Embedded Computing and Mechatronics with the PIC32 Microcontroller

Embedded Computing and Mechatronics with the PIC32 Microcontroller

Kevin M. Lynch
Nicholas Marchuk
Matthew L. Elwin

AMSTERDAM • BOSTON • HEIDELBERG • LONDON
NEW YORK • OXFORD • PARIS • SAN DIEGO
SAN FRANCISCO • SINGAPORE • SYDNEY • TOKYO
Newnes is an imprint of Elsevier

Newnes is an imprint of Elsevier
225 Wyman Street, Waltham, MA 02451, USA
The Boulevard, Langford Lane, Kidlington, Oxford OX5 1GB, UK

British Library Cataloguing in Publication Data
A catalogue record for this book is available from the British Library

Library of Congress Cataloging-in-Publication Data
A catalog record for this book is available from the Library of Congress

For information on all Newnes publications
visit our website at http://store.elsevier.com/

ISBN: 978-0-12-420165-1

Printed and bound in US

Dedication

To Yuko, Erin, and Patrick.

–Kevin M. Lynch

To Mark and Liz.

–Nicholas Marchuk

To Hannah.

–Matthew L. Elwin

Figure Credits

The authors thank the following companies for permission to use their copyrighted images in this book.

* Microchip Technology, Inc. www.microchip.com.

* Digi International, Inc. www.digi.com.

* Pololu Robotics and Electronics. www.pololu.com.

* Digi-Key Electronics. www.digikey.com.

* Advanced Photonix, Inc. www.advancedphotonix.com.

* Contelec AG. www.contelec.ch/en.

* Omega Engineering, Inc., Stamford, CT 06907 USA. www.omega.com.

* Avago Technologies. www.avagotech.com.

* Micro-Measurements, a brand of Vishay Precision Group (VPG), Raleigh, NC, USA. www.vpgsensors.com.

* Maxon Precision Motors. www.maxonmotorusa.com.

* Copley Controls. www.copleycontrols.com.

* H2W Technologies. www.h2wtech.com.

* Hitec RCD USA. www.hitecrcd.com.

Contents

Preface

This book is about the Microchip 32-bit PIC32 microcontroller, its hardware, programming it in C, and interfacing it to sensors and actuators. This book also covers related mechatronics topics such as motor theory, choosing motor gearing, and practical introductions to digital signal processing and feedback control. This book is written for:

- **Anyone starting out with the Microchip PIC32 32-bit microcontroller.** Microchip documentation can be hard to navigate; this is the book we wish we had when we started!
- **The hobbyist ready to explore beyond Arduino.** Arduino software and its large user support community allow you to be up and running quickly with Atmel microcontrollers. But reliance on Arduino software prevents you from fully exploiting or understanding the capability of the microcontroller.
- **Teachers and students in mechatronics.** The exercises, online material, and associated kit are designed to support introductory, advanced, and flipped or online courses in mechatronics.
- **Anyone interested in mechatronics, actuators, sensors, and practical embedded control.**

Contents

This book was written based on the two-quarter mechatronics sequence at Northwestern University, ME 333 Introduction to Mechatronics and ME 433 Advanced Mechatronics. In ME 333, students learn about PIC32 hardware, fundamentals of programming the PIC32 in C, the use of some basic peripherals, and interfacing the PIC32 with sensors and actuators. In ME 433, material from the rest of the book is used as reference by groups working on projects. Students taking the sequence range from sophomores to graduate students. The only prerequisite is introductory circuit analysis and design; experience in C programming is not required. While experience in C would allow faster progression through the material, we decided not to require it, to make the course available to the broad set of students interested in the material. To partially compensate for the wide range of experience in C (from expert to none), we begin ME 333 with an intensive two-week introduction to fundamental C concepts and syntax using the "Crash Course in C" in Appendix A. We also take advantage of student expertise by facilitating peer mentoring.

The goals of this book mirror those of the Northwestern mechatronics sequence:

- to provide the beginner a sound introduction to microcontrollers using the example of the PIC32, a modern 32-bit architecture;
- to do so by first providing an overview of microcontroller hardware, firm in the belief that microcontroller programming is much more grounded when tightly connected to the hardware that implements it;
- to provide a clear understanding of the fundamentals of professional PIC32 programming in C, which builds a foundation for further exploration of the PIC32's capabilities using Microchip documentation and other advanced references;
- to provide reference material and sample code on the major peripherals and special features of the PIC32;
- to instill an understanding of the theory of motor operation and control; and
- to teach how microcontroller peripherals can be used to interface with sensors and motors.

To achieve these goals, the book is divided into five main parts:

1. **Quickstart.** This part (Chapter 1) allows the student to get up and running with the PIC32 quickly.
2. **Fundamentals.** After achieving some early success with the quickstart, the five chapters in Fundamentals (Chapters 2 to 6) examine the PIC32 hardware, the build process in C and the connection of the code to the hardware, the use of libraries, and two important topics for real-time embedded computing: interrupts and the time and space efficiency of code. The time investment in these chapters provides the foundation needed to move quickly through later chapters and to profit from other reference material, like Microchip's PIC32 Reference Manual, Data Sheets, and XC32 C/C++ Compiler User's Guide.
3. **Peripheral Reference.** This part (Chapters 7 to 20) gives details on the operation of the various peripherals on the PIC32, as well as sample code and applications. It is primarily reference material that can be read in any order, though we recommend the first few chapters (digital I/O, counter/timers, output compare, and analog input) be covered in order. The peripheral reference concludes with an introduction to Harmony, Microchip's recent framework for high-level programming of PIC32s.
4. **Mechatronics.** This part (Chapters 21 to 29) focuses on interfacing sensors to a microcontroller, digital signal processing, feedback control, brushed DC motor theory, motor sizing and gearing, control by a microcontroller, and other actuators such as brushless motors, stepper motors, and servo motors.
5. **Appendixes.** The appendixes cover background topics such as analysis of simple circuits and an introduction to programming in C. We have our students first get used to writing C programs on their laptops, and compiling with `gcc`, before moving on to programming a microcontroller.

In ME 333, we cover the crash course in C; the Quickstart; the Fundamentals; select topics from the Peripheral Reference (digital I/O, counter/timers, output compare/PWM, and analog input); and simple sensor interfacing, DC motor theory, motor sizing and gearing, and control of a DC motor from the Mechatronics part. Other chapters are used for reference in ME 433 and other projects that students undertake.

Choices made in this book

We made several choices about how to teach mechatronics in ME 333, and those choices are reflected in this book. Our choices are based on the desire to expose our students to the topics they will need to integrate sensors and actuators and microcontrollers professionally, subject to the constraint that most students will take only one or two courses in mechatronics. Our choices are based on what we believe to be the smallest building blocks that a mechatronics engineer needs to know about. For example, we do not attempt to teach microcontroller architecture at the level that a computer engineer might learn it, since a mechatronics engineer is not likely to design a microcontroller. On the other hand, we also do not rely on software and hardware abstractions that keep the budding mechatronics engineer at arm's length from concepts needed to progress beyond the level of a hobbyist. With that philosophy in mind, here are some of the choices made for ME 333 and this book:

- *Microcontrollers vs. sensors and actuators.* Mechatronics engineering integrates sensors, actuators, and microcontrollers. Handing a student a microcontroller development board and sample code potentially allows the course to focus on the sensors and actuators part. In ME 333, however, we opted to make understanding the hardware and software of the microcontroller approximately 50% of the course. This choice recognizes the fundamental role microcontrollers play in mechatronics, and that mechatronics engineers must be comfortable with programming.
- *Choice of microcontroller manufacturer.* There are many microcontrollers on the market, with a wide variety of features. Manufacturers include Microchip, Atmel, Freescale, Texas Instruments, STMicroelectronics, and many others. In particular, Atmel microcontrollers are used in Arduino boards. Arduinos are heavily used by hobbyists and in K-12 and university courses in large part due to the large online user support community and the wide variety of add-on boards and user-developed software libraries. In this book, we opt for the commercially popular Microchip PIC microcontrollers, and we avoid the high-level software abstractions synonymous with Arduino. (Arduinos are used in other Northwestern courses, particularly those focusing on rapid product prototyping with little mechatronics design.)
- *Choice of a particular microcontroller model.* Microchip's microcontroller line consists of hundreds of different models, including 8-bit, 16-bit, and 32-bit architectures. We have chosen a modern 32-bit architecture. And instead of trying to write a book that deals with all PIC32 models, which includes six different families of PIC32s as of this writing (see

Appendix C), we focus on one particular model: the PIC32MX795F512H. The reasons for this choice are (a) it is a powerful chip with plenty of peripherals and memory (128 KB data RAM and 512 KB program flash), and, more importantly, (b) focusing on a single chip allows us to be concrete in the details of its operation. This is especially important when learning how the hardware relates to the software. (One of the reasons Microchip's documentation is difficult to read, and is so full of exceptions and special cases, is that it is written to be general to all PIC32s in the case of the Reference Manual, or all PIC32s in a specific family in the case of the Data Sheets.) Once the reader has learned about the operation of a specific PIC32, it is not too difficult to learn about the differences for a different PIC32 model.

- *Programming language: C++ vs. C vs. assembly.* C++ is a relatively high-level language, C is lower level, and assembly is lower still. We choose to program in C because of the portability of the language, while staying relatively close to the assembly language level and minimizing abstractions introduced by C++.
- *Integrated Development Environment vs. command line.* MPLAB X is Microchip's Integrated Development Environment (IDE) for developing software for PICs. So why do we avoid using it in this book? Because we feel that it hides key steps in understanding how the code you write turns into an executable for the PIC32. In this book, code is written in a text editor and the C compiler is invoked at the command line. There are no hidden steps. Once the reader has mastered the material in the first few chapters of this book, MPLAB will no longer be mysterious.
- *Use of the Harmony software vs. ignoring it.* Microchip provides an extensive library of middleware, device drivers, system services, and other software to support all of their PIC32 models. One goal of this software is to allow you to write programs that are portable across different PIC32 models. To achieve this, however, a significant amount of abstraction is introduced, separating the code you write from the actual hardware implementation. This is bad pedagogically as you learn about the PIC32. Instead, we use low-level software commands to control the PIC32's peripherals, reinforcing the hardware documentation in this book and in the Data Sheet and Reference Manual. Only with the more complicated peripherals do we use the Harmony software, specifically for USB, in Chapter 20.
- *Sample code vs. writing it yourself.* The usual way to learn to program PIC32s is to take some working sample code and try to modify it to do something else. This is natural, except that if your modified code fails, you are often left with no idea what to do. In this book we provide plenty of sample code, but we also focus on the fundamentals of programming the PIC32 so that you learn to write code from scratch as well as strategies to debug if things go wrong (Figure 0.1).

The philosophy represented by the choices above can be summed up succinctly: There should be no magic steps! You should know how and why the code you write works, and how it

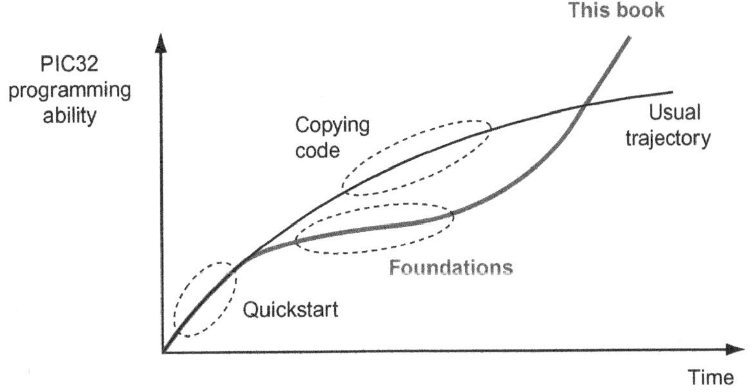

Figure 0.1
The trajectory of PIC32 programming ability vs. time for the usual "copy and modify" approach vs. the foundational approach in this book. The crossover should occur at only a few weeks!

connects to the hardware. You should not be simply modifying opaque and abstract code, compiling with a mysterious IDE, and hoping for the best.

The NU32 development board

The NU32 development board was created to support this book. If you do not have the board, you can still learn a lot about how a PIC32 works from reading this book. We highly recommend that you get the NU32 board and the kit of mechatronics parts, however, to allow you to work through the examples in the book.

In keeping with the "no magic" philosophy, the primary function of the NU32 is to break out the pins of the PIC32MX795F512H to a solderless prototyping breadboard, to allow easy wiring to the pins. Otherwise we try to keep the board as bare bones and inexpensive as possible, leaving external circuits to the reader. To allow you to get up and running as quickly as possible, though, the board does provide a few devices external to the PIC32: two LEDs and two buttons for simple user interaction; a 3.3 V regulator (to provide power to the PIC32) and a 5 V regulator (to provide a commonly needed voltage); a resonator to provide a clock signal; and a USB-to-UART chip that simplifies communication between the user's computer and the PIC32.

The PIC32 on the NU32 comes with a bootloader program pre-installed, allowing you to program the PIC32 with just a USB cable. The NU32 can also be programmed directly using a programmer device, like the PICkit 3. This is covered in Chapter 3.6.

How to use this book in a course

Mechatronics is fundamentally an integrative discipline, requiring knowledge of microcontrollers, programming, circuit design, sensors, signal processing, feedback control, and motors. This book contains a practical introduction to these topics.

Recognizing that most students take no more than one or two courses in mechatronics, however, this book does not delve deeply into the mathematical theory underlying topics such as linear systems, circuit analysis, signal processing, or control theory.[1] Instead, a course based on this book is meant to motivate further theoretical study in these disciplines by exposing students to their practical applications.

As a result, students need only a basic background in circuits and programming to be able to take a course based on this book. At Northwestern, this means that students take ME 333 as early as their sophomore year. ME 333 is an intense 11-week quarter, covering, in order:

- Appendix A, a Crash Course in C. (Approximately 2 weeks.)
- Chapters 1–6, fundamentals of hardware and software of the PIC32 microcontroller. (Approximately 3 weeks.)
- Chapters 7–10, covering digital input and output, counter/timers, output compare/PWM, and analog input. These chapters are primarily used as reference in the context of the following assignment.
- Chapters 23 and 24, on feedback control and PI control of the brightness of an LED using a phototransistor for feedback. This project is the students' first significant project using the PIC32 for embedded control. It also serves as a warmup for the final project. (Approximately 2 weeks.)
- Chapter 25 on theory and experimental characterization of a brushed DC motor. (Approximately 1 week.)
- Introduction to encoders and current sensing in Chapter 21 and all of Chapters 27 and 28 on DC motor control. Chapter 27 introduces all the hardware and software elements of a professional DC motor control system, including a nested-loop control system with an outer-loop motion controller and an inner-loop current controller. Chapter 28 is a chapter-long project that applies the ideas, leading the student through a significant software design project to develop a motor control system that interfaces with a menu system in MATLAB. This "capstone" project is motivated by professional motor amplifier design and integrates the student's knowledge of the PIC32, C programming, brushed DC motors, feedback control, and the interfacing of sensors and actuators with a PIC32. (Approximately 3 weeks.)

[1] Because other courses generally do not cover the operation of motors, this book goes into greater detail on motor theory.

This is a very full quarter, which would be less intense if students were required to know C before taking the course.

ME 333 at Northwestern is taught as a flipped class. Students watch videos that support the text on their own time, then work on assignments and projects during class time while the instructor and TAs circulate to help answer questions. Students bring their laptops and portable mechatronics kits to every class. This kit includes an inexpensive function generator and oscilloscope, the nScope, that uses their laptop as the display. Thus ME 333 does not use a lab facility; students use the classroom and their own dorm rooms. Students work and learn together during classes, but each student completes her own assignment individually. The follow-on course ME 433 focuses on more open-ended mechatronics projects in teams and makes extensive use of a mechatronics lab that is open to students 24/7.

For a 15-week semester, good additions to the course would be two weeks on different sensor technologies (Chapter 21) and digital signal processing of sensor data (Chapter 22). Another week should also be devoted to the final motor control project (Chapter 28), to allow students to experiment with various extensions. Time permitting, other common actuators (e.g., steppers, RC servos, and brushless motors) could be covered in Chapter 29.

For a two-quarter or two-semester sequence, the second course could focus on open-ended team design projects, similar to ME 433 at Northwestern. The book then serves as a reference. Other appropriate material includes chapters on communication protocols and supporting PIC32 peripherals (e.g., UART, SPI, I^2C, USB, and CAN).

Website, videos, and flipped classrooms

The book's website, `www.nu32.org`, has links to downloadable data sheets, sample code, PCB layouts and schematics, chapter extensions, errata, and other useful information and updates. This website also links to short videos that summarize many of the chapters. These videos can be used to flip a traditional classroom, as in ME 333, allowing students to watch the lectures at home and to use class time to ask questions and work on projects.

Other PIC32 references

One goal of this book is to organize Microchip reference material in a logical way, for the beginner. Another goal is to equip the reader to be able to parse Microchip documentation. This ability allows the reader to continue to develop her PIC32 programming abilities beyond the limits of this book. The reader should download and have at the ready the first two references below; the others are optional. The readings are summarized in Figure 0.2.

- **The PIC32 Reference Manual.** The Reference Manual sections describe software and hardware for all PIC32 families and models, so they can sometimes be confusing in their

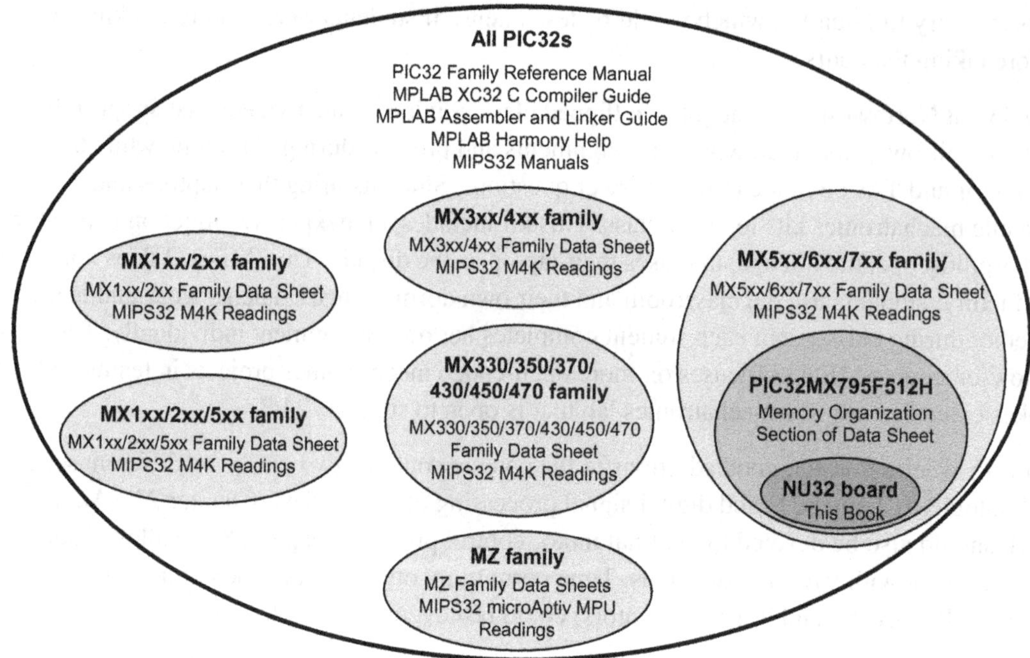

Figure 0.2
Other reference reading and the PIC32s they apply to.

generality. Nevertheless, they are a good source for understanding the functions of the PIC32 in more detail. Some of the sections, particularly the later ones, focus on the PIC32MZ family and are not relevant to the PIC32MX795F512H.

- **The PIC32MX5xx/6xx/7xx Family Data Sheet.** This Data Sheet provides details specific to the PIC32MX5xx/6xx/7xx family. In particular, the Memory Organization section of the Data Sheet clarifies which special function registers (SFRs) are included on the PIC32MX795F512H, and therefore which Reference Manual functions are available for that model.
- **(Optional) The Microchip MPLAB XC32 C Compiler User's Guide** and **The Assembler, Linker, and Utilities User's Guide.** These come with your XC32 C compiler installation, so no need to download separately.
- **(Optional) MPLAB Harmony Help.** This documentation, which comes with the Harmony installation, can be helpful once you start writing more complex code that uses the Harmony software.
- **(Optional) MIPS32 Architecture for Programmers manuals and other MIPS32 documentation.** If you are curious about the MIPS32 M4K CPU, which is used on the PIC32MX795F512H, and its assembly language instruction set, you can find references online.

Acknowledgments

This book has benefited from the feedback of many students and teaching assistants in ME 333 at Northwestern over the years, particularly Alex Ansari, Philip Dames, Mike Hwang, Andy Long, David Meyer, Ben Richardson, Nelson Rosa, Jarvis Schultz, Jian Shi, Craig Shultz, Matt Turpin, and Zack Woodruff.

PART I

Quickstart

CHAPTER 1

Quickstart

Edit, compile, run, repeat: familiar to generations of C programmers, this mantra applies to programming in C, regardless of platform. Architecture, program loading, input and output: these details differ between your computer and the PIC32. Architecture refers to processor type: your computer's x86-64 CPU and the PIC32's MIPS32 CPU understand different machine code and therefore require different compilers. Your computer's operating system allows you to seamlessly run programs; the PIC32's *bootloader* writes programs it receives from your computer to flash memory and executes them when the PIC32 resets.[1] You interact directly with your computer via the screen and keyboard; you interact indirectly with the PIC32 using a *terminal emulator* to relay information between your computer and the microcontroller. As you can see, programming the PIC32 requires attention to details that you probably ignore when programming your computer.

Armed with an overview of the differences between computer programming and microcontroller programming, you are ready to get your hands dirty. The rest of this chapter will guide you through gathering the hardware and installing the software necessary to program the PIC32. You will then verify your setup by running two programs on the PIC32. By the end of the chapter, you will be able to compile and run programs for the PIC32 (almost) as easily as you compile and run programs for your computer!

Throughout this book, we will refer to "the PIC32." Although there are many PIC32 models, for us "the PIC32" is shorthand for the PIC32MX795F512H. While most of the concepts in this book apply to many PIC32 models, you should be aware that some of the details differ between models. (See Appendix C for a discussion of the differences.) Further, we refer to the PIC32MX795F512H as it is configured on the NU32 development board; in particular, it is powered by 3.3 V and is clocked by a system clock and a peripheral bus clock at 80 MHz. You will learn more about these details in Chapter 2.

---

1 Your computer also has a bootloader. It runs when you turn the computer on and loads the operating system. Also, operating systems are available for the PIC32, but we will not use them in this book.

Embedded Computing and Mechatronics with the PIC32 Microcontroller. http://dx.doi.org/10.1016/B978-0-12-420165-1.00001-9

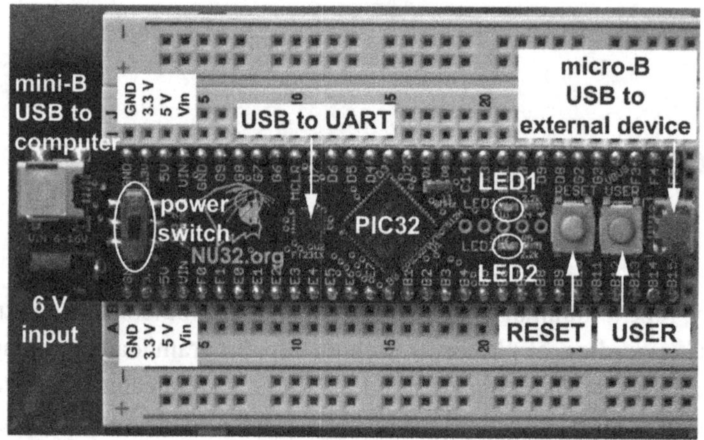

Figure 1.1
A photo of the NU32 development board mounted on a solderless breadboard.

1.1 What You Need

This section explains the hardware and software that you need to program the PIC32. Links to purchase the hardware and download the free software are provided at the book's website, `www.nu32.org`.

1.1.1 Hardware

Although PIC32 microcontrollers integrate many devices on a single chip, they also require external circuitry to function. The NU32 development board, shown in Figure 1.1, provides this circuitry and more: buttons, LEDs, breakout pins, a USB port, and a virtual USB serial port. The examples in this book assume that you use this board. You will also need the following hardware:

1. **Computer with a USB port.** The host computer is used to create PIC32 programs. The examples in this book work with the Linux, Windows, and Mac operating systems.
2. **USB A to mini-B cable.** This cable carries signals between the NU32 board and your computer.
3. **AC/DC adapter (6 V).** This cable provides power to the PIC32 and NU32 board.

1.1.2 Software

Programming the PIC32 requires various software. You should be familiar with some of the software from programming your computer in C; if not, refer to Appendix A.

For your convenience, we have aggregated the software you need at the book's website. You should download and install all of the following software.

1. **The command prompt** allows you to control your computer using a text-based interface. This program, `cmd.exe` on Windows, `Terminal` on Mac, and `bash` on Linux, comes with your operating system so you should not need to install it. See Appendix A for more information about the command line.
2. **A text editor** allows you to create text files, such as those containing C source code. See Appendix A for more information.
3. **A native C compiler** converts human-readable C source code files into machine code that your computer can execute. We suggest the free GNU compiler, `gcc`, which is available for Windows, Mac, and Linux. See Appendix A for more information.
4. **Make** simplifies the build process by automatically executing the instructions required to convert source code into executables. After manually typing all of the commands necessary create your first program, you will appreciate `make`.
5. **The Microchip XC32 compiler** converts C source files into machine code that the PIC32 understands. This compiler is known as a *cross compiler* because it runs on one processor architecture (e.g., x86-64 CPU) and creates machine code for another (e.g., MIPS32). This compiler installation also includes C libraries to help you control PIC32-specific features. Note where you install the compiler; we will refer to this directory as `<xc32dir>`. If you are asked during installation whether you would like to add XC32 to your path variable, do so.
6. **MPLAB Harmony** is Microchip's collection of libraries and drivers that simplify the task of writing code targeting multiple PIC32 models. We will use this library only in Chapter 20; however, you should install it now. Note the installation directory, which we will refer to as `<harmony>`.
7. **The FTDI Virtual COM Port Driver** allows you to use a USB port as a "virtual serial communication (COM) port" to talk to the NU32 board. This driver is already included with most Linux distributions, but Windows and Mac users may need to install it.
8. **A terminal emulator** provides a simple interface to a COM port on your computer, sending keyboard input to the PIC32 and displaying output from the PIC32. For Linux/Mac, you can use the built-in `screen` program. For Windows, we recommend you download `PuTTY`. Remember where you install `PuTTY`; we refer to this directory as `<puttyPath>`.
9. **The PIC32 quickstart code** contains source code and other support files to help you program the PIC32. Download `PIC32quickstart.zip` from the book's website, extract it, and put it in a directory that you create. We will refer to this directory as `<PIC32>`. In `<PIC32>` you will keep the quickstart code, plus all of the PIC32 code you write, so make sure the directory name makes sense to you. For example, depending on your operating system, `<PIC32>` could be `/Users/kevin/PIC32` or `C:\Users\kevin\Documents\PIC32`. In `<PIC32>`, you should have the following three files and one directory:

- `nu32utility.c`: a program for your computer, used to load PIC32 executable programs from your computer to the PIC32
- `simplePIC.c, talkingPIC.c`: PIC32 sample programs that we will test in this chapter
- `skeleton`: a directory containing
  - `Makefile`: a file that will help us compile future PIC32 programs
  - `NU32.c`, `NU32.h`: a library of useful functions for the NU32 board
  - `NU32bootloaded.ld`: a linker script used when compiling programs for the PIC32

We will learn more about each of these shortly.

You should now have code in the following directories (if you are a Windows user, you will also have `PuTTY` in the directory `<puttyPath>`):

- `<xc32dir>`. The Microchip XC32 compiler. **You will never modify code in this directory.** Microchip wrote this code, and there is no reason for you to change it. Depending on your operating system, your `<xc32dir>` could look something like the following:
  - `/Applications/microchip/xc32`
  - `C:\Program Files (x86)\Microchip\xc32`
- `<harmony>`. Microchip Harmony. **You will never modify code in this directory.** Depending on your operating system, your `<harmony>` could look something like the following:
  - `/Users/kevin/microchip/harmony`
  - `C:\microchip\harmony`
- `<PIC32>`. Where PIC32 quickstart code, and code you will write, is stored, as described above.

Now that you have installed all of the necessary software, it is time to program the PIC32. By following these instructions, not only will you run your first PIC32 program, you will also verify that all of the software and hardware is functioning properly. Do not worry too much about what all the commands mean, we will explain the details in subsequent chapters.

> **Notation:** Wherever we write `<something>`, replace it with the value relevant to your computer. On Windows, use a backslash (`\`) and on Linux/Mac use a slash (`/`) to separate the directories in a path. At the command line, place paths that contain spaces between quotation marks (i.e., `"C:\Program Files"`). Enter the text following a > at the command line. Use a single line, even if the command spans multiple lines in the book.

1.2 Compiling the Bootloader Utility

The bootloader utility, located at `<PIC32>/nu32utility.c`, sends compiled code to the PIC32. To use the bootloader utility you must compile it. Navigate to the `<PIC32>` directory by typing:

```
> cd <PIC32>
```

Verify that `<PIC32>/nu32utility.c` exists by executing the following command, which lists all the files in a directory:

- **Windows**
  ```
  > dir
  ```
- **Linux/Mac**
  ```
  > ls
  ```

Next, compile the bootloader utility using the native C compiler `gcc`:

- **Windows**
  ```
  > gcc nu32utility.c -o nu32utility -lwinmm
  ```
- **Linux/Mac**
  ```
  > gcc nu32utility.c -o nu32utility
  ```

When you successfully complete this step the executable file `nu32utility` will be created. Verify that it exists by listing the files in `<PIC32>`.

1.3 Compiling Your First Program

The first program you will load onto your PIC32 is `<PIC32>/simplePIC.c`, which is listed below. We will scrutinize the source code in Chapter 3, but reading it now will help you understand how it works. Essentially, after some setup, the code enters an infinite loop that alternates between delaying and toggling two LEDs. The delay loops infinitely while the USER button is pressed, stopping the toggling.

Code Sample 1.1 `simplePIC.c`**. Blinking Lights on the NU32, Unless the** `USER` **Button Is Pressed.**

```
#include <xc.h>          // Load the proper header for the processor

void delay(void);

int main(void) {
  TRISF = 0xFFFC;        // Pins 0 and 1 of Port F are LED1 and LED2.  Clear
                         // bits 0 and 1 to zero, for output.  Others are inputs.
  LATFbits.LATF0 = 0;    // Turn LED1 on and LED2 off.  These pins sink current
  LATFbits.LATF1 = 1;    // on the NU32, so "high" (1) = "off" and "low" (0) = "on"

  while(1) {
    delay();
    LATFINV = 0x0003;    // toggle LED1 and LED2; same as LATFINV = 0x3;
  }
  return 0;
}
```

```
void delay(void) {
  int j;
  for (j = 0; j < 1000000; j++) { // number is 1 million
    while(!PORTDbits.RD7) {
        ;   // Pin D7 is the USER switch, low (FALSE) if pressed.
    }
  }
}
```

To compile this program you will use the xc32-gcc cross compiler, which compiles code for the PIC32's MIPS32 processor. This compiler and other Microchip tools are located at <xc32dir>/<xc32ver>/bin, where <xc32ver> refers to the XC32 version (e.g., v1.40). To find <xc32ver> list the contents of the Microchip XC32 directory, e.g.,

```
> ls <xc32dir>
```

The subdirectory displayed is your <xc32ver> value. If you happen to have installed two or more versions of XC32, you will always use the most recent version (the largest version number).

Next you will compile simplePIC.c and create the executable *hex file*. To do this, you first create the simplePIC.elf file and then you create the simplePIC.hex file. (This two-step process will be discussed in greater detail in Chapter 3.) Issue the following commands from your <PIC32> directory (where simplePIC.c is), being sure to replace the text between the <> with the values appropriate to your system. Remember, if the paths contain spaces, you must surround them with quotes (i.e., "C:\Program Files\xc32\v1.40\bin\xc32-gcc").

```
> <xc32dir>/<xc32ver>/bin/xc32-gcc -mprocessor=32MX795F512H
    -o simplePIC.elf -Wl,--script=skeleton/NU32bootloaded.ld simplePIC.c
> <xc32dir>/<xc32ver>/bin/xc32-bin2hex simplePIC.elf
```

The -Wl is "-W ell" not "-W one." You can list the contents of <PIC32> to make sure both simplePIC.elf and simplePIC.hex were created. The hex file contains MIPS32 machine code in a format that the bootloader understands, allowing it to load your program onto the PIC32.

If, when you installed XC32, you selected to have XC32 added to your path, then in the two commands above you could have simply typed

```
> xc32-gcc -mprocessor=32MX795F512H
    -o simplePIC.elf -Wl,--script=skeleton/NU32bootloaded.ld simplePIC.c
> xc32-bin2hex simplePIC.elf
```

and your operating system would be able to find xc32-gcc and xc32-bin2hex without needing the full paths to them.

1.4 Loading Your First Program

Loading a program onto the PIC32 from your computer requires communication between the two devices. When the PIC32 is powered and connected to a USB port, your computer creates a new serial communication (COM) port. Depending on your specific system setup, this COM port will have different names. Therefore, we will determine the name of your COM port through experimentation. First, with the PIC32 unplugged, execute the following command to enumerate the current COM ports, and note the names that are listed:

- **Windows**:
  `> mode`
- **Mac**:
  `> ls /dev/tty.*`
- **Linux**:
  `> ls /dev/ttyUSB*`

Next, plug the NU32 board into the wall using the AC adapter, turn the power switch on, and verify that the red "power" LED illuminates. Connect the USB cable from the NU32's mini-B USB jack (next to the power jack) to a USB port on the host computer. Repeat the steps above, and note that a new COM port appears. If it does not appear, make sure that you installed the FTDI driver from Section 1.1.2. The name of the port will differ depending on the operating system; therefore we have listed some typical names:

- **Windows**: `COM4`
- **Mac**: `/dev/tty.usbserial-DJ00DV5V`
- **Linux**: `/dev/ttyUSB0`

Your computer, upon detecting the NU32 board, has created this port. Your programs and the bootloader use this port to communicate with your computer.

After identifying the COM port, place the PIC32 into *program receive mode*. Locate the RESET button and the USER button on the NU32 board (Figure 1.1). The RESET button is immediately above the USER button on the bottom of the board (the power jack is the board's top). Press and hold both buttons, release RESET, and then release USER. After completing this sequence, the PIC32 will flash LED1, indicating that it has entered program receive mode.

Assuming that you are still in the `<PIC32>` directory, start the loading process by typing

- **Windows**
  `nu32utility <COM> simplePIC.hex`
- **Linux/Mac**
  `> ./nu32utility <COM> simplePIC.hex`

where <COM> is the name of your COM port.[2] After the utility finishes, LED1 and LED2 will flash back and forth. Hold USER and notice that the LEDs stop flashing. Release USER and watch the flashing resume. Turn the PIC32 off and then on. The LEDs resume blinking because you have written the program to the PIC32's nonvolatile flash memory. Congratulations, you have successfully programmed the PIC32!

1.5 Using make

As you just witnessed, building an executable for the PIC32 requires several steps. Fortunately, you can use make to simplify this otherwise tedious and error-prone procedure. Using make requires a Makefile, which contains instructions for building the executable. We have provided a Makefile in <PIC32>/skeleton. Prior to using make, you need to modify <PIC32>/skeleton/Makefile so that it contains the paths and COM port specific to your system.

Aside from the paths you have already used, you need your terminal emulator's location, <termEmu>, and the Harmony version, <harmVer>. On Windows, <termEmu> is <puttyPath>/putty.exe and for Linux/Mac, <termEmu> is screen. To find Harmony's version, <harmVer>, list the contents of the <harmony> directory. Edit <PIC32>/skeleton/Makefile and update the first five lines as indicated below.

```
XC32PATH=<xc32dir>/<xc32ver>/bin
HARMONYPATH=<harmony>/<harmVer>
NU32PATH=<PIC32>
PORT=<COM>
TERMEMU=<termEmu>
```

In the Makefile, do not surround paths with quotation marks, even if they contain spaces.

If your computer has more than one USB port, you should always use the same USB port to connect your NU32. Otherwise, the name of the COM port may change, requiring you to edit the Makefile again.

After saving the Makefile, you can use the skeleton directory to easily create new PIC32 programs. The skeleton directory contains not only the Makefile, but also the NU32 library (NU32.h and NU32.c), and the linker script NU32bootloaded.ld, all of which will be used extensively throughout the book. The Makefile automatically compiles and links every .c file in the directory into a single executable; therefore, your project directory should contain all the C files you need and none that you do not want!

2 Windows: Write the ports as \\.\COMx rather than COMx. Linux: To avoid needing to execute commands as root, add yourself to the group that owns the COM port (e.g., uucp).

Each new project you create will have its own directory in <PIC32>, e.g., <PIC32>/<projectdir>. We now explain how to use the <PIC32>/skeleton directory to create a new project, using <PIC32>/talkingPIC.c as an example. For this example, we will name the project talkingPIC, so <projectdir> is talkingPIC. By following this procedure, you will have access to the NU32 library and will be able to avoid repeating the previous setup steps. Make sure you are in the <PIC32> directory, then copy the <PIC32>/skeleton directory to the new project directory:

- **Windows**
```
> mkdir <projectdir>
> copy skeleton\*.* <projectdir>
```
- **Linux/Mac**
```
> cp -R skeleton <projectdir>
```

Now copy the project source files, in this case just talkingPIC.c, to <PIC32>/<projectdir>, and change to that directory:

- **Windows**
```
> copy talkingPIC.c <projectdir>
> cd <projectdir>
```
- **Linux/Mac**
```
> cp talkingPIC.c <projectdir>
> cd <projectdir>
```

Before explaining how to use make, we will examine talkingPIC.c, which accepts input from and prints output to a terminal emulator running on the host computer. These capabilities facilitate user interaction and debugging. The source code for talkingPIC.c is listed below:

Code Sample 1.2 talkingPIC.c. **The PIC32 Echoes Any Messages Sent to It from the Host Keyboard Back to the Host Screen.**

```
#include "NU32.h"          // constants, funcs for startup and UART

#define MAX_MESSAGE_LENGTH 200

int main(void) {
  char message[MAX_MESSAGE_LENGTH];

  NU32_Startup(); // cache on, interrupts on, LED/button init, UART init
  while (1) {
    NU32_ReadUART3(message, MAX_MESSAGE_LENGTH);  // get message from computer
    NU32_WriteUART3(message);                     // send message back
    NU32_WriteUART3("\r\n");                      // carriage return and newline
    NU32_LED1 = !NU32_LED1;                       // toggle the LEDs
    NU32_LED2 = !NU32_LED2;
  }
  return 0;
}
```

The NU32 library function NU32_ReadUART3 allows the PIC32 to read data sent from your computer's terminal emulator. The function NU32_WriteUART3 sends data from your PIC32 to be displayed by the terminal emulator.

Now that you know how talkingPIC.c works, it is time see it in action. First, make sure you are in the <projectdir>. Next, build the project using make.

```
> make
```

This command compiles and assembles all .c files into .o object files, links them into a single out.elf file, and turns that out.elf file into an executable out.hex file. You can do a directory listing to see all of these files.

Next, put the PIC32 into program receive mode (use the RESET and USER buttons) and execute

```
> make write
```

to invoke the bootloader utility nu32utility and program the PIC32 with out.hex. When LED1 stops flashing, the PIC32 has been programmed.

In summary, to create a new project and program the PIC32, you (1) create the project directory <PIC32>/<projectdir>; (2) copy the contents of <PIC32>/skeleton to this new directory; (3) create the source code (talkingPIC.c in this case) in <projectdir>; (4) build the executable by executing make in <projectdir>; and (5) use the RESET and USER buttons to put the PIC32 in program receive mode and execute make write from <projectdir>. To modify the program, you simply edit the source code and repeat steps (4) and (5) above. In fact, you can skip step (4), since make write also builds the executable before loading it onto the PIC32.

Now, to communicate with talkingPIC, you must connect to the PIC32 using your terminal emulator. Recall that the terminal emulator communicates with the PIC32 using <COM>. Enter the following command:

- **Windows**
  `<puttyPath>\putty -serial <COM> -sercfg 230400,R`
- **Linux/Mac**
  `screen <COM> 230400,crtscts`

PuTTY will launch in a new window, whereas screen will use the command prompt window. The number 230400 in the above commands is the baud, the speed at which the PIC32 and computer communicate, and the other parameter enables hardware flow control (see Chapter 11 for details).

After connecting, press RESET to restart the program. Start typing, and notice that no characters appear until you hit `ENTER`. This behavior may seem strange, but it occurs because the terminal emulator only displays the text it receives from the PIC32. The PIC32 does not send any text to your computer until it receives a special *control character*, which you generate by pressing `ENTER`.[3]

For example, if you type `Hello! ENTER`, the PIC32 will receive `Hello!\r`, write `Hello!\r\n` to the terminal emulator, and wait for more input.

When you are done conversing with the PIC32, you can exit the terminal emulator. To exit `screen` type

```
CTRL-a k y
```

Note that `CTRL` and `a` should be pressed simultaneously. To exit `PuTTY` make sure the command prompt window is focused and type

```
CTRL-c
```

Rather than memorizing these rather long commands to connect to the serial port, you can use the `Makefile`. To connect `PuTTY` to the PIC32 type

```
> make putty
```

To use `screen` type

```
> make screen
```

Your system is now configured for PIC32 programming. Although the build process may seem opaque, do not worry. For now it is only important that you can successfully compile programs and load them onto the PIC32. Later chapters will explain the details of the build process.

1.6 Chapter Summary

- To start a new project, copy the `<PIC32>/skeleton` directory to a new location, `<projectdir>`, and add your source code.
- From the directory `<projectdir>`, use `make` to build the executable.

[3] Depending on the terminal emulator, `ENTER` may generate a carriage return (`\r`), newline (`\n`) or both. The terminal emulator typically moves the cursor to the leftmost column when it receives a `\r` and to the next line when it receives a `\n`.

- Put the PIC32 into program receive mode by pressing the USER and RESET buttons simultaneously, then releasing the RESET button, and finally releasing the USER button. Then use `make write` to load your program.
- Use a terminal emulator to communicate with programs running on the PIC32. Typing `make putty` or `make screen` from `<projectdir>` will launch the appropriate terminal emulator and connect it to the PIC32.

Further Reading

Embedded computing and mechatronics with the PIC32 microcontroller website. http://www.nu32.org.

PART II

Fundamentals

CHAPTER 2

Hardware

Microcontrollers power the modern world: from cars to microwaves to toys. These tiny microchips integrate every component of a modern computer—a central processing unit (CPU), random access memory (RAM), nonvolatile memory (flash), and peripherals—into a single chip. Although microcontrollers have significantly less processing power than their personal computer counterparts, they are also much smaller, cost less, and use less power. Additionally, their peripherals—devices that connect the CPU with the physical world—allow software to interact with circuits directly: flashing lights, moving motors, and reading sensors.

Companies including (but certainly not limited to) Microchip, Atmel, Freescale, Texas Instruments, and STMicroelectronics manufacture an overwhelming array of microcontrollers with vastly different specifications. Rather than attempt to discuss microcontrollers generally, we focus on the PIC32MX795F512H (which we usually abbreviate as PIC32). With a fast processor, ample memory, and numerous peripherals, the PIC32MX795F512H is excellent for learning about microcontrollers and completing embedded control projects. Much of what you learn about the PIC32MX795F512H also applies more generally to the PIC32MX family of microcontrollers, and the broader concepts translate to microcontrollers more generally. Appendix C describes the differences between the PIC32MX795F512H and other PIC32 models.

2.1 The PIC32

2.1.1 Pins, Peripherals, and Special Function Registers (SFRs)

The PIC32 requires a supply voltage between 2.3 and 3.6 V and features a maximum CPU clock frequency of 80 MHz, 512 KB of program memory (flash), and 128 KB of data memory (RAM). Its peripherals include a 10-bit analog-to-digital converter (ADC), many digital I/O pins, USB 2.0, Ethernet, two CAN modules, four I^2C and three SPI synchronous serial communication modules, six UARTs for asynchronous serial communication, five 16-bit counter/timers (configurable to give two 32-bit timers and one 16-bit timer), five pulse-width modulation outputs, and several pins that can generate interrupts based on external signals. Whew. Do not worry if you do not know what all of these peripherals do, much of this book is dedicated to explaining them.

Embedded Computing and Mechatronics with the PIC32 Microcontroller. http://dx.doi.org/10.1016/B978-0-12-420165-1.00002-0

Pins connect the peripherals to the outside world. To cram so much functionality into only 64 pins, many serve multiple functions. See the pinout diagram for the PIC32MX795F512H (Figure 2.1). For example, pin 12 can be an analog input, a comparator input, a change notification input (which can generate an interrupt when an input changes state), or a digital input or output.

Table 2.1 summarizes some of the major pin functions. Other pin functions can be found in the PIC32MX5xx/6xx/7xx Data Sheet.

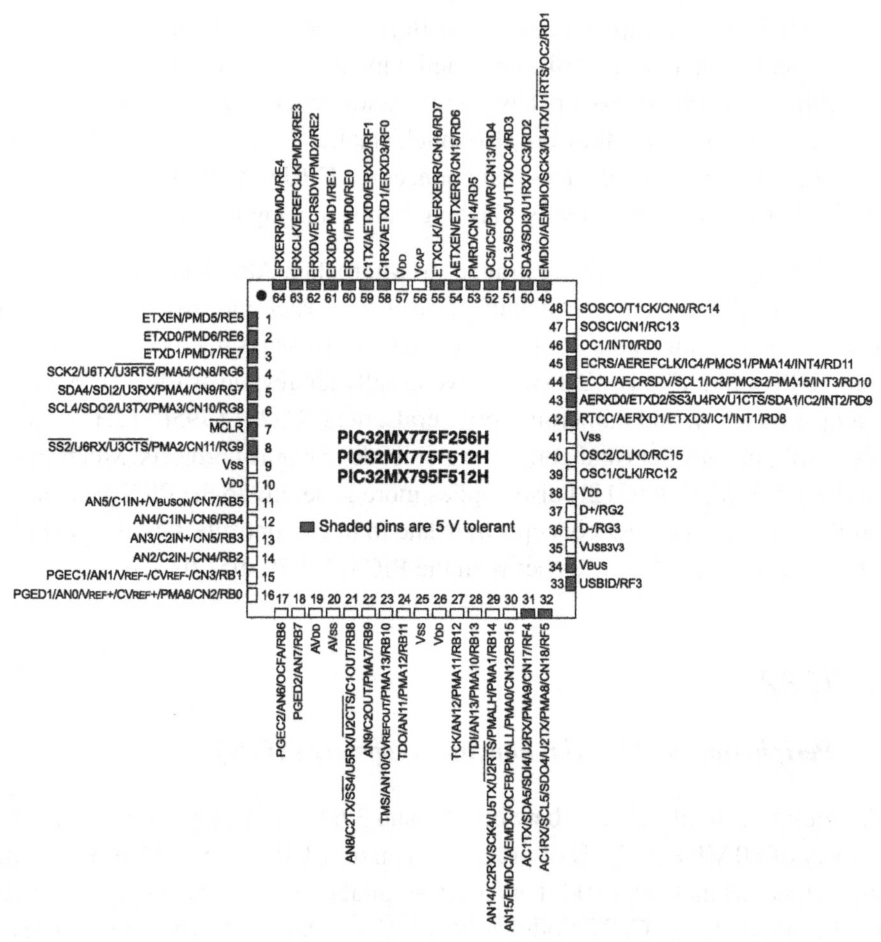

Figure 2.1

The pinout of the PIC32MX795F512H, the microcontroller used on the NU32 development board.

Table 2.1: Some of the pin functions on the PIC32

Pin Label	Function
ANx (x = 0 to 15)	Analog-to-digital (ADC) inputs
AVDD, AVSS	Positive supply and ground reference for ADC
CxIN-, CxIN+, CxOUT (x = 1, 2)	Comparator negative and positive input and output
CxRX, CxTx (x = 1, 2)	CAN receive and transmit pins
CLKI, CLKO	Clock input and output (for particular clock modes)
CNx (x = 0 to 18)	Change notification; voltage changes on these pins can generate interrupts
CVREF-, CVREF+, CVREFOUT	Comparator reference voltage low and high inputs, output
D+, D-	USB communication lines
ENVREG	Enable for on-chip voltage regulator that provides 1.8 V to internal core (on the NU32 board it is set to VDD to enable the regulator)
ICx (x = 1 to 5)	Input capture pins for measuring frequencies and pulse widths
INTx (x = 0 to 4)	Voltage changes on these pins can generate interrupts
$\overline{\text{MCLR}}$	Master clear reset pin, resets PIC when low
OCx (x = 1 to 5)	Output compare pins, usually used to generate pulse trains (pulse-width modulation) or individual pulses
OCFA, OCFB	Fault protection for output compare pins; if a fault occurs, they can be used to make OC outputs be high impedance (neither high nor low)
OSC1, OSC2	Crystal or resonator connections for different clock modes
PMAx (x = 0 to 15)	Parallel master port address
PMDx (x = 0 to 7)	Parallel master port data
PMENB, PMRD, PMWR	Enable and read/write strobes for parallel master port
Rxy (x = B to G, y = 0 to 15)	Digital I/O pins
RTCC	Real-time clock alarm output
SCLx, SDAx (x = 1, 3, 4, 5)	I^2C serial clock and data input/output for I^2C synchronous serial communication modules
SCKx, SDIx, SDOx (x = 2 to 4)	Serial clock, serial data in, out for SPI synchronous serial communication modules
$\overline{\text{SSx}}$ (x = 2 to 4)	Slave select (active low) for SPI communication
T1CK	Input pin for counter/timer 1 when counting external pulses
$\overline{\text{UxCTS}}$, $\overline{\text{UxRTS}}$, UxRX, UxTX (x = 1 to 6)	UART clear to send, request to send, receive input, and transmit output for UART modules
VDD	Positive voltage supply for peripheral digital logic and I/O pins (3.3 V on NU32)
VDDCAP	Capacitor filter for internal 1.8 V regulator when ENVREG enabled
VDDCORE	External 1.8 V supply when ENVREG disabled
VREF-, VREF+	Can be used as negative and positive limit for ADC
VSS	Ground for logic and I/O
VBUS	Monitors USB bus power
VUSB	Power for USB transceiver
USBID	USB on-the-go (OTG) detect

See Section 1 of the Data Sheet for more information.

Which function a particular pin actually serves is determined by *Special Function Registers* (SFRs). Each SFR is a 32-bit word that sits at a memory address. The values of the SFR bits, 0 (cleared) or 1 (set), control the functions of the pins as well as other PIC32 behavior.

For example, pin 51 in Figure 2.1 can be OC4 (output compare 4) or RD3 (digital I/O number 3 on port D). If we wanted to use pin 51 as a digital output we would set the SFRs that control this pin to disable the OC4 functionality and enable RD3 as a digital output. The Data Sheet explains the memory addresses and meanings of the SFRs. Be careful, because it includes information for many different PIC32 models. Looking at the Data Sheet section on Output Compare reveals that the 32-bit SFR named "OC4CON" determines whether OC4 is enabled. Specifically, for bits numbered 0-31, we see that bit 15 is responsible for enabling or disabling OC4. We refer to this bit as OC4CON⟨15⟩. If it is cleared (0), OC4 is disabled, and if it is set (1), OC4 is enabled. So we clear this bit to 0. (Bits can be "cleared to 0" or simply "cleared," or "set to 1" or simply "set.") Now, referring to the I/O Ports section of the Data Sheet, we see that the input/output direction of Port D is controlled by the SFR TRISD, and bits 0-11 correspond to RD0-RD11. Bit 3 of the SFR TRISD, i.e., TRISD⟨3⟩, should be cleared to 0 to make RD3 (pin 51) a digital output.

According to the Memory Organization section of the Data Sheet, OC4CON⟨15⟩ is cleared by default on reset, so it is not necessary for our program to clear OC4CON⟨15⟩. On the other hand, TRISD⟨3⟩ is set to 1 on reset, making pin 51 a digital input by default, so the program must clear TRISD⟨3⟩. For safety, all pins are inputs on reset to prevent the PIC32 from imposing an unwanted voltage on external circuitry.

In addition to setting the behavior of the pins, SFRs are the primary means of communication between the PIC32's CPU and its peripherals. You can think of a peripheral, such as a UART communication peripheral, as an independent circuit on the same chip as the CPU. Your program, running on the CPU, configures behavior of this circuit (such as the speed of UART communication) by writing bits to one or more SFRs which are read by the peripheral circuit. The CPU sends data to the peripheral (e.g., data to be sent by the UART) by writing to SFRs, and the CPU receives data from the peripheral (e.g., data received by the UART) by reading SFRs controlled by the peripheral.

We will see and use SFRs repeatedly as we learn about the PIC32.

2.1.2 PIC32 Architecture

Peripherals

Figure 2.2 depicts the PIC32's architecture. Of course there is a CPU, program memory (flash), and data memory (RAM). Perhaps most interesting to us, though, are the *peripherals*, which make microcontrollers useful for embedded control. We briefly discuss the available peripherals here; subsequent chapters cover them in detail. The peripherals are listed roughly in top to bottom, left to right order, as they appear in Figure 2.2.

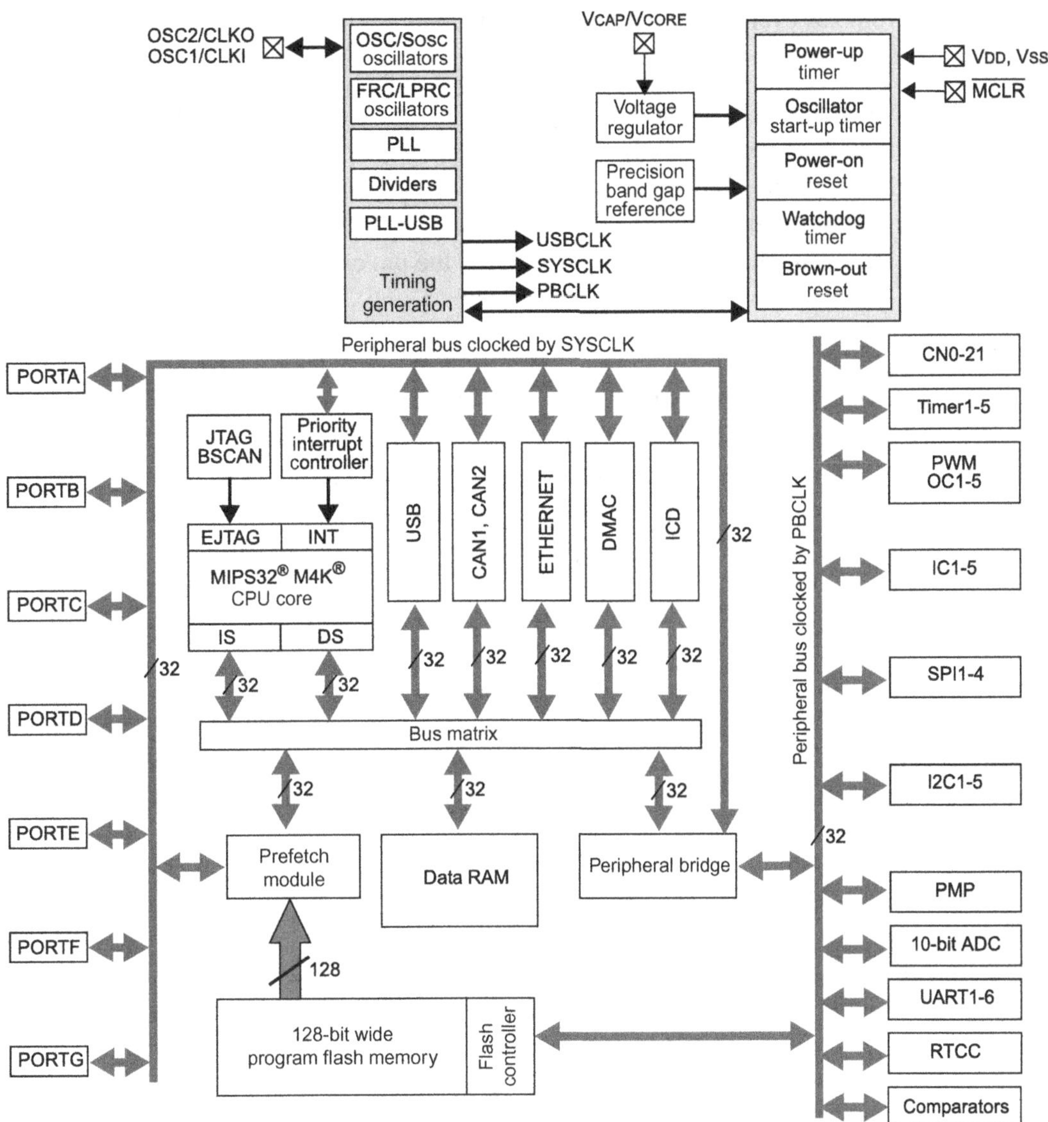

Figure 2.2

The PIC32MX5XX/6XX/7XX architecture. The PIC32MX795F512H is missing the digital I/O PORTA and has only 19 change notification inputs, 3 SPI modules, and 4 I^2C modules.

Digital input and output

Digital I/O ports (PORTB to PORTG on the PIC32MX795F512H) allow you to read or output a digital voltage. A digital I/O pin configured as an input can detect whether the input voltage is low or high. On the NU32, the PIC32 is powered by 3.3 V, so voltages close to 0 V are considered low and those close to 3.3 V are considered high. Some input pins can tolerate up to 5.5 V, while voltages over 3.3 V on other pins could damage the PIC32 (see Figure 2.1 for the pins that can tolerate 5.5 V).

A digital I/O pin configured as an output can produce a voltage of 0 or 3.3 V. An output pin can also be configured as *open drain*. In this configuration, the pin is connected by an external pull-up resistor to a voltage of up to 5.5 V. This allows the pin's output transistor to either sink current (to pull the voltage down to 0 V) or turn off (allowing the voltage to be pulled up as high as 5.5 V), increasing the range of output voltages the pin can produce.

Universal Serial Bus

The Universal Serial Bus (USB) is an asynchronous communication protocol heavily used by computers and other devices. One master communicates with one or more slaves over a four-line bus: +5 V, ground, D+, and D− (differential data signals). The PIC32 has a single USB peripheral implementing USB 2.0 full-speed and low-speed options, and can communicate at theoretical data rates of up to several megabits per second.

Controller area network

Controller area network (CAN) is pervasive in industrial and automotive applications, where electrical noise can be problematic. CAN allows many devices to communicate over a single two-wire bus. Data rates of up to 1 megabit per second are possible. The CAN peripheral uses an external transceiver chip to convert between signals on the bus and signals that the PIC32 can process. The PIC32 contains two CAN modules.

Ethernet

The Ethernet module allows the PIC32 to connect to the Internet. It uses an external physical layer protocol transceiver (PHY) chip and direct memory access (DMA) to offload the heavy processing requirements of Ethernet communication from the CPU. The NU32 board does not include a PHY chip.

DMA controller

The direct memory access (DMA) controller (DMAC) transfers data without involving the CPU. For example, DMA can allow an external device to dump data through a UART directly into PIC32 RAM.

In-Circuit Debugger

The In-Circuit Debugger (ICD) is used by Microchip debugging tools to control the PIC32's operation during debugging.

Watchdog timer

If the watchdog timer (WDT) is used by your program, your program must periodically reset a counter. Otherwise, when the counter reaches a specified value, the PIC32 will reset. The WDT allows the PIC32 to recover from an unexpected state or infinite loop caused by software errors.

Change notification

A change notification (CN) pin can be used to generate an interrupt when the input voltage changes from low to high or vice-versa. The PIC32 has 19 change notification pins (CN0 to CN18).

Counter/timers

The PIC32 has five 16-bit counters/timers (Timer1 to Timer5). A counter counts the number of pulses of a signal. If the pulses occur at a regular frequency, the count can be used as a time; hence timers are just counters with inputs at a fixed frequency. Microchip uniformly refers to these devices as "timers", so we adopt that terminology from now on. Each timer can count from 0 up to $2^{16} - 1$, or any preset value less than $2^{16} - 1$ that we choose, before rolling over. Timers can count external events, such as pulses on the T1CK pin, or internal pulses on the peripheral bus clock. Two 16-bit timers can be configured to make a single 32-bit timer. Two different pairs of timers can be combined, yielding one 16-bit and two 32-bit timers.

Output compare

The five output compare (OC) pins (OC1 to OC5) are used to generate a single pulse of specified duration, or a continuous pulse train of specified duty cycle and frequency. They work with timers to generate pulses with precise timing. Output compare is commonly used to generate PWM (pulse-width modulated) signals that can control motors or be low-pass filtered to create a specified analog voltage output. (You cannot output an arbitrary analog voltage from the PIC32.)

Input capture

The five input capture (IC) pins (IC1 to IC5) store the current time, as measured by a timer, when an input changes. Thus, this peripheral allows precise measurements of input pulse widths and signal frequencies.

Serial Peripheral Interface

The PIC32 has three Serial Peripheral Interface (SPI) peripherals (SPI2 to SPI4). The SPI bus provides a method for synchronous serial communication between a master device (typically a microcontroller) and one or more slave devices. The interface typically requires four communication pins: a clock (SCK), data in (SDI), data out (SDO), and slave select (SS). Communication can occur at up to tens of megabits per second.

Inter-integrated circuit

The PIC32 has four inter-integrated circuit (I^2C) modules (I2C1, I2C3, I2C4, I2C5). I^2C (pronounced "I squared C") is a synchronous serial communication standard (like SPI) that allows several devices to communicate over only two wires. Any device can be the master and control communication at any given time. The maximum data rate is less than for SPI, usually 100 or 400 kilobits per second.

Parallel master port

The parallel master port (PMP) module is used to read data from and write data to external parallel devices. Parallel communication allows multiple data bits to be transferred simultaneously, but each bit requires its own wire.

Analog input

The PIC32 has one analog-to-digital converter (ADC), but 16 different pins can be connected to it, allowing up to 16 analog voltage values (typically sensor inputs) to be monitored. The ADC can be programmed to continuously read data from a sequence of input pins, or to read a single value. Input voltages must be between 0 and 3.3 V. The ADC has 10 bits of resolution, allowing it to distinguish $2^{10} = 1024$ different voltage levels. Conversions are theoretically possible at a maximum rate of 1 million samples per second.

Universal asynchronous receiver/transmitter

The PIC32 has six universal asynchronous receiver transmitter (UART) modules (UART1 to UART6). These peripherals provide another method for serial communication between two devices. Unlike synchronous serial protocols such as SPI, the UART has no clock line; rather the devices communicating each have their own clocks that must operate at the same frequency. Each of the two devices participating in UART communication has, at minimum, a receive (RX) and transmit (TX) line. Often request to send (RTS) and clear to send (CTS) lines are used as well, allowing the devices to coordinate when to send data. Typical data rates are 9600 bits per second (9600 baud) up to hundreds of thousands of bits per second. The `talkingPIC.c` program uses a UART configured to operate at 230,400 baud to communicate with your computer.

Real-time clock and calendar

The real-time clock and calendar (RTCC) module maintains accurate time, in seconds, minutes, days, months, and years, over extended periods of time.

Comparators

The PIC32 has two comparators, each of which compares two analog input voltages and determines which is larger. A configurable internal voltage reference may be used in the comparison, or even output to a pin, resulting in a limited-resolution digital-to-analog converter.

Other components

Note that the peripherals are on two different buses: one is clocked by the system clock SYSCLK, and the other is clocked by the peripheral bus clock PBCLK. A third clock, USBCLK, is used for USB communication. The timing generation block that creates these clock signals and other elements of the architecture in Figure 2.2 are briefly described below.

CPU

The central processing unit runs everything. It fetches program instructions over its "instruction side" (IS) bus, reads data over its "data side" (DS) bus, executes the instructions, and writes the results over the DS bus. The CPU can be clocked by SYSCLK at up to 80 MHz, meaning it can execute one instruction every 12.5 ns. The CPU is capable of multiplying a 32-bit integer by a 16-bit integer in one cycle, or a 32-bit integer by a 32-bit integer in two cycles. There is no floating point unit (FPU), so floating point math is carried out by software algorithms, making floating point operations much slower than integer math.

The CPU is the MIPS32® M4K® microprocessor core, licensed from Imagination Technologies. The CPU operates at 1.8 V (provided by a voltage regulator internal to the PIC32, as it's used on the NU32 board). The interrupt controller, discussed below, can notify the CPU about external events.

Bus matrix

The CPU communicates with other units through the 32-bit bus matrix. Depending on the memory address specified by the CPU, the CPU can read data from, or write data to, program memory (flash), data memory (RAM), or SFRs. The memory map is discussed in Section 2.1.3.

Interrupt controller

The interrupt controller presents "interrupt requests" to the CPU. An interrupt request (IRQ) may be generated by a variety of sources, such as a changing input on a change notification pin or by the elapsing of a specified time on one of the timers. If the CPU accepts the request, it will suspend whatever it is doing and jump to an *interrupt service routine* (ISR), a function defined in the program. After completing the ISR, program control returns to where it was suspended. Interrupts are an extremely important concept in embedded control and are discussed thoroughly in Chapter 6.

Memory: Program flash and data RAM

The PIC32 has two types of memory: flash and RAM. Flash is generally more plentiful on PIC32's (e.g., 512 KB flash vs. 128 KB RAM on the PIC32MX795F512H), nonvolatile (meaning that its contents are preserved when powered off, unlike RAM), but slower to read and write than RAM. Your program is stored in flash memory and your temporary data is stored in RAM. When you power cycle the PIC32, your program is still there but your data in RAM is lost.[1]

Because flash is slow, with a max access speed of 30 MHz for the PIC32MX795F512H, reading a program instruction from flash may take three CPU cycles when operating at 80 MHz (see Electrical Characteristics in the Data Sheet). The prefetch cache module (described below) can minimize or eliminate the need for the CPU to wait for program instructions to load from flash.

Prefetch cache module

You might be familiar with the term *cache* from your web browser. Your browser's cache stores recent documents or pages you have accessed, so the next time you request them, your browser can provide a local copy immediately, instead of waiting for the download.

The *prefetch cache module* operates similarly—it stores recently executed program instructions, which are likely to be executed again soon (as in a program loop), and, in linear code with no branches, it can even run ahead of the current instruction and predictively *prefetch* future instructions into the cache. In both cases, the goal is to have the next instruction requested by the CPU already in the cache. When the CPU requests an instruction, the cache is first checked. If the instruction at that memory address is in the cache (a cache *hit*), the prefetch module provides the instruction to the CPU immediately. If there is a *miss*, the slower load from flash memory begins.

In some cases, the prefetch module can provide the CPU with one instruction per cycle, hiding the delays due to slow flash access. The module can cache all instructions in small program loops, so that flash memory does not have to be accessed while executing the loop. For linear code, the 128-bit wide data path between the prefetch module and flash memory allows the prefetch module to run ahead of execution despite the slow flash load times.

The prefetch cache module can also store constant data.

Clocks and timing generation

There are three clocks on the PIC32: SYSCLK, PBCLK, and USBCLK. USBCLK is a 48 MHz clock used for USB communication. SYSCLK clocks the CPU at a maximum

[1] It is also possible to store program instructions in RAM, and to store data in flash, but we ignore that for now.

frequency of 80 MHz, adjustable down to 0 Hz. Higher frequency means more calculations per second but higher power usage (approximately proportional to frequency). PBCLK is used by many peripherals, and its frequency is set to SYSCLK's frequency divided by 1, 2, 4, or 8. You might want to set PBCLK's frequency lower than SYSCLK's if you want to save power. If PBCLK's frequency is less than SYSCLK's, then programs with back-to-back peripheral operations will cause the CPU to wait a few cycles before issuing the second peripheral command to ensure that the first one has completed.

All clocks are derived either from an oscillator internal to the PIC32 or an external resonator or oscillator provided by the user. High-speed operation requires an external circuit, so the NU32 provides an external 8 MHz resonator as a clock source. The NU32 software sets the PIC32's configuration bits (see Section 2.1.4) to use a phase-locked loop (PLL) on the PIC32 to multiply this frequency by a factor of 10, generating a SYSCLK of 80 MHz. The PBCLK is set to the same frequency. The USBCLK is also derived from the 8 MHz resonator by multiplying the frequency by 6.

2.1.3 The Physical Memory Map

The CPU accesses peripherals, data, and program instructions in the same way: by writing a memory address to the bus. The PIC32's memory addresses are 32-bits long, and each address refers to a byte in the *memory map*. Thus, the PIC32's memory map consists of 4 GB (four gigabytes, or 2^{32} bytes). Of course most of these addresses are meaningless; there are far more addresses than needed.

The PIC32's memory map consists of four main components: RAM, flash, peripheral SFRs that we write to (to control the peripherals or send outputs) or read from (to get sensor input, for example), and *boot flash*. Of these, we have not yet seen "boot flash." This extra flash memory, 12 KB on the PIC32MX795F512H, contains program instructions that are executed immediately upon reset.[2] The boot flash instructions typically perform PIC32 initialization and then call the program installed in program flash. For the PIC32 on the NU32 board, the boot flash contains a "bootloader" program that communicates with your computer when you load a new program on the PIC32 (see Chapter 3).

The following table illustrates the PIC32's *physical* memory map. It consists of a block of "RAMsize" bytes of RAM (128 KB for the PIC32MX795F512H), "flashsize" bytes of flash (512 KB for the PIC32MX795F512H), 1 MB for the peripheral SFRs, and "bootsize" for the boot flash (12 KB for the PIC32MX795F512H):

---

[2] The last four 32-bit words of the boot flash memory region are Device Configuration Registers (see Section 2.1.4).

Physical Memory Start Address	Size (bytes)	Region
0x00000000	RAMsize (128 KB)	Data RAM
0x1D000000	flashsize (512 KB)	Program Flash
0x1F800000	1 MB	Peripheral SFRs
0x1FC00000	bootsize (12 KB)	Boot Flash

The memory regions are not contiguous. For example, the first address of program flash is 480 MB after the first address of data RAM. An attempt to access an address between the data RAM segment and the program flash segment would generate an error.

It is also possible to allocate a portion of RAM to hold program instructions.

In Chapter 3, when we discuss programming the PIC32, we will introduce the *virtual* memory map and its relationship to the physical memory map.

2.1.4 Configuration Bits

The last four 32-bit words of the boot flash are the Device Configuration Registers, DEVCFG0 to DEVCFG3, containing the *configuration bits*. The values in these configuration bits determine important properties of how the PIC32 will function. You can learn more about configuration bits in the Special Features section of the Data Sheet. For example, DEVCFG1 and DEVCFG2 contain configuration bits that determine the frequency multiplier converting the external resonator frequency to the SYSCLK frequency, as well as bits that determine the ratio between the SYSCLK and PBCLK frequencies. On the NU32 board (below), the PIC32's configuration bits were programmed along with the bootloader.

2.2 The NU32 Development Board

The NU32 development board is shown in Figure 1.1, and the pinout is given in Table 2.2. The NU32 board provides easy breadboard access to most of the PIC32MX795F512H's 64 pins. The NU32 acts like a big 60-pin DIP (dual in-line package) chip and plugs into a standard prototyping breadboard as shown in Figure 1.1. More details and the latest information on the NU32 can be found on the book's website.

Beyond simply breaking out the pins, the NU32 provides many features that make it easy to get started with the PIC32. For example, to power the PIC32, the NU32 provides a barrel jack that accepts a 1.35 mm inner diameter, 3.5 mm outer diameter center-positive power plug. The plug should provide 1 A at DC 6 V or more. The PIC32 requires a supply voltage VDD between 2.3 and 3.6 V, and the NU32 provides a 3.3 V voltage regulator providing a stable voltage source for the PIC32 and other electronics on board. Since it is often convenient to have a 5 V supply available, the NU32 also has a 5 V regulator. The power plug's raw input

Table 2.2: The NU32 pinout (in gray, with power jack at top) with PIC32MX795F512H pin numbers

Function	PIC32			PIC32	Function
GND		GND	GND		GND
3.3 V		3.3 V	3.3 V		3.3 V
5 V		5 V	5 V		5 V
VIN		VIN	VIN		VIN
C1RX/RF0	√ 58	**F0**	GND		GND
C1TX/RF1	√ 59	**F1**	**G9**	8 √	U6RX/$\overline{\text{U3CTS}}$/PMA2/CN11/RG9
PMD0/RE0	√ 60	E0	**G8**	6 √	SCL4/SDO2/U3TX/PMA3/CN10/RG8
PMD1/RE1	√ 61	E1	**G7**	5 √	SDA4/SDI2/U3RX/PMA4/CN9/RG7
PMD2/RE2	√ 62	E2	**G6**	4 √	SCK2/U6TX/$\overline{\text{U3RTS}}$/PMA5/CN8/RG6
PMD3/RE3	√ 63	E3	**$\overline{\text{MCLR}}$**	7 √	$\overline{\text{MCLR}}$
PMD4/RE4	√ 64	E4	**D7**	55 √	CN16/RD7
PMD5/RE5	√ 1	E5	D6	54 √	CN15/RD6
PMD6/RE6	√ 2	E6	D5	53 √	PMRD/CN14/RD5
PMD7/RE7	√ 3	E7	D4	52 √	OC5/IC5/PMWR/CN13/RD4
AN0/PMA6/CN2/RB0	16	B0	D3	51 √	SCL3/SDO3/U1TX/OC4/RD3
AN1/CN3/RB1	15	B1	D2	50 √	SDA3/SDI3/U1RX/OC3/RD2
AN2/C2IN-/CN4/RB2	14	B2	D1	49 √	SCK3/U4TX/$\overline{\text{U1RTS}}$/OC2/RD1
AN3/C2IN+/CN5/RB3	13	B3	D0	46 √	OC1/INT0/RD0
AN4/C1IN-/CN6/RB4	12	B4	C14	48	T1CK/CN0/RC14
AN5/C1IN+/CN7/RB5	11	B5	C13	47	CN1/RC13
AN6/OCFA/RB6	17	B6	D11	45 √	IC4/PMA14/INT4/RD11
AN7/RB7	18	B7	D10	44 √	SCL1/IC3/PMA15/INT3/RD10
AN8/C2TX/U5RX/$\overline{\text{U2CTS}}$/RB8	21	B8	D9	43 √	U4RX/$\overline{\text{U1CTS}}$/SDA1/IC2/INT2/RD9
AN9/PMA7/RB9	22	B9	D8	42 √	IC1/INT1/RD8
AN10/PMA13/RB10	23	B10	G2	37	D+/RG2
AN11/PMA12/RB11	24	B11	G3	36	D−/RG3
AN12/PMA11/RB12	27	B12	VBUS	34 √	VBUS
AN13/PMA10/RB13	28	B13	F3	33 √	USBID/RF3
AN14/C2RX/SCK4/U5TX/$\overline{\text{U2RTS}}$/ PMA1/RB14	29	B14	F4	31 √	SDA5/SDI4/U2RX/PMA9/CN17/RF4
AN15/OCFB/PMA0/CN12/RB15	30	B15	F5	32 √	SCL5/SDO4/U2TX/PMA8/CN18/RF5

Pins marked with a √ are 5.5 V tolerant. Not all pin functions are listed; see Figure 2.1 or the PIC32 Data Sheet. Board pins in **bold** should only be used with care, as they are shared with other functions on the NU32. In particular, the NU32 pins G6, G7, G8, G9, F0, F1, D7, and MCLR should be considered outputs during normal usage. The value of $\overline{\text{MCLR}}$ is determined by the $\overline{\text{MCLR}}$ button on the NU32; the value of D7 is determined by the USER button; F0 and F1 are used by the PIC32 as digital outputs to control LED1 and LED2 on the NU32, respectively; and G6 through G9 are used by the PIC32's UART3 for communication with the host computer through the mini-B USB jack.

voltage Vin and ground, as well as the regulated 3.3 V and 5 V supplies, are made available to the user as illustrated in Figure 1.1. The power jack is directly connected to the Vin and GND pins so you could power the NU32 by putting Vin and GND on these pins directly and not connecting the power jack.

The 3.3 V regulator provides up to 800 mA and the 5 V regulator provides up to 1 A of current, provided the power supply can source that much current. In practice you should stay well under each of these limits. For example, you should not plan to draw more than 200-300 mA or so from the NU32. Even if you use a higher-current power supply, such as a battery, you should respect these limits, as the current has to flow through the relatively thin traces of the PCB. It is also not recommended to use high voltage supplies greater than 9 V or so, as the regulators will heat up.

Since motors tend to draw lots of current (even small motors may draw hundreds of milliamps up to several amps), do not try to power them from the NU32. Use a separate battery or power supply instead.

In addition to the voltage regulators, the NU32 provides an 8 MHz resonator as the source of the PIC32's 80 MHz clock signal. It also has a mini-B USB jack to connect your computer's USB port to a USB-to-UART FTDI chip that allows your PIC32 to use its UART to communicate with your computer.

A USB micro-B jack is provided to allow the PIC32 to speak USB to another external device, like a smartphone.

The NU32 board also has a power switch which connects or disconnect the input power supply to the voltage regulators, and two LEDs and two buttons (labeled USER and RESET) allowing very simple input and output. The two LEDs, LED1 and LED2, are connected at one end by a resistor to 3.3 V and the other end to digital outputs RF0 and RF1, respectively, so that they are off when those outputs are high and on when they are low. The USER and RESET buttons are attached to the digital input RD7 and $\overline{\text{MCLR}}$ pins, respectively, and both buttons are configured to give 0 V to these inputs when pressed and 3.3 V otherwise. See Figure 2.3.

Because pins RG6 through RG9, RF0, RF1, and RD7 on the PIC32 are used for UART communication with the host computer, LEDs, and the USER button, other potential functions of these pins are not available if you would like to use the communication, LEDs, and USER button. In particular:

- UART6 is unavailable (conflicts with pins RG6 and RG9). Since UART3 is used for communication with the host computer, this leaves UART1, UART2, UART4, and UART5 for your programs.
- SPI2 is unavailable (conflicts with pins RG6 and RG7). This leaves SPI3 and SPI4.
- I2C4 is unavailable (conflicts with pins RG7 and RG8). This leaves I2C1, I2C3, and I2C5.

- The default CAN1 pins C1RX and C1TX cannot be used (they conflict with pins RF0 and RF1), but the configuration bit FCANIO in DEVCFG3 has been cleared to zero on the NU32, thereby setting CAN1 to use the alternate pins AC1RX (RF4) and AC1TX (RF5). Therefore no CAN module is lost.
- Media-independent interface (MII) Ethernet is unavailable (conflicts with pins RD7, RF0, and RF1). The PIC32 can implement Ethernet communication using either the MII or the reduced media-independent interface (RMII), and RMII Ethernet communication, which uses many fewer pins than MII, is still available on the NU32.
- Several change notification and digital I/O pins are unavailable, but many more remain.

In all, very little functionality is unavailable due to connections on the NU32, and advanced users can find ways to bypass even these limitations.

Although the NU32 comes with a bootloader installed in its flash memory, you have the option to use a programmer to install a standalone program. The five plated through-holes on the USB board align with the pins of devices such as the PICkit 3 programmer (Figure 2.4).

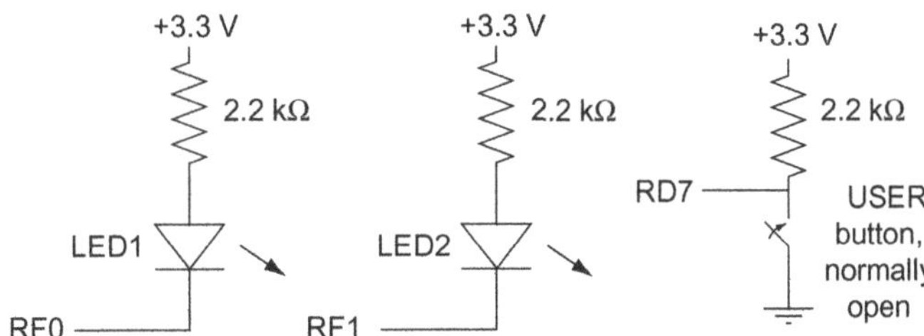

Figure 2.3
The NU32 connection of the PIC32 pins RF0, RF1, and RD7 to LED1, LED2, and the USER button, respectively.

Figure 2.4
Attaching the PICkit 3 programmer to the NU32 board.

2.3 Chapter Summary

- The PIC32 features a 32-bit data bus and a CPU capable of performing some 32-bit operations in a single clock cycle.
- In addition to nonvolatile flash program memory and RAM data memory, the PIC32 provides peripherals particularly useful for embedded control, including analog inputs, digital I/O, PWM outputs, counter/timers, inputs that generate interrupts or measure pulse widths or frequencies, and pins for a variety of communication protocols, including USB, Ethernet, CAN, I^2C, and SPI.
- The functions performed by the pins and peripherals are determined by Special Function Registers. SFRs are also used for communication back and forth between the CPU and peripherals.
- The PIC32 has three main clocks: the SYSCLK that clocks the CPU, the PBCLK that clocks peripherals, and the USBCLK that clocks USB communication.
- Physical memory addresses are specified by 32 bits. The physical memory map contains four regions: data RAM, program flash, SFRs, and boot flash. RAM can be accessed in one clock cycle, while flash access may be slower. The prefetch cache module can be used to minimize delays in accessing program instructions.
- Four 32-bit configuration words, DEVCFG0 to DEVCFG3, set important behavior of the PIC32. For example, these configuration bits determine how an external clock frequency is multiplied or divided to create the PIC32 clocks.
- The NU32 development board provides voltage regulators for power, includes a resonator for clocking, breaks out the PIC32 pins to a solderless breadboard, provides a couple of LEDs and buttons for simple input and output, and simplifies communication with the PIC32 via your computer's USB port.

2.4 Exercises

You will need to refer to the PIC32MX5XX/6XX/7XX Data Sheet and PIC32 Reference Manual to answer some questions.

1. Search for a listing of PIC32 products on Microchip's webpage, showing the specifications of all the PIC32 models.
   a. Find PIC32s that meet the following specs: at least 128 KB of flash, at least 32 KB of RAM, and at least 80 MHz max CPU speed. What is the cheapest PIC32 that meets these specs, and what is its volume price? How many ADC, UART, SPI, and I^2C channels does it have? How many timers?
   b. What is the cheapest PIC32 overall? How much flash and RAM does it have, and what is its maximum clock speed?
   c. Among all PIC32s with 512 KB flash and 128 KB RAM, which is the cheapest? How does it differ from the PIC32MX795F512H?

2. Based on C syntax for bitwise operators and bit-shifting, calculate the following and give your results in hexadecimal.
   a. `0x37 | 0xA8`
   b. `0x37 & 0xA8`
   c. `~0x37`
   d. `0x37>>3`
3. Describe the four functions that pin 12 of the PIC32MX795F512H can have. Is it 5 V tolerant?
4. Referring to the Data Sheet section on I/O Ports, what is the name of the SFR you have to modify if you want to change pins on PORTC from output to input?
5. The SFR CM1CON controls comparator behavior. Referring to the Memory Organization section of the Data Sheet, what is the reset value of CM1CON in hexadecimal?
6. In one sentence each, without going into detail, explain the basic function of the following items shown in the PIC32 architecture block diagram Figure 2.2: SYSCLK, PBCLK, PORTA to PORTG (and indicate which of these can be used for analog input on the NU32's PIC32), Timer1 to Timer5, 10-bit ADC, PWM OC1-5, Data RAM, Program Flash Memory, and Prefetch Cache Module.
7. List the peripherals that are *not* clocked by PBCLK.
8. If the ADC is measuring values between 0 and 3.3 V, what is the largest voltage difference that it may not be able to detect? (It's a 10-bit ADC.)
9. Refer to the Reference Manual chapter on the Prefetch Cache. What is the maximum size of a program loop, in bytes, that can be completely stored in the cache?
10. Explain why the path between flash memory and the prefetch cache module is 128 bits wide instead of 32, 64, or 256 bits.
11. Explain how a digital output could be configured to swing between 0 and 4 V, even though the PIC32 is powered by 3.3 V.
12. PIC32's have increased their flash and RAM over the years. What is the maximum amount of flash memory a PIC32 can have before the current choice of base addresses in the physical memory map (for RAM, flash, peripherals, and boot flash) would have to be changed? What is the maximum amount of RAM? Give your answers in bytes in hexadecimal.
13. Examine the Special Features section of the Data Sheet.
    a. If you want your PBCLK frequency to be half the frequency of SYSCLK, which bits of which Device Configuration Register do you have to modify? What values do you give those bits?
    b. Which bit(s) of which SFR set the watchdog timer to be enabled? Which bit(s) set the postscale that determines the time interval during which the watchdog must be reset to prevent it from restarting the PIC32? What values would you give these bits to enable the watchdog and to set the time interval to be the maximum?

c. The SYSCLK for a PIC32 can be generated several ways, as discussed in the Oscillator chapter in the Reference Manual and the Oscillator Configuration section in the Data Sheet. The PIC32 on the NU32 uses the (external) primary oscillator in HS mode with the phase-locked loop (PLL) module. Which bits of which device configuration register enable the primary oscillator and turn on the PLL module?

14. Your NU32 board provides four power rails: GND, regulated 3.3 V, regulated 5 V, and the unregulated input voltage (e.g., 6 V). You plan to put a load from the 5 V output to ground. If the load is modeled as a resistor, what is the smallest resistance that would be safe? An approximate answer is fine. In a sentence, explain how you arrived at the answer.
15. The NU32 could be powered by different voltages. Give a reasonable range of voltages that could be used, minimum to maximum, and explain the reason for the limits.
16. Two buttons and two LEDs are interfaced to the PIC32 on the NU32. Which pins are they connected to? Give the actual pin numbers, 1-64, as well as the name of the pin function as it is used on the NU32. For example, pin 37 on the PIC32MX795F512H could have the function D+ (USB data line) or RG2 (Port G digital input/output), but only one of these functions could be active at a given time.

Further Reading

PIC32 family reference manual. Section 03: Memory organization. (2010). Microchip Technology Inc.

PIC32 family reference manual. Section 02: CPU for devices with the M4K core. (2012). Microchip Technology Inc.

PIC32 family reference manual. Section 32: Configuration. (2013). Microchip Technology Inc.

PIC32MX5XX/6XX/7XX family data sheet. (2013). Microchip Technology Inc.

CHAPTER 3

Software

In this chapter we explore how a simple C program interacts with the hardware described in the previous chapter. We begin by introducing the virtual memory map and its relationship to the physical memory map. We then use the `simplePIC.c` program from Chapter 1 to explore the compilation process and the XC32 compiler installation.

3.1 The Virtual Memory Map

In the previous chapter we learned about the PIC32's physical memory map, which allows the CPU to access any SFR or any location in data RAM, program flash, or boot flash, using a 32-bit address. The PIC32 does not actually have 2^{32} bytes, or 4 GB, worth of SFRs and memory; therefore, many physical addresses are invalid.

Rather than use physical addresses (PAs), software refers to memory and SFRs using virtual addresses (VAs). The fixed mapping translation (FMT) unit in the CPU converts VAs into PAs using the following formula:

$$PA = VA \ \& \ 0x1FFFFFFF$$

This bitwise AND operation clears the three most significant bits of the address; thus multiple VAs map to the same PA.

If the mapping from the VA to the PA just discards the first three bits, why bother having them? Well, the CPU and the prefetch cache module we learned about in the previous chapter use them. If the first three bits of the virtual address are 0b100 (corresponding to an 8 or 9 as the most significant hex digit of the VA), then the contents of that memory address can be cached. If the first three bits are 0b101 (corresponding to an A or B as the most significant hex digit of the VA), then it cannot be cached. Thus the segment of virtual memory 0x80000000 to 0x9FFFFFFF is cacheable, while the segment 0xA0000000 to 0xBFFFFFFF is noncacheable. The cacheable segment is called KSEG0 (for "kernel segment") and the noncacheable segment is called KSEG1.

Embedded Computing and Mechatronics with the PIC32 Microcontroller. http://dx.doi.org/10.1016/B978-0-12-420165-1.00003-2

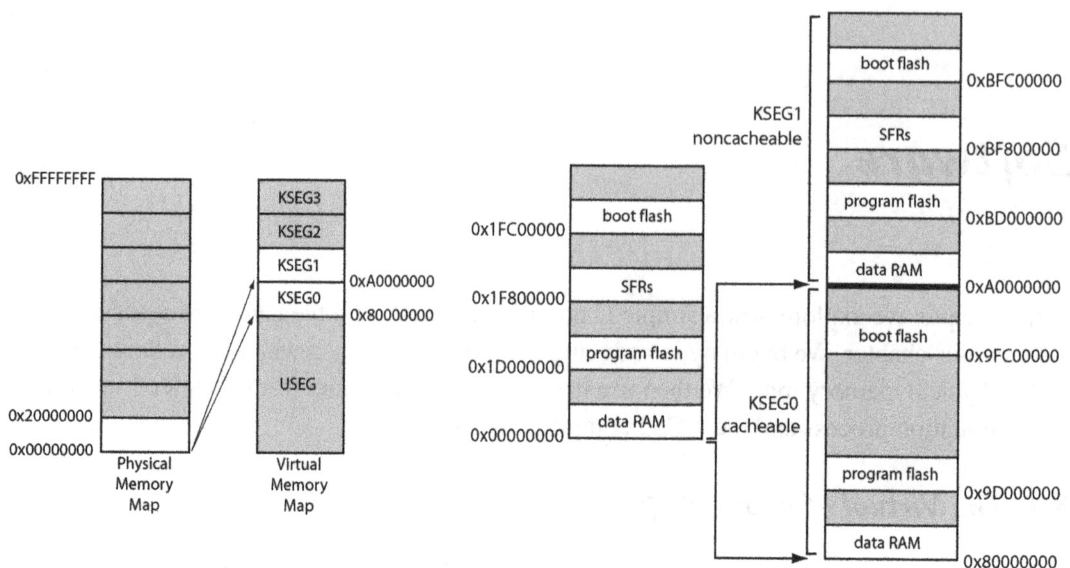

Figure 3.1

(Left) The 4 GB physical and virtual memory maps are divided into 512 MB segments. The mapping of the valid physical memory addresses to the virtual memory regions KSEG0 and KSEG1 is illustrated. We use only KSEG0 and KSEG1, not KSEG2, KSEG3, or the user segment USEG. (Right) Physical addresses mapped to virtual addresses in cacheable memory (KSEG0) and noncacheable memory (KSEG1). Note that SFRs are not cacheable. The last four words of boot flash, 0xBFC02FF0 to 0xBFC02FFF in KSEG1, correspond to the device configuration words DEVCFG0 to DEVCFG3. Memory regions are not drawn to scale.

Figure 3.1 illustrates the relationship between the physical and virtual memory maps. Note that the SFRs are excluded from the KSEG0 cacheable virtual memory segment. SFRs correspond to physical devices (e.g., peripherals); therefore their values cannot be cached. Otherwise, the CPU could read outdated SFR values because the state of the SFR could change between when it was cached and when it was needed by the CPU. For instance, if port B were configured as a digital input port, the SFR PORTB would contain the current input values of some pins. The voltage on these pins could change at any time; therefore, the only way to retrieve a reliable value is to read directly from the SFR rather than from the cache.

Also note that program flash and data RAM can be accessed using either cacheable or noncacheable VAs. Typically, you can ignore this detail because the PIC32 will be configured to access program flash via the cache (since flash memory is slow), and data RAM without the cache (since RAM is fast).

Going forward, we will use virtual addresses like 0x9D000000 and 0xBD000000, and you should realize that these refer to the same physical address. Since virtual addresses start at 0x80000000, and all physical addresses are below 0x20000000, there is no possibility of confusing whether we are talking about a VA or a PA.

3.2 An Example: `simplePIC.c`

Let us build the `simplePIC.c` executable from Chapter 1. For convenience, here is the program again:

Code Sample 3.1 `simplePIC.c`**. Blinking Lights, Unless the USER Button Is Pressed.**

```
#include <xc.h>          // Load the proper header for the processor

void delay(void);

int main(void) {
  TRISF = 0xFFFC;        // Pins 0 and 1 of Port F are LED1 and LED2.  Clear
                         // bits 0 and 1 to zero, for output.  Others are inputs.
  LATFbits.LATF0 = 0;    // Turn LED1 on and LED2 off.  These pins sink current
  LATFbits.LATF1 = 1;    // on the NU32, so "high" (1) = "off" and "low" (0) = "on"

  while(1) {
    delay();
    LATFINV = 0x0003;    // toggle LED1 and LED2; same as LATFINV = 0x3;
  }
  return 0;
}

void delay(void) {
  int j;
  for (j = 0; j < 1000000; j++) { // number is 1 million
    while(!PORTDbits.RD7) {
        ;   // Pin D7 is the USER switch, low (FALSE) if pressed.
    }
  }
}
```

Navigate to the `<PIC32>` directory. Following the same procedure as in Chapter 1.3, build `simplePIC.hex` and load it onto your NU32. We have reprinted the instructions here (you may need to specify the full path to these commands):

```
> xc32-gcc -mprocessor=32MX795F512H
   -o simplePIC.elf -Wl,--script=skeleton/NU32bootloaded.ld simplePIC.c
> xc32-bin2hex simplePIC.elf
> nu32utility <COM> simplePIC.hex
```

When you have the program running, the NU32's two LEDs should alternate on and off and stop while you press the USER button.

Look at the source code: the program refers to SFRs named TRISF, LATFINV, etc. These names align with the SFR names in the Data Sheet and Reference Manual sections on input/output (I/O) ports. We will often consult the Data Sheet and Reference Manual when programming the PIC32. We will explain the use of these SFRs shortly.

3.3 What Happens When You Build?

First, let us begin to understand what happens when you create `simplePIC.hex` from `simplePIC.c`. Refer to Figure 3.2.

First the **preprocessor** removes comments and inserts `#include`d header files. It also handles other preprocessor instructions such as `#define`. You can have multiple `.c` C source files and `.h` header files, but only one C file is allowed to have a `main` function. The other files may contain helper functions. We will learn more about projects with multiple C source files in Chapter 4.

Then the **compiler** turns the C files into MIPS32 assembly language files, machine instructions specific to the PIC32's MIPS32 CPU. Basic C code will not vary between processor architectures, but assembly language may be completely different. These assembly files are readable by a text editor, and it is possible to program the PIC32 directly in assembly language.

The **assembler** turns the assembly files into machine-level *relocatable object code*. This code cannot be inspected with a text editor. The code is called relocatable because the final memory addresses of the program instructions and data used in the code are not yet specified. The **archiver** is a utility that allows you to package several related `.o` object files into a single `.a` library file. We will not be making our own archived libraries, but we will certainly be using `.a` libraries that have already been made by Microchip!

Finally, the **linker** takes one or more object files and combines them into a single executable file, with all program instructions assigned to specific memory locations. The linker uses a linker script that has information about the amount of RAM and flash on your particular PIC32, as well as directions about where in virtual memory to place the data and instructions. The result is an executable and linkable format (`.elf`) file, a standard executable file format. This file contains useful debugging information as well as information that allows tools such as `xc32-objdump` to *disassemble* the file, which converts it back into assembly code (Section 3.8). This extra information adds up; building `simplePIC.c` results in a `.elf` file that is hundreds of kilobytes! A final step creates a stripped-down `.hex` file of less than 10 KB. This `.hex` file is a representation of your executable suitable for sending to the bootloader program on your PIC32 (more on this in the next section) that writes the program into flash on your PIC32.

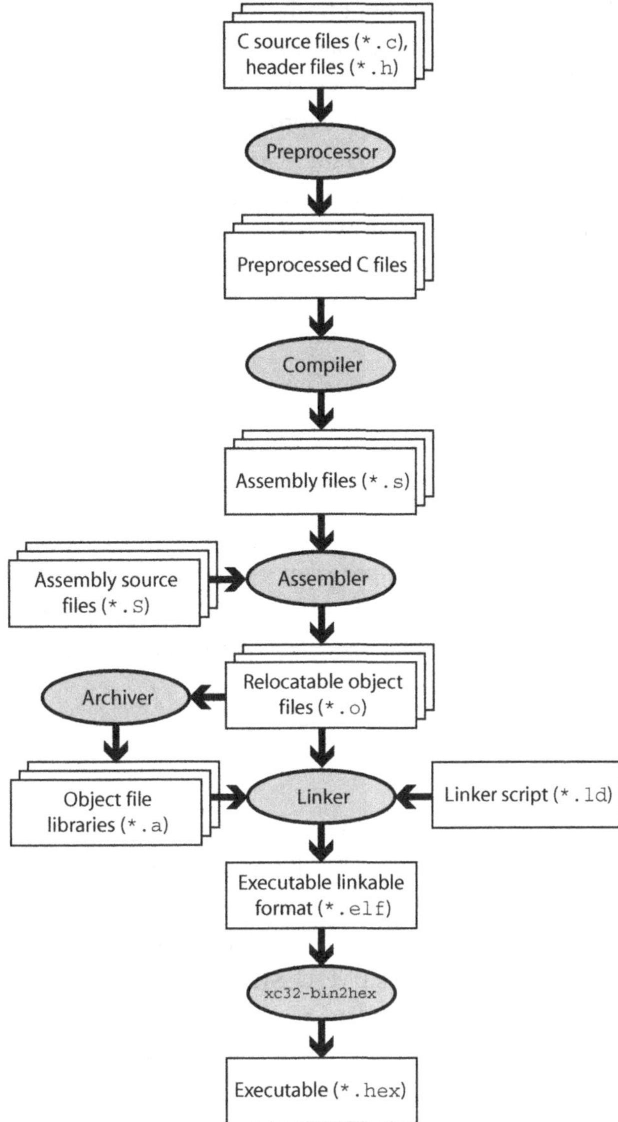

Figure 3.2
The "compilation" process.

Although the entire process consists of several steps, it is often referred to as "compiling" for short. "Building" or "making" is more accurate.

3.4 What Happens When You Reset the PIC32?

Your program is running. You hit the RESET button on the NU32. What happens next?

First the CPU jumps to the beginning of boot flash, address 0xBFC00000, and starts executing instructions.[1] For the NU32, the boot flash contains the *bootloader*, a program used to load other programs onto the PIC32. The bootloader first checks to see if you are pressing the USER button. If so, it knows that you want to reprogram the PIC32, so it attempts to communicate with the bootloader utility (`nu32utility`) on your computer. With communication established, the bootloader receives the executable `.hex` file and writes it to the PIC32's program flash (see Exercise 2). We refer to the virtual address where your program is installed as `_RESET_ADDR`.

> **Note:** The PIC32's reset address 0xBFC00000 is hardwired and cannot be changed. The address where the bootloader writes your program, however, can be changed in software.

Now assume that you were not pressing the USER button when you reset the PIC32. In that case the bootloader jumps to the address `_RESET_ADDR` and begins executing the program you previously installed there. Notice that our program, `simplePIC.c`, is an infinite loop, so it never stops executing, the desired behavior in embedded control. If a program exits, the PIC32 will sit in a tight loop, doing nothing until it is reset. (Interrupts, described in Chapter 6, will continue to execute.)

3.5 Understanding `simplePIC.c`

Let us return to understanding `simplePIC.c`. The `main` function initializes values of TRISF and LATFbits, then enters an infinite `while` loop. Each time through the loop it calls `delay()` and then assigns a value to LATFINV. The `delay` function executes a `for` loop that iterates one million times. During each iteration it enters a `while` loop, which checks the value of (!PORTDbits.RD7). If PORTDbits.RD7 is 0 (FALSE), then the expression (!PORTDbits.RD7) evaluates to TRUE, and the program remains here, doing nothing except checking the expression (!PORTDbits.RD7). When this expression evaluates to FALSE, the `while` loop exits, and the program continues with the `for` loop. After the `for` loop finishes, control returns to `main`.

Special function registers (SFRs)

The main difference between `simplePIC.c` and programs that you may have written for your computer is how it interacts with the outside world. Rather than via keyboard or mouse,

[1] If you are just powering on your PIC32, it will wait a short period while electrical transients die out, clocks synchronize, etc., before jumping to 0xBFC00000.

`simplePIC.c` accesses SFRs like TRISF, LATF, and PORTD, all of which correspond to peripherals. Specifically, TRISF, LATF, and PORTD refer to the digital I/O ports F and D. Digital I/O ports allow the PIC32 to read or set the digital voltage on a pin. To discover what these SFRs control we start by consulting the table in Section 1 of the Data Sheet, which lists the pinout descriptions. For example, we see that port D, with pins named RD0 to RD11, has 12 pins, and port F, with pins RF0, RF1, RF3, RF4, and RF5, has five pins. Port B has 16 pins, labeled RB0 to RB15.

We now turn to the Data Sheet section on I/O Ports for more information. We find that TRISF, short for "tri-state F," controls the direction, input or output, of the pins on port F. Each port F pin has a corresponding bit in TRISF. If this bit is 0, the pin is an output. If the bit is a 1, the pin is an input. ($0 = O_{\text{utput}}$ and $1 = I_{\text{nput}}$.) We can make some pins inputs and some outputs, or we can make them all have the same direction.

If you are curious about which direction the pins are by default, you can consult the Memory Organization section of the Data Sheet. Tables there list the VAs of many of the SFRs, as well as the values they default to upon reset. There are a lot of SFRs! After some searching, you will find that TRISF sits at virtual address 0xBF886140, and its default value upon reset is 0x0000003B. (We have reproduced part of this table for you in Figure 3.3.) In binary, this would be

$$\text{0x0000003B} = 0000\quad 0000\quad 0000\quad 0000\quad 0000\quad 0000\quad 0011\quad 1011.$$

The four most significant hex digits (two bytes, or 16 bits) are all 0. This is because those bits, technically, do not exist. Microchip calls them "unimplemented." No I/O port has more than 16 pins, so we do not need those bits, which are numbered 16-31. (The 32 bits are numbered 0-31.) Of the remaining bits, since the 0th bit (least significant bit) is the rightmost bit, we see that bits 0, 1, 3, 4, and 5 are 1, while the rest are 0. The bits set to 1 correspond precisely to the pins we have available, meaning that they are inputs. (The other pins are unimplemented.). I/O pins are configured as inputs on reset for safety reasons; when we power on the PIC32, each pin will take its default direction before the program can change it. If an output pin were

Virtual address (BF88_#)	Register name(1)	Bit range	Bits																All resets
			31/15	30/14	29/13	28/12	27/11	26/10	25/9	24/8	23/7	22/6	21/5	20/4	19/3	18/2	17/1	16/0	
6140	TRISF	31:16	—	—	—	—	—	—	—	—	—	—	—	—	—	—	—	—	0000
		15:0	—	—	—	—	—	—	—	—	—	—	TRISF5	TRISF4	TRISF3	—	TRISF1	TRISF0	003B

Figure 3.3
The TRISF SFR, taken from the PIC32 Data Sheet.

connected to an external circuit that is also trying to control the voltage on the pin, the two devices would fight each other, with damage to one or the other a possibility. No such problems arise if the pin is configured as an input by default.

So now we understand that the instruction

```
TRISF = 0xFFFC;
```

clears bits 0 and 1, implicitly clears bits 16-31 (which is ignored, since the bits are not implemented), and sets the rest of the bits to 1. It does not matter that we try to set some unimplemented bits to 1; those bits are ignored. The result is that port F pins 0 and 1, or RF0 and RF1 for short, are now outputs.

Our PIC32 C compiler allows the use of binary (base 2) representations of unsigned integers using 0b at the beginning of the number, so if you do not get lost counting bits, you could have equally written

```
TRISF = 0b1111111111111100;
```

or simply

```
TRISF = 0b111100;
```

since no bits are implemented after RF5.

Another option would have been to use the instructions

```
TRISFbits.TRISF0 = 0; TRISFbits.TRISF1 = 0;
```

This allows us to specify individual bits without affecting the other bits. We see this kind of notation later in the program, with LATFbits.LATF0 and LATFbits.LATF1, for example.

The two other basic SFRs in this program are LATF and PORTD. Again consulting the I/O Ports section of the Data Sheet, we see that LATF, short for "latch F," is used to write values to the output pins. Thus

```
LATFbits.LATF1 = 1;
```

sets pin RF1 high. Finally, PORTD contains the digital inputs on the port D pins. (Notice we did not configure port D as input; we relied on the fact that it's the default.) PORTDbits.RD7 is 0 if 0 V is present on pin RD7 and 1 if approximately 3.3 V is present. Note that we use the latch when writing pins and the port when reading pins, for reasons explained in Chapter 7.

Pins RF0, RF1, and RD7 on the NU32

Figure 2.3 shows how pins RF0, RF1 and RD7 are wired on the NU32 board. LED1 (LED2) is on if RF0 (RF1) is 0 and off if it is 1. When the USER button is pressed, RD7 registers a 0, and otherwise it registers a 1.

The result of these electronics and the `simplePIC.c` program is that the LEDs flash alternately, but remain unchanging while you press the USER button.

CLR, SET, and INV SFRs

So far we have ignored the instruction

```
LATFINV = 0x0003;
```

Again consulting the Memory Organization section of the Data Sheet, we see that associated with the SFR LATF are three more SFRs, called LATFCLR, LATFSET, and LATFINV. (Indeed, many SFRs have corresponding CLR, SET, and INV SFRs.) These SFRs are used to easily change some of the bits of LATF without affecting the others. A write to these registers causes a change to LATF's bits, but only in the bits corresponding to bits on the right-hand side that have a value of 1. For example,

```
LATFINV = 0x3;      // flips (inverts) bits 0 and 1 of LATF; all others unchanged
LATFINV = 0b11;     // same as above
LATFSET = 0x9;      // sets bits 0 and 3 of LATF to 1; all others unchanged
LATFCLR = 0x2;      // clears bit 1 of LATF to 0;  all others unchanged
```

A nominally less efficient way to toggle bits 0 and 1 of LATF is

```
LATA5LATFbits.LATF0 = !LATFbits.LATF0; LATFbits.LATF1 = !LATFbits.LATF1;
```

The compiler, however, sometimes optimizes these instructions into the equivalent more efficient operation. We shall look at efficiency in Chapter 5. In most cases, the difference between the methods is negligible so you should access the fields using the bit structures (e.g., LATFbits.LATF0) for code clarity.

You can return to the table in the Data Sheet to see the VAs of the CLR, SET, and INV registers. They are always offset from their base register by 4, 8, and 12 bytes, respectively. Since LATF is at 0xBF886160, LATFCLR, LATFSET, and LATFINV are at 0xBF886164, 0xBF886168, and 0xBF88616C, respectively.

You should now understand how `simplePIC.c` works. But we have ignored the fact that we never declared TRISF, LATFINV, etc., before we started using them. We know you cannot do that in C; these variables must be declared somewhere. The only place they could be declared is in the included file `xc.h`. We have ignored that `#include <xc.h>` statement until now. Time to take a look.[2]

3.5.1 Down the Rabbit Hole

Where do we find `xc.h`? The line `#include <xc.h>` means that the preprocessor will look for `xc.h` in directories specified in the *include path*.

For us, the default include path means that the compiler finds `xc.h` sitting at

```
<xc32dir>/<xc32ver>/pic32mx/include/xc.h
```

You should substitute your install directory in place of `<xc32dir>/<xc32ver>`.

Including `xc.h` gives us access to many data types, variables, and constants that Microchip has provided for our convenience. In particular, it provides variable declarations for SFRs like TRISF, allowing us to access the SFRs from C.

Before we open `xc.h`, let us examine the directory structure of the XC32 compiler installation. There's a lot here! We certainly do not need to understand it all now, but we should get a sense of what's going on. We start at the level of your XC32 install directory and summarize the important nested directories, without being exhaustive.

1. `bin`: Contains the actual executable programs that do the compiling, assembling, linking, etc. For example, `xc32-gcc` is the C compiler.
2. `docs`: Some manuals, including the XC32 C Compiler User's Guide, and other documentation.
3. `examples`: Some sample code.
4. `lib`: Contains some `.h` header files and `.a` library archives containing general C object code.
5. `pic32-libs`: This directory contains the source code (`.c` C files, `.h` header files, and `.S` assembly files) needed to create numerous Microchip-provided libraries. These files are provided for reference and are not included directly in any of your code.
6. `pic32mx`: This directory has several files we are interested in because many of them end up in your project.

[2] Microchip often changes the software it distributes, so there may be differences in details, but the essence of what we describe here will be the same.

a. `lib`: This directory consists mostly of PIC32 object code and libraries that are linked with our compiled and assembled source code. For some of these libraries, source code exists in `pic32-libs`; for others we have only the object code libraries. Some important files in this directory include:

- `proc/32MX795F512H/crt0_mips32r2.o`: The linker combines this object code with your program's object code when it creates the `.elf` file. The linker ensures that this "C Runtime Startup" code is executed first, since it performs various initializations your code needs to run, such as initializing the values of global variables. Different PIC32 models have different versions of this file under the appropriate `proc/<processor>` directory. You can find readable assembly source code at `pic32-libs/libpic32/startup/crt0.S`.
- `libc.a`: Implementations of functions that are part of the C standard library.
- `libdsp.a`: This library contains MIPS implementations of finite and infinite impulse response filters, the fast Fourier transform, and various vector math functions.
- `proc/32MX795F512H/processor.o`: This object file provides the virtual memory addresses for the PIC32's SFRs; each specific model has its own `processor.o` file. We cannot look at it directly with a text editor, but there are utilities that allow us to examine it. For example, from the command line you could use the `xc32-nm` program in the top-level `bin` directory to see all the SFR VAs:

```
> xc32-nm processor.o
bf809040 A AD1CHS
...
bf886140 A TRISF
bf886144 A TRISFCLR
bf88614c A TRISFINV
bf886148 A TRISFSET
...
```

  All of the SFRs are printed out, in alphabetical order, with their corresponding VA. The spacing between SFRs is four, since there are four bytes (32 bits) in an SFR. The "A" means that these are absolute addresses. The linker must use these addresses when making final address assignments because the SFRs are implemented in hardware and cannot be moved! The listing above indicates that TRISF is located at VA 0xBF886140, agreeing with the Memory Organization section of the Data Sheet.
- `proc/32MX795F512H/configuration.data`: This file describes some constants used in setting the configuration bits in DEVCFG0 to DEVCFG3 (Chapter 2.1.4). These bits are set by the bootloader (Section 3.6), so you do not need to worry about them in your programs. It is possible to use a programmer device to load programs onto the PIC32 without having a bootloader pre-installed on the PIC32 (that's how the

bootloader got there in the first place!), in which case you would need to worry about these bits. See Section 3.6 for more information about programs that do not use a bootloader.

b. `include`: This directory contains several `.h` header files.
   - `cp0defs.h`: This file defines constants and macros that allow us to access functions of coprocessor 0 (CP0) on the MIPS32 M4K CPU. In particular, it allows us to read and set the *core timer* clock that ticks once every two SYSCLK cycles using macros like `_CP0_GET_COUNT()` (see Chapters 5 and 6 for more details). More information on CP0 can be found in the "CPU for Devices with the M4K Core" section of the Reference Manual.
   - `sys/attribs.h`: In the directory `sys`, the file `attribs.h` defines the macro syntax `__ISR` that we will use for interrupt service routines starting in Chapter 6.
   - `sys/kmem.h`: Contains macros for converting between physical and virtual addresses.
   - `xc.h`: This is the file we include in `simplePIC.c`. The main purpose of `xc.h` is to include the appropriate processor-specific header file, in our case `include/proc/p32mx795f512h.h`. It does this by checking if `__32MX795F512H__` is defined:

```
#elif defined(__32MX795F512H__)
#include <proc/p32mx795f512h.h>
```

   If you look at the command for compiling `simplePIC.c`, you may notice the option `-mprocessor=32MX795F512H`. This option defines the constant `__32MX795F512H__` to the compiler, allowing `xc.h` to function properly. This file also defines some macros for easily inserting some specific assembly instructions directly from C.
   - `proc/p32mx795f512h.h`: Open this file in your text editor. Whoa! This file is over 40,000 lines long! It must be important. Time to look at it in more detail.

3.5.2 The Header File `p32mx795f512h.h`

The first 31% of `p32mx795f512h.h`, about 14,000 lines, consists of code like this, with line numbers added to the left for reference:

```
extern volatile unsigned int        TRISF __attribute__((section("sfrs")));
typedef union {
  struct {
    unsigned TRISF0:1;   // TRISF0 is bit 0 (1 bit long), interpreted as unsigned int
    unsigned TRISF1:1;   // bits are in order, so the next bit, bit 1, is TRISF1
    unsigned TRISF2:1;   // TRISF2 doesn't actually exist (unimplemented)
    unsigned TRISF3:1;
    unsigned TRISF4:1;
```

```
9       unsigned TRISF5:1;  // later bits are not given names, since they're
                               unimplemented
10    };
11    struct {
12      unsigned w:32;      // w refers to all 32 bits
13    };
14  } __TRISFbits_t;
15  extern volatile __TRISFbits_t TRISFbits __asm__ ("TRISF") __attribute__
    ((section("sfrs")));
16  extern volatile unsigned int        TRISFCLR __attribute__((section("sfrs")));
17  extern volatile unsigned int        TRISFSET __attribute__((section("sfrs")));
18  extern volatile unsigned int        TRISFINV __attribute__((section("sfrs")));
```

The first line, beginning `extern`, declares the variable TRISF as an `unsigned int`. The keyword `extern` means that no RAM has to be allocated for it; memory to hold the variable has been allocated for it elsewhere. In a typical C program, memory for the variable has been allocated by another C file using syntax without the `extern`, like `volatile unsigned int TRISF;`. In this case, however, no RAM has to be allocated for TRISF because it refers to an SFR, not a word in RAM. The `processor.o` file actually defines the VA of the symbol TRISF, as mentioned earlier.

The `volatile` keyword, applied to all the SFRs, means that the value of this variable could change without the CPU knowing it. Thus the compiler should generate assembly code to reload TRISF into the CPU registers every time it is used, rather than assuming that its value is unchanged just because no C code has modified it.

Finally, the `__attribute__` syntax tells the linker that TRISF is in the `sfrs` section of memory.

The next section of code, lines 2-14, defines a new data type called `__TRISFbits_t`. Next, in line 15, a variable named TRISFbits is declared of type `__TRISFbits_t`. Again, since it is an `extern` variable, no memory is allocated, and the `__asm__ ("TRISF")` syntax means that TRISFbits is at the same VA as TRISF.

It is worth understanding the new data type `__TRISFbits_t`. It is a `union` of two `structs`. The `union` means that the two `structs` share the same memory, a 32-bit word in this case. Each `struct` is called a *bit field*, which gives names to specific groups of bits within the 32-bit word. Thus declaring a variable TRISFbits of type `__TRISFbits_t`, and forcing it to be located at the same VA as `TRISF` allows us to use syntax like TRISFbits.TRISF0 to refer to bit 0 of TRISF.

A named set of bits in a bit field need not be one bit long; for example, TRISFbits.w refers to the entire `unsigned int` TRISF, created from all 32 bits. The type `__RTCALRMbits_t` defined earlier in the file by

```
typedef union {
  struct {
    unsigned ARPT:8;
```

```
        unsigned AMASK:4;
    ...
    } __RTCALRMbits_t;
```

has a first field ARPT that is eight bits long and a second field AMASK that is four bits long. Since RTCALRM is a variable of type `__RTCALRMbits_t`, a C statement of the form `RTCALRMbits.AMASK = 0xB` would put the values 1, 0, 1, 1 in bits 11, 10, 9, 8, respectively, of RTCALRM.

After the declaration of TRISF and TRISFbits, lines 16-18 contain declarations of TRISFCLR, TRISFSET, and TRISFINV. These declarations allow `simplePIC.c`, which uses these variables, to compile successfully. When the object code of `simplePIC.c` is linked with the `processor.o` object code, references to these variables are resolved to the proper SFR VAs.

With these declarations in `p32mx795f512h.h`, the `simplePIC.c` statements

```
TRISF = 0xFFFC;
LATFINV = 0x0003;
while(!PORTDbits.RD7)
```

finally make sense; these statements write values to, or read values from, SFRs at VAs specified by `processor.o`. You can see that `p32mx795f512h.h` declares many SFRs, but no RAM has to be allocated for them; they exist at fixed addresses in the PIC32's hardware.

The next 9% of `p32mx795f512h.h` is the `extern` variable declaration of the same SFRs, without the bit field types, for assembly language. The VAs of each of the SFRs is given, making this a handy reference.

Starting at just over 17,000 lines into the file, we see more than 20,000 lines with constant definitions like the following:

```
#define _T1CON_TCS_POSITION                          0x00000001
#define _T1CON_TCS_MASK                              0x00000002
#define _T1CON_TCS_LENGTH                            0x00000001

#define _T1CON_TCKPS_POSITION                        0x00000004
#define _T1CON_TCKPS_MASK                            0x00000030
#define _T1CON_TCKPS_LENGTH                          0x00000002
```

These refer to the Timer1 SFR T1CON. Consulting the information about T1CON in the Timer1 section of the Data Sheet, we see that bit 1, called TCS, controls whether Timer1's clock input comes from the T1CK input pin or from PBCLK. Bits 4 and 5, called TCKPS for "timer clock prescaler," control how many times the input clock has to "tick" before Timer1 is incremented (e.g., TCKPS = 0b10 means there is one clock increment per 64 input ticks). The constants defined above are for convenience in accessing these bits. The `POSITION` constants indicate the least significant bit location in TCS or TCKPS in T1CON—one for TCS and four for TCKPS. The `LENGTH` constants indicate that TCS consists of one bit and TCKPS consists of two bits. Finally, the `MASK` constants can be used to determine the values of the bits we care about. For example:

```
unsigned int tckpsval = (T1CON & _T1CON_TCKPS_MASK) >> _T1CON_TCKPS_POSITION;
// AND MASKing clears all bits except 5 and 4, which are unchanged and shifted to
// positions 1 and 0, so tckpsval now contains the value T1CONbits.TCKPS
```

The definitions of the `POSITION`, `LENGTH`, and `MASK` constants take up most of the rest of the file. Of course, there is also a `T1CONbits` defined that allows you to access these bits directly (e.g., T1CONbits.TCKPS). We recommend that you use this method, as it is typically clearer and less error prone than performing direct bit manipulations.

At the end, some more constants are defined, like below:

```
#define _ADC10
#define _ADC10_BASE_ADDRESS     0xBF809000
#define _ADC_IRQ                33
#define _ADC_VECTOR             27
```

The first is merely a flag indicating to other `.h` and `.c` files that the 10-bit ADC is present on this PIC32. The second indicates the first address of 22 consecutive SFRs related to the ADC (see the Memory Organization section of the Data Sheet). The third and fourth relate to interrupts. The PIC32MX's CPU is capable of being interrupted by up to 96 different events, such as a change of voltage on an input line or a timer rollover event. Upon receiving these interrupts, it can call up to 64 different interrupt service routines, each identified by a "vector" corresponding to its address. These two lines say that the ADC's "interrupt request" line is 33 (out of 0 to 95), and its corresponding interrupt service routine is at vector 27 (out of 0 to 63). Interrupts are covered in Chapter 6.

Finally, `p32mx795f512h.h` concludes by including `ppic32mx.h`, which contains legacy code that is no longer needed but remains for backward compatibility with old programs.

3.5.3 Other Microchip Software: Harmony

Installed in your Harmony directory (`<harmony>`) is an extensive and complex set of libraries and sample code written by Microchip. Because of the complexity and abstraction it introduces, we avoid using Harmony functions until Chapter 20, when our programs are complex enough that low-level access to the peripherals through SFRs becomes more difficult.[3]

[3] Even though most sample code in the book does not use Harmony, we had you install it and record its installation directory in the `Makefile`; this way, your system is prepared for Harmony when we are ready to use it.

3.5.4 The NU32bootloaded.ld Linker Script

To create the executable .hex file, we needed the C source file simplePIC.c and the linker script NU32bootloaded.ld. Examining NU32bootloaded.ld with a text editor, we see the following line near the beginning:

```
INPUT("processor.o")
```

This line tells the linker to load the processor.o file specific to your PIC32. This allows the linker to resolve references to SFRs (declared as extern variables in p32mx795f512h.h) to actual addresses.

The rest of the NU32bootloaded.ld linker script has information such as the amount of program flash and data memory available, as well as the virtual addresses where program elements and global data should be placed. Below is a portion of NU32bootloaded.ld:

```
_RESET_ADDR              = (0xBD000000 + 0x1000 + 0x970);

/*************************************************************************
 * NOTE:  What is called boot_mem and program_mem below do not directly
 * correspond to boot flash and program flash.  For instance, here
 * kseg0_boot_mem and kseg1_boot_mem both live in program flash memory.
 * (We leave the boot flash solely to the bootloader.)
 * The boot_mem names below tell the linker where the startup codes should
 * go (here, in program flash).  The first 0x1000 + 0x970 + 0x490 = 0x1E00 bytes
 * of program flash memory is allocated to the interrupt vector table and
 * startup codes.  The remaining 0x7E200 is allocated to the user's program.
 *************************************************************************/
MEMORY
{
  /* interrupt vector table */
  exception_mem              : ORIGIN = 0x9D000000, LENGTH = 0x1000
  /* Start-up code sections; some cacheable, some not */
  kseg0_boot_mem             : ORIGIN = (0x9D000000 + 0x1000), LENGTH = 0x970
  kseg1_boot_mem             : ORIGIN = (0xBD000000 + 0x1000 + 0x970), LENGTH =
                               0x490
  /* User's program is in program flash, kseg0_program_mem, all cacheable      */
  /* 512 KB flash = 0x80000, or 0x1000 + 0x970 + 0x490 + 0x7E200              */
  kseg0_program_mem    (rx)  : ORIGIN = (0x9D000000 + 0x1000 + 0x970 + 0x490),
                               LENGTH = 0x7E200
  debug_exec_mem             : ORIGIN = 0xBFC02000, LENGTH = 0xFF0
  /* Device Configuration Registers (configuration bits) */
  config3                    : ORIGIN = 0xBFC02FF0, LENGTH = 0x4
  config2                    : ORIGIN = 0xBFC02FF4, LENGTH = 0x4
  config1                    : ORIGIN = 0xBFC02FF8, LENGTH = 0x4
  config0                    : ORIGIN = 0xBFC02FFC, LENGTH = 0x4
  configsfrs                 : ORIGIN = 0xBFC02FF0, LENGTH = 0x10
  /* all SFRS */
  sfrs                       : ORIGIN = 0xBF800000, LENGTH = 0x100000
  /* PIC32MX795F512H has 128 KB RAM, or 0x20000 */
  kseg1_data_mem       (w!x) : ORIGIN = 0xA0000000, LENGTH = 0x20000
}
```

Converting virtual to physical addresses, we see that the cacheable interrupt vector table (we will learn more about this in Chapter 6) in `exception_mem` is placed in a memory region of length 0x1000 bytes beginning at PA 0x1D000000 and running to 0x1D000FFF; cacheable startup code in `kseg0_boot_mem` is placed at PAs 0x1D001000 to 0x1D00196F; noncacheable startup code in `kseg1_boot_mem` is placed at PAs 0x1D001970 to 0x1D001DFF; and cacheable program code in `kseg0_program_mem` is allocated the rest of program flash, PAs 0x1D001E00 to 0x1D07FFFF. This program code includes the code we write plus other code that is linked.

The linker script for the NU32 bootloader placed the bootloader completely in the 12 KB boot flash with little room to spare. Therefore, the linker script for our bootloaded programs should place the programs solely in program flash. Therefore, the `boot_mem` sections above are defined to be in program flash. The label `boot_mem` tells the linker where the startup code should be placed, just as the label `kseg0_program_mem` tells the linker where the program code should be placed. (For the bootloader program, `kseg0_program_mem` was in boot flash.)

If the LENGTH of any given memory region is not large enough to hold all the program instructions or data for that region, the linker will fail.

Upon reset, the PIC32 always jumps to 0xBFC00000, where the first instruction of the startup code for the bootloader resides. The bootloader's final action is to jump to VA 0xBD001970. Since the first instruction in the startup code for our bootloaded program is installed at the first address in `kseg1_boot_mem`, `NU32bootloaded.ld` *must* define the ORIGIN of `kseg1_boot_mem` at this address. This address is also known as `_RESET_ADDR` in `NU32bootloaded.ld`.

3.6 Bootloaded Programs vs. Standalone Programs

Your executable is installed on the PIC32 by another executable: the bootloader. The bootloader has been pre-installed in the boot flash portion of flash memory using an external programming tool such as the PICkit 3. The bootloader, which always runs first when the PIC32 is reset, has already defined some of the behavior of the PIC32, so you did not need to specify it in `simplePIC.c`. Particularly, the bootloader performs tasks such as such as enabling the prefetch cache module, enabling multi-vector interrupts (see Chapter 6), and freeing some pins to be used as general I/O. The bootloader code also defines the PIC32's *configuration bits*. These bits, which control the PIC32's low-level behavior, are located in the last four words of boot flash and are written by the programming tool. When the bootloader was installed, it also set the configuration bits using XC32-specific commands that begin with `#pragma config`. The configuration bits that were set when the bootloader was installed are

```
#pragma config DEBUG    = OFF        // Background Debugger disabled
#pragma config FPLLMUL = MUL_20      // PLL Multiplier: Multiply by 20
#pragma config FPLLIDIV = DIV_2      // PLL Input Divider:  Divide by 2
#pragma config FPLLODIV = DIV_1      // PLL Output Divider: Divide by 1
```

```
#pragma config FWDTEN = OFF            // WD timer: OFF
#pragma config WDTPS = PS4096          // WD period: 4.096 sec
#pragma config POSCMOD = HS            // Primary Oscillator Mode: High Speed xtal
#pragma config FNOSC = PRIPLL          // Oscillator Selection: Primary oscillator
                                          w/ PLL
#pragma config FPBDIV = DIV_1          // Peripheral Bus Clock: Divide by 1
#pragma config UPLLEN = ON             // USB clock uses PLL
#pragma config UPLLIDIV = DIV_2        // Divide 8 MHz input by 2, mult by 12 for
                                          48 MHz
#pragma config FUSBIDIO = ON           // USBID controlled by USB peripheral when it
                                          is on
#pragma config FVBUSONIO = ON          // VBUSON controlled by USB peripheral when it
                                          is on
#pragma config FSOSCEN = OFF           // Disable second osc to get pins back
#pragma config BWP = ON                // Boot flash write protect: ON
#pragma config ICESEL = ICS_PGx2       // ICE pins configured on PGx2
#pragma config FCANIO = OFF            // Use alternate CAN pins
#pragma config FMIIEN = OFF            // Use RMII (not MII) for ethernet
#pragma config FSRSSEL = PRIORITY_6 // Shadow Register Set for interrupt priority 6
```

The directives above

- disable some debugging features;
- turn the PIC32's watchdog timer off and set its period (see Chapter 17);
- configure the PIC32's clock generation circuit to take the external 8 MHz resonator signal, divide its frequency by 2, input the divided frequency into a phase-locked loop (PLL) that multiplies the frequency by 20, and divide the PLL's output frequency by 1, creating a SYSCLK of $8/2 \times 20/1$ MHz = 80 MHz;
- set the PBCLK frequency to be SYSCLK divided by 1 (80 MHz);
- use a PLL to generate the 48 MHz USBCLK by first dividing the 8 MHz signal frequency by 2 before multiplying by a fixed factor of 12;
- allow two pins to be controlled by the USB peripheral when USB is enabled;
- disable the secondary oscillator (this could provide an alternative clock source for power-saving modes or as a backup);
- prevent the boot flash from being written when a program is running;
- connect the CAN modules to the alternate pins instead of the default pins;
- configure the ethernet module to use the reduced media-independent interface (RMII); and
- set the shadow register set to be used for interrupts of priority level 6 (see Chapter 6).

Remember, these bits are set by the programming tool, so they are only stored when the bootloader is written to the PIC32; using these commands in a bootloaded program has no effect. The file `pic32-libs/proc/32MX795F512H/configuration.data` contains definitions for the values you can use in the `#pragma config` directives. The Data Sheet and Configuration section of the Reference Manual have more details about the configuration bits.

If you decide not to use a bootloader, and instead use a programming tool like the PICkit 3 (Figure 2.4) to install a standalone program, you must set the configuration bits, enable the prefetch cache module, perform other configuration tasks, and use the default linker script. If you use the NU32 library you need not write this code: `NU32.c` contains the necessary configuration bit settings (which have no effect for a bootloaded program) and `NU32_Startup` performs the necessary setup tasks (which are redundant but harmless for a bootloaded program).[4] To use the default linker script (a copy of which is located at `pic32-libs/proc/32MX795F512H/p32MX795F512H.ld`) when you build a program using `make` (Chapter 1.5), change the line `LINKSCRIPT="NU32bootloaded.ld"` to `LINKSCRIPT=` in the `Makefile`.

After building a standalone hex file, you must load it onto the PIC32 using a programming tool. The easiest method for loading a hex file is to use the MPLAB X IDE. In the IDE, create a new "precompiled" project, selecting your processor model, programming tool, and hex file. Next, hit "run," and the hex file will be written to the PIC32. Remember, the PIC32 must be powered for it to be programmed.

3.7 Build Summary

Recall that what we colloquially refer to as "compiling" actually consists of multiple steps. You initiated these steps by invoking the compiler, `xc32-gcc`, at the command line:

```
> xc32-gcc -mprocessor=32MX795F512H
   -o simplePIC.elf -Wl,--script=skeleton/NU32bootloaded.ld simplePIC.c
```

This step creates the `.elf` file, which then needs to be converted into a `.hex` file that the bootloader understands:

```
> xc32-bin2hex simplePIC.elf
```

The compiler requires multiple *command line options* to work. It accepts arguments, as detailed in the XC32 Users Manual, and some important ones are displayed by typing `xc32-gcc --help`. The arguments we used were

- `-mprocessor=32MX795F512H`: Tells the compiler what PIC32MX model to target. This also causes the compiler to define `__32MX795F512H__` so that the processor model can be detected in header files such as `xc.h`.
- `-o simplePIC.elf`: Specifies that the final output will be named `simplePIC.elf`.

---

4 Technically, performing the configuration tasks a second time in NU32 startup wastes an iota of program memory and computation time, but it allows you to use the same code in both bootloaded and standalone modes without modification.

- `-Wl`: Tells the compiler that what follows are a comma-separated list of options for the linker.
- `--script=skeleton/NU32bootloaded.ld`: A linker option that specifies the linker script to use.
- `simplePIC.c`: The C files that you want compiled and linked are listed. In this case the whole program is in just one file.

Another option that may be useful when exploring what the compiler does is `-save-temps`. This option will save all of the intermediate files generated during the build process, allowing you to examine them.

Here is what happens when you build and load `simplePIC.c`.

- **Preprocessing.** The preprocessor (`xc32-cpp`), among other duties, handles include files. By including `xc.h` at the beginning of your program, we get access to variables for all the SFRs. The output of the preprocessor is a `.i` file, which by default is not saved.
- **Compiling.** After the preprocessor, the compiler (`xc32-gcc`) turns your C code into assembly language specific to the PIC32. For convenience, (`xc32-gcc`) automatically invokes the other commands required in the build process. The result of the compilation step is an assembly language `.S` file, containing a human-readable version of instructions specific to a MIPS32 processor. This output is also not saved by default.
- **Assembling.** The assembler (`xc32-as`) converts the human-readable assembly code into object files (`.o`) that contain machine code. These files cannot be executed directly, however, because addresses have not been resolved. This step yields `simplePIC.o`
- **Linking.** The object code `simplePIC.o` is linked with the `crt0_mips32r2.o` C run-time startup library, which performs functions such as initializing global variables, and the `processor.o` object code, which contains the SFR VAs. The linker script `NU32bootloaded.ld` provides information to the linker on the allowable absolute virtual addresses for the program instructions and data, as required by the bootloader and the specific PIC32 model. Linking yields a self-contained executable in `.elf` format.
- **Hex file.** The `xc32-bin2hex` utility converts `.elf` files into `.hex` files. The `.hex` is a different format for the executable from the `.elf` file that the bootloader understands and can load into the PIC32's program memory.
- **Installing the program.** The last step is to use the NU32 bootloader and the host computer's bootloader utility to install the executable. By resetting the PIC32 while holding the USER button, the bootloader enters a mode where it tries to communicate with the bootload communication utility on the host computer. When it receives the executable from the host, it writes the program instructions to the virtual memory addresses specified by the linker. Now every time the PIC32 is reset without holding the USER button, the bootloader exits and jumps to the newly installed program.

3.8 Useful Command Line Utilities

The `bin` directory of the XC32 installation contains several useful command line utilities. These utilities can be used directly at the command line and many are invoked by the `Makefile`. We have already seen the first two of these utilities, as described in Section 3.7:

xc32-gcc The XC32 version of the `gcc` compiler is used to compile, assemble, and link, creating the executable `.elf` file.
xc32-bin2hex Converts a `.elf` file into a `.hex` file suitable for placing directly into PIC32 flash memory.
xc32-ar The archiver can be used to create an archive, list the contents of an archive, or extract object files from an archive. An archive is a collection of `.o` files that can be linked into a program. Example uses include:

```
xc32-ar -t lib.a          // list the object files in lib.a
                             (in current directory)
xc32-ar -x lib.a code.o   // extract code.o from lib.a to the current directory
```

xc32-as The assembler.
xc32-ld This is the actual linker called by `xc32-gcc`.
xc32-nm Prints the symbols (e.g., global variables) in an object file. Examples:

```
xc32-nm processor.o       // list the symbols in alphabetical order
xc32-nm -n processor.o    // list the symbols in numerical order of their VAs
```

xc32-objdump Displays the assembly code corresponding to an object or `.elf` file. This process is called *disassembly*. Example:

```
xc32-objdump -S file.elf > file.dis    // send output to the file file.dis
```

xc32-readelf Displays a lot of information about the `.elf` file. Example:

```
xc32-readelf -a filename.elf   // output is dominated by SFR definitions
```

These utilities correspond to standard "GNU binary utilities" of the same name without the preceding `xc32-`. To learn the options available for a command called `xc32-cmdname`, you can type `xc32-cmdname --help` or read about them in the XC32 compiler reference manual.

3.9 Chapter Summary

OK, that's a lot to digest. Do not worry, you can view much of this chapter as reference material; you do not have to memorize it to program the PIC32!

- Software refers almost exclusively to the virtual memory map. Virtual addresses map directly to physical addresses by PA = VA & 0x1FFFFFFF.
- Building an executable `.hex` file from a source file consists of the following steps: preprocessing, compiling, assembling, linking, and converting the `.elf` file to a `.hex` file.

- Including the file `xc.h` gives our program access to variables, data types, and constants that significantly simplify programming by allowing us to access SFRs easily from C code without needing to specify addresses directly.
- The included file `pic32mx/include/proc/p32mx795f512h.h` contains variable declarations, like TRISF, that allow us to read from and write to the SFRs. We have several options for manipulating these SFRs. For TRISF, for example, we can directly assign the bits with `TRISF=0x3`, or we can use bitwise operations like `&` and `|`. Many SFRs have associated CLR, SET, and INV registers which can be used to efficiently clear, set, or invert certain bits. Finally, particular bits or groups of bits can be accessed using bit fields. For example, we access bit 3 of TRISF using TRISFbits.TRISF3. The names of the SFRs and bit fields follow the names in the Data Sheet (particularly the Memory Organization section) and Reference Manual.
- All programs are linked with `pic32mx/lib/proc/32MX795F512H/crt0_mips32r2.o` to produce the final `.hex` file. This C run-time startup code executes first, doing things like initializing global variables in RAM, before jumping to the `main` function. Other linked object code includes `processor.o`, with the VAs of the SFRs.
- Upon reset, the PIC32 jumps to the boot flash address 0xBFC00000. For a PIC32 with a bootloader, the `crt0_mips32r2` of the bootloader is installed at this address. When the bootloader completes, it jumps to an address where the bootloader has previously installed a bootloaded executable.
- When the bootloader was installed with a device programmer, the programmer set the Device Configuration Registers. In addition to loading or running executables, the bootloader enables the prefetch cache module and minimizes the number of CPU wait cycles for instructions to load from flash.
- A bootloaded program is linked with a custom linker script, like `NU32bootloaded.ld`, to make sure the flash addresses for the instructions do not conflict with the bootloader's, and to make sure that the program is placed at the address where the bootloader jumps.

3.10 Exercises

1. Convert the following virtual addresses to physical addresses, and indicate whether the address is cacheable or not, and whether it resides in RAM, flash, SFRs, or boot flash. (a) 0x80000020. (b) 0xA0000020. (c) 0xBF800001. (d) 0x9FC00111. (e) 0x9D001000.
2. Look at the linker script used with programs for the NU32. Where does the bootloader install your program in virtual memory? (Hint: look at the `_RESET_ADDR`.)
3. Refer to the Memory Organization section of the Data Sheet and Figure 2.1.
   a. Referring to the Data Sheet, indicate which bits, 0-31, can be used as input/outputs for each of Ports B through G. For the PIC32MX795F512H in Figure 2.1, indicate which pin corresponds to bit 0 of port E (this is referred to as RE0).

b. The SFR INTCON refers to "interrupt control." Which bits, 0-31, of this SFR are unimplemented? Of the bits that are implemented, give the numbers of the bits and their names.

4. Modify `simplePIC.c` so that both lights are on or off at the same time, instead of opposite each other. Turn in only the code that changed.
5. Modify `simplePIC.c` so that the function `delay` takes an `int cycles` as an argument. The `for` loop in `delay` executes `cycles` times, not a fixed value of 1,000,000. Then modify `main` so that the first time it calls `delay`, it passes a value equal to `MAXCYCLES`. The next time it calls `delay` with a value decreased by `DELTACYCLES`, and so on, until the value is less than zero, at which time it resets the value to `MAXCYCLES`. Use `#define` to define the constants `MAXCYCLES` as 1,000,000 and `DELTACYCLES` as 100,000. Turn in your code.
6. Give the VAs and reset values of the following SFRs. (a) I2C3CON. (b) TRISC.
7. The `processor.o` file linked with your `simplePIC` project is much larger than your final `.hex` file. Explain how that is possible.
8. The building of a typical PIC32 program makes use of a number of files in the XC32 compiler distribution. Let us look at a few of them.
   a. Look at the assembly startup code `pic32-libs/libpic32/startup/crt0.S`. Although we are not studying assembly code, the comments help you understand what the startup code does. Based on the comments, you can see that this code clears the RAM addresses where uninitialized global variables are stored, for example. Find and list the line(s) of code that call the user's `main` function when the C runtime startup completes.
   b. Using the command `xc32-nm -n processor.o`, give the names and addresses of the five SFRs with the highest addresses.
   c. Open the file `p32mx795f512h.h` and go to the declaration of the SFR SPI2STAT and its associated bit field data type __SPI2STATbits_t. How many bit fields are defined? What are their names and sizes? Do these coincide with the Data Sheet?
9. Give three C commands, using TRISDSET, TRISDCLR, and TRISDINV, that set bits 2 and 3 of TRISD to 1, clear bits 1 and 5, and flip bits 0 and 4.

Further Reading

MPLAB XC32 C/C++ compiler user's guide. (2012). Microchip Technology Inc.
MPLAB XC32 linker and utilities user guide. (2013). Microchip Technology Inc.
PIC32 family reference manual. Section 32: Configuration. (2013). Microchip Technology Inc.

Further Reading

CHAPTER 4

Using Libraries

You have used libraries all your life—well, at least as long as you have programmed in C. Want to display text on the screen? printf. What about determining the length of a string? strlen. Need to sort an array? qsort. You can find these functions, along with numerous others, in the C standard library. A *library* consists of a collection object files (.o), that have been combined into an archive file (.a): for example, the C standard library libc.a. Using a library requires you to include the associated header files (.h) and link with the archive file. The header file (e.g., stdio.h) declares the functions, constants, and data types used by the library while the archive file contains function implementations. Libraries make it easy to share code between multiple projects without needing to repeatedly compile the code.

In addition to the C standard library, Microchip provides some other libraries specific to programming PIC32s. In Chapter 3, we learned about the header file xc.h which includes the processor-specific header pic32mx795f512h.h, providing us with definitions for the SFRs. The "archive" file for this library is processor.o.[1] Microchip also provides a higher-level framework called Harmony, which contains libraries and other source code to help you create code that works with multiple PIC32 models; we use Harmony later in this book.

Libraries can also be distributed as source code: for example, the NU32 library consists of <PIC32>/skeleton/NU32.h and <PIC32>/skeleton/NU32.c. To use libraries distributed as source code you must include the library header files, compile your source code and the library code, and link the resulting object files. You can link as many object files as you want, as long as they do not declare the same symbols (e.g., two C files in one project cannot both have a main function).

The NU32 library provides initialization and communication functions for the NU32 board. The talkingPIC.c code in Chapter 1 uses the NU32 library, as will most of the examples throughout the book. Let us revisit talkingPIC.c, and examine how it includes libraries during the build process.

---

[1] The library consists of only one object file so Microchip did not create an archive, which holds multiple object files.

Embedded Computing and Mechatronics with the PIC32 Microcontroller. http://dx.doi.org/10.1016/B978-0-12-420165-1.00004-4

4.1 Talking PIC

In the previous chapter, to keep things as simple as possible, we built the executable from `simplePIC.c` by directly issuing the `xc32-gcc` and `xc32-bin2hex` commands at the command line. In this chapter, and all future chapters, we use the `Makefile` with `make` to build the executable, as with `talkingPIC.c` in Chapter 1.

Recall from Chapter 1 that the `Makefile` compiles and links all `.c` files in the directory. Since the project directory `<PIC32>/talkingPIC` contains `NU32.c`, this file was compiled along with `talkingPIC.c`. To see how this process works, we examine the commands that `make` issues to build your project.

Navigate to where you created talkingPIC in Chapter 1 (`<PIC32>/talkingPIC`). Issue the following command:

```
> make clean
```

This command removes the files created when you originally built the project, so we can start fresh. Next, issue the `make` command to build the project. Notice that it issues commands similar to:

```
> xc32-gcc -g -O1 -x c -c -mprocessor=32MX795F512H -o talkingPIC.o talkingPIC.c
> xc32-gcc -g -O1 -x c -c -mprocessor=32MX795F512H -o NU32.o NU32.c
> xc32-gcc -mprocessor=32MX795F512H -o out.elf  talkingPIC.o  NU32.o
      -Wl,--script="NU32bootloaded.ld",-Map=out.map
> xc32-bin2hex out.elf
> xc32-objdump -S out.elf > out.dis
```

The first two commands compile the C files necessary to create `talkingPIC` using the following options:

- `-g`: Include debugging information, extra data added into the object file that helps us to inspect the generated files later.
- `-O1`: Sets optimization level one. We discuss optimization in Chapter 5.
- `-x c`: Tells the compiler to treat input files as C language files. Typically the compiler can detect the proper language based on the file extension, but we use this here to be certain.
- `-c`: Compile and assemble only, do not link. The output of this command is just an object (`.o`) file because the linker is not invoked to create an `.elf` file.

Thus the first two commands create two object files: `talkingPIC.o`, which contains the `main` function, and `NU32.o`, which includes helper functions that `talkingPIC.c` calls. The third command tells the compiler to invoke the linker, because all the "source" files specified are actually object (`.o`) files. We do not invoke the linker `xc32-ld` directly because the compiler automatically tells the linker to link against some standard libraries that we need. Notice that `make` always names its output `out.elf`, regardless of what you name the source files.

Some additional options that `make` provides to the linker are specified after the `Wl` flag:

- `--script`: Tells the linker to use the `NU32bootloaded.ld` linker script.
- `-Map`: This option is passed to the linker and tells it to produce a map file, which details the program's memory usage. Chapter 5 explains map files.

The next command produces the hex file. The final line, `xc32-objdump`, disassembles `out.elf`, saving the results in `out.dis`. This file contains interspersed C code and assembly instructions, allowing you to inspect the assembly instructions that the compiler produces from your C code.

4.2 The NU32 Library

The NU32 library provides several functions that make programming the PIC32 easier. Not only does `talkingPIC.c` use this library, but so do most examples in this book. The `<PIC32>/skeleton` directory contains the NU32 library files, `NU32.c` and `NU32.h`; you copy this directory to create a new project. The `Makefile` automatically links all files in the directory, thus `NU32.c` will be included in your project. By writing `#include "NU32.h"` at the beginning of the program, we can access the library. We list `NU32.h` below:

Code Sample 4.1 `NU32.h`. **The NU32 Header File.**

```
#ifndef NU32__H__
#define NU32__H__

#include <xc.h>                     // processor SFR definitions
#include <sys/attribs.h>            // __ISR macro

#define NU32_LED1 LATFbits.LATF0    // LED1 on the NU32 board
#define NU32_LED2 LATFbits.LATF1    // LED2 on the NU32 board
#define NU32_USER PORTDbits.RD7     // USER button on the NU32 board
#define NU32_SYS_FREQ 80000000ul    // 80 million Hz

void NU32_Startup(void);
void NU32_ReadUART3(char * string, int maxLength);
void NU32_WriteUART3(const char * string);

#endif // NU32__H__
```

The `NU32__H__` include guard, consisting of the first two lines and the last line, ensure that `NU32.h` is not included twice when compiling any single C file. The next two lines include Microchip-provided headers that you would otherwise need to include in most programs. The next three lines define aliases for SFRs that control the two LEDs (`NU32_LED1` and `NU32_LED2`) and the USER button (`NU32_USER`) on the NU32 board. Using these aliases allows us to write code like

```
int button = NU32_USER; // button now has 0 if pressed, 1 if not
NU32_LED1 = 0;          // turn LED1 on
NU32_LED2 = 1;          // turn LED2 off
```

which is easier than remembering which PIC32 pin is connected to these devices. The header also defines the `NU32_SYS_FREQ` constant, which contains the frequency, in Hz, at which the PIC32 operates. The rest of `NU32.h` consists of function prototypes, described below.

void NU32_Startup(void) Call `NU32_Startup()` at the beginning of `main` to configure the PIC32 and the NU32 library. You will learn about the details of this function as the book progresses, but here is an overview. First, the function configures the prefetch cache module and flash wait cycles for maximum performance. Next, it configures the PIC32 for multi-vector interrupt mode. Then it disables JTAG debugging so that the associated pins are available for other functions. The pins RF0 and RF1 are then configured as digital outputs, to control LED1 and LED2. The function then configures UART3 so that the PIC32 can communicate with your computer. Configuring UART3 allows you to use `NU32_WriteUART3()` and `NU32_ReadUART3()` to send strings between the PIC32 and the computer. The communication occurs at 230,400 baud (bits per second), with eight data bits, no parity, one stop bit, and hardware flow control with CTS/RTS. We discuss the details of UART communication in Chapter 11. Finally, it enables interrupts (see Chapter 6). You may notice that these tasks, such as configuring the prefetch cache, are also performed by the bootloader. We do this because `NU32_Startup()` also works with standalone code, in which case these actions would be required, not redundant.

void NU32_ReadUART3(char * string, int maxLength) This function takes a character array `string` and a maximum input length `maxLength`. It fills `string` with characters received from the host via UART3 until a newline `\n` or carriage return `\r` is received. If the string exceeds `maxLength`, the new characters wrap around to the beginning of the string. Note that this function will not exit unless it receives a `\n` or a `\r`.

Example:

```
char message[100] = {}, str[100] = {};
int i = 0;
NU32_ReadUART3(message, 100);
sscanf(message, "%s %d", str, &i);  // if message is expected to have a string and int
```

void NU32_WriteUART3(const char * string) This function sends a string over UART3. The function does not complete until the transmission has finished. Thus, if the host computer is not reading the UART and asserting flow control, the function will wait to send its data.

Example:

```
char msg[100] = {};
sprintf(msg,"The value is %d.\r\n",22);
NU32_WriteUART3(msg);
```

4.3 Bootloaded Programs

Throughout the rest of this book, all C files with a `main` function will begin with

```
#include "NU32.h"          // constants, funcs for startup and UART
```

and the first line of code (other than local variable definitions) in `main` will be

```
NU32_Startup();
```

While other C files and header files might include `NU32.h` to gain access to its contents and function prototypes, no file except the C file with the `main` function should call `NU32_Startup()`.

For bootloaded programs, the configuration bits are set by the bootloader. However, `NU32.c` includes the configuration bit settings. This provides a convenient reference and also allows you to use the same code for both bootloaded and standalone applications (see Chapter 3.6).

4.4 An LCD Library

Dot matrix LCD screens are inexpensive portable devices that can display information to the user. LCD screens often come with an integrated controller that simplifies communication with the LCD. We now discuss a library that allows the PIC32 to control a Hitachi HD44780 (or compatible) LCD controller connected to a 16x2 LCD screen.[2] You can purchase the screen and controller as a pre-built module. The data sheet for this controller is available on the book's website.

The HD44780 has 16 pins: ground (GND), power (VCC), contrast (VO), backlight anode (A), backlight cathode (K), register select (RS), read/write (RW), enable strobe (E), and 8 data pins (D0-D7). We show the pins below.

1	2	3	4	5	6	7	8	9	10	11	12	13	14	15	16
GND	VCC	VO	RS	R/W	E	D0	D1	D2	D3	D4	D5	D6	D7	A	K

Connect the LCD as shown in Figure 4.1.

The LCD is powered by VCC (5 V) and GND. The resistors R1 and R2 determine the LCD's brightness and contrast, respectively. Good guesses for these values are R1 = 100 Ω and R2 = 1000 Ω, but you should consult the data sheet and experiment. The remaining pins are for communication. The R/W pin controls the communication direction. From the PIC32's perspective, R/W = 0 means write and R/W = 1 means read. The RS pin indicates whether the

[2] Many LCD controllers are compatible with the HD44780. We used the Samsung KS006U for the examples in this chapter.

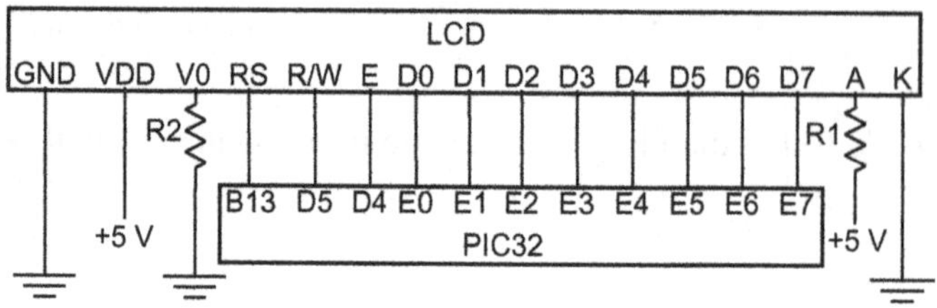

Figure 4.1
Circuit diagram for the LCD.

PIC32 is sending data (e.g., text) or a command (e.g., clear screen). The pins D0-D7 carry the actual data between the devices; after setting data on these pins the PIC32 pulses the enable strobe (E) signal to tell the LCD that the data is ready. For every pulse of E, the LCD receives or sends eight bits of data simultaneously (in *parallel*). We delve into this parallel communication scheme more deeply in Chapter 14, where we discuss the parallel master port (PMP), the peripheral that coordinates the signals between the PIC32 and the LCD.

Now we present the LCD library by looking at its interface. The LCD controller has many features, such as the ability to horizontally scroll text, display custom characters, display a larger font on a single line, and display a cursor. The LCD library contains many functions that enable access to these features; however, we only discuss the basics.

Code Sample 4.2 `LCD.h`**. The LCD Library Header File.**

```
#ifndef LCD_H
#define LCD_H
// LCD control library for Hitachi HD44780-compatible LCDs.

void LCD_Setup(void);                         // Initialize the LCD
void LCD_Clear(void);                         // Clear the screen, return to position (0,0)
void LCD_Move(int line, int col);             // Move position to the given line and column
void LCD_WriteChar(char c);                   // Write a character at the current position
void LCD_WriteString(const char * string);   // Write string starting at current position
void LCD_Home(void);                          // Move to (0,0) and reset any scrolling
void LCD_Entry(int id, int s);                // Control display motion after sending a char
void LCD_Display(int d, int c, int b);        // Turn display on/off and set cursor settings
void LCD_Shift(int sc, int rl);               // Shift the position of the display
void LCD_Function(int n, int f);              // Set number of lines (0,1) and the font size
void LCD_CustomChar(unsigned char val, const char data[7]); // Write custom char to CGRAM
void LCD_Write(int rs, unsigned char db70);  // Write a command to the LCD
void LCD_CMove(unsigned char addr);           // Move to the given address in CGRAM
unsigned char LCD_Read(int rs);               // Read a value from the LCD
#endif
```

LCD_Setup(void) Initializes the LCD, putting it into two-line mode and clearing the screen. You should call this at the beginning of `main()`, after you call `NU32_Startup()`.
LCD_Clear(void) Clears the screen and returns the cursor to line zero, column zero.
LCD_Move(int line, int col) Causes subsequent text to appear at the given line and column. After calling `LCD_Setup()`, the LCD has two lines and 16 columns. Remember, just like C arrays, numbering starts at zero!
LCD_WriteChar(unsigned char s) Write a character to the current cursor position. The cursor position will then be incremented.
LCD_WriteString(const char * str) Displays the string, starting at the current position. Remember, the LCD does not understand control characters like `'\n'`; you must use `LCD_Move` to access the second line.

The program `LCDwrite.c` uses both the NU32 and LCD libraries to accept a string from your computer and write it to the LCD. To build the executable, copy the `<PIC32>/skeleton` directory and then add the files `LCDwrite.c`, `LCD.c`, and `LCD.h`. After building, loading, and running the program, open the terminal emulator. You can now converse with your LCD! The terminal emulator will ask

```
What do you want to write?
```

If you respond `Echo!!`, the LCD prints

```
Echo!!__________
___Received_1___
```

where the underscores represent blank spaces. As you send more strings, the Received number increments. The code is given below.

Code Sample 4.3 `LCDwrite.c`**. Takes Input from the User and Prints It to the LCD Screen.**

```
#include "NU32.h"          // constants, funcs for startup and UART
#include "LCD.h"

#define MSG_LEN 20

int main() {
  char msg[MSG_LEN];
  int nreceived = 1;

  NU32_Startup();          // cache on, interrupts on, LED/button init, UART init

  LCD_Setup();

  while (1) {
    NU32_WriteUART3("What do you want to write? ");
```

```
        NU32_ReadUART3(msg, MSG_LEN);                 // get the response
        LCD_Clear();                                  // clear LCD screen
        LCD_Move(0,0);
        LCD_WriteString(msg);                         // write msg at row 0 col 0
        sprintf(msg, "Received %d", nreceived);      // display how many messages received
        ++nreceived;
        LCD_Move(1,3);
        LCD_WriteString(msg);                         // write new msg at row 1 col 3
        NU32_WriteUART3("\r\n");
    }
    return 0;
}
```

4.5 Microchip Libraries

Microchip provides several libraries for PIC32s. Understanding these libraries is rather confusing (as we began to see in Chapter 3), partially because they are written to support many PIC32 models, and partially because of the requirement to maintain backwards compatibility, so that code written years ago does not become obsolete with new library releases.

Historically, people primarily programmed microcontrollers in assembly language, where the interaction between the code and the hardware is quite direct: typically the CPU executes one assembly instruction per clock cycle, without any hidden steps. For complex software projects, however, assembly language becomes cumbersome because it is processor-specific and lacks convenient higher-level constructs.

The C language, although still low-level, provides some portability and abstraction. Much of your C code will work for different microcontrollers with different CPUs, provided you have a compiler for the particular CPU. Still, if your code directly manipulates a particular SFR that does not exist on another microcontroller model, portability is broken.

Microchip's recent software release, Harmony addresses this issue by providing functions that allow your code to work for many PIC32 models. In a simplified hierarchical view, the user's application may call Microchip *middleware* libraries, which provide a high level of abstraction and keep the user somewhat insulated from the hardware details. The middleware libraries may interface with lower-level *device drivers*. Device drivers may interface with still lower-level *peripheral libraries*. These peripheral libraries then, finally, read or write the SFRs associated with your particular PIC32.

Our philosophy is to stay close to the hardware, similar to assembly language programming, but with the benefits of the easier higher-level C language. This approach allows you to directly translate from the PIC32 hardware documentation to C code because the SFRs are accessed from C using the same names as the hardware documentation. If unsure of how to access an SFR from C code, open the processor-specific header file `<xc32dir>/<xc32ver>/pic32mx/proc/p32mx795f512h.h`, search for the SFR name, and read the declarations related to that SFR. Overall, we believe that this low-level approach to programming the PIC32 should provide you with a strong foundation in microcontroller programming. Additionally, after programming using SFRs directly, you should be able to understand the documentation for any Microchip-provided software and, if you desire, use it in your own projects. Finally, we believe that programming at the SFR level translates better to other microcontrollers: Harmony is Microchip-specific, but concepts such as SFRs are widespread.

4.6 Your Libraries

Now that you have seen how some libraries function, you can create your own libraries. As you program, try to think about the interconnections between parts of your code. If you find that some functions are independent of other functions, you may want to code them in separate `.c` and `.h` files. Splitting projects into multiple files that contain related functions helps increase program modularity. By leaving some definitions out of the header file and declaring functions and variables in your C code as `static` (meaning that they cannot be used outside the C file), you can hide the implementation details of your code from other code. Once you divide your code into independent modules, you can think about which of those modules might be useful in other projects; these files can then be used as libraries.

4.7 Chapter Summary

- A library is a `.a` archive of `.o` object files and associated `.h` header files that give programs access to function prototypes, constants, macros, data types, and variables associated with the library. Libraries can also be distributed in source code form and need not be compiled into archive format prior to being used; in this way they are much like code that you write and split amongst multiple C files. We often call a "library" a `.c` file and its associated `.h` file.
- For a project with multiple C files, each C file is compiled and assembled independently with the aid of its included header files. Compiling a C file does not require the actual definitions of helper functions in other helper C files; only the prototypes are needed. The function calls are resolved to the proper virtual addresses when the multiple objects are linked. If multiple object files have functions with the same name, and these functions are not `static` (private) to the particular file, the linker will fail.
- The NU32 library provides functions for initializing the PIC32 and communicating with the host computer. The LCD library provides functions to write to a 16×2 character dot matrix LCD screen.

4.8 Exercises

1. Identify which functions, constants, and global variables in `NU32.c` are private to `NU32.c` and which are meant to be used in other C files.

2. You will create your own libraries.
   a. Remove the comments from `invest.c` in Appendix A. Now modify it to work on the NU32 using the NU32 library. You will need to replace all instances of `printf` and `scanf` with appropriate combinations of `sprintf`, `sscanf`, `NU32_ReadUART3` and `NU32_WriteUART3`. Verify that you can provide data to the PIC32 with your keyboard and display the results on your computer screen. Turn in your code for all the files, with comments where you altered the input and output statements.
   b. Split `invest.c` into two C files, `main.c` and `helper.c`, and one header file, `helper.h`. `helper.c` contains all functions other than `main`. Which constants, function prototypes, data type definitions, etc., should go in each file? Build your project and verify that it works. For the safety of future `helper` library users, put an include guard in `helper.h`. Turn in your code and a separate paragraph justifying your choice for where to put the various definitions.
   c. Break `invest.c` into three files: `main.c`, `io.c`, and `calculate.c`. Any function which handles input or output should be in `io.c`. Think about which prototypes, data types, etc., are needed for each C file and come up with a good choice of a set of header files and how to include them. Again, for safety, use include guards in your header files. Verify that your code works. Turn in your code and a separate paragraph justifying your choice of header files.
3. When you try to build and run a program, you could run into (at least) three different kinds of errors: a compiler error, a linker error, or a run-time error. A compiler or linker error would prevent the building of an executable, while a run-time error would only become evident when the program does not behave as expected. Say you are building a program with no global variables and two C files, exactly one of which has a `main()` function. For each of the three types of errors, give simple code that would lead to it.
4. Write a function, `void LCD_ClearLine(int ln)`, that clears a single line of the LCD (either line zero or line one). You can clear a line by writing enough space (`' '`) characters to fill it.
5. Write a function, `void LCD_print(const char *)`, that writes a string to the LCD and interprets control characters. The function should start writing from position (0,0). A carriage return (`'\r'`) should reset the cursor to the beginning of the line, and a line feed (`'\n'`) should move the cursor to the other line.

Further Reading

32-Bit language tools libraries. (2012). Microchip Technology Inc.
HD44780U (LCD-II) dot matrix liquid crystal display controller/driver. HITACHI.
KS0066U 16COM/40SEG driver and controller for dot matrix LCD. Samsung.

CHAPTER 5

Time and Space

How long does your program take to execute? How much RAM does it require? How much flash does it occupy? Fast processing speeds and plentiful memory have reduced the importance of these questions for programmers of personal computers. After all, if a process takes a few microseconds longer or uses a few extra megabytes of RAM, the user may not notice the difference. On embedded systems, however, efficiency is important: processors are relatively slow, control commands must execute in a timely manner, and RAM and flash are precious resources. Additionally, specific timing requirements may be imposed on your system due to physics: for example, you might need to provide commands to a motor at regular, timely intervals to control it properly. If your code is too slow, it may fail to accomplish its purpose.

Writing efficient code is a balancing act. Code can be time-efficient (runs fast), RAM-efficient (uses less RAM), flash-efficient (has a smaller executable size), but perhaps most importantly, programmer-time-efficient (minimizes the time needed to write and debug the code, or for a future programmer to understand and modify it). Often these interests compete with each other. Some XC32 compiler options reflect this trade-off, allowing you to explicitly make space-time tradeoffs.[1] For example, the compiler could "unroll" loops. If a loop is known to be executed 20 times, for example, instead of using a small piece of code, incrementing a counter, and checking to see if the count has reached 20, the compiler could simply write the same block of code 20 times. This may save some execution time (no counter increments, no conditional tests, no branches) at the expense of using more flash to store the program.

This chapter explains some of the tools available for understanding the time and space consumed by your program. These tools will not only help you squeeze the most out of your PIC32; by writing more efficient code you can often choose a less expensive microcontroller. More importantly, though, you will better understand how your software works.

[1] Some options are not available in the free version of the compiler.

Embedded Computing and Mechatronics with the PIC32 Microcontroller. http://dx.doi.org/10.1016/B978-0-12-420165-1.00005-6

5.1 Compiler Optimization

The XC32 compiler provides five levels of optimization. Their availability depends on whether you have a license for the free version of the compiler, the Standard version, or the Pro version:

Version	Label	Description
All	00	no optimization
All	01	level 1: attempts to reduce both code size and execution time
Standard, Pro	02	level 2: further reduces code size and execution time beyond 01
Pro	03	level 3: maximum optimization for speed
Pro	0s	maximum optimization for code size

The greater the optimization, the longer it takes the compiler to produce the assembly code. You can learn more about compiler optimization in the XC32 C/C++ Compiler User's Guide.

When you issue a `make` command with the `Makefile` from the quickstart code, you see that the compiler is invoked with optimization level `01`, using commands like

```
xc32-gcc -g -01 -x c ...
```

`-g -01 -x c` are compiler flags set in the variable `CFLAGS` in the `Makefile`. The `-01` means that optimization level 1 is being requested.

In this chapter, we examine the assembly code that the compiler produces from your C code. The mapping between your C code and the assembly code is relatively direct when no optimization is used, but is less clear when optimization is invoked. (We will see an example of this in Section 5.2.3.) To create clearer assembly code, we will find it useful to be able to `make` files with no optimization. You can override the compilation flags by specifying the `CFLAGS` `Makefile` variable at the command line:

```
> make CFLAGS="-g -x c"
```

or

```
> make write CFLAGS="-g -x c"
```

Alternatively, you could edit the `Makefile` line to remove the `-01` where `CFLAGS` is defined and just use `make` and `make write` as usual.

In these examples, since no optimization level is specified, the default (no optimization) is applied. Unless otherwise specified, all examples in this chapter assume that no optimization is applied.

5.2 Time and the Disassembly File

5.2.1 Timing Using a Stopwatch (or an Oscilloscope)

A direct way to time something is to toggle a digital output and look at that digital output using an oscilloscope or stopwatch. For example:

```
  ...                       // digital output RF0 has been high for some time
LATFCLR = 0x1;              // clear RF0 to 0 (turn on NU32 LED1)
  ...                       // some code you want to time
LATFSET = 0x1;              // set RF0 to 1 (turn off LED1)
```

The time that RF0 is low (or LED1 is on) approximates the execution time of the code.

If the duration is too short to measure with your scope or stopwatch, you could modify the code to something like

```
  ...                         // digital output RF0 has been high for some time
LATFCLR = 0x1;                // clear RF0 to 0 (turn on NU32 LED1)
for (i=0; i<1000000; i++) {   // modify 1,000,000 as appropriate for you
  ...                         // some code you want to time
}
LATFSET = 0x1;                // set RF0 to 1 (turn off LED1)
```

Then you can divide the total time by 1,000,000.[2] Remember, however, that the `for` loop introduces additional overhead: for example, instructions to increment the counter and check the inequality. We will examine the overhead in Section 5.2.3. If the code you want to time uses only a few assembly instructions, then the time you actually measure will be dominated by the implementation of the `for` loop.

5.2.2 Timing Using the Core Timer

A more accurate time can be obtained using a timer onboard the PIC32. The NU32's PIC32 has six timers: a 32-bit *core* timer, associated with the MIPS32 CPU, and five 16-bit *peripheral* timers. We can use the core timer for pure timing operations, leaving the much more flexible peripheral timers available for other tasks (see Chapter 8). The core timer increments once for every two ticks of SYSCLK. For a SYSCLK of 80 MHz, the timer increments every 25 ns. Because the timer is 32 bits, it rolls over every $2^{32} \times 25$ ns $= 107$ s.

---

[2] If you use optimization in compiling your program, however, the compiler might recognize that you are not doing anything with the results of the loop, and not generate assembly code for the loop at all! You can place a `_nop()` macro in the loop to force it to remain. The `_nop()` inserts a `nop` assembly instruction that does nothing but the compiler cannot remove it.

The include file <cp0defs.h> contains two macros for accessing the core timer: _CP0_GET_COUNT() and _CP0_SET_COUNT(val). When you include NU32.h, it includes <xc.h> which, in turn, includes <cp0defs.h>, so you typically already have access to the required macros. The macro _CP0_GET_COUNT() returns the current core timer count while _CP0_SET_COUNT(val) sets the count to val.

```
unsigned int elapsedticks, elapsedns;

_CP0_SET_COUNT(0);                  // set the core timer counter to 0
  ...                               // some code you want to time
elapsedticks = _CP0_GET_COUNT();    // read the core timer
elapsedns = elapsedticks * 25;      // duration in ns, for 80 MHz SYSCLK
```

If the core timer is being used to time different things, do not reset the counter to zero. Instead, read the value initial at the start of the timing, then the value final at the end, and subtract.

```
unsigned int initial, final, elapsed;
initial = _CP0_GET_COUNT();         // read the initial time
  ...                               // some code you want to time
final = _CP0_GET_COUNT();           // the end duration
elapsed = final - initial;          // total elapsed time, in ticks
```

The above code works even if final is less than initial due to a single timer rollover.[3] If the timer rolls over twice the answer will be incorrect, but your code will have taken longer than 107 s.

5.2.3 Disassembling Your Code

By looking at the assembly code the compiler produces, you can determine approximately how long your code takes to execute. Fewer instructions mean faster code.

In Chapter 3.5, we claimed that the code

```
LATFINV = 0x3;
```

is more efficient than

```
LATFbits.LATF0 = !LATFbits.LATF0; LATFbits.LATF1 = !LATFbits.LATF1;
```

[3] Unsigned arithmetic actually computes $(a \odot b) \bmod 2^N$, where N is the number of bits in the type and $\odot$ represents an operator such as $+$ or $-$. Thus, unsigned subtraction computes the distance between the two numbers, modulo 2^N.

Let us examine that claim by looking at the assembly code of the following program. This program delays by executing a `for` loop 50 million times, then toggles RF1 (LED2 on the NU32).

Code Sample 5.1 `timing.c`. **RF1 Toggles (LED2 on the NU32 Flashes).**

```
#include "NU32.h"          // constants, functions for startup and UART
#define DELAYTIME 50000000 // 50 million

void delay(void);
void toggleLight(void);

int main(void) {
  NU32_Startup(); // cache on, min flash wait, interrupts on, LED/button init, UART init

  while(1) {
    delay();
    toggleLight();
  }
}

void delay(void) {
  int i;
  for (i = 0; i < DELAYTIME; i++) {
      ; //do nothing
  }
}

void toggleLight(void) {
  LATFINV = 0x2; // invert LED2 (which is on port F1)
  // LATFbits.LATF1 = !LATFbits.LATF1;
}
```

Put `timing.c` in a project directory together with the `Makefile`, `NU32.c`, `NU32.h`, and `NU32bootloaded.ld`, and nothing else. Then `make` with no optimization, as described above. The `Makefile` automatically disassembles the `out.elf` file to the file `out.dis`, but if you wanted to do it manually, you could type

```
> xc32-objdump -S out.elf > out.dis
```

Open `out.dis` in a text editor. You will see a listing showing the assembly code corresponding to `out.hex`. The file interleaves your C code and the assembly code it generated.[4] Each assembly line has the actual virtual address where the assembly instruction is placed in memory, the 32-bit machine instruction, and the equivalent human-readable (if you know assembly!) assembly code. Let us look at the segment of the listing corresponding to the command `LATFINV = 0x2`. You should see something like

[4] Sometimes the output from the `xc32-objdump` duplicates part of the C code, due to the way in which the C gets translated into assembly.

```
  LATFINV = 0x2; // invert the LED2 (which is on port F1)
9d00221c: 3c02bf88  lui v0,0xbf88
9d002220: 24030002  li v1,2
9d002224: ac43616c  sw v1,24940(v0)
  // LATFbits.LATF1 = !LATFbits.LATF1;
```

We see that the `LATFINV = 0x2` instruction has expanded to three assembly statements. Without going into detail, the `li` instruction stores the base-10 value 2 (or hex 0x2) in the CPU register `v1`, which is then written by the `sw` command to the memory address corresponding to LATFINV (`v0`, or 0xBF88, is bits 16-31 of the address, and the base-10 value 24940, or hex 0x616C, is bits 0-15).[5]

If instead we comment out the `LATFINV = 0x2`; command and replace it with the bit manipulation version, we get the following disassembly:

```
  // LATFINV = 0x2; // invert the LED2 (which is on port F1)
  LATFbits.LATF1 = !LATFbits.LATF1;
9d00221c: 3c02bf88  lui v0,0xbf88
9d002220: 8c426160  lw v0,24928(v0)
9d002224: 30420002  andi v0,v0,0x2
9d002228: 2c420001  sltiu v0,v0,1
9d00222c: 304400ff  andi a0,v0,0xff
9d002230: 3c03bf88  lui v1,0xbf88
9d002234: 90626160  lbu v0,24928(v1)
9d002238: 7c820844  ins v0,a0,0x1, 0x1
9d00223c: a0626160  sb v0,24928(v1)
```

The bit manipulation version requires nine assembly statements. Basically the value of LATF is being copied to a CPU register, manipulated, then stored back in LATF. In contrast, with the LATFINV syntax, there is no copying the values of LATFINV back and forth.

Although one method of manipulating the SFR bit appears three times slower than the other, we do not yet know how many CPU cycles each consumes. Assembly instructions are generally performed in a single clock cycle, but there is still the question of whether the CPU is getting one instruction per cycle (due to the slow program flash.) We will investigate further by manipulating the prefetch cache module in Section 5.2.4. For now, though, we time the 50 million iteration delay loop. Here is the disassembly for `delay()`, with comments added to the right:

```
void delay(void) {
9d0021c0: 27bdfff0  addiu sp,sp,-16     // manipulate the stack pointer on ...
9d0021c4: afbe000c  sw s8,12(sp)        // ... entering the function (see text)
9d0021c8: 03a0f021  move s8,sp
  int i;
  for (i = 0; i < DELAYTIME; i++) {
9d0021cc: afc00000  sw zero,0(s8)       // initialization of i in RAM to 0
```

[5] You can refer to the MIPS32 documentation if interested.

```
9d0021d0: 0b400879  j 9d0021e4          // jump to 9d0021e4 (skip adding 1 to i),
9d0021d4: 00000000  nop                 // but "no operation" executed first
9d0021d8: 8fc20000  lw v0,0(s8)         // start of loop; load RAM i into
                                           register v0
9d0021dc: 24420001  addiu v0,v0,1       // add 1 to v0 ...
9d0021e0: afc20000  sw v0,0(s8)         // ... and store it to i in RAM
9d0021e4: 8fc30000  lw v1,0(s8)         // load i into register v1
9d0021e8: 3c0202fa  lui v0,0x2fa        // load the upper 16 bits and ...
9d0021ec: 3442f080  ori v0,v0,0xf080    // ... lower 16 bits of 50,000,000
                                           into v0
9d0021f0: 0062102a  slt v0,v1,v0        // store "true" (1) in v0 if v1 < v0
9d0021f4: 1440fff8  bnez v0,9d0021d8    // if v0 not equal to 0, branch to top of
                                           loop,
9d0021f8: 00000000  nop                 // but branch "delay slot" is executed
                                           first
      ; //do nothing
  }
}
9d0021fc: 03c0e821  move sp,s8          // manipulate the stack pointer on exiting
9d002200: 8fbe000c  lw s8,12(sp)
9d002204: 27bd0010  addiu sp,sp,16
9d002208: 03e00008  jr ra               // jump to return address ra stored by jal,
9d00220c: 00000000  nop                 // but jump delay slot is executed first
```

There are nine instructions in the delay loop itself, starting with `lw v0,0(s8)` and ending with the next `nop`. When the LED turns on, these instructions are carried out 50 million times, and then the LED turns off. (There are additional instructions to set up the loop and increment the counter, but the duration of these is negligible compared to the 50 million executions of the loop.) So if one instruction is executed per cycle, we would predict the light to stay on for approximately 50 million $\times$ 9 instructions $\times$ 12.5 ns/instruction = 5.625 s. When we time by a stopwatch, we get about 6.25 s, which implies ten CPU (SYSCLK) cycles per loop. So our cache module has the CPU executing one assembly instruction almost every cycle.

In the code above there are two "jumps" (`j` for "jump" to the specified address and `jr` for "jump register" to jump to the address in the return address register `ra`, which was set by the calling function) and one "branch" (`bnez` for "branch if not equal to zero"). For MIPS32, the command after a jump or branch is executed before the jump actually occurs. This next command is said to be in the "delay slot" for the jump or branch. In all three delay slots in this code is a `nop` command, which stands for "no operation."

You might notice a few ways you could have written the assembly code for the delay function to use fewer assembly commands. Certainly one of the advantages of coding directly in assembly is that you have direct control of the processor instructions. The disadvantage, of course, is that MIPS32 assembly is a much lower-level language than C, requiring significantly more knowledge of MIPS32 from the programmer. Until you have already invested a great deal of time learning the assembly language, programming in assembly fails the "programmer-time-efficient" criterion! (Not to mention that `delay()` was designed to waste time, so no need to minimize assembly lines!)

You may have also noticed, in the disassembly of `delay()`, the manipulation of the *stack pointer* (`sp`) upon entering and exiting the function. The *stack* is an area of RAM that holds temporary local variables and parameters. When a function is called, its parameters and local variables are "pushed" onto the stack. When the function exits, the local variables are "popped" off of the stack by moving the stack pointer back to its original position before the function was called. A *stack overflow* occurs if there is not enough RAM available for the stack to hold all the local variables defined in currently-called functions. We will see the stack again in Section 5.3.

The overhead due to passing parameters and manipulating the stack pointer upon entering and exiting a function should not discourage you from writing modular code. Function call overhead should only concern you when you need to squeeze a final few nanoseconds out of your program execution time.

Finally, if you revert back to the `LATFINV` method for toggling the LED and compile `timing.c` with optimization level 1 (the optimization flag `-O1`), you see that `delay()` is optimized to

```
void delay(void) {
9d00211c: 3c0202fa   lui v0,0x2fa           // load the upper 16 bits and ...
9d002120: 3442f080   ori v0,v0,0xf080       // ... lower 16 bits of 50,000,000 into v0
9d002124: 2442ffff   addiu v0,v0,-1         // subtract 1 from v0
  int i;
  for (i = 0; i < DELAYTIME; i++) {
9d002128: 1440ffff   bnez v0,9d002128       // if v0 !=0, branch back to the same line,
9d00212c: 2442ffff   addiu v0,v0,-1         // but before branch, subtract 1 from v0
      ; //do nothing
  }
}
9d002130: 03e00008   jr ra                  // jump to return address ra stored by jal
9d002134: 00000000   nop                    // no operation in jump delay slot
```

No local variables are stored in RAM, and there is no stack pointer manipulation upon entering and exiting the function. The counter variable is simply stored in a CPU register. The loop itself has only two lines instead of nine, and it has been designed to count down from 49,999,999 to zero instead of counting up. The branch delay slot is actually used to implement the counter update instead of having a wasted `nop` cycle.

More importantly, however, `delay()` is never called by the assembly code for `main` in our `-O1` optimized code! The compiler has recognized that `delay()` does not do anything.[6] As a result, the LED toggles so quickly that you cannot see it by eye. The LED just looks dim.[7]

---

6 A better optimization would not have produced code for `delay` at all, reducing flash usage.

7 To prevent `delay()` from being optimized away, we could have added a "no operation" `_nop();` command inside the delay loop. Or we could have accessed a `volatile` variable inside the loop. Or we could have polled the core timer to implement a desired delay.

5.2.4 The Prefetch Cache Module

In the previous section, we saw that our `timing.c` program executed approximately one assembly instruction every clock cycle. We achieved this performance because the bootloader (and `NU32_Startup()`) enabled the prefetch cache module and selected the minimum number of CPU wait cycles for instructions loading from flash.[8]

We can disable the prefetch cache module to observe its effect on the program `timing.c`. The prefetch cache module performs two primary tasks: (1) it keeps recent instructions in the cache, ready if the CPU requests the instruction at that address again (allowing the cache to completely store small loops); and, (2) for linear code, it retrieves instructions ahead of the current execution location, so they are ready when needed (prefetch). We can disable each of these functions separately, or we can disable both.

Let us start by disabling both. Modify `timing.c` in Code Sample 5.1 by adding

```
// Turn off function (1), storing recent instructions in cache
__builtin_mtc0(_CP0_CONFIG, _CP0_CONFIG_SELECT, 0xa4210582);
CHECONCLR = 0x30;  // Turn off function (2), prefetch
```

right after `NU32_Startup()` in `main`. Everything else stays the same. The first line modifies a CPU register, preventing the prefetch cache module from storing recent instructions in cache. As for the second line, consulting the section on the prefetch cache module in the Reference Manual, we see that bits 4 and 5 of the SFR CHECON determine whether instructions are prefetched, and that clearing both bits disables predictive prefetch.

Recompiling `timing.c` with no compiler optimizations and rerunning, we find that the LED stays on for approximately 17 s, compared to approximately 6.25 s before. This corresponds to 27 SYSCLK cycles per delay loop, which we saw earlier has nine assembly commands. These numbers make sense—since the prefetch cache is completely disabled, it takes three CPU cycles (one request cycle plus two wait cycles) for each instruction to get from flash to the CPU.

If we comment out the second line, so that (1) the cache of recent instructions is off but (2) the prefetch is enabled, and rerun, we find that the LED stays on for about 8.1 s, or 13 SYSCLK cycles per loop, a small penalty compared to our original performance of 10 cycles. The prefetch is able to run ahead to grab future instructions, but it cannot run past the `for` loop conditional statement, since it does not know the outcome of the test.

---

[8] The number of "wait cycles" is the number of extra cycles the CPU must wait for instructions to finish loading from flash if they are not cached. Since the PIC32's flash operates at a maximum of 30 MHz and the CPU operates at 80 MHz, the number of wait cycles is configured as two in the bootloader and `NU32_Startup()`, to allow three total cycles for a flash instruction to load. Fewer wait cycles would result in errors and more wait cycles would slow performance unnecessarily.

Finally, if we comment out the first line but leave the second line uncommented, so that (1) the cache of recent instructions is on but (2) the prefetch is disabled, we recover our original performance of approximately 6.25 s or 10 SYSCLK cycles per loop. The reason is that the entire loop is stored in the cache, so prefetch is not necessary.

5.2.5 Math

For real-time systems, it is often critical to perform mathematical operations as quickly as possible. Mathematical expressions should be coded to minimize execution time. We will delve into the speed of various math operations in the Exercises, but here are a few rules of thumb for efficient math:

- There is no floating point unit on the PIC32MX, so all floating point math is carried out in software. Integer math is much faster than floating point math. If speed is an issue, perform all math as integer math, scaling the variables as necessary to maintain precision, and only convert to floating point when needed.
- Floating point division is slower than multiplication. If you will be dividing by a fixed value many times, consider taking the reciprocal of the value once and then using multiplication thereafter.
- Functions such as trigonometric functions, logarithms, square roots, etc. in the math library are generally slower to evaluate than arithmetic functions. Their use should be minimized when speed is an issue.
- Partial results should be stored in variables for future use to avoid performing the same computation multiple times.

5.3 Space and the Map File

The previous section focused on execution time. We now examine how much program memory (flash) and data memory (RAM) our programs use.

The linker allocates virtual addresses in program flash for all program instructions, and virtual addresses in data RAM for all global variables. The rest of RAM is allocated to the *heap* and the *stack*.

The heap is memory set aside to hold dynamically allocated memory, as allocated by `malloc` and `calloc`. These functions allow you to, for example, create an array whose length is determined at runtime, rather than specifying a (possibly space-wasteful) fixed-sized array in advance.

The stack holds temporary local variables used by functions. When a function is called, space on the stack is allocated for its local variables. When the function exits, the local variables are discarded and the space is made available again by moving the stack pointer. The stack grows "down" from the end of RAM—as local variables are "pushed" onto the stack, the stack pointer address decreases, and when local variables are "popped" off the stack after exiting a function, the stack pointer address increases. (See the assembly listing for `delay()` in `timing.c` in Section 5.2.3 for an example of moving the stack pointer when a function is called and when it exits.)

If your program attempts to put too many local variables on the stack (stack overflow), the error will not appear until run time. The linker does not catch this error because it does not explicitly set aside space for temporary local variables; it assumes they will be handled by the stack.

To further examine how memory is allocated, we can ask the linker to create a "map" file when it creates the `.elf` file. The map file indicates where instructions are placed in program memory and where global variables are placed in data memory. Your `Makefile` automatically creates an `out.map` file for you by including the `-Map` option to the linker command:

```
> xc32-gcc [details omitted] -Wl,--script="NU32bootloaded.ld",-Map="out.map"
```

The map file can be opened with a text editor.

Let us examine the `out.map` file for `timing.c` as shown in Code Sample 5.1, and compiled with no optimizations. Here's an edited portion of this rather large file:

```
Microchip PIC32 Memory-Usage Report

kseg0 Program-Memory Usage
section                      address  length [bytes]      (dec)  Description
-------                   ----------  -------------------------  -----------
.text                     0x9d001e00           0x2b4        692  App's exec code
.text.general_exception   0x9d0020b4            0xdc        220
.text                     0x9d002190            0xac        172  App's exec code
.text.main_entry          0x9d00223c            0x54         84
.text._bootstrap_except   0x9d002290            0x48         72
.text._general_exceptio   0x9d0022d8            0x48         72
.vector_default           0x9d002320            0x48         72
.text                     0x9d002368            0x5c         92  App's exec code
.dinit                    0x9d0023c4            0x10         16
.text._on_reset           0x9d0023d4            0x8          8
.text._on_bootstrap       0x9d0023dc            0x8          8
      Total kseg0_program_mem used  :           0x5e4        1508  0.3% of 0x7e200

kseg0 Boot-Memory Usage
section                      address  length [bytes]      (dec)  Description
-------                   ----------  -------------------------  -----------
         Total kseg0_boot_mem used  :               0          0  <1% of 0x970

Exception-Memory Usage
section                      address  length [bytes]      (dec)  Description
-------                   ----------  -------------------------  -----------
.app_excpt                0x9d000180            0x10         16  General-Exception
.vector_0                 0x9d000200             0x8          8  Interrupt Vector 0
.vector_1                 0x9d000220             0x8          8  Interrupt Vector 1
```

```
[[[ ... omitting long list of vectors ...]]]

.vector_51               0x9d000860            0x8          8  Interrupt Vector 51
         Total exception_mem used    :       0x1b0        432  10.5% of 0x1000

kseg1 Boot-Memory Usage
section                     address  length [bytes]      (dec)  Description
-------                  ----------  -------------------------  -----------
.reset                   0xbd001970          0x1f4        500  Reset handler
.bev_excpt               0xbd001cf0           0x10         16  BEV-Exception
        Total kseg1_boot_mem used    :       0x204        516  44.2% of 0x490
        --------------------------------------------------------------------------
        Total Program Memory used    :       0x998       2456  0.5% of 0x80000
        --------------------------------------------------------------------------
```

The kseg0 program memory usage report tells us that 1508 (or `0x5e4`) bytes are used for the main part of our program. The first entry is denoted `.text`, and holds program instructions. It is the largest single section, using 692 bytes, described as `App's exec code`, and installed starting at VA 0x9d001e00. Searching for this address in the map file, we see that this is the code for `NU32.o`, the object code associated with the NU32 library.

Subsequent sections of kseg0 program memory (some also denoted as `.text`), are packed tightly and in order of decreasing section size. The next section is `.text.general_exception`, which corresponds to a routine that is called when the CPU encounters certain types of "exceptions" (run-time errors). This code was linked from `pic32mx/lib/libpic32.a`. The next `.text` section, also labeled `App's exec code`, is the object code `timing.o` and is 172 (or `0xac`) bytes long. Searching for `timing.o` we find the following text:

```
.text           0x9d002190      0xac
 .text          0x9d002190      0xac timing.o
                0x9d002190               main
                0x9d0021c0               delay
                0x9d002210               toggleLight
```

Our functions `main`, `delay`, and `toggleLight` of `timing.o` are stored consecutively in memory. The addresses agree with our disassembly file from Section 5.2.3.

Continuing, the kseg0 boot memory report indicates that no code is placed in this memory region. The exception memory report indicates that placeholders for instructions corresponding to interrupts occupy 432 bytes. Finally, the kseg1 boot memory report indicates that the C runtime startup code installed reset functions that occupy 516 bytes. The address of the `.reset` section is the address that the bootloader (already installed in the 12 KB boot flash) jumps to.

In all, 2456 bytes of the 512 KB of program memory are used.

Continuing further in the map file, we see

```
kseg1 Data-Memory Usage
section                   address  length [bytes]      (dec)  Description
-------                ----------  -------------------------  -----------
         Total kseg1_data_mem used  :           0          0  <1% of 0x20000
        -----------------------------------------------------------------------
            Total Data Memory used  :           0          0  <1% of 0x20000
        -----------------------------------------------------------------------


Dynamic Data-Memory Reservation
section                   address  length [bytes]      (dec)  Description
-------                ----------  -------------------------  -----------
heap                   0xa0000008           0          0  Reserved for heap
stack                  0xa0000020     0x1ffd8     131032  Reserved for stack
```

There are no global variables, so no kseg1 data memory is used. The heap size is zero, so essentially all data memory is reserved for the stack.

Now let us modify our program by adding some useless global variables, just to see what happens to the map file. Let us add the following lines just before `main`:

```
char my_cat_string[] = "2 cats!";
int my_int = 1;
char my_message_string[] = "Here's a long message stored in a character array.";
char my_small_string[6], my_big_string[97];
```

Rebuilding and examining the new map file, we see the following for the data memory report:

```
kseg1 Data-Memory Usage
section                   address  length [bytes]      (dec)  Description
-------                ----------  -------------------------  -----------
.sdata                 0xa0000000         0xc         12  Small init data
.sbss                  0xa000000c         0x6          6  Small uninit data
.bss                   0xa0000014        0x64        100  Uninitialized data
.data                  0xa0000078        0x34         52  Initialized data
         Total kseg1_data_mem used  :        0xaa        170  0.1% of 0x20000
        -----------------------------------------------------------------------
            Total Data Memory used  :        0xaa        170  0.1% of 0x20000
        -----------------------------------------------------------------------
```

Our global variables now occupy 170 bytes of data RAM. The global variables have been placed in four different data memory sections, depending on whether the variable is small or large (according to a command line option or `xc32-gcc` default) and whether it is initialized:

section name	data type	variables stored there
`.sdata`	small initialized data	`my_cat_string`, `my_int`
`.sbss`	small uninitialized data	`my_small_string`
`.bss`	larger uninitialized data	`my_big_string`
`.data`	larger initialized data	`my_message_string`

Searching for the `.sdata` section further in the map file, we see

```
.sdata          0xa0000000          0xc timing.o
                0xa0000000                  my_cat_string
                0xa0000008                  my_int
                0xa000000c                  _sdata_end = .
```

Even though the string `my_cat_string` uses only seven bytes, the variable `my_int` starts eight bytes after the start of `my_cat_string`. This gap occurs because variables are aligned on four-byte boundaries, meaning that their addresses are evenly divisible by four. Similarly, the strings `my_message_string`, `my_small_string`, and `my_big_string` occupy memory to the next four-byte boundary. Due to data alignment, a five-byte string uses the same amount of memory as an eight-byte string.

Apart from the addition of these sections to the data memory usage report, we see that the global variables reduce the data memory available for the stack, and the `.dinit` (global data initialization, from the C runtime startup code) section of the kseg0 program memory report has grown to 112 bytes, meaning that our total kseg0 program memory used is now 1604 bytes instead of 1508.

Now let us change the definition of `my_cat_string` to put the qualifier `const` in front of it, becoming

```
const char my_cat_string[] = "2 cats!";
```

This makes the array a constant; `my_cat_string` cannot be changed later in the program. Global variables declared with the `const` qualifier are placed in flash rather than RAM (this behavior is XC32 specific). Building again and examining the map file, we see the following changes in kseg0 program memory usage

```
kseg0 Program-Memory Usage
section                     address  length [bytes]      (dec)  Description
-------                  ----------  -------------------------  -----------
  [[[ .dinit has shrunk by 16 bytes; no initialization code for my_cat_string ]]]
.dinit                   0x9d00223c            0x60         96
  [[[ a new eight-byte section, .rodata, has been added to hold my_cat_string ]]]
.rodata                  0x9d002424             0x8          8  Read-only const
```

and the following change in kseg1 data memory usage

```
kseg1 Data-Memory Usage
section                     address  length [bytes]      (dec)  Description
-------                  ----------  -------------------------  -----------
  [[[ .sdata has shrunk since RAM no longer holds my_cat_string ]]]
.sdata                   0xa0000000             0x4          4  Small init data
```

The new section in kseg0 program memory, `.rodata` (read-only data), contains eight bytes to hold `my_cat_string`. This constant character array is stored in flash memory at the address 0x9d002424. Correspondingly, the `.sdata` section in RAM (kseg1 data memory) has dropped by eight bytes, since the eight-byte `my_cat_string` is no longer a variable that has to be stored in RAM. Finally, the `.dinit` section in flash program memory has shrunk by 16 bytes, since we no longer need assembly code to initialize `my_cat_string`.

One last change. Let us move the definition

```
const char my_cat_string[] = "2 cats!";
```

inside the `main` function, so that `my_cat_string` is now local to `main`. Building the program again, we find only one change: one `.text` section in the kseg0 program memory report has grown by 24 bytes.

```
kseg0 Program-Memory Usage
section                    address  length [bytes]      (dec)  Description
-------                 ----------  -------------------------  -----------
  [[[ this .text section has grown by 24 bytes ]]]
.text                   0x9d002190            0xc4        196  App's exec code
```

Examining the disassembly file, we see that six lines of assembly code were added in `main` that copy the string to RAM. Since this is a local variable, the local copy uses the stack, and therefore there is no memory allocated for it in the kseg1 data memory report.

Finally, we might wish to reserve some RAM for a heap for dynamic memory allocation using `malloc` or `calloc`. By default, the heap size is set to zero. To set a nonzero heap size, we can pass a linker option to `xc32-gcc`:

```
xc32-gcc [details omitted] -Wl,--script="NU32bootloaded.ld",-Map="out.map",--defsym=_
min_heap_size=4096
```

This defines a heap of 4 KB. After building, the map file shows

```
Dynamic Data-Memory Reservation
section                    address  length [bytes]      (dec)  Description
-------                 ----------  -------------------------  -----------
heap                    0xa00000a8          0x1000       4096  Reserved for heap
stack                   0xa00010c0         0x1ef30     126768  Reserved for stack
```

The heap is allocated at low RAM addresses, close after the global variables, starting in this case at address 0xa00000a8. The stack occupies most of the rest of RAM.

For most embedded applications, there is no need to use a heap.

5.4 Chapter Summary

- The CPU's core timer increments once every two ticks of the SYSCLK, or every 25 ns for an 80 MHz SYSCLK. The commands `_CP0_SET_COUNT(0)` and `unsigned int dt = _CP0_GET_COUNT()` can be used to measure the execution time of the code in between to within a few SYSCLK cycles.
- To generate a disassembly listing at the command line, use `xc32-objdump -S filename.elf > filename.dis`.
- With the prefetch cache module fully enabled, your PIC32 should be able to execute an assembly instruction nearly every cycle. The prefetch allows instructions to be fetched in advance for linear code, but the prefetch cannot run past conditional statements. For small loops, the entire loop can be stored in the cache.
- The linker assigns specific program flash VAs to all program instructions and data RAM VAs to all global variables. The rest of RAM is allocated to the heap, for dynamic memory allocation, and to the stack, for function parameters and temporary local variables. The heap is zero bytes by default.
- A map file provides a detailed summary of memory usage. To generate a map file at the command line, use the `-Map` option to the linker, e.g.,

  ```
  xc32-gcc [details omitted] -Wl,-Map="out.map"
  ```

- Global variables can be initialized (assigned a value when they are defined) or uninitialized. Initialized global variables are stored in RAM memory sections `.data` and `.sdata` and uninitialized globals are stored in RAM memory sections `.bss` and `.sbss`. Sections beginning with `.s` mean that the variables are "small." When the program is executed, initialized global variables are assigned their values by C runtime startup code, and uninitialized global variables are set to zero.
- Global variables are packed tightly at the beginning of data RAM, 0xA0000000. The heap comes immediately after. The stack begins at the high end of RAM and grows "down" toward lower RAM addresses. Stack overflow occurs if the stack pointer attempts to move into an area reserved for the heap or global variables.

5.5 Exercises

Unless otherwise specified, compile with no optimizations for all problems.

1. Describe two examples of how you can write code differently to either make it execute faster or use less program memory.
2. Compile and run `timing.c`, Code Sample 5.2, with no optimizations (`make CFLAGS="-g -x c"`). With a stopwatch, verify the time taken by the delay loop. Do your results agree with Section 5.2.3?

3. To write time-efficient code, it is important to understand that some mathematical operations are faster than others. We will look at the disassembly of code that performs simple arithmetic operations on different data types. Create a program with the following local variables in `main`:

```
char c1=5, c2=6, c3;
int i1=5, i2=6, i3;
long long int j1=5, j2=6, j3;
float f1=1.01, f2=2.02, f3;
long double d1=1.01, d2=2.02, d3;
```

   Now write code that performs add, subtract, multiply, and divide for each of the five data types, i.e., for `char`s:

```
c3 = c1+c2;
c3 = c1-c2;
c3 = c1*c2;
c3 = c1/c2;
```

   Build the program with no optimization and look at the disassembly. For each of the statements, you will notice that some of the assembly code involves simply loading the variables from RAM into CPU registers and storing the result (also in a register) back to RAM. Also, while some of the statements are completed by a few assembly commands in sequence, others result in a jump to a software subroutine to complete the calculation. (These subroutines are provided with our C installation and included in the linking process.) Answer the following questions.

   a. Which combinations of data types and arithmetic functions result in a jump to a subroutine? From your disassembly file, copy the C statement and the assembly commands it expands to (including the jump) for one example.
   b. For those statements that do not result in a jump to a subroutine, which combination(s) of data types and arithmetic functions result in the fewest assembly commands? From your disassembly, copy the C statement and its assembly commands for one of these examples. Is the smallest data type, `char`, involved in it? If not, what is the purpose of extra assembly command(s) for the `char` data type vs. the `int` data type? (Hint: the assembly command `ANDI` takes the bitwise AND of the second argument with the third argument, a constant, and stores the result in the first argument. Or you may wish to look up a MIPS32 assembly instruction reference.)
   c. Fill in the following table. Each cell should have two numbers: the number of assembly commands for the specified operation and data type, and the ratio of this number (greater than or equal to 1.0) to the smallest number of assembly commands in the table. For example, if addition of two `int`s takes four assembly commands, and this is the fewest in the table, then the entry in that cell would be 1.0 (4). This has been filled in below, but you should change it if you get a different result. If a statement results in a jump to a subroutine, write J in that cell.

	char	int	long long	float	long double
+		1.0 (4)			
−					
*					
/					

d. From the disassembly, find out the name of any math subroutine that has been added to your assembly code. Now create a map file of the program. Where are the math subroutines installed in virtual memory? Approximately how much program memory is used by each of the subroutines? You can use evidence from the disassembly file and/or the map file. (Hint: You can search backward from the end of your map file for the name of any math subroutines.)

4. Let us look at the assembly code for bit manipulation. Create a program with the following local variables:

```
unsigned int u1=33, u2=17, u3;
```

and look at the assembly commands for the following statements:

```
u3 = u1 & u2;      // bitwise AND
u3 = u1 | u2;      // bitwise OR
u3 = u2 << 4;      // shift left 4 spaces, or multiply by 2^4 = 16
u3 = u1 >> 3;      // shift right 3 spaces, or divide by 2^3 = 8
```

How many commands does each use? For unsigned integers, bit-shifting left and right make for computationally efficient multiplies and divides, respectively, by powers of 2.

5. Use the core timer to calculate a table similar to that in Exercise 3, except with entries corresponding to the actual execution time in terms of SYSCLK cycles. So if a calculation takes 15 cycles, and the fastest calculation is 10 cycles, the entry would be 1.5 (15). This table should contain all 20 entries, even for those that jump to subroutines. (Note: subroutines often have conditional statements, meaning that the calculation could terminate faster for some operands than for others. You can report the results for the variable values given in Exercise 3.)
To minimize uncertainty due to the setup and reading time of the core timer, and the fact that the timer only increments once every two SYSCLK cycles, each math statement could be repeated ten or more times (no loops) between setting the timer to zero and reading the timer. The average number of cycles, rounded down, should be the number of cycles for each statement. Use the NU32 communication routines, or any other communication routines, to report the answers back to your computer.

6. Certain math library functions can take quite a bit longer to execute than simple arithmetic functions. Examples include trigonometric functions, logarithms, square roots, etc. Make a program with the following local variables:

```
float f1=2.07, f2;            // four bytes for each float
long double d1=2.07, d2;      // eight bytes for each long double
```

Also be sure to put `#include <math.h>` at the top of your program to make the math function prototypes available.

a. Using methods similar to those in Exercise 5, measure how long it takes to perform each of `f2 = cosf(f1)`, `f2 = sqrtf(f1)`, `d2 = cos(d1)`, and `d2 = sqrt(d1)`.
b. Copy and paste the disassembly from a `f2 = cosf(f1)` statement and a `d2 = cos(d1)` statement into your solution set and compare them. Based on the comparison of the assembly codes, comment on the advantages and disadvantages of using the eight-byte `long double` floating point representation compared to the four-byte `float` representation when you compute a cosine with the PIC32 compiler.
c. Make a map file for this program, and search for the references to the math library `libm.a` in the map file. There are several `libm.a` files in your C installation, but which one was used by the linker when you built your program? Give the directory.

7. Explain what stack overflow is, and give a short code snippet (not a full program) that would result in stack overflow on your PIC32.
8. In the map file of the original `timing.c` program, there are several `App's exec code`, one corresponding to `timing.o`. Explain briefly what each of the others are for. Provide evidence for your answer from the map file.
9. Create a map file for `simplePIC.c` from Chapter 3. (a) How many bytes does `simplePIC.o` use? (b) Where are the functions `main` and `delay` placed in virtual memory? Are instructions at these locations cacheable? (c) Search the map file for the `.reset` section. Where is it in virtual memory? Is it consistent with your `NU32bootloaded.ld` linker file? (d) Now augment the program by defining `short int`, `long int`, `long long int`, `float`, `double`, and `long double` global variables. Provide evidence from the map file indicating how much memory each data type uses.
10. Assume your program defines a global `int` array `int glob[5000]`. Now what is the maximum size of an array of `int`s that you can define as a local variable?
11. Provide global variable definitions (not an entire program) so that the map file has data sections `.sdata` of 16 bytes, `.sbss` of 24 bytes, `.data` of 0 bytes, and `.bss` of 200 bytes.
12. If you define a global variable and you want to set its initial value, is it "better" to initialize it when the variable is defined or to initialize it in a function? Explain any pros and cons.
13. The program `readVA.c` (Code Sample 5.2) prints out the contents of the 4-byte word at any virtual address, provided the address is word-aligned (i.e., evenly divisible by four). This allows you to inspect anything in the virtual memory map (Chapter 3), including the representations of variables in data RAM, program instructions in flash or boot flash, SFRs, and configuration bits. You can use code like this in conjunction with the map and disassembly files to better understand the code produced by a build. Here's a sample of the program's output to the terminal:

```
Enter the start VA (e.g., bd001970) and # of 32-bit words to print.
Listing 4 32-bit words starting from address bd001970.
 ADDRESS  CONTENTS
bd001970  0f40065e
bd001974  00000000
bd001978  401a6000
bd00197c  7f5a04c0
```

a. Build and run the program. Check its operation by consulting your disassembly file and confirming that 32-bit program instructions (in flash) listed there match the output of the program. Confirm that you get the same results whether you reference the same physical memory address using a cacheable or noncacheable virtual address. What happens if you specify a virtual address that is not divisible by four?
b. Examine the map file. At what virtual address in RAM is the variable `val` stored? Confirm its value using the program.
c. The values of the configuration bits (the four words DEVCFG0 to DEVCFG3, see Chapter 2.1.4) were set by the preinstalled bootloader. Use the program to print the values of these device configuration registers.
d. Consulting the map or disassembly file, what is the address of the last instruction in the program? Use the program to provide a listing of the instructions from a few addresses before to a few addresses after the last instruction. Addresses that do not have an instruction were erased when the program was loaded by the bootloader, but no instructions were written there. Knowing that, and by looking at your program's output, what value does an erased flash byte have?
e. Modify the program so the `unsigned int`s are defined as local to `main`, so that they are on the stack. Since `val` is no longer given a specific address by the linker, you do not find it in the map file. Use your program to find the address in RAM where `val` is stored.

Code Sample 5.2 `readVA.c`**. Code to Inspect the Virtual Memory Map.**

```
#include "NU32.h"
#define MSGLEN 100

// val is an initialized global; you can find it in the memory map with this program
char msg[MSGLEN];
unsigned int *addr;
unsigned int k = 0, nwords = 0, val = 0xf01dable;

int main(void) {

  NU32_Startup();
  while (1) {
    sprintf(msg, "Enter the start VA (e.g., bd001970) and # of 32-bit words to print: ");
    NU32_WriteUART3(msg);

    NU32_ReadUART3(msg,MSGLEN);
    sscanf(msg,"%x %d",&addr, &nwords);
```

```
    sprintf(msg,"\r\nListing %d 32-bit words starting from VA %08x.\r\n",nwords,addr);
    NU32_WriteUART3(msg);

    sprintf(msg," ADDRESS  CONTENTS\r\n");
    NU32_WriteUART3(msg);

    for (k = 0; k < nwords; k++) {
      sprintf(msg,"%08x  %08x\r\n", addr,*addr); // *addr is the 32 bits starting at addr
      NU32_WriteUART3(msg);
      ++addr;                     // addr is an unsigned int ptr so it increments by 4 bytes
    }
  }
  return 0;
}

// handle cpu exceptions, such as trying to read from a bad memory location
void _general_exception_handler(unsigned cause, unsigned status)
{
  unsigned int exccode = (cause & 0x3C) >> 2; // the exccode is reason for the exception
  // note: see PIC32 Family Reference Manual Section 03 CPU M4K Core for details
  // Look for the Cause register and the Status Register
  NU32_WriteUART3("Reset the PIC32 due to general exception.\r\n");
  sprintf(msg,"cause 0x%08x (EXCCODE = 0x%02x), status 0x%08x\r\n",cause,exccode,status);
  NU32_WriteUART3(msg);
  while(1) {
    ;
  }
}
```

Further Reading

MIPS32 M4K processor core software user's manual (2.03 ed.). (2008). MIPS Technologies.
PIC32 family reference manual. Section 04: Prefectch cache module. (2011). Microchip Technology Inc.

CHAPTER 6

Interrupts

Interrupts allow the PIC32 to respond to important events, even when performing other tasks. For example, perhaps the PIC32 is in the midst of a time-consuming calculation when the user presses a button. If software waited for the calculation to complete before checking the button state it could introduce a delay or even miss the button press altogether. To avoid this fate, we can have the button press generate an interrupt, or *interrupt request* (IRQ). When an IRQ occurs, the CPU pauses its current computation and jumps to a special routine called an *interrupt service routine* (ISR). Once the ISR has completed, the CPU returns to its original task. Interrupts appear frequently in real-time embedded control systems, and can arise from many different events. This chapter describes how the PIC32 handles interrupts and how you can implement your own ISRs.

6.1 Overview

Interrupts can be generated by the processor, peripherals, and external inputs. Example events include

- a digital input changing its value,
- information arriving on a communication port,
- the completion of a task that a peripheral was executing,
- the elapsing of a specified amount of time.

For example, to guarantee performance in real-time control applications, sensors must be read and new control signals calculated at a known fixed rate. For a robot arm, a common control loop frequency is 1 kHz. So we could configure one of the PIC32's timers to roll over every 80,000 ticks (or 1 ms at 12.5 ns per tick). This rollover event generates an interrupt that calls the feedback control ISR, which reads sensors and produces output. In this case, we would have to ensure that the control ISR always executes in less than 1 ms (you could measure the time using the core timer).

Pretend that the PIC32 is controlling a robot arm: a control loop running in a timer ISR holds the arm at a specified position. When the PIC32 receives data over the UART, a communication interrupt triggers another ISR, which parses the data and sets a new desired

Embedded Computing and Mechatronics with the PIC32 Microcontroller. http://dx.doi.org/10.1016/B978-0-12-420165-1.00006-8

angle for the robot arm. What happens if the PIC32 is executing the control ISR when a communication interrupt occurs? Alternatively, what happens if the PIC32 is executing the communication ISR and a control interrupt occurs? Each interrupt has a configurable *priority* that we can use to decide which ISR receives precedence. If a high priority interrupt occurs while a low priority ISR is executing, the CPU will jump to the high priority ISR, complete it, and then return to finish the low priority ISR. If a low priority interrupt occurs while a high priority ISR is executing, the low priority ISR remains pending until the high priority ISR finishes executing. When the high priority ISR finishes, the CPU jumps to the low priority ISR. Mainline code (i.e., any code that is not in an ISR) has the lowest priority and will usually be preempted by any interrupt.[1]

So, for our robot arm example, what should the relative priorities of the interrupts be? Assuming that communication is slow and lacks precise timing requirements, we should give the control loop (timer) ISR higher priority than the communication (UART) ISR. This scheme would prevent the control loop ISR from being preempted, helping to ensure the stability of the robot arm. We would then have to ensure that the control ISR executes fast enough to allow time for communication and other processes.

Every time an interrupt is generated, the CPU must save the contents of the internal CPU registers, called the "context," to the stack (data RAM). It then uses the registers while running the ISR. After the ISR completes, it copies the context from RAM back to its registers, restoring the previous CPU state and allowing it to continue where it left off before the interrupt. The copying of register data back and forth between the registers and RAM is called "context save and restore."

6.2 Details

The address of an ISR in virtual memory is determined by the *interrupt vector* associated with the IRQ. The CPU of the PIC32MX supports up to 64 unique interrupt vectors (and therefore 64 ISRs). For `timing.c` in Chapter 5.3, the virtual addresses of the interrupt vectors can be seen in this edited exception memory listing from the map file (an interrupt is also known as an "exception"):

```
.vector_0          0x9d000200          0x8          8  Interrupt Vector 0
.vector_1          0x9d000220          0x8          8  Interrupt Vector 1

  [[[ ... snipping long list of vectors ...]]]

.vector_51         0x9d000860          0x8          8  Interrupt Vector 51
```

[1] Interrupts can have the same priority as mainline code, in which case they will never execute, even when enabled.

If an ISR had been written for the core timer (interrupt vector 0), the code at 0x9D000200 would contain a jump to the location in program memory that actually holds the ISR.

Although the PIC32MX can have only 64 interrupt vectors, it has up to 96 events (or IRQs) that generate an interrupt. Therefore, some of the IRQs share the same interrupt vector and ISR.

Before interrupts can be used, the CPU has to be enabled to process them in either "single vector mode" or "multi-vector mode." In single vector mode, all interrupts jump to the same ISR. This is the default setting on reset of the PIC32. In multi-vector mode, different interrupt vectors are used for different IRQs. We use multi-vector mode, which is set by the bootloader (and `NU32_Startup()`).

With interrupts enabled, the CPU jumps to an ISR when three conditions are satisfied: (1) the specific IRQ has been enabled by setting a bit to 1 in the SFR IECx (one of three Interrupt Enable Control SFRs, with x equal to 0, 1, or 2); (2) an event causes a 1 to be written to the corresponding bit of the SFR IFSx (Interrupt Flag Status); and (3) the priority of the interrupt vector, as represented in the SFR IPCy (one of 16 Interrupt Priority Control SFRs, y = 0 to 15), is greater than the current priority of the CPU. If the first two conditions are satisfied, but not the third, the interrupt waits until the CPU's current priority drops.

The "x" in the IECx and IFSx SFRs above can be 0, 1, or 2, corresponding to (3 SFRs) × (32 bits/SFR) = 96 interrupt sources. The "y" in IPCy takes values 0-15, and each of the IPCy registers contains the priority level for four different interrupt vectors, i.e., (16 SFRs) × (four vectors per register) = 64 interrupt vectors. The priority level for each of the 64 vectors is represented by five bits: three indicating the priority (taking values 0 to 7, or 0b000 to 0b111; an interrupt with priority of 0 is effectively disabled) and two bits indicating the subpriority (taking values 0 to 3). Thus each IPCy has 20 relevant bits—five for each of the four interrupt vectors—and 12 unused bits. For easier access from your code, you can access these SFRs using the structures `IECxbits`, `IFSxbits`, and `IPCybits`.

The list of interrupt sources (IRQs) and their corresponding bit locations in the IECx and IFSx SFRs, as well as the bit locations in IPCy of their corresponding interrupt vectors, are given in Table 6.1, reproduced from the Interrupts section of the Data Sheet. Consulting Table 6.1, we see that the change notification's (CN) interrupt has x = 1 (for the IRQ) and y = 6 (for the vector), so information about this interrupt is stored in IFS1, IEC1, and IPC6. Specifically, IEC1⟨0⟩ is its interrupt enable bit, IFS1⟨0⟩ is its interrupt flag status bit, IPC6⟨20:18⟩ are the three priority bits for its interrupt vector, and IPC6⟨17:16⟩ are the two subpriority bits.

As mentioned earlier, some IRQs share the same vector. For example, IRQs 26, 27, and 28, each corresponding to UART1 events, all share vector number 24. Priorities and subpriorities are associated with interrupt vectors, not IRQs.

Table 6.1: Interrupt IRQ, vector, and bit location

Interrupt Source[a]	IRQ Number	Vector Number	Interrupt Bit Location			
			Flag	Enable	Priority	Sub-Priority
			Highest Natural Order Priority			
CT—Core Timer Interrupt	0	0	IFS0<0>	IEC0<0>	IPC0<4:2>	IPC0<1:0>
CS0—Core Software Interrupt 0	1	1	IFS0<1>	IEC0<1>	IPC0<12:10>	IPC0<9:8>
CS1—Core Software Interrupt 1	2	2	IFS0<2>	IEC0<2>	IPC0<20:18>	IPC0<17:16>
INTO—External Interrupt 0	3	3	IFS0<3>	IEC0<3>	IPC0<28:26>	IPC0<25:24>
T1—Timer1	4	4	IFS0<4>	IEC0<4>	IPC1<4:2>	IPC1<1:0>
IC1—Input Capture 1	5	5	IFS0<5>	IEC0<5>	IPC1<12:10>	IPC1<9:8>
OC1—Output Compare 1	6	6	IFS0<6>	IEC0<6>	IPC1<20:18>	IPC1<17:16>
INT1—External Interrupt 1	7	7	IFS0<7>	IEC0<7>	IPC1<28:26>	IPC1<25:24>
T2—Timer2	8	8	IFS0<8>	IEC0<8>	IPC2<4:2>	IPC2<1:0>
IC2—Input Capture 2	9	9	IFS0<9>	IEC0<9>	IPC2<12:10>	IPC2<9:8>
OC2—Output Compare 2	10	10	IFS0<10>	IEC0<10>	IPC2<20:18>	IPC2<17:16>
INT2—External Interrupt 2	11	11	IFS0<11>	IEC0<11>	IPC2<28:26>	IPC2<25:24>
T3—Timer3	12	12	IFS0<12>	IEC0<12>	IPC3<4:2>	IPC3<1:0>
IC3—Input Capture 3	13	13	IFS0<13>	IEC0<13>	IPC3<12:10>	IPC3<9:8>
OC3—Output Compare 3	14	14	IFS0<14>	IEC0<14>	IPC3<20:18>	IPC3<17:16>
INT3—External Interrupt 3	15	15	IFS0<15>	IEC0<15>	IPC3<28:26>	IPC3<25:24>
T4—Timer4	16	16	IFS0<16>	IEC0<16>	IPC4<4:2>	IPC4<1:0>
IC4—Input Capture 4	17	17	IFS0<17>	IEC0<17>	IPC4<12:10>	IPC4<9:8>
OC4—Output Compare 4	18	18	IFS0<18>	IEC0<18>	IPC4<20:18>	IPC4<17:16>
INT4—External Interrupt 4	19	19	IFS0<19>	IEC0<19>	IPC4<28:26>	IPC4<25:24>
T5—Timer5	20	20	IFS0<20>	IEC0<20>	IPC5<4:2>	IPC5<1:0>
IC5—Input Capture 5	21	21	IFS0<21>	IEC0<21>	IPC5<12:10>	IPC5<9:8>
OC5—Output Compare 5	22	22	IFS0<22>	IEC0<22>	IPC5<20:18>	IPC5<17:16>
SPI1E—SPI1 Fault	23	23	IFS0<23>	IEC0<23>	IPC5<28:26>	IPC5<25:24>
SPI1RX—SPI1 Receive Done	24	23	IFS0<24>	IEC0<24>	IPC5<28:26>	IPC5<25:24>
SPI1TX—SPI1 Transfer Done	25	23	IFS0<25>	IEC0<25>	IPC5<28:26>	IPC5<25:24>
U1E—UART1 Error SPI3E—SPI3 Fault I2C3B—I2C3 Bus Collision Event	26	24	IFS0<26>	IEC0<26>	IPC6<4:2>	IPC6<1:0>
U1RX—UART1 Receiver SPI3RX—SPI3 Receive Done I2C3S—I2C3 Slave Event	27	24	IFS0<27>	IEC0<27>	IPC6<4:2>	IPC6<1:0>

U1TX—UART1 Transmitter SPI3TX—SPI3 Transfer Done I2C3M—I2C3 Master Event	28	24	IFS0<28>	IEC0<28>	IPC6<4:2>	IPC6<1:0>
I2C1B—I2C1 Bus Collision Event	29	25	IFS0<29>	IEC0<29>	IPC6<12:10>	IPC6<9:8>
I2C1S—I2C1 Slave Event	30	25	IFS0<30>	IEC0<30>	IPC6<12:10>	IPC6<9:8>
I2C1M—I2C1 Master Event	31	25	IFS0<31>	IEC0<31>	IPC6<12:10>	IPC6<9:8>
CN—Input Change Interrupt	32	26	IFS1<0>	IEC1<0>	IPC6<20:18>	IPC6<17:16>
AD1—ADC1 Convert Done	33	27	IFS1<1>	IEC1<1>	IPC6<28:26>	IPC6<25:24>
PMP—Parallel Master Port	34	28	IFS1<2>	IEC1<2>	IPC7<4:2>	IPC7<1:0>
CMP1—Comparator Interrupt	35	29	IFS1<3>	IEC1<3>	IPC7<12:10>	IPC7<9:8>
CMP2—Comparator Interrupt	36	30	IFS1<4>	IEC1<4>	IPC7<20:18>	IPC7<17:16>
U3E—UART2A Error SPI2E—SPI2 Fault I2C4B—I2C4 Bus Collision Event	37	31	IFS1<5>	IEC1<5>	IPC7<28:26>	IPC7<25:24>
U3RX—UART2A Receiver SPI2RX—SPI2 Receive Done I2C4S—I2C4 Slave Event	38	31	IFS1<6>	IEC1<6>	IPC7<28:26>	IPC7<25:24>
U3TX—UART2A Transmitter SPI2TX—SPI2 Transfer Done IC4M—I2C4 Master Event	39	31	IFS1<7>	IEC1<7>	IPC7<28:26>	IPC7<25:24>
U2E—UART3A Error SPI4E—SPI4 Fault I2C5B—I2C5 Bus Collision Event	40	32	IFS1<8>	IEC1<8>	IPC8<4:2>	IPC8<1:0>
U2RX—UART3A Receiver SPI4RX—SPI4 Receive Done I2C5S—I2C5 Slave Event	41	32	IFS1<9>	IEC1<9>	IPC8<4:2>	IPC8<1:0>
U2TX—UART3A Transmitter SPI4TX—SPI4 Transfer Done IC5M—I2C5 Master Event	42	32	IFS1<10>	IEC1<10>	IPC8<4:2>	IPC8<1:0>
I2C2B—I2C2 Bus Collision Event	43	33	IFS1<11>	IEC1<11>	IPC8<12:10>	IPC8<9:8>
I2C2S—I2C2 Slave Event	44	33	IFS1<12>	IEC1<12>	IPC8<12:10>	IPC8<9:8>
I2C2M—I2C2 Master Event	45	33	IFS1<13>	IEC1<13>	IPC8<12:10>	IPC8<9:8>
FSCM—Fail-Safe Clock Monitor	46	34	IFS1<14>	IEC1<14>	IPC8<20:18>	IPC8<17:16>
RTCC—Real-Time Clock and Calendar	47	35	IFS1<15>	IEC1<15>	IPC8<28:26>	IPC8<25:24>
DMA0—DMA Channel 0	48	36	IFS1<16>	IEC1<16>	IPC9<4:2>	IPC9<1:0>
DMA1—DMA Channel 1	49	37	IFS1<17>	IEC1<17>	IPC9<12:10>	IPC9<9:8>

Continued

Table 6.1: (b) Interrupt IRQ, vector, and bit location - Cont'd

Interrupt Source[a]	IRQ	Vector	Interrupt Bit Location			
	Number	Number	Flag	Enable	Priority	Sub-Priority
Highest Natural Order Priority						
DMA2—DMA Channel 2	50	38	IFS1<18>	IEC1<18>	IPC9<20:18>	IPC9<17:16>
DMA3—DMA Channel 3	51	39	IFS1<19>	IEC1<19>	IPC9<28:26>	IPC9<25:24>
DMA4—DMA Channel 4	52	40	IFS1<20>	IEC1<20>	IPC10<4:2>	IPC10<1:0>
DMA5—DMA Channel 5	53	41	IFS1<21>	IEC1<21>	IPC10<12:10>	IPC10<9:8>
DMA6—DMA Channel 6	54	42	IFS1<22>	IEC1<22>	IPC10<20:18>	IPC10<17:16>
DMA7—DMA Channel 7	55	43	IFS1<23>	IEC1<23>	IPC10<28:26>	IPC10<25:24>
FCE—Flash Control Event	56	44	IFS1<24>	IEC1<24>	IPC11<4:2>	IPC11<1:0>
USB—USB Interrupt	57	45	IFS1<25>	IEC1<25>	IPC11<12:10>	IPC11<9:8>
CAN1—Control Area Network 1	58	46	IFS1<26>	IEC1<26>	IPC11<20:18>	IPC11<17:16>
CAN2—Control Area Network 2	59	47	IFS1<27>	IEC1<27>	IPC11<28:26>	IPC11<25:24>
ETH—Ethernet Interrupt	60	48	IFS1<28>	IEC1<28>	IPC12<4:2>	IPC12<1:0>
IC1E—Input Capture 1 Error	61	5	IFS1<29>	IEC1<29>	IPC1<12:10>	IPC1<9:8>
IC2E—Input Capture 2 Error	62	9	IFS1<30>	IEC1<30>	IPC2<12:10>	IPC2<9:8>
IC3E—Input Capture 3 Error	63	13	IFS1<31>	IEC1<31>	IPC3<12:10>	IPC3<9:8>
IC4E—Input Capture 4 Error	64	17	IFS2<0>	IEC2<0>	IPC4<12:10>	IPC4<9:8>
IC4E—Input Capture 5 Error	65	21	IFS2<1>	IEC2<1>	IPC5<12:10>	IPC5<9:8>
PMPE—Parallel Master Port Error	66	28	IFS2<2>	IEC2<2>	IPC7<4:2>	IPC7<1:0>
U4E—UART4 Error	67	49	IFS2<3>	IEC2<3>	IPC12<12:10>	IPC12<9:8>
U4RX—UART4 Receiver	68	49	IFS2<4>	IEC2<4>	IPC12<12:10>	IPC12<9:8>
U4TX—UART4 Transmitter	69	49	IFS2<5>	IEC2<5>	IPC12<12:10>	IPC12<9:8>
U6E—UART6 Error	70	50	IFS2<6>	IEC2<6>	IPC12<20:18>	IPC12<17:16>
U6RX—UART6 Receiver	71	50	IFS2<7>	IEC2<7>	IPC12<20:18>	IPC12<17:16>
U6TX—UART6 Transmitter	72	50	IFS2<8>	IEC2<8>	IPC12<20:18>	IPC12<17:16>
U5E—UART5 Error	73	51	IFS2<9>	IEC2<9>	IPC12<28:26>	IPC12<25:24>
U5RX—UART5 Receiver	74	51	IFS2<10>	IEC2<10>	IPC12<28:26>	IPC12<25:24>
U5TX—UART5 Transmitter	75	51	IFS2<11>	IEC2<11>	IPC12<28:26>	IPC12<25:24>
(Reserved)	—	—	—	—	—	—
Lowest Natural Order Priority						

[a]Not all interrupt sources are available on all PIC32s.

If the CPU is currently processing an ISR at a particular priority level, and it receives an interrupt request for a vector (and therefore ISR) at the same priority, it will complete its current ISR before servicing the other IRQ, regardless of the subpriority. When the CPU has multiple IRQs pending at a higher priority than its current operating level, the CPU first processes the IRQ with the highest priority level. If multiple IRQs sharing the highest priority are pending, the CPU processes them based on their subpriority. If interrupts have the same priority and subpriority, then their priority is resolved using the "natural order priority" given in Table 6.1, where vectors earlier in Table 6.1 have higher priority.

If the priority of an interrupt vector is zero, then the interrupt is effectively disabled.[2] There are seven enabled priority levels.

Every ISR should clear the interrupt flag (clear the appropriate bit of IFSx to zero), indicating that the interrupt has been serviced. By doing so, after the ISR completes, the CPU is free to return to the program state when the ISR was called. If the interrupt flag is not cleared, then the interrupt will be triggered immediately upon exiting the ISR.

When configuring an interrupt, you set a bit in IECx to 1 indicating the interrupt is enabled (all bits are set to zero upon reset) and assign values to the associated IPCy priority bits. (These priority bits default to zero upon reset, which will keep the interrupt disabled.) You generally never write code setting an IFSx bit to 1.[3] Instead, when you set up the device that generates the interrupt (e.g., a UART or timer), you configure it to set the interrupt flag IFSx upon the appropriate event.

The shadow register set

The PIC32MX's CPU provides an internal *shadow register set* (SRS), which is a full extra set of registers. You can use these extra registers to eliminate the time needed for context save and restore. When processing an ISR using the SRS, the CPU switches to this extra set of internal registers. When it finishes the ISR, it switches back to its original register set, without needing to save and restore them. We see examples using the shadow register set in Section 6.4.

The Device Configuration Register DEVCFG3 determines which priority level is assigned to the shadow register set. The preprocessor command

```
#pragma config FSRSSEL = PRIORITY_6
```

implemented in `NU32.c` and the NU32 bootloader allows only priority level 6 to use the shadow register set. To change this setting you need to either use a standalone program or modify and re-install the bootloader.

---

[2] One reason for giving an interrupt priority zero is to prevent the ISR from triggering while still having the interrupt source set the IRQ flag, allowing you to monitor the interrupt without it preempting other code.

[3] Setting an IFSx bit to one would cause the interrupt to be pending, just as if hardware had set the flag.

External interrupt inputs

The PIC32 has five digital inputs (INT0 to INT4) that can generate interrupts on rising or falling signal edges. The flag and enable status bits are IFS0 and IEC0, respectively, at bits 3, 7, 11, 15, and 19 for INT0, INT1, INT2, INT3, and INT4, respectively. The priority and subpriority bits are in IPCy⟨28:26⟩ and IPCy⟨25:24⟩ for the input INTy. The SFR INTCON bits 0-4 determine whether the associated interrupt is triggered on a falling edge (bit cleared to 0) or rising edge (bit set to 1). From C, you can access these bits as `IFS0bits.INTxIF`, `IEC0bits.INTxIE`, `IPCxbits.INTxIP`, and `IPCxbits.INTxIS`, where `x` is 0 to 4. The interrupt vector number for each external interrupt is stored as the constant `_EXTERNAL_x_VECTOR`.

Special Function Registers

The SFRs associated with interrupts are summarized below. We omit some fields: for full details, consult the Interrupt Controller section of the Data Sheet, the Interrupt IRQ, Vector, and Bit Location table reproduced earlier (Table 6.1), or the Interrupt section of the Reference Manual.

INTCON The interrupt control SFR determines the interrupt controller mode and the behavior of the five external interrupt pins INT0 to INT4.

- INTCON⟨12⟩, or INTCONbits.MVEC: Set to enable multi-vector mode. The bootloader (and `NU32_Startup()`) sets this bit, and you will probably always use this mode.
- INTCON⟨x⟩, for x = 0 to 4, or INTCONbits.INTxEP: Determines whether the given external interrupt (INTx) triggers on a rising or falling edge.
  - 1 External interrupt pin x triggers on a rising edge.
  - 0 External interrupt pin x triggers on a falling edge.

INTSTAT The interrupt status SFR is read-only and contains information about the latest IRQ given to the CPU when in single vector mode. We will not need it.

IPTMR The interrupt proximity timer SFR can be used to implement a delay to queue up interrupt requests before presenting them to the CPU. For example, upon receiving an interrupt request, the timer starts counting clock cycles, queuing up any subsequent interrupt requests, until IPTMR cycles have passed. By default, this timer is turned off by INTCON, and we will leave it that way.

IECx, x = 0, 1, or 2 Three 32-bit interrupt enable control SFRs for up to 96 interrupt sources. Setting to 1 enables the given interrupt, clearing to 0 disables it.

IFSx, x = 0, 1, or 2 The three 32-bit interrupt flag status SFRs represent the status of up to 96 interrupt sources. Setting to one requests a given interrupt, and clearing to 0 indicates that no interrupt is requested. Typically, peripherals set the appropriate IFSx bit in response to an event, and user software clears the IFSx bit from within the ISR. The CPU services pending interrupts (those whose IFSx and IECx bits are set) in priority order.

IPCy, y = 0 to 15 Each of the 16 interrupt priority control SFRs contains 5-bit priority and subpriority values for 4 different interrupt vectors (64 vectors total). Interrupts will not be serviced unless the CPU's current priority is less than the interrupt's priority. When the CPU services an interrupt, it sets its current priority to that of the interrupt, and when the ISR completes, its previous priority is restored.

6.3 Steps to Configure and Use an Interrupt

The bootloader (and `NU32_Startup()`) enables the CPU to receive interrupts, setting multi-vector mode by setting INTCONbits.MVEC to 1. After being in the correct mode, there are seven steps to configure and use an interrupt. We recommend your program execute steps 2-7 in the order given below. The details of the syntax are left to the examples in Section 6.4.

1. Write an ISR with a priority level 1-7 using the syntax

   ```
   void __ISR(vector_number, IPLnXXX) interrupt_name(void) { ... }
   ```

   where `vector_number` is the interrupt vector number, `n`=1 to 7 is the priority, `XXX` is either `SOFT` or `SRS`, and `interrupt_name` can be anything but should describe the ISR. `SOFT` uses software context save and restore, and `SRS` uses the shadow register set. (The bootloader on the NU32 allows only priority level 6 to use the SRS, so if you use the SRS, you should use the syntax `IPL6SRS`.) No subpriority is specified in the ISR function. The ISR should clear the appropriate interrupt flag IFSxbit.
2. Disable interrupts at the CPU to prevent spurious generation of interrupts while you are configuring. Although interrupts are disabled by default on reset, `NU32_Startup()` enables them. To disable all interrupts you can use the special compiler instruction `__builtin_disable_interrupts()`.
3. Configure the device (e.g., peripheral) to generate interrupts on the appropriate event. This procedure involves configuring the SFRs of the particular peripheral.
4. Configure the interrupt priority and subpriority in IPCy. **The IPCy priority should match the priority of the ISR defined in Step 1.**
5. Clear the interrupt flag status bit to 0 in IFSx.
6. Set the interrupt enable bit to 1 in IECx.
7. Re-enable interrupts at the CPU. You can use the compiler instruction `__builtin_enable_interrupts()`.

6.4 Sample Code

6.4.1 Core Timer Interrupt

Let us toggle a digital output once per second based on an interrupt from the CPU's core timer. First, we store a value in the CPU's CP0_COMPARE register. Whenever the core timer counter equals this value (CP0_COMPARE), an interrupt is generated. In the interrupt service routine, we reset the core timer counter to 0. Since the core timer runs at half the frequency of the system clock, setting CP0_COMPARE to 40,000,000 toggles the digital output once per second.

To view the ISR in action we toggle LED2 on the NU32 board. We shall use priority level 6, subpriority 0, and the shadow register set.

Code Sample 6.1 `INT_core_timer.c`. **A Core Timer Interrupt Using the Shadow Register Set.**

```
#include "NU32.h"          // constants, funcs for startup and UART
#define CORE_TICKS 40000000 // 40 M ticks (one second)

void __ISR(_CORE_TIMER_VECTOR, IPL6SRS) CoreTimerISR(void) {  // step 1: the ISR
  IFS0bits.CTIF = 0;                  // clear CT int flag IFS0<0>, same as IFS0CLR=0x0001
  LATFINV = 0x2;                      // invert pin RF1 only
  _CP0_SET_COUNT(0);                  // set core timer counter to 0
  _CP0_SET_COMPARE(CORE_TICKS);       // must set CP0_COMPARE again after interrupt
}

int main(void) {
  NU32_Startup(); // cache on, min flash wait, interrupts on, LED/button init, UART init

  __builtin_disable_interrupts();     // step 2: disable interrupts at CPU
  _CP0_SET_COMPARE(CORE_TICKS);       // step 3: CP0_COMPARE register set to 40 M
  IPC0bits.CTIP = 6;                  // step 4: interrupt priority
  IPC0bits.CTIS = 0;                  // step 4: subp is 0, which is the default
  IFS0bits.CTIF = 0;                  // step 5: clear CT interrupt flag
  IEC0bits.CTIE = 1;                  // step 6: enable core timer interrupt
  __builtin_enable_interrupts();      // step 7: CPU interrupts enabled

  _CP0_SET_COUNT(0);                  // set core timer counter to 0

  while(1) { ; }
  return 0;
}
```

Following our seven steps to use an interrupt, we have:

Step 1. The ISR.

```
void __ISR(_CORE_TIMER_VECTOR, IPL6SRS) CoreTimerISR(void) {  // step 1: the ISR
  IFS0bits.CTIF = 0;             // clear CT int flag IFS0<0>, same as IFS0CLR=0x0001
  ...
}
```

We can name the ISR anything, so we name it `CoreTimerISR`. The `__ISR` syntax is XC32-specific (not a C standard) and tells the compiler and linker that this function should be treated as an interrupt handler.[4] The two arguments are the interrupt vector number for the core timer, called `_CORE_TIMER_VECTOR` (defined as 0 in `p32mx795f512h.h`, which agrees with Table 6.1), and the interrupt priority level. The interrupt priority level is specified using the syntax `IPLnSRS` or `IPLnSOFT`, where `n` is 1 to 7, `SRS` indicates that the shadow register set should be used, and `SOFT` indicates that software context save and restore should be used. Use `IPL6SRS` if you'd like to use the shadow register set, as in this example, since the device configuration registers on the NU32's PIC32 specify priority level 6 for the shadow register set. You do not specify subpriority in the ISR.

The ISR should clear the interrupt flag in IFS0⟨0⟩ (`IFS0bits.CTIF`), because, according to the table of interrupts (Table 6.1), this bit corresponds to the core timer interrupt. We also need to write to CP0_COMPARE to clear the interrupt, an action specific to the core timer (many interrupts have peripheral-specific actions that are required to clear the interrupt).

Step 2. Disabling interrupts. Since `NU32_Startup()` enables interrupts, we disable them before configuring the core timer interrupt.

```
__builtin_disable_interrupts();   // step 2: disable interrupts at CPU
```

Disabling interrupts before configuring the device that generates interrupts is good general practice, to avoid unwanted interrupts during configuration. In many cases it is not strictly necessary, however.

Step 3. Configuring the core timer to interrupt.

```
_CP0_SET_COMPARE(CORE_TICKS);     // step 3: CP0_COMPARE register set to 40 M
```

This line sets the core timer's CP0_COMPARE value so that an interrupt is generated when the core timer counter reaches `CORE_TICKS`. If the interrupt were to be generated by a peripheral, we would consult the appropriate book chapter or the Reference Manual, to set the SFRs to generate an IRQ on the appropriate event.

Step 4. Configuring interrupt priority.

```
IPC0bits.CTIP = 6;                // step 4: interrupt priority
IPC0bits.CTIS = 0;                // step 4: subp is 0, which is the default
```

These two commands set the appropriate bits in IPCy (y = 0, according to Table 6.1). Consulting the file `p32mx795f512h.h` or the Memory Organization section of the Data Sheet

[4] `__ISR` is a macro defined in `<sys/attribs.h>`, which `NU32.h` includes.

shows us that the core timer's priority and subpriority bits of IPC0 are called IPC0bits.CTIP and IPC0bits.CTIS, respectively. Alternatively, we could have used any other means to manipulate the bits IPC0⟨4:2⟩ and IPC0⟨1:0⟩, as indicated in Table 6.1, while leaving all other bits unchanged. We prefer using the bit-field `structs` because it is the most readable method. The priority must agree with the ISR priority. It is unnecessary to set the subpriority, which defaults to zero.

Step 5. Clearing the interrupt flag status bit.

```
IFS0bits.CTIF = 0;                // step 5: clear CT interrupt flag
```

This command clears the appropriate bit in IFSx (x = 0 here). A less readable but possibly more efficient alternative would be `IFS0CLR = 1`, to clear the zeroth bit of IFS0.

Step 6. Enabling the core timer interrupt.

```
IEC0bits.CTIE = 1;                // step 6: enable core timer interrupt
```

This command sets the appropriate bit in IECx (x = 0 here). An alternative would be `IEC0SET = 1` to set the zeroth bit of IEC0.

Step 7. Re-enable interrupts at the CPU.

```
__builtin_enable_interrupts();    // step 7: CPU interrupts enabled
```

This compiler built-in function generates an assembly instruction that allows the CPU to process interrupts.

6.4.2 External Interrupt

Code Sample 6.2 causes the NU32's LEDs to illuminate briefly, on a falling edge of external interrupt input pin INT0. You can find the IRQ associated with INT0, and the flag, enable, priority, and subpriority bits in Table 6.1. In this example we use interrupt priority level 2, with software context save and restore.

You can test this program with the NU32 by connecting a wire from the D7/USER pin to the D0/INT0 pin. Pressing the USER button creates a falling edge on digital input RD7 (see the wiring diagram in Figure 2.3) and therefore INT0, which causes the LEDs to flash. You might also notice the issue of switch *bounce*: when you release the button, nominally creating a rising edge, you might see the LEDs flash again. The extra flash occurs because of chattering when mechanical contact between two conductors is established or broken, causing spurious rising and falling edges. Reading reliably from mechanical switches requires a *debouncing* circuit or software; see Exercise 16.

Code Sample 6.2 `INT_ext_int.c`. **Using an External Interrupt to Flash LEDs on the NU32.**

```
#include "NU32.h"          // constants, funcs for startup and UART

void __ISR(_EXTERNAL_0_VECTOR, IPL2SOFT) Ext0ISR(void) { // step 1: the ISR
  NU32_LED1 = 0;                  // LED1 and LED2 on
  NU32_LED2 = 0;
  _CP0_SET_COUNT(0);

  while(_CP0_GET_COUNT() < 10000000) { ; } // delay for 10 M core ticks, 0.25 s

  NU32_LED1 = 1;                  // LED1 and LED2 off
  NU32_LED2 = 1;
  IFS0bits.INT0IF = 0;            // clear interrupt flag IFS0<3>
}

int main(void) {
  NU32_Startup(); // cache on, min flash wait, interrupts on, LED/button init, UART init
  __builtin_disable_interrupts(); // step 2: disable interrupts
  INTCONbits.INT0EP = 0;          // step 3: INT0 triggers on falling edge
  IPC0bits.INT0IP = 2;            // step 4: interrupt priority 2
  IPC0bits.INT0IS = 1;            // step 4: interrupt priority 1
  IFS0bits.INT0IF = 0;            // step 5: clear the int flag
  IEC0bits.INT0IE = 1;            // step 6: enable INT0 by setting IEC0<3>
  __builtin_enable_interrupts();  // step 7: enable interrupts
                                  // Connect RD7 (USER button) to INT0 (D0)
  while(1) {
      ; // do nothing, loop forever
  }

  return 0;
}
```

6.4.3 Speedup Due to the Shadow Register Set

This example measures the amount of time it takes to enter and exit an ISR using context save and restore vs. the SRS. We write two identical ISRs; the only difference is that one uses `IPL6SOFT` and the other uses `IPL6SRS`. The two ISRs are based on the external interrupts INT0 and INT1, respectively. To get precise timing, however, we trigger interrupts in software by setting the appropriate bit of IFS0.

After setting up the interrupts, the program `INT_timing.c` enters an infinite loop. First the core timer is reset to zero, then the interrupt flag is set for INT0. The `main` function then waits until the ISR clears the flag. First the ISR for INT0 records the core timer counter. Next it toggles LED2 and clears the interrupt flag. Finally it logs the time again. After the interrupt exits the `main` function logs the time when control is returned. The timing results are written back to the host computer over the UART. The ISR for INT1 is timed in a similar manner.

These are the results (which are repeated over and over):

```
IPL6SOFT in  19 out  26 total  40 time 1000 ns
 IPL6SRS in  17 out  24 total  37 time  925 ns
```

For context save and restore, it takes 19 core clock ticks (about 38 SYSCLK ticks) to begin executing statements in the ISR; the last ISR statement completes about 7 (14) ticks later; and finally control is returned to `main` approximately 40 (80) total ticks, or 1000 ns, after the interrupt flag is set. For the SRS, the first ISR statement is executed after about 17 (34) ticks; the ISR runs in an identical 7 (14) ticks; and a total of approximately 37 (74) ticks, or 925 ns, elapse between the time the interrupt flag is set and control is returned to `main`.

Although the exact timing may be different for other ISRs and `main` functions, depending on the register context that must be saved and restored (which depends on what the code does), we can make some general observations:

- The ISR is not entered immediately after the flag is set. It takes time to respond to the interrupt request, and instructions in `main` may be executed after the flag is set.
- The SRS reduces the time needed to enter and exit the ISR, approximately 75 ns total in this case.
- Simple ISRs can be completed less than a microsecond after the interrupt event occurs.

The sample code is below. We note that this example also serves to demonstrate different methods for setting bit fields in ISRs. We hope that after viewing this code you will agree that using the bit-field `structs` makes the code easier to read and understand.

Code Sample 6.3 `INT_timing.c`. **Timing the Shadow Register Set vs. Typical Context Save and Restore.**

```
#include "NU32.h"          // constants, funcs for startup and UART
#define DELAYTIME 40000000 // 40 million core clock ticks, or 1 second

void delay(void);

static volatile unsigned int Entered = 0, Exited = 0;  // note the qualifier "volatile"

void __ISR(_EXTERNAL_0_VECTOR, IPL6SOFT) Ext0ISR(void) {
  Entered = _CP0_GET_COUNT();        // record time ISR begins
  IFS0CLR = 1 << 3;                  // clear the interrupt flag
  NU32_LED2 = !NU32_LED2;            // turn off LED2
  Exited = _CP0_GET_COUNT();         // record time ISR ends
}

void __ISR(_EXTERNAL_1_VECTOR, IPL6SRS) Ext1ISR(void) {
  Entered = _CP0_GET_COUNT();        // record time ISR begins
```

```
  IFS0CLR = 1 << 7;                // clear the interrupt flag
  NU32_LED2 = !NU32_LED2;          // turn on LED2
  Exited = _CP0_GET_COUNT();       // record time ISR ends
}

int main(void) {
  unsigned int dt = 0;
  unsigned int encopy, excopy;     // local copies of globals Entered, Exited
  char msg[128] = {};

  NU32_Startup(); // cache on, min flash wait, interrupts on, LED/button init, UART init
  __builtin_disable_interrupts();  // step 2: disable interrupts at CPU
  INTCONSET = 0x3;                 // step 3: INT0 and INT1 trigger on rising edge
  IPC0CLR = 31 << 24;              // step 4: clear 5 priority and subp bits for INT0
  IPC0 |= 24 << 24;                // step 4: set INT0 to priority 6 subpriority 0
  IPC1CLR = 0x1F << 24;            // step 4: clear 5 priority and subp bits for INT1
  IPC1 |= 0x18 << 24;              // step 4: set INT1 to priority 6 subpriority 0
  IFS0bits.INT0IF = 0;             // step 5: clear INT0 flag status
  IFS0bits.INT1IF = 0;             // step 5: clear INT1 flag status
  IEC0SET = 0x88;                  // step 6: enable INT0 and INT1 interrupts
  __builtin_enable_interrupts();   // step 7: enable interrupts
  while(1) {
    delay();                       // delay, so results sent back at reasonable rate
    _CP0_SET_COUNT(0);             // start timing
    IFS0bits.INT0IF = 1;           // artificially set the INT0 interrupt flag
    while(IFS0bits.INT0IF) {
        ;                          // wait until the ISR clears the flag
    }
    dt = _CP0_GET_COUNT();         // get elapsed time
    __builtin_disable_interrupts(); // good practice before using vars shared w/ISR
    encopy = Entered;              // copy the shared variables to local copies ...
    excopy = Exited;               // ... so the time interrupts are off is short
    __builtin_enable_interrupts(); // turn interrupts back on quickly!
    sprintf(msg,"IPL6SOFT in %3d out %3d total %3d time %4d ns\r\n"
        ,encopy,excopy,dt,dt*25);
    NU32_WriteUART3(msg);          // send times to the host

    delay();                       // same as above, except for INT1
    _CP0_SET_COUNT(0);
    IFS0bits.INT1IF = 1;           // trigger INT1 interrupt
    while(IFS0bits.INT1IF) {
        ;                          // wait until the ISR clears the flag
    }
    dt = _CP0_GET_COUNT();
    __builtin_disable_interrupts();
    encopy = Entered;
    excopy = Exited;
    __builtin_enable_interrupts();
    sprintf(msg," IPL6SRS in %3d out %3d total %3d time %4d ns\r\n"
        ,encopy,excopy,dt,dt*25);
    NU32_WriteUART3(msg);
  }
  return 0;
}

void delay() {
  _CP0_SET_COUNT(0);
  while(_CP0_GET_COUNT() < DELAYTIME) {
      ;
  }
}
```

6.4.4 Sharing Variables with ISRs

Code Sample 6.3 was the first to share variables between an ISR and other functions. Namely, `Entered` and `Exited` are used in both ISRs as well as `main`. Sharing data between ISRs and mainline code is one area where using global variables is required, even though they should be generally avoided. We use the `static` keyword so at least, in a larger project, the variables are limited in scope to the file in which they are declared.

This code demonstrates two good practices when sharing variables with ISRs:

(1) Using the type qualifier `volatile`

By putting the qualifier `volatile` in front of the type in the global variable definition

```
static volatile unsigned int Entered = 0, Exited = 0;
```

we tell the compiler that external processes (namely, an ISR that may be triggered at an unknown time) may read or write the variables at any time. Therefore, any optimizations the compiler performs should not take shortcuts in generating assembly code associated with a `volatile` variable. For example, if you had code of the form

```
static int i = 0;    // global variable shared by functions and an ISR

void myFunc(void) {
  i = 1;
  // some other code that doesn't use or affect i
  i = 2;
  // some code that uses i
}
```

a compiler running optimizations might not generate any code for `i = 1` at all, believing that the value 1 for `i` is never used. If an external interrupt triggered during execution of the code between `i = 1` and `i = 2`, however, and the ISR used the value of `i`, it would use the wrong value (perhaps the originally initialized value `i = 0`).

To correct this problem, the declaration of the global variable `i` should be

```
static volatile int i = 0;
```

The `volatile` qualifier ensures that the compiler will emit full assembly code for any reads or writes of `i`. The compiler does not assume anything about the value of `i` or whether it is changed or used by processes that it does not know about. This is why all SFRs are declared as `volatile` in `p32mx795f512h.h`; their values can be changed by processes external to the CPU.

(2) Enabling and disabling interrupts

Consider a scenario where the mainline code and an ISR share a 64-bit `long double` variable. To load the variable into two of the CPU's 32-bit registers, one assembly instruction first loads the most significant 32 bits into one register. Then the process is interrupted, the ISR modifies the variable in RAM, and control returns to the main code. At that point, the next assembly instruction loads the lower 32 bits of the new value of the `long double` in RAM into the other CPU register. Now the CPU registers have neither the correct variable value from before nor after the ISR.

To prevent data corruption like this, interrupts can be disabled before reading or writing the shared variables, then re-enabled afterward. If an IRQ is generated during the time that the CPU is ignoring interrupts, the IRQ will simply wait until the CPU is accepting interrupts again.

Interrupts should not be disabled for long, as this defeats the purpose of interrupts. In the sample code, the time that interrupts are disabled is minimized by simply copying the shared variables to local copies during this period, rather than performing time-consuming operations with them. This avoids having the interrupts disabled during the `sprintf` command, which can take many CPU cycles.

In many cases it is not necessary to disable interrupts before using shared variables. (For example, it was not necessary in the sample code above.) Determining whether such precautions are necessary sometimes depends not only on the algorithms you employ but how those algorithms translate into assembly instructions. Disabling interrupts is usually the simpler and safer option.[5] As long as you limit the code that executes while interrupts are disabled, you will delay any interrupts that would have occurred by at most a few hundred nanoseconds; most applications can tolerate such delays.

6.5 Chapter Summary

- The CPU of the PIC32MX supports 96 interrupt requests (IRQs) and 64 interrupt vectors, and therefore up to 64 interrupt service routines (ISRs). Therefore, some IRQs share the same ISR. For example, all IRQs related to UART1 (data received, data transmitted, error) have the same interrupt vector.
- The PIC32 can be configured to operate in single vector mode (all IRQs result in a jump to the same ISR) or in multi-vector mode. The bootloader (and `NU32_Startup()`) puts the NU32 in multi-vector mode.

[5] Errors caused by incorrect sharing of data between interrupts and mainline code are called *race conditions*. Race conditions are notoriously difficult to discover and fix; they may only appear intermittently because they closely depend on timing.

- Priorities and subpriorities are associated with interrupt vectors, and therefore ISRs, not IRQs. The priority of a vector is defined in an SFR IPCy, y = 0 to 15. In the definition of the associated ISR, the same priority `n` should be specified using `IPLnSOFT` or `IPLnSRS`. `SOFT` indicates that software context save and restore is performed, while `SRS` means that the shadow register set is used instead, reducing ISR entry and exit time. The PIC32 on the NU32 is configured by its device configuration registers to make the SRS available only at priority level 6, so the SRS can only be used with `IPL6SRS`.
- When an interrupt is generated, it is serviced immediately if its priority is higher than the current priority. Otherwise it waits until the current ISR is finished.
- In addition to configuring the CPU to accept interrupts, enabling specific interrupts, and setting their priority, the specific peripherals (such as counter/timers, UARTs, change notification pins, etc.) must be configured to generate interrupt requests on the appropriate events. These configurations are left for the chapters covering those peripherals.
- The seven steps to use an interrupt, after putting the CPU in multi-vector mode, are: (1) write the ISR; (2) disable interrupts; (3) configure a device or peripheral to generate interrupts; (4) set the ISR priority and subpriority; (5) clear the interrupt flag; (6) enable the IRQ; and (7) enable interrupts at the CPU.
- If a variable is shared with an ISR, it is a good idea to (1) define that variable with the type qualifier `volatile` (also use `static` unless you have good reason not to) and (2) turn off interrupts before reading or writing it if there is a danger the process could be interrupted. If interrupts are disabled, they should be disabled for as short a period as possible.

6.6 Exercises

1. Interrupts can be used to implement a fixed frequency control loop (e.g., 1 kHz). Another method for executing code at a fixed frequency is *polling*: you can keep checking the core timer, and when some number of ticks has passed, execute the control routine. Polling can also be used to check for changes on input pins and other events. Give pros and cons (if any) of using interrupts vs. polling.
2. You are watching TV. Give an analogy to an IRQ and ISR for your mental attention in this situation. Also give an analogy to polling.
3. What is the relationship between an interrupt vector and an ISR? What is the maximum number of ISRs that the PIC32 can handle?
4. (a) What happens if an IRQ is generated for an ISR at priority level 4, subpriority level 2 while the CPU is in normal execution (not executing an ISR)? (b) What happens if that IRQ is generated while the CPU is executing a priority level 2, subpriority level 3 ISR? (c) What happens if that IRQ is generated while the CPU is executing a priority level 4, subpriority level 0 ISR? (d) What happens if that IRQ is generated while the CPU is executing a priority level 6, subpriority level 0 ISR?

5. An interrupt asks the CPU to stop what it's doing, attend to something else, and then return to what it was doing. When the CPU is asked to stop what it's doing, it needs to remember "context" of what it was working on, i.e., the values currently stored in the CPU registers. (a) Assuming no shadow register set, what is the first thing the CPU must do before executing the ISR and the last thing it must do upon completing the ISR? (b) How does using the shadow register set change the situation?
6. What is the peripheral and interrupt vector number associated with IRQ 35? What are the SFRs and bit numbers controlling its interrupt enable, interrupt flag status, and priority and subpriority? Does IRQ 35 share the interrupt vector with any other IRQ?
7. What peripherals and IRQs are associated with interrupt vector 24? What are the SFRs and bit numbers controlling the priority and subpriority of the vector and the interrupt enable and flag status of the associated IRQs?
8. For the problems below, use only the SFRs IECx, IFSx, IPCy, and INTCON, and their CLR, SET, and INV registers (do not use other registers, nor the bit fields as in IFS0bits.INT0IF). Give valid C bit manipulation commands to perform the operations without changing any uninvolved bits. Also indicate, in English, what you are trying to do, in case you have the right idea but wrong C statements. Do not use any constants defined in Microchip XC32 files; just use numbers.
   a. Enable the Timer2 interrupt, set its flag status to 0, and set its vector's priority and subpriority to 5 and 2, respectively.
   b. Enable the Real-Time Clock and Calendar interrupt, set its flag status to 0, and set its vector's priority and subpriority to 6 and 1, respectively.
   c. Enable the UART4 receiver interrupt, set its flag status to 0, and set its vector's priority and subpriority to 7 and 3, respectively.
   d. Enable the INT2 external input interrupt, set its flag status to 0, set its vector's priority and subpriority to 3 and 2, and configure it to trigger on a rising edge.
9. Edit Code Sample 6.3 so that each line correctly uses the "bits" forms of the SFRs. In other words, the left-hand sides of the statements should use a form similar to that used in step 5, except using INTCONbits, IPC0bits, and IEC0bits.
10. Consulting the `p32mx795f512h.h` file, give the names of the constants, and the numerical values, associated with the following IRQs: (a) Input Capture 5. (b) SPI3 receive done. (c) USB interrupt.
11. Consulting the `p32mx795f512h.h` file, give the names of the constants, and the numerical values, associated with the following interrupt vectors: (a) Input Capture 5. (b) SPI3 receive done. (c) USB interrupt.
12. True or false? When the PIC32 is in single vector interrupt mode, only one IRQ can trigger an ISR. Explain your answer.
13. Give the numerical value of the SFR INTCON, in hexadecimal, when it is configured for single vector mode using the shadow register set; and external interrupt input INT3

triggers on a rising edge while the rest of the external inputs trigger on a falling edge. The Interrupt Proximity Timer bits are left as the default.

14. So far we have only seen interrupts generated by the core timer and the external interrupt inputs, because we first have to learn something about the other peripherals to complete Step 3 of the seven-step interrupt setup procedure. Let us jump ahead and see how the Change Notification peripheral could be configured in Step 3. Consulting the Reference Manual chapter on I/O Ports, name the SFR and bit number that has to be manipulated to enable Change Notification pins to generate interrupts.
15. Build `INT_timing.c` and open its disassembly file `out.dis` with a text editor. Starting at the top of the file, you see the startup code inserted by `crt0.o`. Continuing down, you see the "bootstrap exception" section `.bev_excpt`, which handles any exceptions that might occur while executing boot code; the "general exception" section `.app_excpt`, which handles any serious errors the CPU encounters (such as attempting to access an invalid memory address) (Table 6.1); and finally the interrupt vector sections, labeled `.vector_x`, where `x` can take values from 0 to 51 (12 of the possible 64 vectors are not used by the PIC32MX). Each of these exception vectors simply jumps to another address. (Note that `j`, `jal`, and `jr` are all jump statements in assembly. Jumps are not executed immediately; the next assembly statement, in the *jump delay slot*, executes before the jump completes. The jump `j` jumps to the address specified. `jal` jumps to the address specified, usually corresponding to a function, and stores in a CPU register `ra` a return address two instructions [eight bytes] later. `jr` jumps to an address stored in a register, often `ra` to return from a function.)
    a. What addresses do the `.vector_x` sections jump to? What is installed at these addresses?
    b. Find the `Ext0ISR` and `Ext1ISR` functions. How many assembly commands are before the first `_CP0_GET_COUNT()` command in each function? How many assembly commands are after the last `_CP0_GET_COUNT()` command in each function? What is the purpose of the commands that account for the majority of the difference in the number of commands? (Note that `sw`, short for "store word," copies a 32-bit CPU register to RAM, and `lw`, short for "load word," copies a 32-bit word from RAM to a CPU register.) Explain why the two functions are different even though their C code is essentially identical.
16. Modify Code Sample 6.2 so the USER button is debounced. How can you change the ISR so the LEDs do not flash if the falling edge comes at the beginning of a very brief, spurious down pulse? Verify that your solution works. (Hint: Any real button press should last much more than 10 ms, while the mechanical bouncing period of any decent switch should be much less than 10 ms. See also Chapter B.2.1 for a hardware solution to debouncing.)
17. Using your solution for debouncing the USER button (Exercise 16), write a stopwatch program using an ISR based on INT2. Connect a wire from the USER button pin to the INT2 pin so you can use the USER button as your timing button. Using the NU32

library, your program should send the following message to the user's screen: `Press the USER button to start the timer.` When the USER button has been pressed, it should send the following message: `Press the USER button again to stop the timer.` When the user presses the button again, it should send a message such as `12.505 seconds elapsed.` The ISR should either (1) start the core timer at 0 counts or (2) read the current timer count, depending on whether the program is in the "waiting to begin timing" state or the "timing state." Use priority level 6 and the shadow register set. Verify that the timing is accurate. The stopwatch only has to be accurate for periods of less than the core timer's rollover time.

You could also try using polling in your `main` function to write out the current elapsed time (when the program is in the "timing state") to the user's screen every second so the user can see the running time.

18. Write a program identical to the one in Exercise 17, but using a 16×2 LCD screen for output instead of the host computer's display.
19. Write a program that interrupts at a frequency defined interactively by the user. The `main` function is an infinite loop that uses the NU32 library to ask the user to specify the integer variable `InterruptPeriod`. If the user enters a number greater than an appropriate minimum and less than an appropriate maximum, this becomes the number of core clock ticks between core timer interrupts. The ISR simply toggles the LEDs, so the `InterruptPeriod` is visible. Set the vector priority to 3 and subpriority to 0.
20. (a) Write a program that has two ISRs, one for the core timer and one for the debounced input INT2. The core timer interrupts every 4 s, and the ISR simply turns on LED1 for 2 s, turns it off, and exits. The INT2 interrupt turns LED2 on and keeps it on until the user releases the button. Choose interrupt priority level 1 for the core timer and 5 for INT2. Run the program, experiment with button presses, and see if it agrees with what you expect. (b) Modify the program so the two priority levels are switched. Run the program, experiment with button presses, and see if it agrees with what you expect.
21. A CPU run-time error, such as attempting to access an invalid memory address, generates a general exception. As with an interrupt, program execution jumps to a new function, in this case called `_gen_exception`. In turn, this function calls the function `_general_exception_context` which calls `_general_exception_handler`. You have the option to use the Microchip default general exception handler, or you can write your own, as in Sample Code 5.2 `readVA.c` in Chapter 5, the only sample code in this book that defines a general exception handler. Looking at the disassembly file for any program that uses the Microchip default general exception handler, what does the program do after the software debug breakpoint (`sdbbp`)?

Further Reading

PIC32 family reference manual. Section 08: Interrupts. (2013). Microchip Technology Inc.

PART III

Peripheral Reference

CHAPTER 7

Digital Input and Output

Digital inputs and outputs (DIO) are the simplest interfaces between the PIC32 and other electronics, sensors, and actuators. The PIC32 has many DIO pins, each of which normally has one of two states: high (1) or low (0). These states usually correspond to 3.3 V or 0 V.[1] The DIO peripheral also handles change notification (CN) interrupts, which happen when the input changes on at least one of up to 19 digital inputs.

7.1 Overview

The PIC32 offers many DIO pins, arranged into "ports" B through G on the PIC32MX795F512H. The pins are labeled Rxy, where x is the port letter and y is the pin number. For example, pin 5 of port B is named RB5. (We often omit the R, referring to pin RB5 as B5, when there is no possibility of confusion.) Port B is a full 16-bit port, with pins 0-15, and port B can also be used for analog input (Chapter 10). Other ports have a smaller number of pins, not necessarily sequentially numbered; for example, port G has RG2, RG3, and RG6-RG9. All pins labeled Rxy can be used for input or output, except for RG2 and RG3, pins that are shared with USB communication lines and are input only. For more details on the available pin numbers, see Section 1 of the Data Sheet.

The PIC32 has two types of digital outputs: buffered and open drain. Usually, outputs are buffered and can drive the associated pin to either 0 V or 3.3 V. Some pins, however, can be configured with open drains. Their pins can be driven to 0 V or to a high impedance state often called "floating" or "high-Z." When floating, the output is effectively disconnected, allowing you to attach an external "pull-up" resistor from the output to a positive voltage. Then, when the output floats, the external pull-up resistor connects the pin to the positive voltage (Figure 7.1). The positive voltage must be less than either 3.3 V or 5 V, depending on the pin (see, e.g., Figure 2.1 and Table 2.2 to determine which pins can tolerate 5 V).

A pin configured as an output should not source or sink more than about 10 mA. For example, it would be a mistake to connect a digital output to a 10 Ω resistor to ground. In this case,

[1] Technically, there are tolerances associated with these voltages. According to the Electrical Characteristics section of the Data Sheet, input voltages below about 0.5 V are treated as logic low, and input voltages above 2.1 V are treated as logic high.

Embedded Computing and Mechatronics with the PIC32 Microcontroller. http://dx.doi.org/10.1016/B978-0-12-420165-1.00007-X

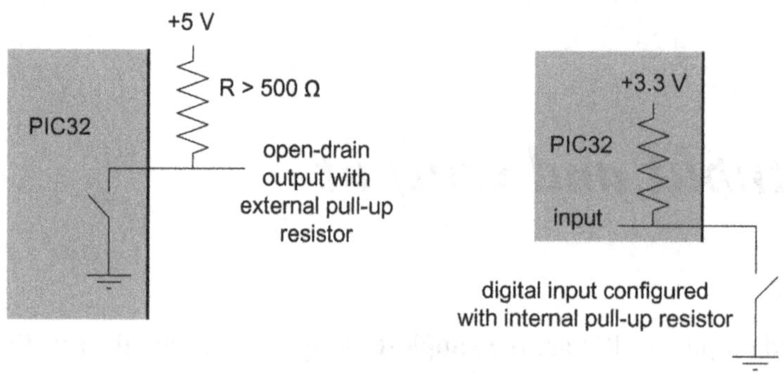

Figure 7.1

Left: A digital output configured as an open-drain output with external pull-up resistor to 5 V. (This should only be done for 5 V tolerant pins.) The resistor should allow no more than 10 mA to flow into the PIC32 when the PIC32 holds the output low. If the LAT bit controlling the pin has the value 1, the internal switch is open, and the output reads 5 V. If the bit is a 0, internal switch is closed and the output reads 0 V. Right: A digital input configured with an internal pull-up resistor allows a simple open-close switch to yield digital high when the switch is open and digital low when the switch is closed.

creating a digital high output of 3.3 V would require 3.3 V/10 Ω = 330 mA, which a digital output cannot source. Be careful; trying to source or sink too much current from a pin may damage your PIC32!

An input pin will read low, or 0, if the input voltage is close to zero, and it will read high, or 1, if it is close to 3.3 V. Some pins tolerate inputs up to 5 V. Some input pins, those that can also be used for "change notification" (labeled CNy, $y = 0$ to 18, spread out over several of the ports), can be configured with an internal pull-up resistor to 3.3 V. If configured this way, the input will read "high" if it is disconnected (Figure 7.1). This is useful for interfacing with simple buttons, which either connect the input to ground or leave it floating (e.g., the USER button in Figure 2.3).[2] Otherwise, if an input pin is not connected to anything, we cannot be certain what the input will read.

Input pins have fairly high input impedance—very little current will flow in or out of an input pin. Therefore, connecting an external circuit to an input pin should have little effect on the behavior of the external circuit.

Up to 19 inputs can be configured as change notification inputs. When enabled, the change notification pins generate interrupts if their digital input state changes. The ISR must then read from the ports configured with change notification, or else future input changes will not result

[2] According to the PIC32MX575/675/695/775/795 Family Silicon Errata and Data Sheet Clarification, an internal pull-up resistor is not guaranteed to make *external* devices read the pin as high, and even the PIC32 may not read the pin as high if an external load causes it to source more than 50 μA.

in an interrupt. The ISR can compare the new port values to the previous values to determine which specific input has changed.

Microchip recommends that unused digital I/O pins be configured as outputs and driven to a logic low state, though this is not required. All pins are configured as inputs by default. This safety feature prevents the PIC32 from imposing unwanted voltages on attached circuitry before your program begins executing. The pins on port B are shared with the with the analog-to-digital converter (ADC) and default to analog inputs, unless explicitly set as digital pins.

7.2 Details

TRISx, x = B to G These tri-state SFRs determine which port x DIO pins are inputs and which are outputs. Bit y corresponds to the port's pin y (i.e., Rxy). Bits can be accessed individually by using TRISxbits.TRISxy. For example, TRISDbits.TRISD5 = 0 makes RD5 an output (0 = Output), and TRISDbits.TRISD5 = 1 makes RD5 an input (1 = Input). Bits of TRISx default to 1 on reset.

LATx, x = B to G A write to the latch chooses the output for pins configured as digital outputs. (Pins configured as inputs ignore the write.) Bit y correspond to that port's pin (i.e., Rxy). For example, if TRISD = 0x0000, making all port D pins outputs, then LATD = 0x00FF sets pins RD0-RD7 high and other RD pins low. Individual pins can be referenced using LATxbits.LATxy, where y is the pin number. For example, LATDbits.LATD11 = 1 sets pin RD11 high. A write of 1 to an open-drain output sets the output to floating, while a write of 0 makes the output to low.

PORTx, x = B to G PORTx returns the current digital value for DIO pins on port x configured as inputs. Bit y corresponds to pin Rxy. Individual pins can be accessed as PORTxbits.RDy; for example, PORTDbits.RD6 returns the digital input for RD6.

ODCx, x = B to G The open-drain configuration SFR determines whether outputs are open drain or not. Individual bits can be accessed using ODCxbits.ODCxy. For example, if TRISbits.TRISD8 = 0, making RD8 an output, then ODCDbits.ODCD8 = 1 configures RD8 as an open-drain output, and ODCDbits.ODCD8 = 0 configures RD8 as a typical buffered output. The reset default for all bits is 0.

AD1PCFG The pin configuration bits in this register determine whether port B's pins are analog or digital inputs. See Chapter 10 for information about analog inputs.

AD1PCFG⟨x⟩, or AD1PCFGbits.PCFGx, x = 0 to 15, controls whether pin ANx (equivalently RBx) is an analog input: 0 = analog input, 1 = digital pin.

Each of the lower 16 bits of this SFR correspond to a port B pin. On reset, they are zero, meaning that all port B pins are analog inputs by default. Therefore, to use a port B pin as a digital input, you must explicitly set the appropriate pin in AD1PCFG.

For example, to use port B, pin 2 (RB2) as a digital input, set AD1PCFGbits.PCFG2 (AD1PCFG⟨2⟩) to one. This extra configuration step applies only to port B because the other ports do not overlap with analog inputs.

CNPUE Change notification pull-up enable allows you to enable an internal pull-up resistor on the change notification pins (CN0-CN18). Each bit in CNPUE⟨18:0⟩, when set, enables the pull-up, and when clear, disables it. Bit x corresponds to pin CNx. Individual bits can be accessed using CNPUEbits.CNPUEx. For example, if CNPUEbits.CNPUE2 = 1, then CN2/RB0 has the internal pull-up resistor enabled, and if CNPUEbits.CNPUE2 = 0, then it is disabled. The reset default for all bits is 0 so the pull-up resistors are disabled. An internal pull-up resistor can be convenient because it ensures that the input pin is in a known state when disconnected.

Latches vs. ports: What is the difference between the latch (LATx) and port (PORTx) SFRs? The PORTx SFRs correspond to voltages on the pins while LATx SFRS correspond to what the pin should output if configured as an output. When you read from LATx you are actually reading the last value you wanted to put out on the port, not the pins' actual state. Therefore, to read pins, use PORTx, and to write digital outputs, use LATx.

Using the INV, CLR, and SET SFRs vs. directly manipulating bits: To directly set a bit, say bit 4 of LATF, you could use either `LATFSET = 0b10000` (equivalently `LATFSET = 0x10`) or you could use `LATFbits.LATF4 = 1`, based on the bit field names in the `p32mx795f512h.h` file. The former approach is atomic—it executes in a single assembly statement. The latter approach causes the CPU to copy LATF to a CPU register, set bit 4 of the CPU's copy, and write the result back to LATF. While this takes more assembly commands, we generally recommend that you use this approach instead of using the SFR's SET, CLR, and INV registers. The resulting syntax is clearer (essentially self-documenting by the name of the bit field) and less error prone.

7.2.1 Change Notification

Change notification interrupts can be generated on pins CN0-CN18 and are triggered when the input state on the pin changes. The relevant SFRs are:

CNPUE Change notification pull-up enable. See above (it is listed there because pull-ups can be useful even if no change notification is used).

CNCON The change notification control SFR enables CN interrupts if CNCON⟨15⟩ (CNCONbits.ON) equals 1. The default is 0.

CNEN A particular pin CNx can generate a change notification interrupt if CNEN⟨x⟩ (CNENbits.CNENx) is 1. Otherwise it is not included in the change notification.

The relevant interrupt bit fields and constants are:

IFS1⟨0⟩ or IFS1bits.CNIF: Interrupt status flag for change notification, set when the interrupt is pending.

IEC1⟨0⟩ or IEC1bits.CNIE: Interrupt enable flag for change notification, set to enable the interrupt.

IPC6⟨20:18⟩ or IPC6bits.CNIP: Change notification interrupt priority.
IPC6⟨17:16⟩ or IPC6bits.CNIS: Change notification sub-priority.
Vector Number: The change notification uses interrupt vector 26 or `_CHANGE_NOTICE_VECTOR`, as defined in `p32mx795f512h.h`.

A recommended procedure for enabling the CN interrupt:

1. Write an ISR using the vector `_CHANGE_NOTICE_VECTOR` (26). It should clear IFS1bits.CNIF and read the pins involved in the CN scan to re-enable the interrupt.
2. Disable all interrupts using `__builtin_disable_interrupts()`.
3. Set CNENbits.CNENx to one for each pin x that you want included in the change notification, and set CNCONbits.ON to one.
4. Choose the interrupt priority IPC6bits.CNIP and subpriority IPC6bits.CNIS. The priority should match that used in the definition of the ISR.
5. Clear the interrupt flag status IFS1bits.CNIF.
6. Enable the CN interrupt by setting IEC1bits.CNIE to one.
7. Enable interrupts using `__builtin_enable_interrupts()`.

7.3 Sample Code

Our first program, `simplePIC.c`, demonstrated the use of two digital outputs (to control two LEDs) and one digital input (the USER button).

The example below configures the following inputs and outputs:

- Pins RB0 and RB1 are digital inputs with internal pull-up resistors enabled.
- Pins RB2 and RB3 are digital inputs without pull-up resistors.
- Pins RB4 and RB5 are buffered digital outputs.
- Pins RB6 and RB7 are open-drain digital outputs.
- Pins AN8-AN15 are analog inputs.
- RF4 is a digital input with an internal pull-up resistor.
- Change notification is enabled on pins RB0 (CN2), RF4 (CN17), and RF5 (CN18). Since both ports B and F are involved in the change notification, both ports must be read inside the ISR to allow the interrupt to be re-enabled. The ISR toggles one of the NU32 LEDs to indicate that a change has been noticed on pin RB0, RF4, or RF5.

Code Sample 7.1 `DIO_sample.c`. **Digital Input, Output, and Change Notification.**

```
#include "NU32.h"          // constants, funcs for startup and UART

volatile unsigned int oldB = 0, oldF = 0, newB = 0, newF = 0; // save port values

void __ISR(_CHANGE_NOTICE_VECTOR, IPL3SOFT) CNISR(void) { // INT step 1
  newB = PORTB;           // since pins on port B and F are being monitored
```

```
    newF = PORTF;             // by CN, must read both to allow continued functioning
                              // ... do something here with oldB, oldF, newB, newF ...
    oldB = newB;              // save the current values for future use
    oldF = newF;
    LATBINV = 0xF0;           // toggle buffered RB4, RB5 and open-drain RB6, RB7
    NU32_LED1 = !NU32_LED1;   // toggle LED1
    IFS1bits.CNIF = 0;        // clear the interrupt flag
}

int main(void) {
  NU32_Startup(); // cache on, min flash wait, interrupts on, LED/button init, UART init

  AD1PCFG = 0x00FF;           // set B8-B15 as analog in, 0-7 as digital pins
  TRISB = 0xFF0F;             // set B4-B7 as digital outputs, 0-3 as digital inputs
  ODCBSET = 0x00C0;           // set ODCB bits 6 and 7, so RB6, RB7 are open drain outputs
  CNPUEbits.CNPUE2 = 1;       // CN2/RB0 input has internal pull-up
  CNPUEbits.CNPUE3 = 1;       // CN3/RB1 input has internal pull-up
  CNPUEbits.CNPUE17 = 1;      // CN17/RF4 input has internal pull-up
                              // due to errata internal pull-ups may not result in a logic 1

  oldB = PORTB;               // bits 0-3 are relevant input
  oldF = PORTF;               // pins of port F are inputs, by default
  LATB = 0x0050;              // RB4 is buffered high, RB5 is buffered low,
                              // RB6 is floating open drain (could be pulled to 3.3 V by
                              // external pull-up resistor), RB7 is low

  __builtin_disable_interrupts(); // step 1: disable interrupts
  CNCONbits.ON = 1;               // step 2: configure peripheral: turn on CN
  CNENbits.CNEN2 = 1;             // Use CN2/RB0 as a change notification
  CNENbits.CNEN17 = 1;            // Use CN17/RF4 as a change notification
  CNENbits.CNEN18 = 1;            // Use CN18/RF5 as a change notification

  IPC6bits.CNIP = 3;              // step 3: set interrupt priority
  IPC6bits.CNIS = 2;              // step 4: set interrupt subpriority
  IFS1bits.CNIF = 0;              // step 5: clear the interrupt flag
  IEC1bits.CNIE = 1;              // step 6: enable the CN interrupt
  __builtin_enable_interrupts();  // step 7: CPU enabled for mvec interrupts

  while(1) {
      ;                           // infinite loop
  }
  return 0;
}
```

7.4 Chapter Summary

- The PIC32 has several DIO ports, labeled with the letters B through G. Only port B has all 16 pins. Almost every pin can be configured as a digital input or a digital output. Some outputs can be configured to be open-drain.
- By default, port B inputs are configured as analog inputs. To use these pins as digital inputs you must set the corresponding bits in AD1PCFG to one.
- Nineteen pins can be configured as change notification pins (CN0-CN18). For those pins that are enabled as change notification pins, any change of input state generates an interrupt. To re-enable the interrupt, the ISR should clear the IRQ flag and read the ports with pins involved in the change notification.

- The change notification pins offer an optional internal pull-up resistor so that the input registers as high when it is left floating. These pull-up resistors can be used regardless of whether change notification is enabled for the pin. The internal pull-up resistor allows simple interfacing with push-buttons, for example.

7.5 Exercises

1. True or false? If an input pin is not connected to anything, it always reads digital low.
2. You are configuring port B to receive analog and digital inputs and to write digital output. Here is how you would like to configure the port. (Pin x corresponds to RBx.)
   - Pin 0 is an analog input.
   - Pin 1 is a "typical" buffered digital output.
   - Pin 2 is an open-drain digital output.
   - Pin 3 is a "typical" digital input.
   - Pin 4 is a digital input with an internal pull-up resistor.
   - Pins 5-15 are analog inputs.
   - Pin 3 is monitored for change notification, and the change notification interrupt is enabled.

   Questions:
   a. Which digital pin would most likely have an external pull-up resistor? What would be a reasonable range of resistances to use? Explain what factors set the lower bound on the resistance and what factors set the upper bound on the resistance.
   b. To achieve the configuration described above, give the eight-digit hex values you should write to AD1PCFG, TRISB, ODCB, CNPUE, CNCON, and CNEN. (Some of these SFRs have unimplemented bits 16-31; write 0 for those bits.)

Further Reading

PIC32 family reference manual. Section 12: I/O ports. (2011). Microchip Technology Inc.

CHAPTER 8

Counter/Timers

Counters count rising edges of a pulse train. The pulses may come from the internal peripheral bus clock or external sources. If a fixed frequency clock produces the pulses, counters become timers (the count represents a time). Therefore, the words "counter" and "timer" are often used interchangeably. Because Microchip's documentation refers to these devices as "timers," we adopt that terminology. Timers can generate interrupts after a preset number of pulses has been counted or on the falling edge of an external pulse whose duration is being timed. These timers differ from the core timer introduced in Chapter 5 because they are peripherals rather than part of the MIPS32 CPU.

8.1 Overview

The PIC32 is equipped with five 16-bit peripheral timers named Timerx, where x is 1 to 5. A timer increments on the rising edge of a clock signal, which may come from the PBCLK or from an external source of pulses. The input for an external source for Timerx is pin TxCK. For the 64-pin PIC32MX795F512H on the NU32 board, only Timer1 is equipped with a pin for an external input (T1CK). The 100-pin version of this chip (the PIC32MX795F512L) has an external input pin for all five timers.

A prescaler $N \geq 1$ determines how many clock pulses must be received before the timer increments. If the prescaler is set to $N = 1$, the timer increments on every clock rising edge; if it is set to $N = 8$, it increments every eighth rising edge. The clock source type (internal or external) and the prescaler value is chosen by setting the value of the SFR TxCON.

Each 16-bit timer can count from 0 up to a period $P \leq 2^{16} - 1 = 65{,}535 = \text{0xFFFF}$. The current count is stored in the SFR TMRx and the value of P can be chosen by writing to the period register SFR PRx. When the timer reaches the value P, a *period match* occurs, and after N more pulses are received, the counter "rolls over" to 0. If the input to the timer is the 80 MHz PBCLK, with 12.5 ns between rising edges, then the time between rollovers is $T = (P + 1) \times N \times 12.5$ ns. Choosing $N = 8$ and $P = 9{,}999$, we get $T = 1$ ms, and changing N to 64 gives $T = 8$ ms. By configuring the timer to use the PBCLK as input and to generate an interrupt when a period match occurs, the timer can implement a function that runs at a fixed frequency (a controller, for example) (see Figure 8.1).

Embedded Computing and Mechatronics with the PIC32 Microcontroller. http://dx.doi.org/10.1016/B978-0-12-420165-1.00008-1

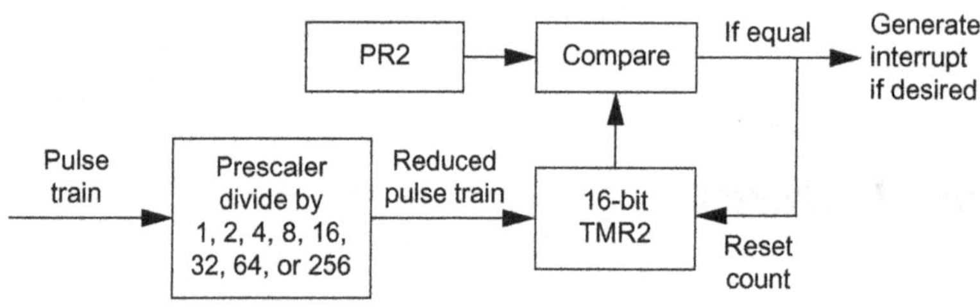

Figure 8.1
Simplified block diagram for a typical use of the 16-bit Timer2. The pulse train feeds a prescaler, which produces one output pulse for every N input pulses ($N = 1, 2, 4, 8, 16, 32, 64$, or 256). The TMR2 SFR stores the count of these pulses. TMR2 resets to zero on the first pulse in the reduced pulse train after TMR2 matches the period register PR2. By default, PR2 is 0xFFFF, so TMR2 counts up to $2^{16} - 1$ before rolling over to zero. Timers 3, 4, and 5 are similar to Timer 2, but Timer 1 can only have prescaler values $N = 1, 8, 64$, or 256.

If the period P is zero, then once the count reaches zero it will never increment again (it keeps rolling over). No interrupt can be generated by a period match if $P = 0$.

The PIC32 has two types of timers: Type A and Type B, each with slightly different features (explained shortly). Timer1 is Type A and Timer2 to Timer5 are Type B. The timers can be used in the following modes, chosen by the SFR TxCON:

Counting PBCLK pulses. In this mode, the timer counts PBCLK pulses, so the count corresponds to an elapsed time. This mode is often used to generate interrupts at a desired frequency by appropriate setting of N and P. It can also be used to time the duration of code, like how we used the core timer in Chapter 5. A peripheral timer, however, can increment once every N PBCLK cycles, including $N = 1$, not just every 2 SYSCLK cycles.

Synchronous counting of external pulses. For Timer1, an external pulse source is connected to the pin T1CK. The timer count increments after each rising edge of the external source. This mode is called "synchronous" because timer increments are synchronized to the PBCLK; the timer does not actually increment until the first rising edge of PBCLK after the rising edge of the external source. If the external pulses are too fast, the timer will not accurately count them. According to the Electrical Characteristics section of the Data Sheet, the duration of the high and low portions of a pulse should be at least 37.5 ns each.

Asynchronous counting of external pulses (Type A Timer1 only). The Type A pulse counting circuit can be configured to increment independently of the PBCLK, allowing it to count even when the PBCLK is not operating, such as in the power-saving Sleep mode. If a period match occurs, Timer1 can generate an interrupt and wake up the PIC32. When used in asynchronous mode, the timer can count pulses with high and low durations as low as 10 ns each.

Timing the duration of an external pulse. Also called "gated accumulation mode." For Timer1, when the input on external pin T1CK goes high, the counter starts incrementing according to the PBCLK and the prescaler N. When the input drops low, the count stops. The timer can also generate an interrupt when the input drops low.

Important differences between Timer1 (Type A) and Timer2 to Timer5 (Type B) are:

- Only Timer1 can count external pulses on the PIC32MX795F512H.
- Timer1 can have prescalers $N = 1, 8, 64$, and 256, while Timer2 to Timer5 can have prescalers $N = 1, 2, 4, 8, 16, 32, 64$, and 256.
- Timer2 and Timer3 can be chained together to make a single 32-bit timer, called Timer23. Timer4 and Timer5 can similarly be used to make a single 32-bit timer, called Timer45. These combined timers allow counts of up to $2^{32} - 1$, or over 4 billion. When two timers are used to make Timerxy (x < y), Timerx is called the "master" and Timery is the "slave"—only the prescaler and mode information in TxCON are relevant, while those fields in TyCON are ignored. When Timerx rolls over from $2^{16} - 1$ to 0, it sends a clock pulse to increment Timery. In Timerxy mode, the 16 bits of TMRy are also stored in the most significant 16 bits of the SFR TMRx, so the 32 bits of TMRx contain the full count of Timerxy. Similarly, the 32 bits of PRx contain the 32-bit period match value Pxy. The interrupt associated with a period match (or a falling input in gated accumulation mode) is actually generated by Timery, so interrupt settings should be chosen for Timery's IRQ and vector.

Timers are used in conjunction with digital waveform generation by the Output Compare peripheral (Chapter 9) and in timing digital input waveforms by the Input Capture peripheral (Chapter 15). A timer can also be used to repeatedly trigger analog to digital conversions (Chapter 10).

8.2 Details

The following SFRs are associated with the timers. All SFRs default to 0x0000, except the PRx SFRs, which default to 0xFFFF. First, we describe settings common to both Type A and Type B timers.

TxCON, x = 1 to 5 The Timerx control SFR configures the behavior of Timerx. Important bits common to both Type A and Type B timers include

TxCON⟨15⟩, or TxCONbits.ON: Enables and disables the timer.

1 Timerx enabled.

0 Timerx disabled.

TxCON⟨7⟩, or TxCONbits.TGATE: Sets gated accumulation mode, which can be used to time the duration of an external pulse. In gated accumulation mode the timer starts counting when an external signal goes high and stops when it goes low.

1 Gated accumulation mode enabled.
0 Gated accumulation mode disabled.

Gated accumulation also requires TxCONbits.TCS = 0 (below).

TxCON⟨1⟩, or TxCONbits.TCS: Determines whether the timer uses an external clock source or PBCLK. On the PIC32MX795F512H, only Timer1 has a pin for an external clock source.

1 Timerx uses the signal on TxCK as an external pulse source.
0 PBCLK provides the pulse source.

The Type A timer, Timer1, also has the relevant fields:

T1CON The control register for Timer1. Has the same fields as above; here we describe the fields specific to Timer1.

T1CON⟨5:4⟩, or T1CONbits.TCKPS: Sets the prescaler ratio. The prescaler determines how many pulses are required to increment the timer count. Type A timers have fewer prescaler ratios than Type B timers.

0b11 Prescaler of 1:256.
0b10 Prescaler of 1:64.
0b01 Prescaler of 1:8.
0b00 Prescaler of 1:1.

T1CON⟨2⟩, or T1CONbits.TSYNC: Determines whether external clock inputs are synchronized to PBCLK. When synchronized, the timer counts external rising edges on the PBCLK ticks. When asynchronous, every external pulse is registered immediately (subject to timing requirements specified in the Data Sheet). Only Type A timers can count pulses asynchronously.

1 External counting is synchronized.
0 External counting is asynchronous.

The Type B timers, Timer2 to Timer5, have their own type-specific TxCON fields.

TxCON, x = 2 to 5 Here we describe the fields in TxCON specific to Type B timers.

TxCON⟨6:4⟩, or TxCONbits.TCKPS: Sets the prescaler ratio. There are more choices than for Type A timers.

0b111 Prescaler of 1:256.
0b110 Prescaler of 1:64.
0b101 Prescaler of 1:32.
0b100 Prescaler of 1:16.
0b011 Prescaler of 1:8.
0b010 Prescaler of 1:4.
0b001 Prescaler of 1:2.
0b000 Prescaler of 1:1.

TxCON⟨3⟩, or TxCONbits.T32: This bit is only relevant for x = 2 and 4 (Timer2 and Timer4). When set, Timerx and Timery are chained together to make the 32 bit timer called Timerxy (y = x + 1). When in 32-bit mode, TyCON settings are ignored, TMRy is enabled, and its clock value comes from the rollover of TMRx after it hits 0xFFFF. Interrupts are generated by Timery, but the timer's full 32-bit value and 32-bit rollover count are accessed via TMRx and PRx. When TxCONbits.T32 is zero, Timerx and Timery operate as independent 16-bit timers.

1 Use Timer23 (if x = 2) or Timer45 (if x = 4) as 32-bit timers.

0 User Timerx as a 16-bit timer.

The following SFRs apply to each Timerx, x = 1 to 5.

TMRx, x = 1 to 5 TMRx⟨15:0⟩ stores the 16-bit count of Timerx. TMRx resets to 0 on the next count after TMRx reaches the number stored in PRx. This rollover process is called a period match. In Timerxy 32-bit mode, TMRx contains the 32-bit value of the chained timer, and period match occurs when TMRx = PRx.

PRx, x = 1 to 5 PRx⟨15:0⟩ contains the maximum value of the count of TMRx before it resets to zero on the next count. An interrupt can be generated on this period match. In Timerxy 32-bit timer mode, PRx contains the 32-bit value of the period *P*xy. Interrupts are generated as if Timery triggered the interrupt.

The timer can generate an interrupt on the falling edge of the gate input when it is in gated mode (TxCONbits.TCS = 0 and TxCONbits.TGATE = 1). Otherwise, it can generate an interrupt whenever a period match occurs.

The relevant interrupt flags are shown in Table 8.1. To enable the interrupt for Timerx, the interrupt enable bit IEC0bits.TxIE must be set. The interrupt flag bit IFS0bits.TxIF should be cleared and the priority and subpriority bits IPCxbits.TxIP and IPCxbits.TxIS must be written. In 32-bit Timerxy mode, interrupts are generated by Timery; interrupt settings for Timerx are ignored.

Table 8.1: Vectors and bits relevant to timer interrupts

IRQ Source	Vector	Flag	Enable	Priority	Subpriority
Timer1	4 _TIMER_1_VECTOR	IFS0⟨4⟩ IFS0bits.T1IF	IEC0⟨4⟩ IEC0bits.T1IE	IPC1⟨4:2⟩ IPC1bits.T1IP	IPC1⟨1:0⟩ IPC1bits.T1IS
Timer2	8 _TIMER_2_VECTOR	IFS0⟨8⟩ IFS0bits.T2IF	IEC0⟨8⟩ IEC0bits.T2IE	IPC2⟨4:2⟩ IPC2bits.T2IP	IPC2⟨1:0⟩ IPC2bits.T2IS
Timer3	12 _TIMER_3_VECTOR	IFS0⟨12⟩ IFS0bits.T3IF	IEC0⟨12⟩ IEC0bits.T3IE	IPC3⟨4:2⟩ IPC3bits.T3IP	IPC3⟨1:0⟩ IPC3bits.T3IS
Timer4	16 _TIMER_4_VECTOR	IFS0⟨16⟩ IFS0bits.T4IF	IEC0⟨16⟩ IEC0bits.T4IE	IPC4⟨4:2⟩ IPC4bits.T4IP	IPC4⟨1:0⟩ IPC4bits.T4IS
Timer5	20 _TIMER_5_VECTOR	IFS0⟨20⟩ IFS0bits.T5IF	IEC0⟨20⟩ IEC0bits.T5IE	IPC5⟨4:2⟩ IPC5bits.T5IP	IPC5⟨1:0⟩ IPC5bits.T5IS

8.3 Sample Code

8.3.1 A Fixed Frequency ISR

To create a 5 Hz ISR with an 80 MHz PBCLK, the interrupt must be triggered every 16 million PBCLK cycles. The highest a 16-bit timer can count is $2^{16} - 1$. Instead of wasting two timers to make a 32-bit timer with a prescaler $N = 1$, let us use a single 16-bit timer with a prescaler $N = 256$. We shall use Timer1. We should choose PR1 to satisfy

```
16,000,000 = (PR1 + 1) x 256
```

that is, PR1 = 62,499. The ISR in the following code toggles a digital output at 5 Hz, creating a 2.5 Hz square wave (a flashing LED on the NU32).

Code Sample 8.1 `TMR_5Hz.c`. **Timer1 Toggles RF0 Five Times a Second (LED1 on the NU32 Flashes).**

```
#include "NU32.h"          // constants, functions for startup and UART

void __ISR(_TIMER_1_VECTOR, IPL5SOFT) Timer1ISR(void) {  // INT step 1: the ISR
  LATFINV = 0x1;                    // toggle RF0 (LED1)
  IFS0bits.T1IF = 0;                // clear interrupt flag
}

int main(void) {
  NU32_Startup(); // cache on, min flash wait, interrupts on, LED/button init, UART init

  __builtin_disable_interrupts(); // INT step 2: disable interrupts at CPU
                                  // INT step 3: setup peripheral
  PR1 = 62499;                    //             set period register
  TMR1 = 0;                       //             initialize count to 0
  T1CONbits.TCKPS = 3;            //             set prescaler to 256
  T1CONbits.TGATE = 0;            //             not gated input (the default)
  T1CONbits.TCS = 0;              //             PCBLK input (the default)
  T1CONbits.ON = 1;               //             turn on Timer1
  IPC1bits.T1IP = 5;              // INT step 4: priority
  IPC1bits.T1IS = 0;              //             subpriority
  IFS0bits.T1IF = 0;              // INT step 5: clear interrupt flag
  IEC0bits.T1IE = 1;              // INT step 6: enable interrupt
  __builtin_enable_interrupts();  // INT step 7: enable interrupts at CPU
  while (1) {
      ;                           // infinite loop
  }
  return 0;
}
```

8.3.2 Counting External Pulses

The following code uses the 16-bit Timer1 to count the rising edges on the input T1CK. The 32-bit Timer45 creates an interrupt at 2 kHz to toggle a digital output, generating a 1 kHz pulse train on RD1 that acts as input to T1CK. Although a 16-bit timer can certainly generate a 2 kHz interrupt, we use a 32-bit timer just to show the configuration. In Chapter 9 we will learn about the Output Compare peripheral, a better way to use a timer to create more flexible waveforms.

To create an IRQ every 0.5 ms (2 kHz), we use a prescaler $N = 1$ and a period match PR4 = 39,999, so

```
(PR4 + 1) x N x 12.5 ns = 0.5 ms
```

The code below displays to your computer's screen the amount of time that has elapsed since the PIC32 was reset, in milliseconds. If you wait 65 s, you will see Timer1 roll over.

Code Sample 8.2 `TMR_external_count.c`**. Timer45 Creates a 1 kHz Pulse Train on RD1, and These External Pulses Are Counted by Timer1. The Elapsed Time Is Periodically Reported Back to the Host Computer Screen.**

```
#include "NU32.h"          // constants, functions for startup and UART

void __ISR(_TIMER_5_VECTOR, IPL4SOFT) Timer5ISR(void) {  // INT step 1: the ISR
  LATDINV = 0x02;                                         // toggle RD1
  IFS0bits.T5IF = 0;                                      // clear interrupt flag
}

int main(void) {
  char message[200] = { };
  int i = 0;

  NU32_Startup();                  // cache on, interrupts on, LED/button init, UART init
  __builtin_disable_interrupts(); // INT step 2: disable interrupts

  TRISDbits.TRISD1 = 0;            // make D1 an output. connect D1 to T1CK (C14)!

                                   // configure Timer1 to count external pulses.
                                   // The remaining settings are left at their defaults
  T1CONbits.TCS = 1;               // count external pulses
  PR1 = 0xFFFF;                    // enable counting to max value of 2^16 - 1
  TMR1 = 0;                        // set the timer count to zero
  T1CONbits.ON = 1;                // turn Timer1 on and start counting

                                   // 1 kHz pulses with 2 kHz interrupt from Timer45
  T4CONbits.T32 = 1;               // INT step 3: set up Timers 4 and 5 as 32-bit Timer45
  PR4 = 39999;                     //             rollover at 40,000; 80MHz/40k = 2 kHz
  TMR4 = 0;                        //             set the timer count to zero
  T4CONbits.ON = 1;                //             turn the timer on
  IPC5bits.T5IP = 4;               // INT step 4: priority for Timer5 (int goes with T5)
  IFS0bits.T5IF = 0;               // INT step 5: clear interrupt flag
  IEC0bits.T5IE = 1;               // INT step 6: enable interrupt
  __builtin_enable_interrupts();   // INT step 7: enable interrupts at CPU

  while (1) {
                                   // display the elapsed time in ms
    sprintf(message,"Elapsed time:  %u ms\r\n", TMR1);
    NU32_WriteUART3(message);
```

```
    for(i = 0; i < 10000000; ++i){// loop to delay printing
      _nop();                       // include nop so loop is not optimized away
    }
  }
  return 0;
}
```

8.3.3 Timing the Duration of an External Pulse

In this last example we modify our previous code to use Timer45 to toggle the digital output RD1 every 100 ms, creating a 5 Hz square wave. These pulses are timed by Timer1 in gated accumulation mode. The accumulated count begins when the T1CK input from RD1 goes high and stops when the T1CK input drops low. The falling edge triggers an ISR that displays the Timer1 count to the screen and resets the timer. You should find that the measured time is very close to 100 ms, as expected.

Code Sample 8.3 `TMR_pulse_duration.c`**. Timer45 Creates a Series of 100 ms Pulses on RD1. These Pulses Are Input to T1CK and Timer1 Measures Their Duration in Gated Accumulation Mode.**

```
#include "NU32.h"          // constants, functions for startup and UART

void __ISR(_TIMER_5_VECTOR, IPL4SOFT) Timer5ISR(void) {  // INT step 1: the ISR
  LATDINV = 0x02;                    // toggle RD1
  IFS0bits.T5IF = 0;                 // clear interrupt flag
}

void __ISR(_TIMER_1_VECTOR, IPL3SOFT) Timer1ISR(void) { // INT step 1: the ISR
  char msg[100] = { };
  sprintf(msg,"The count was %u, or %10.8f seconds.\r\n", TMR1, TMR1/312500.0);
  NU32_WriteUART3(msg);
  TMR1 = 0;                          // reset Timer1
  IFS0bits.T1IF = 0;                 // clear interrupt flag
}

int main(void) {
  NU32_Startup();                    // cache on, interrupts on, LED/button init, UART init

  __builtin_disable_interrupts(); // INT step 2: disable interrupts

  TRISDbits.TRISD1 = 0;              // make D1 an output. connect D1 to T1CK (C14)

                                     // INT step 3
  T1CONbits.TGATE = 1;               //             Timer1 in gated accumulation mode
  T1CONbits.TCKPS = 3;               //             1:256 prescale ratio
  T1CONbits.TCS = 0;
  PR1 = 0xFFFF;                      //             use the full period of Timer1
  T1CONbits.TON = 1;                 //             turn Timer1 on

  T4CONbits.T32 = 1;                 // for T45:    enable 32 bit mode Timer45
  PR4 = 7999999;                     //             set PR so timer rolls over at 10 Hz
  TMR4 = 0;                          //             initialize count to 0
  T4CONbits.TON = 1;                 //             turn Timer45 on
```

```
    IPC5bits.T5IP = 4;              // INT step 4: priority for Timer5 (int for Timer45)
    IPC1bits.T1IP = 3;              //             priority for Timer1
    IFS0bits.T5IF = 0;              // INT step 5: clear interrupt flag for Timer45
    IFS0bits.T1IF = 0;              //             clear interrupt flag for Timer1
    IEC0bits.T5IE = 1;              // INT step 6: enable interrupt for Timer45
    IEC0bits.T1IE = 1;              //             enable interrupt for Timer1
    __builtin_enable_interrupts();  // INT step 7: enable interrupts at the CPU

    while (1) {
      :
    }
    return 0;
}
```

8.4 Chapter Summary

- The PIC32 timers can be used to generate fixed-frequency interrupts, count external pulses, and time the duration of external pulses. Additionally, the Type A Timer1 can asynchronously count external pulses even when the PIC32 is in Sleep mode, while the Type B timers Timer2 and Timer3 can be chained to make the 32-bit timer Timer23. Similarly, Timer4 and Timer5 can be chained to make the 32-bit timer Timer45.
- For a 32-bit timer Timerxy, the timer configuration information in TxCON is used (TyCON is ignored), and the interrupt enable, flag status, and priority bits are configured for Timery (this information for Timerx is ignored). The 32-bit Timerxy count is held in TMRx and the 32-bit period match value is held in PRx.
- A timer can generate an interrupt when either the external pulse being timed falls low (gated accumulation mode) or the count reaches a value stored in a period register (period match).

8.5 Exercises

1. Assume PBCLK is running at 80 MHz. Give the four-digit hex values for T3CON and PR3 so that Timer3 is enabled, has a 1:64 prescaler, and rolls over (generates an interrupt) every 16 ms.
2. Using a 32-bit timer (Timer23 or Timer45) to count rising edges from the 80 MHz PBCLK, what is the longest duration you can time, in seconds, before the timer rolls over? (Use the prescaler that maximizes this time.)

Further Reading

PIC32 family reference manual. Section 14: Timers. (2013). Microchip Technology Inc.

CHAPTER 9

Output Compare

The output compare (OC) peripheral sets the state of an output pin based on the value of a timer. Output compare can be used to generate a single pulse of specified duration or a continuous pulse train. Either mode of operation can generate an interrupt when the value of the output pin changes.

Like most microcontrollers, the PIC32 cannot output an arbitrary analog voltage because it lacks a true digital-to-analog converter (DAC) (see Chapter 16 for details about the PIC32's limited analog output capability). By generating a pulse train, the output compare can be used to generate a time-based analog output. The analog value is proportional to the *duty cycle* of the pulse train: the percentage of the period that the signal is high. Generating a signal whose value is determined by the duty cycle is called "pulse width modulation" (PWM) (see Figure 9.1). High-frequency PWM signals can be low-pass filtered to create a true analog output. PWM signals are also commonly used as input to H-bridge amplifiers that drive motors.

9.1 Overview

The PIC32's five OC peripherals can be configured to operate in seven different modes. In every mode, the module uses either the count of the 16-bit timer Timer2 or Timer3, or the count of the 32-bit timer Timer23, depending on the OC control SFR OCxCON (where x = 1 to 5 refers to the particular output compare module). We call the timer used by an output compare module Timery, where y = 2, 3, or 23. You must configure Timery with its own prescaler and period register, which influences the OC peripheral's behavior.

The OC peripheral's operating modes consist of three "single compare" modes, two "dual compare" modes, and two PWM modes. In the three single compare modes, the Timery count TMRy is compared to the value in the OCx count SFR OCxR. In the "driven high" single compare mode, the OCx output is initially driven low when OCx is enabled, and then transitions to high when TMRy first matches OCxR. In the "driven low" single compare mode, the OCx output is initially driven high and then transitions to low on the first TMRy match. In the "toggle" single compare mode, the OCx output is initially driven low and then toggles on each TMRy match. This toggle mode generates a continuous pulse train.

Embedded Computing and Mechatronics with the PIC32 Microcontroller. http://dx.doi.org/10.1016/B978-0-12-420165-1.00009-3

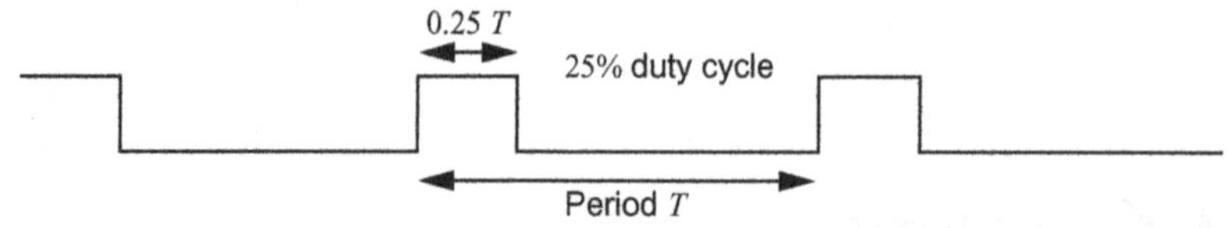

Figure 9.1
A PWM waveform. The duty cycle is the percentage of a period that the signal is high.

In the dual compare modes, TMRy is compared to two values, OCxR and OCxRS. When TMRy matches OCxR the output is driven high, and when it matches OCxRS it is driven low. Depending on a bit in OCxCON, either a single pulse or continuous pulse train is produced.

The two PWM modes create continuous pulse trains. Each pulse begins (is set high) at the rollover of Timery, as set by the period register PRy. The output is then set low when the timer count reaches OCxR. To change the value of OCxR, the user's program may alter the value in OCxRS at any time. This value will then be transferred to OCxR at the beginning of the new time period. The duty cycle of the pulse train, as a percentage, is

```
duty cycle = OCxR/(PRy + 1) x 100%.
```

One of the two PWM modes offers the use of a fault protection input. If chosen, the OCFA input pin (corresponding to OC1 through OC4) or the OCFB input pin (corresponding to OC5) must be high for PWM to operate. If the pin drops to logic low, corresponding to some external fault condition, the PWM output will be high impedance (like an open-drain output, effectively disconnected) until the fault condition is removed and the PWM mode is reset by a write to OCxCON.

9.2 Details

The output compare modules are controlled by the following SFRs. The OCxCON SFRs default to 0x0000 on reset; the OCxR and OCxRS SFR values are unknown after reset.

OCxCON, x = 1 to 5 This output compare control SFR determines the operating mode of OCx.

OCxCON⟨15⟩, or OCxCONbits.ON: Enables and disables the output compare module.

- 1 Output compare enabled.
- 0 Output compare disabled.

OCxCON⟨5⟩, or OCxCONbits.OC32: Determines which timer to use.

- 1 Use the 32-bit timer Timer23.
- 0 Use a 16-bit timer, either Timer2 or Timer3.

OCxCON⟨4⟩, or OCxCONbits.OCFLT: The read-only PWM fault condition status bit. If a fault has occurred you must reset the PWM module by writing to OCxCONbits.OCM (assuming the external fault condition has been removed).

1 PWM fault has occurred.
0 No fault has occurred.

OCxCON⟨3⟩, or OCxCONbits.OCTSEL: This timer select bit chooses the timer used for comparison. If using the 32-bit Timer23, then this bit is ignored.

1 Use Timer3.
0 Use Timer2.

OCxCON⟨2:0⟩, or OCxCONbits.OCM: These three bits determine the operating mode:

0b111 PWM mode with fault pin enabled. OCx is set high on the timer rollover, then set low when the timer value matches OCxR. The SFR OCxRS can be altered at any time, and its value is copied to OCxR at the beginning of the next timer period.[1] The duty cycle of the PWM signal is

$$\text{OCxR}/(\text{PRy} + 1) \times 100\%, \tag{9.1}$$

where PRy is the period register of the timer.
If the fault pin (OCFA for OC1-OC4 and OCFB for OC5) drops low, the read-only fault status bit OCxCONbits.OCFLT is set to 1, the OCx output is set to high impedance, and an interrupt is generated if the interrupt enable bit is set. The fault condition is cleared and PWM resumes once the fault pin goes high and the OCxCONbits.OCM bits are rewritten.
You can use the fault pin with an Emergency Stop button that is normally high but drops low when the user presses it. If the OCx output is driving an H-bridge that powers a motor, setting the OCx output to high impedance will signal the H-bridge to stop sending current to the motor. An emergency stop button will likely have other requirements (such as also physically cutting power to the motor), depending on your application.

0b110 PWM mode with fault pin disabled. Identical to above, except without the fault pin.

0b101 Dual compare mode, continuous output pulses. When the module is enabled, OCx is driven low. OCx is driven high on a match with OCxR and driven low on a match with OCxRS. The process repeats, creating an output pulse train. An interrupt can be generated when OCx is driven low.

0b100 Dual compare mode, single output pulse. Same as above, except the OCx pin will remain low after the match with OCxRS until the OC mode is changed or the module is disabled.

0b011 Single compare mode, continuous pulse train. When the module is enabled, OCx is driven low. The output is toggled on all future matches with OCxR, until the mode is changed or the module disabled. Each toggle can generate an interrupt.

[1] Initialize OCxR before enabling the OC module to handle the first PWM cycle. After enabling the OC module, OCxR is read-only.

0b010 Single compare mode, single high pulse. When the module is enabled, OCx is driven high. OCx will be driven low and an interrupt optionally generated on a match with OCxR. OCx remains low until the mode is changed or the module disabled.

0b001 Single compare mode, single low pulse. When the module is enabled, OCx is driven low. OCx will be driven high and an interrupt optionally generated on a match with OCxR. OCx will remain high until the mode is changed or the module disabled.

0b000 The output compare module is disabled but still drawing current, unless OCxCONbits.ON = 0.

OCxR, x = 1 to 5 If OCxCONbits.OC32 = 1, then all 32-bits of OCxR are compared against Timer23's 32-bit count. Otherwise, only OCxR⟨15:0⟩ is compared to the 16-bit count of Timer2 or Timer3, depending on OCxCONbits.OCTSEL.

OCxRS, x = 1 to 5 In dual compare mode, if OCxCONbits.OC32 = 1, then all 32-bits of OCxRS are compared against Timer23's 32-bit count. Otherwise, only OCxRS⟨15:0⟩ is compared to the 16-bit counter Timer2 or Timer3, depending on OCxCONbits.OCTSEL. In PWM mode, the value of this register is transferred into OCxR at the beginning of each period; therefore, modifying this register sets the next duty cycle. This SFR is unused in the single compare modes.

Timer2, Timer3, or Timer23 (depending on OCxCONbits.OC32 and OCxCONbits.OCTSEL) must be separately configured. Output compare modules do not affect the behavior of the timers; they simply compare the timer count to values in OCxR and OCxRS and alter the digital output OCx on match events.

The interrupt flag status and enable bits for OCx are IFS0bits.OCxIF and IEC0bits.OCxIE, and the priority and subpriority bits are IPCxbits.OCxIP and IPCxbits.OCxIS.

PWM modes

The Output Compare modes you are most likely to use are the PWM modes. They can be used to drive H-bridges powering motors or to continuously transmit analog values represented by the duty cycle. Microchip often equates "duty cycle" to the duration OCxR of the high portion of the PWM waveform, but it is more standard to refer to the duty cycle as a percentage, 0 to 100%. A plot of a PWM waveform is shown in Figure 9.2.

9.3 Sample Code

9.3.1 Generating a Pulse Train with PWM

Below is sample code using OC1 with Timer2 to generate a 10 kHz PWM signal, initially at 25% duty cycle and then changed to 50% duty cycle. The fault pin is not used.

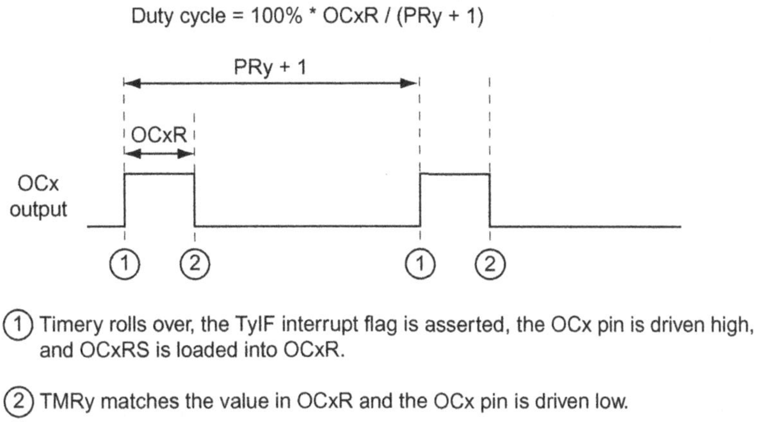

Figure 9.2
A PWM waveform from OCx using Timery as the time base.

Code Sample 9.1 `OC_PWM.c`. **Generating 10 kHz PWM with 50% Duty Cycle.**

```
#include "NU32.h"          // constants, functions for startup and UART

int main(void) {
  NU32_Startup();          // cache on, interrupts on, LED/button init, UART init

  T2CONbits.TCKPS = 2;     // Timer2 prescaler N=4 (1:4)
  PR2 = 1999;              // period = (PR2+1) * N * 12.5 ns = 100 us, 10 kHz
  TMR2 = 0;                // initial TMR2 count is 0
  OC1CONbits.OCM = 0b110;  // PWM mode without fault pin; other OC1CON bits are defaults
  OC1RS = 500;             // duty cycle = OC1RS/(PR2+1) = 25%
  OC1R = 500;              // initialize before turning OC1 on; afterward it is read-only
  T2CONbits.ON = 1;        // turn on Timer2
  OC1CONbits.ON = 1;       // turn on OC1

  _CP0_SET_COUNT(0);       // delay 4 seconds to see the 25% duty cycle on a 'scope
  while(_CP0_GET_COUNT() < 4 * 40000000) {
    ;
  }
  OC1RS = 1000;            // set duty cycle to 50%
  while(1) {
    ;                      // infinite loop
  }
  return 0;
}
```

9.3.2 *Analog Output*

DC analog output

Low-pass filtering a high-frequency, constant duty cycle PWM signal can create an approximately constant analog output. The low-pass filter essentially time-averages the high and low voltages of the waveform,

```
average voltage = duty cycle * 3.3 V
```

assuming that the output compare module swings between 0 and 3.3 V (the range may actually be a bit less).

There are many ways to build circuits to low-pass filter a signal, including *active* filter circuits using op amps. Here we focus on a simple *passive* RC filter, shown in Figure 9.3 and described in Appendix B.2. The voltage V_C across the capacitor C is the output of the filter. When R is zero, the output compare module attempts to source or sink enough current to allow the capacitor voltage to exactly track the nominal PWM square wave, and there is no "averaging" effect. As the resistance R is increased, however, the resistor increasingly limits the current I available to charge or discharge the capacitor, meaning that the capacitor's voltage changes more and more slowly, according to the relationship $\mathrm{d}V_C/\mathrm{d}t = I/C$.

The charging and discharging of the capacitor, and its relationship to the product *RC*, is shown in Figure 9.4. RC low-pass filters are discussed in more detail in Appendix B.2.

In Figure 9.4, the RC filter voltage variation during one PWM cycle is rather large. To reduce this variation, we would choose a larger product *RC* by increasing the resistance *R* and/or capacitance *C*. The drawback of a large *RC* is that the filter's average output voltage changes slowly in response to a change in the PWM duty cycle. While this is not an issue if the desired

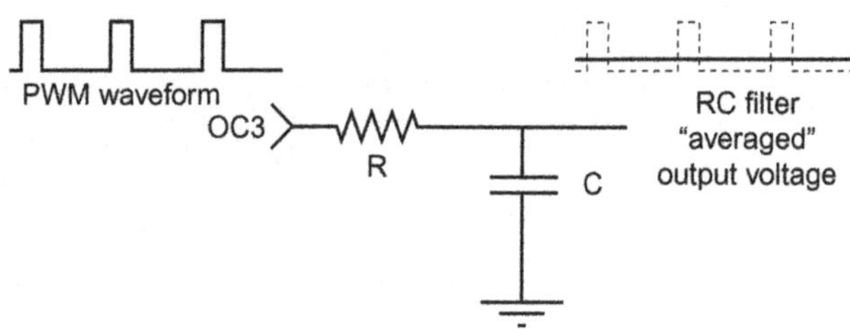

Figure 9.3
An RC low-pass filter "averaging" the PWM output from OC3.

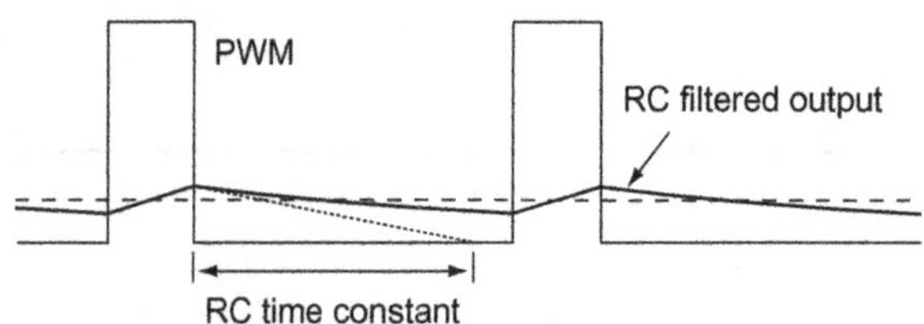

Figure 9.4
A close-up of the PWM, the RC filter output (with RC charging/discharging time constant illustrated), and the true time-averaged output (dashed). If the variation in the RC filtered output is unacceptably large, a larger value of RC should be chosen.

analog voltage is DC (constant), it is an issue if we want the analog voltage to vary in time, as discussed below.

The PWM OCxR value can range from 0 to PRy + 1, where PRy is the period register of the Timery base for the OCx module. This means that PRy + 2 different average voltage levels are achievable.

Code Sample 9.2 generates a PWM signal at 78.125 kHz with a duty cycle determined by OC3R in the range 0-1024. The timer base is Timer2. With an RC filter with a suitably large time constant attached to OC3, the voltage across the capacitor reflects the DC analog voltage requested by the user.

Code Sample 9.2 `OC_analog_out.c`. **Using Timer2, OC3, and an RC Low-pass Filter to Create Analog Output.**

```
#include "NU32.h"          // constants, functions for startup and UART

#define PERIOD 1024        // this is PR2 + 1
#define MAXVOLTAGE 3.3     // corresponds to max high voltage output of PIC32

int getUserPulseWidth(void) {
  char msg[100] = {};
  float f = 0.0;

  sprintf(msg, "Enter the desired voltage, from 0 to %3.1f (volts): ", MAXVOLTAGE);
  NU32_WriteUART3(msg);

  NU32_ReadUART3(msg,10);
  sscanf(msg, "%f", &f);

  if (f > MAXVOLTAGE) {   // clamp the input voltage to the appropriate range
    f = MAXVOLTAGE;
  } else if (f < 0.0) {
    f = 0.0;
  }

  sprintf(msg, "\r\nCreating %5.3f volts.\r\n", f);
  NU32_WriteUART3(msg);
  return PERIOD * (f / MAXVOLTAGE);  // convert volts to counts
}

int main(void) {
  NU32_Startup();          // cache on, interrupts on, LED/button init, UART init

  PR2 = PERIOD - 1;        // Timer2 is OC3's base, PR2 defines PWM frequency, 78.125 kHz
  TMR2 = 0;                // initialize value of Timer2
  T2CONbits.ON = 1;        // turn Timer2 on, all defaults are fine (1:1 divider, etc.)
  OC3CONbits.OCTSEL = 0;   // use Timer2 for OC3
  OC3CONbits.OCM = 0b110;  // PWM mode with fault pin disabled
  OC3CONbits.ON = 1;       // Turn OC3 on
  while (1) {
    OC3RS = getUserPulseWidth();
  }
  return 0;
}
```

Time-varying analog output

Suppose we want to create a sinusoidal analog output voltage, such as

$$V_a(t) = 1.65\ \text{V} + A \sin(2\pi f_a t),$$

by changing the duty cycle of the PWM. The frequency of this desired analog output is f_a. Now we have three relevant frequencies: the PWM frequency f_{PWM}, the RC filter cutoff frequency $f_c = 1/(2\pi RC)$ (Appendix B.2.2), and the desired analog voltage frequency f_a. Examining the frequency response of the low-pass RC filter in Figure B.9(a), we can adopt the following rules of thumb for choosing these three frequencies:

- $f_{PWM} \geq 100 f_c$: The PWM waveform consists of a DC component, a base frequency at f_{PWM}, and higher harmonics to create the square wave output. According to the gain response of the filter, only about 1% of the magnitude of the PWM frequency component at $100 f_c$ makes it through the RC filter.
- $f_c \geq 10 f_a$: Again consulting the RC filter frequency response, we see that signals at frequencies ten times less than f_c are relatively unaffected by the RC filter: the phase delay is only a few degrees and the gain is nearly 1.

For example, if the PWM is at 100 kHz, then we might choose an RC filter cutoff frequency of 1 kHz, and the highest frequency analog output we should expect to be able to create would be 100 Hz. In other words, we can vary the PWM duty cycle through a full sinusoid (e.g., from 50% duty cycle to 100% duty cycle to 0% duty cycle and back to 50% duty cycle) 100 times per second.[2] If the desired analog output is not sinusoidal, then it should be a sum of signals at frequencies less than 100 Hz.

The maximum possible PWM frequency is determined by the 80 MHz PBCLK and the number of bits of resolution we require for the analog output. For example, if we want 8 bits of resolution in the analog output levels, this means we need $2^8 = 256$ different PWM duty cycles. Therefore the maximum PWM frequency is 80 MHz/256 = 312.5 kHz.[3] On the other hand, if we require $2^{10} = 1024$ voltage levels, the maximum PWM frequency is 78.125 kHz. Thus there is a fundamental trade-off between the voltage resolution and the maximum PWM frequency (and therefore the maximum analog output frequency f_a). While higher resolution analog output is generally desirable, there are limits to the value of increasing resolution beyond a certain point, because the device receiving the analog input may have a limit to its analog input sensing resolution and the transmission lines for the analog signal may be subject to electromagnetic noise that creates voltage variations larger than the analog output resolution.

---

[2] Note that this creates a signal that is the sum of a 100 Hz sinusoid with a duty cycle amplitude equal to 50% plus a DC (zero frequency) component of amplitude equal to 50% duty cycle.

[3] Technically this yields 257 possible duty cycle levels, since OCxR = 0 corresponds to 0% duty cycle and OCxR = 256 corresponds to 100% duty cycle.

9.4 Chapter Summary

- Output compare modules pair with Timer2, Timer3, or the 32-bit Timer23 to generate a single timed pulse or a continuous pulse train with controllable duty cycle. Microcontrollers commonly control motors using pulse-width modulation (PWM) to drive H-bridge amplifiers that power the motors.
- Low-pass filtering of PWM signals, perhaps using an RC filter with a cutoff frequency $f_c = 1/(2\pi RC)$, allows the generation of analog outputs. There is a fundamental tradeoff between the resolution of the analog output and the maximum possible frequency component f_a of the generated analog signal. If the PWM frequency is f_{PWM}, then generally the frequencies should satisfy $f_{PWM} \gg f_c \gg f_a$.

9.5 Exercises

1. Enforce the constraints $f_{PWM} \geq 100 f_c$ and $f_c \geq 10 f_a$. Given that PBCLK is 80 MHz, provide a formula for the maximum f_a given that you require n bits of resolution in your DC analog voltage outputs. Provide a formula for RC in terms of n.
2. You will use PWM and an RC low-pass filter to create a time-varying analog output waveform that is the sum of a constant offset and two sinusoids of frequency f and kf, where k is an integer greater than 1. The PWM frequency will be 10 kHz and f satisfies 50 Hz $\geq f \geq$ 10 Hz. Use OC1 and Timer2 to create the PWM waveform, and set PR2 to 999 (so the PWM waveform is 0% duty cycle when OC1R = 0 and 100% duty cycle when OC1R = 1000). You can break this program into the following pieces:
   a. Write a function that forms a sampled approximation of a single period of the waveform

$$V_{out}(t) = C + A_1 \sin(2\pi f t) + A_2 \sin(2\pi k f t + \phi),$$

   where the constant C is 1.65 V (half of the full range 0 to 3.3 V), A_1 is the amplitude of the sinusoid at frequency f, A_2 is the amplitude of the sinusoid at frequency kf, and ϕ is the phase offset of the higher frequency component. Typically values of A_1 and A_2 would be 1 V or less so the analog output is not saturated at 0 or 3.3 V. The function takes f, A_1, k, A_2, and ϕ as input and creates an array `dutyvec`, of appropriate length, where each entry is a value 0 to 1000 corresponding to the voltage range 0 to 3.3 V. Each entry of `dutyvec` corresponds to a time increment of 1/10 kHz = 0.1 ms, and `dutyvec` holds exactly one cycle of the analog waveform, meaning that it has n = 10 kHz/f elements. A MATLAB implementation is given below. You can experiment plotting waveforms or just use the function for reference. A reasonable call of the function is `signal(20, 0.5, 2, 0.25, 45)`, where the phase 45 is in degrees. An example waveform is shown in Figure 9.5.

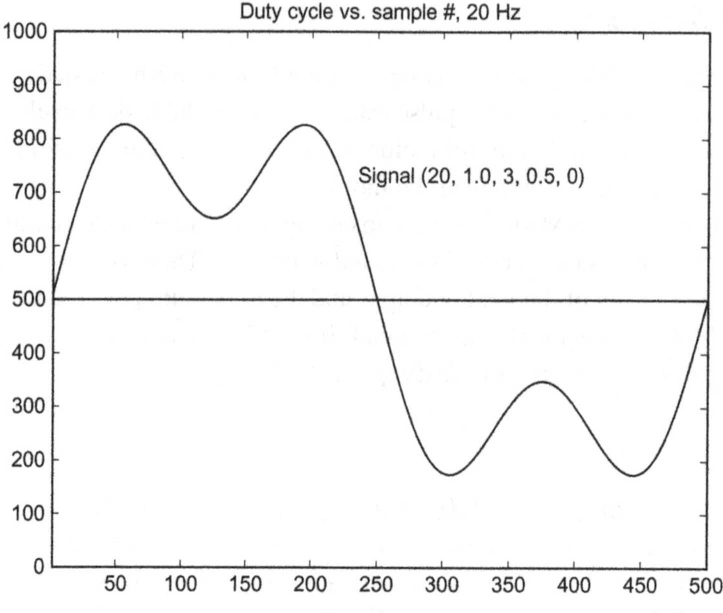

Figure 9.5
An example analog output waveform from Exercise 2, plotted as the duration 0 to 1000 of the high portion of the PWM waveform, which has a period of 1000.

```
function signal(BASEFREQ,BASEAMP,HARMONIC,HARMAMP,PHASE)

% This function calculates the sum of two sinusoids of different
% frequencies and populates an array with the values.  The function
% takes the arguments
%
% * BASEFREQ:  the frequency of the low frequency component (Hz)
% * BASEAMP:   the amplitude of the low frequency component (volts)
% * HARMONIC:  the other sinusoid is at HARMONIC*BASEFREQ Hz; must be
%              an integer value > 1
% * HARMAMP:   the amplitude of the other sinusoid (volts)
% * PHASE:     the phase of the second sinusoid relative to
%              base sinusoid (degrees)
%
% Example matlab call:  signal(20,1,2,0.5,45);

% some constants:

MAXSAMPS = 1000;      % no more than MAXSAMPS samples of the signal
ISRFREQ = 10000;      % frequency of the ISR setting the duty cycle; 10kHz

% Now calculate the number of samples in your representation of the
% signal; better be less than MAXSAMPS!

numsamps = ISRFREQ/BASEFREQ;
if (numsamps>MAXSAMPS)
```

```
  disp('Warning: too many samples needed; choose a higher base freq.');
  disp('Continuing anyway.');
end
numsamps = min(MAXSAMPS,numsamps);  % continue anyway

ct_to_samp = 2*pi/numsamps;         % convert counter to time
offset = 2*pi*(PHASE/360);          % convert phase offset to signal counts

for i=1:numsamps  % in C, we should go from 0 to NUMSAMPS-1
  ampvec(i) = BASEAMP*sin(i*ct_to_samp) + ...
              HARMAMP*sin(HARMONIC*i*ct_to_samp + offset);
  dutyvec(i) = 500 + 500*ampvec(i)/1.65; % duty cycle values,
                                         % 500 = 1.65 V is middle of 3.3V
                                         % output range
  if (dutyvec(i)>1000) dutyvec(i)=1000;
  end
  if (dutyvec(i)<0) dutyvec(i)=0;
  end
end

% ampvec is in volts; dutyvec values are in range 0...1000

plot(dutyvec);
hold on;
plot([1 1000],[500 500]);
axis([1 numsamps 0 1000]);
title(['Duty Cycle vs. sample #, ',int2str(BASEFREQ),' Hz']);
hold off;
```

b. Write a function using the NU32 library that prompts the user for A_1, A_2, k, f, and ϕ. The array `dutyvec` is then updated based on the input.
c. Use Timer2 and OC1 to create a PWM signal at 10 kHz. Enable the Timer2 interrupt, which generates an IRQ at every Timer2 rollover (10 kHz). The ISR for Timer2 should update the PWM duty cycle with the next entry in the `dutyvec` array. When the last element of the `dutyvec` array is reached, wrap around to the beginning of `dutyvec`. Use the shadow register set for the ISR.
d. Choose reasonable values for *RC* for your RC filter. Justify your choice.
e. The main function of your program should sit in an infinite loop, asking the user for new parameters. In the meantime, the old waveform continues to be "played" by the PWM. For the values given in Figure 9.5, use your oscilloscope to confirm that your analog waveform looks correct.

Further Reading

PIC32 family reference manual. Section 16: Output compare. (2011). Microchip Technology Inc.

CHAPTER 10

Analog Input

The PIC32 has one analog-to-digital converter (ADC) that, through the use of multiplexers, can sample the analog voltage from 16 pins (Port B). Typically used with sensors that produce analog voltages, the ADC can capture nearly one million readings per second. The ADC has 10-bit resolution, which means it distinguishes $2^{10} = 1024$ voltage values, usually in the range from 0 to 3.3 V, yielding approximately 3 mV resolution. For higher resolution analog inputs, you can use an external chip and communicate with it using SPI (Chapter 12) or I^2C (Chapter 13).

10.1 Overview

Analog to digital conversion is a multi-step process. First the voltage on the appropriate pin must be routed to an internal differencing amplifier, which outputs the difference between the pin voltage and a reference voltage. Next, the voltage difference is sampled and held by an internal capacitor. Finally, the ADC converts the voltage on the capacitor into a 10-bit binary number.

Figure 10.1 shows a block diagram of the ADC, adapted from the Reference Manual. First we must determine which signals feed the differencing amp, which is located near the middle of Figure 10.1. Control logic (determined by SFRs) selects the differencing amp's + input from the analog pins AN0 to AN15 and the − input from either AN1 or V_{REFL}, a selectable reference voltage.[1] For proper operation, the − input voltage V_{INL} should be less than or equal to the + input voltage V_{INH}.

The differencing amp sends the difference of the two input voltages, $V_{SHA} = V_{INH} - V_{INL}$, to the Sample and Hold Amplifier (SHA). During the *sampling* (or *acquisition*) stage, a 4.4 pF internal holding capacitor charges or discharges to hold the voltage difference V_{SHA}. Once the sampling period has ended, the SHA is disconnected from the inputs, allowing V_{SHA} to remain constant during the *conversion* stage, even if the input voltages change.

[1] This reference V_{REFL} can be chosen to be either V_{REF-}, a voltage provided on an external pin, or AV_{SS}, the PIC32's GND line, also known as V_{SS}.

Embedded Computing and Mechatronics with the PIC32 Microcontroller. http://dx.doi.org/10.1016/B978-0-12-420165-1.00010-X

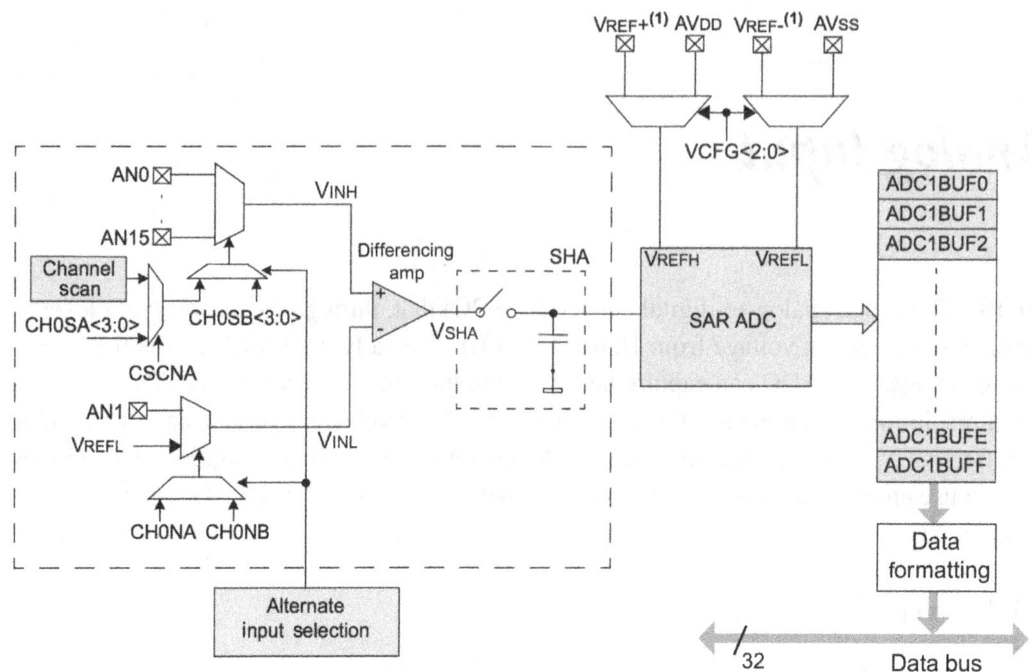

Figure 10.1
A simplified schematic of the ADC module.

The Successive Approximation Register (SAR) converts V_{SHA} to a 10-bit result depending on the low (V_{REFL}) and high (V_{REFH}) reference voltages: $1024 \times V_{SHA}/(V_{REFH} - V_{REFL})$, rounded to the nearest integer between 0 and 1023. (See the Reference Manual for more details on the ADC transfer function.) The 10-bit conversion result is written to the buffer ADC1BUF which is read by your program. If you do not read the result right away, ADC1BUF can store up to 16 results (in the SFRs ADC1BUF0, ADC1BUF1, ..., ADC1BUFF) before the ADC begins overwriting old results.[2]

Sampling and conversion timing

The two main stages of an ADC read are sampling/acquisition and conversion. During the sampling stage, we must allow sufficient time for the internal holding capacitor to converge to the difference $V_{INH} - V_{INL}$. According to the Electrical Characteristics section of the Data Sheet, this time is 132 ns when the SAR ADC uses the external voltage references V_{REF-} and

[2] The ADC1BUFx buffers are not contiguous in memory. Each buffer is four bytes long, but they are 16 bytes apart.

V_{REF+} as its low and high references. The minimum sampling time is 200 ns when using AV_{SS} and AV_{DD} as the low and high references.

Once the sampling stage finishes, the SAR begins the conversion process, using successive approximation to find the digital representation of the voltage. This method uses a binary search, iteratively comparing V_{SHA} to test voltages produced by an internal digital-to-analog converter (DAC). The DAC converts 10-bit numbers into test voltages between V_{REFL} and V_{REFH}: 0x000 produces V_{REFL} and 0x3FF produces V_{REFH}. During the first iteration, the DAC's test value is 0x200 = 0b1000000000, which produces a voltage in the middle of the reference voltage range. If V_{SHA} is greater than this DAC voltage, the first result bit is one, otherwise it is zero. On the second cycle, the DAC's most significant bit is set to the first test's result and the second most significant bit is set to 1. The comparison is performed and the second result bit determined. The process continues until all 10 bits of the result are determined. The entire process requires 10 cycles, plus 2 more, for a total of 12 ADC clock cycles.

The ADC clock is derived from PBCLK. According to the Electrical Characteristics section of the Data Sheet, the ADC clock period (Tad) must be at least 65 ns to allow enough time to convert a single bit. The ADC SFR AD1CON3 allows us to choose the ADC clock period as $2 \times k \times Tpb$, where Tpb is the PBCLK period and k is any integer from 1 to 256. Since Tpb is 12.5 ns for the NU32, to meet the 65 ns specification, the smallest value we can choose is $k = 3$, or $Tad = 75$ ns.

The minimum time between samples is the sum of the sampling time and the conversion time. If configured to sample automatically, we must choose the sampling time to be an integer multiple of Tad. The shortest time we can choose is $2 \times Tad = 150$ ns to satisfy the 132 ns minimum sampling time. Thus the fastest we can read from an analog input is

```
minimum read time = 150 ns + 12 * 75 ns = 1050 ns
```

or just over 1 μs. We can, theoretically, read the ADC at almost one million samples per second (1 MHz).

Multiplexers

Two multiplexers determine which analog input pins to connect to the differencing amp. These two multiplexers are called MUX A and MUX B. MUX A is the default active multiplexer, and the SFR AD1CON3 contains CH0SA bits that determine which of AN0 to AN15 is connected to the + input and CH0NA bits that determine which of AN1 and V_{REF-} is connected to the − input. It is possible to alternate between MUX A and MUX B, but you are unlikely to need this feature.

Options

The ADC peripheral provides a bewildering array of options, some of which are described here. No need to remember them all! The sample code provides a good starting point.

- Data format: The result of the conversion is stored in a 32-bit word, and it can be represented as a signed integer, unsigned integer, fractional value, etc. Typically we use either 16-bit or 32-bit unsigned integers.
- Sampling and conversion initiation events: Sampling can be initiated by a software command or immediately after the previous conversion has completed (auto sample). Conversion can be initiated by a software command, the expiration of a specified sampling period (auto convert), a period match with Timer3, or a signal change on the INT0 pin. If sampling and conversion happen automatically (i.e., not through software commands), the conversion results are placed in the ADC1BUF at successively higher addresses (ADC1BUF0 to ADC1BUFF) before returning to the first address in ADC1BUF.
- Input scan and alternating modes: You can read one analog input at a time, scan through a list of inputs (using MUX A), or alternate between two inputs (one from MUX A and one from MUX B).
- Voltage reference: The ADC normally uses reference voltages of 0 and 3.3 V (the power rails of the PIC32); therefore, a reading of 0x000 corresponds to 0 V and a reading of 0x3FF corresponds to 3.3 V. If you are interested in a different voltage range—say 1.0 V to 2.0 V—you can configure the ADC so that 0x000 corresponds to 1.0 V and 0x3FF corresponds to 2.0 V, giving you better resolution: (2 V – 1 V)/1024 = 1 mV resolution. You supply alternate voltage references on pins V_{REF-} and V_{REF+}. The voltages provided must be between 0 and 3.3 V.
- Unipolar differential mode: Any of the analog inputs ANx (x = 2 to 15, e.g., AN5) can be compared to AN1, allowing you to read the voltage difference between ANx and AN1. The voltage on ANx should be greater than the voltage on AN1.
- Interrupts: An interrupt may be generated after a specified number of conversions. The number of conversions per interrupt also determines which ADC1BUFx buffer is used, even if you do not enable the interrupt. Conversion results are placed in successively higher numbered ADC1BUFx buffers (i.e., the first conversion goes in ADC1BUF0, the next in ADC1BUF1, etc.). When the interrupt triggers, the current buffer wraps around to ADC1BUF0 (or, in dual buffer mode, ADC1BUF8, see below). So if you set the ADC to interrupt on every conversion (the default), the results will always be stored in ADC1BUF0.
- ADC clock period: The ADC clock period Tad can range from 2 times the PB clock period up to 512 times the PB clock period, in integer multiples of two. Tad must be long enough to convert a single bit (65 ns according to the Electrical Characteristics section of the Data Sheet). You may also choose Tad to be the period of the ADC internal RC clock.

- Dual buffer mode: When an ADC conversion finishes, the result is written into the output buffer ADC1BUFx (x = 0x0 to 0xF). The ADC can be configured to write a series of conversions into a sequence of output buffers. The first conversion is stored in ADC1BUF0, the second in ADC1BUF1, etc. After a series of conversions, an interrupt flag is set, indicating that the results are available for the program to read. The next set of conversions starts over at ADC1BUF0; if the program is slow to read the results, the next conversions may overwrite the previous results. To help with this scenario, the 16 ADC1BUFx buffers can be split into two 8-word groups: one in which the current conversions are written, and one from which the program should read the results. The first conversion sequence starts writing at ADC1BUF0, the next starts at ADC1BUF8, and the starting buffers alternate from there.

10.2 Details

The operation of the ADC peripheral is determined by the following SFRs, all of which default to all zeros on reset.

AD1PCFG Only the least significant 16 bits are relevant. If a bit is 0, the associated pin on port B is configured as an analog input. If a bit is 1, it is digital I/O. The analog input pins AN0 to AN15 correspond to the port B pins RB0 to RB15.

AD1CON1 One of three main ADC control registers: controls the output format and conversion and sampling methods.

AD1CON1⟨15⟩ or AD1CON1bits.ON: Enables and disables the ADC.

1 The ADC is enabled.

0 The ADC is disabled.

AD1CON1⟨10:8⟩ or AD1CON1bits.FORM: Determines the data output format. We usually use either

0b100 32-bit unsigned integer

0b000 16-bit unsigned integer (the default).

AD1CON1⟨7:5⟩ or AD1CON1bits.SSRC: Determines what begins the conversion process. The two most common methods are

0b111 Auto conversion. The conversion begins as soon as sampling ends. Hardware automatically clears AD1CON1bits.SAMP to begin the conversion.

0b000 Manual conversion. You must clear AD1CONbits.SAMP to start the conversion.

AD1CON1⟨2⟩ or AD1CON1bits.ASAM: Determines whether another sample occurs immediately after conversion.

1 Use auto sampling. Sampling starts after the last conversion is finished. Hardware automatically sets AD1CON1bits.SAMP.

0 Use manual sampling. Sampling begins when the user sets AD1CON1bits.SAMP.

AD1CON1⟨1⟩ or AD1CON1bits.SAMP: Indicates whether the sample and hold amplifier (SHA) is sampling or holding. When auto sampling is disabled (AD1CON1bits.ASAM = 0), set this bit to initiate sampling. When using manual conversion (AD1CON1bits.SSRC = 0) clear this bit to zero to start conversion.

1 The SHA is sampling. Setting this bit initiates sampling when in manual sampling mode (AD1CON1bits.ASAM = 0).

0 The SHA is holding. Clearing this bit begins conversion when in manual conversion mode (AD1CON1bits.SSRC = 0).

AD1CON1⟨0⟩ or AD1CON1bits.DONE: Indicates whether a conversion is occurring. When using automatic sampling, hardware clears this bit automatically.

1 The analog-to-digital conversion is finished.

0 The analog-to-digital conversion is either pending or has not begun.

AD1CON2 Determines voltage reference sources, input pin selections, and the number of conversions per interrupt.

AD1CON2⟨15:13⟩ or AD1CON2bits.VCFG: Determines the voltage reference sources for the V_{REFH} and V_{REFL} inputs to the SAR. These references determine what voltage a given reading corresponds to: 0x000 corresponds to V_{REFL} and 0x3FF corresponds to V_{REFH}.

0b000 Use the internal references: V_{REFH} is 3.3 V and V_{REFL} is 0 V.

0b001 Use an external reference for V_{REFH} and an internal reference for V_{REFL}: V_{REFH} is the voltage on the V_{REF+} pin and V_{REFL} is 0 V.

0b010 Use an internal reference for V_{REFH} and an external reference for V_{REFL}: V_{REFH} is 3.3 V and V_{REFL} is the voltage on the V_{REF-} pin.

0b011 Use external references: V_{REFH} is the voltage on the V_{REF+} pin and V_{REFL} is the voltage on the V_{REF-} pin.

AD1CON2⟨10⟩ or AD1CON2bits.CSNA: Control scanning of inputs. The pins to scan are selected by AD1CSSL.

1 Scan inputs. Each subsequent sample will be from a different pin, selected by AD1CSSL, wrapping around to the beginning when the last pin is reached.

0 Do not scan inputs. Only one input is used.

AD1CON2⟨7⟩ or AD1CON2bits.BUFS: Used only in split buffer mode (AD1CON2bits.BUFM = 1). Indicates which buffer the ADC is currently filling.

1 The ADC is filling buffers ADC1BUF8 to ADC1BUFF, so the user should read from buffers ADC1BUF0 to ADC1BUF7.

0 The ADC is filling buffers ADC1BUF0 to ADC1BUF7, so the user should read from buffers ADC1BUF8 to ADC1BUFF.

AD1CON2⟨5:2⟩ or AD1CON2bits.SMPI: The number of sample/conversion sequences per interrupt is AD1CON2bits.SMPI + 1. In addition to determining when the interrupt occurs, these bits also determine how many conversions must occur before the ADC starts storing data in the first buffer (or the alternate first buffer when AD1CON2bits.BUFM = 1). For example, if AD1CON2bits.SMPI = 1 then there will be two conversions per interrupt. The first conversion will be stored in AD1BUF0 and the second in AD1BUF1. After the second conversion the ADC interrupt flag will be set. The next conversion will be stored in AD1BUF0 if AD1CON2bits.BUFM = 0 or AD1BUF8 if AD1CON2bits.BUFM = 1.

AD1CON2⟨1⟩ or AD1CON2bits.BUFM: Determines if the ADC buffer is split into two 8-word buffers or is used as a single 16-word buffer.

1 The ADC buffer is split into two 8-word buffers, ADC1BUF0 to ADC1BUF7 and ADC1BUF8 to ADC1BUFF. Data is alternatively stored in the lower and upper buffers, every AD1CON2bits.SMPI + 1 sample/conversion sequences.

0 The ADC buffer is used as a single 16 word buffer, ADC1BUF0 to ADC1BUFF.

AD1CON3 Controls settings for the ADC clock and other ADC timing settings. Determines Tad, the ADC clock period.

AD1CON3⟨12:8⟩ or AD1CON3bits.SAMC: Determines the length of auto sampling time, in Tad. Can be set anywhere from 1 Tad to 31 Tad; however, sampling requires at least 132 ns.

AD1CON3⟨7:0⟩ or AD1CON3bits.ADCS: Determines the length of Tad, in terms of the peripheral bus clock period Tpb, according to the formula

$$\text{Tad} = 2 \times \text{Tpb} \times (\text{AD1CON3bits.ADCS} + 1). \tag{10.1}$$

Tad must be at least 65 ns, so with an 80 MHz peripheral bus clock frequency, the minimum value for AD1CON3bits.ADCS is 2, which yields a 75 ns Tad by the equation above.

AD1CHS This SFR determines which pins will be sampled (the "positive" inputs) and what they will be compared to (i.e., V_{REFL} or AN1). When in scan mode, the sample pins specified in this SFR are ignored. There are two multiplexers available, MUX A and MUX B; we focus on the settings for MUX A.

AD1CHS⟨23⟩ or AD1CHSbits.CH0NA: Determines the negative input for MUX A. When MUX A is selected (the default), this input is the negative input to the differencing amplifier.

1 The negative input is the pin AN1.

0 The negative input is V_{REFL}. V_{REFL} is determined by AD1CON2bits.VCFG.

AD1CHS⟨19:16⟩ or AD1CHSbits.CH0SA: Determines the positive input to MUX A. When MUX A is selected (the default), this input is the positive input to the differencing amplifier. The value of this field determines which ANx pin is used. For example, if AD1CHS.CHOSA = 6 than AN6 is used.

AD1CSSL Bits set to 1 in this SFR indicate which analog inputs will be sampled in scan mode (if AD1CON2 has configured the ADC for scan mode). Inputs will be scanned from lower number inputs to higher numbers. Bit x corresponds to an ANx. Individual bits can be accessed using AD1CSSLbits.CSSLx.

Apart from these SFRs, the ADC module has bits associated with the ADC interrupt in IFS1bits.AD1IF (IFS1⟨1⟩), IEC1bits.AD1IE (IEC1⟨1⟩), IPC6bits.AD1IP (IPC6⟨28:26⟩), and IPC6bits.AD1IS (IPC6⟨25:24⟩). The interrupt vector is 27, also known as `_ADC_VECTOR`.

10.3 Sample Code

10.3.1 Manual Sampling and Conversion

There are many ways to read the analog inputs, but the sample code below is perhaps the simplest. This code reads in analog inputs AN14 and AN15 every half second and sends their values to the user's terminal. It also logs the time it takes to do the two samples and conversions, which is a bit under 5 μs total. In this program we set the ADC clock period Tad to be $6 \times \text{Tpb} = 75$ ns, and the acquisition time to be at least 250 ns. There are two places in this program where we wait and do nothing: during the sampling and during the conversion. If speed were an issue, we could use more advanced settings to let the ADC work in the background and interrupt when samples are ready.

In the exercises you will write code to initiate the conversion automatically, rather than manually as in the sample code below.

Code Sample 10.1 `ADC_Read2.c`. **Reading Two Analog Inputs with Manual Initialization of Sampling and Conversion.**

```
#include "NU32.h"          // constants, functions for startup and UART

#define VOLTS_PER_COUNT (3.3/1024)
#define CORE_TICK_TIME 25    // nanoseconds between core ticks
#define SAMPLE_TIME 10       // 10 core timer ticks = 250 ns
#define DELAY_TICKS 20000000 // delay 1/2 sec, 20 M core ticks, between messages

unsigned int adc_sample_convert(int pin) { // sample & convert the value on the given
                                           // adc pin the pin should be configured as an
                                           // analog input in AD1PCFG
    unsigned int elapsed = 0, finish_time = 0;
    AD1CHSbits.CH0SA = pin;                // connect chosen pin to MUXA for sampling
    AD1CON1bits.SAMP = 1;                  // start sampling
    elapsed = _CP0_GET_COUNT();
    finish_time = elapsed + SAMPLE_TIME;
    while (_CP0_GET_COUNT() < finish_time) {
      ;                                   // sample for more than 250 ns
    }
    AD1CON1bits.SAMP = 0;                 // stop sampling and start converting
    while (!AD1CON1bits.DONE) {
      ;                                   // wait for the conversion process to finish
    }
    return ADC1BUF0;                      // read the buffer with the result
}

int main(void) {
  unsigned int sample14 = 0, sample15 = 0, elapsed = 0;
```

```
  char msg[100] = {};

  NU32_Startup();                  // cache on, interrupts on, LED/button init, UART init
  AD1PCFGbits.PCFG14 = 0;                   // AN14 is an adc pin
  AD1PCFGbits.PCFG15 = 0;                   // AN15 is an adc pin
  AD1CON3bits.ADCS = 2;                     // ADC clock period is Tad = 2*(ADCS+1)*Tpb =
                                            //                           2*3*12.5ns = 75ns
  AD1CON1bits.ADON = 1;                     // turn on A/D converter
  while (1) {
    _CP0_SET_COUNT(0);                      // set the core timer count to zero
    sample14 = adc_sample_convert(14);      // sample and convert pin 14
    sample15 = adc_sample_convert(15);      // sample and convert pin 15
    elapsed = _CP0_GET_COUNT();             // how long it took to do two samples
                                            // send the results over serial
    sprintf(msg, "Time elapsed: %5u ns  AN14: %4u (%5.3f volts)"
                 "  AN15: %4u (%5.3f volts) \r\n",
                 elapsed * CORE_TICK_TIME,
                 sample14, sample14 * VOLTS_PER_COUNT,
                 sample15, sample15 * VOLTS_PER_COUNT);
    NU32_WriteUART3(msg);
    _CP0_SET_COUNT(0);                      // delay to prevent a flood of messages
    while(_CP0_GET_COUNT() < DELAY_TICKS) {
      ;
    }
  }
  return 0;
}
```

If AN14 is connected to 0 V and AN15 is connected to 3.3 V, typical output of the program repeats the following two lines,

```
  ...
Time elapsed:  4550 ns  AN14:    0 (0.000 volts)  AN15: 1023 (3.297 volts)
Time elapsed:  4675 ns  AN14:    0 (0.000 volts)  AN15: 1023 (3.297 volts)
  ...
```

indicating that the two conversions take less than 5 μs with some minor variation each time through the loop.

10.3.2 Maximum Possible Sample Rate

The program `ADC_max_rate.c` reads from a single analog input, AN2, at the maximum speed that fits the PIC32 Electrical Characteristics and the 80 MHz PBCLK (Tpb = 12.5 ns). We choose

```
Tad = 6 * Tpb = 75 ns
```

as the smallest time that is an even integer multiple of Tpb and greater than the 65 ns required in the Electrical Characteristics section of the Data Sheet. We choose the sample time to be

```
Tsamp = 2 * Tad = 150 ns,
```

the smallest integer multiple of Tad that meets the minimum spec of 132 ns in the Data Sheet.[3] The ADC is configured to auto-sample and auto-convert eight samples and then generate an interrupt. The ISR reads eight samples from ADCBUF0 to ADCBUF7 or from ADCBUF8 to ADCBUFF while the ADC fills the other eight-word section. The ISR must finish reading one eight-word section before the other eight-word section is filled. Otherwise, the ADC results will start overwriting unread results.

After reading 1000 samples, the ADC interrupt is disabled to free the CPU from servicing the ISR. The program writes the data and average sample/conversion times to the user's terminal.

High-speed sampling requires pin V_{REF-} (RB1) to be connected to ground and pin V_{REF+} (RB0) to be connected to 3.3 V. These pins are the external low and high voltage references for analog input. Technically, the Reference Manual states that V_{REF-} should be attached to ground through a 10 ohm resistor and V_{REF+} should be attached to two capacitors in parallel to ground (0.1 μF and 0.01 μF) as well as a 10 ohm resistor to 3.3 V.

To provide input to the ADC, we configure OC1 to output a 5 kHz 25% duty cycle square wave. The program also uses Timer45 to time the duration between ISR entries. The first 1000 analog input samples are written to the screen, as well as the time they were taken, confirming that the samples correspond to 889 kHz sampling of a 5 kHz 25% duty cycle waveform. The ISR that reads eight samples from ADCBUF also toggles an LED once every million times it is entered, allowing you to measure the time it takes to acquire eight million samples with a stopwatch (about 9 s).

Code Sample 10.2 `ADC_max_rate.c`. **Reading a Single Analog Input at the Maximum Possible Rate to Meet the Electrical Characteristics Section of the Data Sheet, Given That the PBCLK Is 80 MHz.**

```
// ADC_max_rate.c
//
// This program reads from a single analog input, AN2, at the maximum speed
// that fits the PIC32 Electrical Characteristics and the 80 MHz PBCLK
// (Tpb = 12.5 ns).  The input to AN2 is a 5 kHz 25% duty cycle PWM from
// OC1.  The results of 1000 analog input reads is sent to the user's
// terminal.  An LED on the NU32 also toggles every 8 million samples.
//
// RB1/VREF- must be connected to ground and RB0/VREF+ connected to 3.3 V.
//

#include "NU32.h"                    // constants, functions for startup and UART

#define NUM_ISRS 125                 // the number of 8-sample ISR results to be printed
#define NUM_SAMPS (NUM_ISRS*8)       // the number of samples stored
#define LED_TOGGLE 1000000           // toggle the LED every 1M ISRs (8M samples)
```

[3] The Electrical Characteristics section of the Data Sheet lists 132 ns as the minimum sampling time for an analog input from a source with 500 Ω output impedance. If the source has a much lower output impedance, you may be able to reduce the sampling time below 132 ns.

```
// these variables are static because they are not needed outside this C file
// volatile because they are written to by ISR, read in main

static volatile int storing = 1;    // if 1, currently storing data to print; if 0, done
static volatile unsigned int trace[NUM_SAMPS];   // array of stored analog inputs
static volatile unsigned int isr_time[NUM_ISRS]; // time of ISRs from Timer45

void __ISR(_ADC_VECTOR, IPL6SRS) ADCHandler(void) { // interrupt every 8 samples
  static unsigned int isr_counter = 0; // the number of times the isr has been called
                                       // "static" means the variable maintains its value
                                       // in between function (ISR) calls
  static unsigned int sample_num = 0;  // current analog input sample number

  if (isr_counter <= NUM_ISRS) {
    isr_time[isr_counter] = TMR4;      // keep track of Timer45 time the ISR is entered
  }

  if (AD1CON2bits.BUFS) {              // 1=ADC filling BUF8-BUFF, 0=filling BUF0-BUF7
    trace[sample_num++] = ADC1BUF0;    // all ADC samples must be read in, even
    trace[sample_num++] = ADC1BUF1;    // if we don't want to store them, so that
    trace[sample_num++] = ADC1BUF2;    // the interrupt can be cleared
    trace[sample_num++] = ADC1BUF3;
    trace[sample_num++] = ADC1BUF4;
    trace[sample_num++] = ADC1BUF5;
    trace[sample_num++] = ADC1BUF6;
    trace[sample_num++] = ADC1BUF7;
  }
  else {
    trace[sample_num++] = ADC1BUF8;
    trace[sample_num++] = ADC1BUF9;
    trace[sample_num++] = ADC1BUFA;
    trace[sample_num++] = ADC1BUFB;
    trace[sample_num++] = ADC1BUFC;
    trace[sample_num++] = ADC1BUFD;
    trace[sample_num++] = ADC1BUFE;
    trace[sample_num++] = ADC1BUFF;
  }
  if (sample_num >= NUM_SAMPS) {
    storing = 0;                       // done storing data
    sample_num = 0;                    // reset sample number
  }
  ++isr_counter;                       // increment ISR count
  if (isr_counter == LED_TOGGLE) {     // toggle LED every 1M ISRs (8M samples)
    LATFINV = 0x02;
    isr_counter = 0;                   // reset ISR counter
  }

  IFS1bits.AD1IF = 0;                  // clear interrupt flag
}

int main(void) {
  int i = 0, j = 0, ind = 0;        // variables used for indexing
  float tot_time = 0.0;             // time between 8 samples
  char msg[100] ={};                // buffer for writing messages to uart
  unsigned int prev_time = 0;       // used for calculating time differences

  NU32_Startup(); // cache on, min flash wait, interrupts on, LED/button init, UART init

  __builtin_disable_interrupts(); // INT step 2: disable interrupts
```

```
                                  // configure OC1 to use T2 to make 5 kHz 25% DC
  PR2 = 15999;                    // (15999+1)*12.5ns = 200us period = 5kHz
  T2CONbits.ON = 1;               // turn on Timer2
  OC1CONbits.OCM = 0b110;         // OC1 is PWM with fault pin disabled
  OC1R = 4000;                    // hi for 4000 counts, lo for rest (25% DC)
  OC1RS = 4000;
  OC1CONbits.ON = 1;              // turn on OC1

                                  // set up Timer45 to count every pbclk cycle
  T4CONbits.T32 = 1;              // configure 32-bit mode
  PR4 = 0xFFFFFFFF;               // rollover at the maximum possible period, the default
  T4CONbits.TON = 1;              // turn on Timer45

                                  // INT step 3: configure ADC generating interrupts
  AD1PCFGbits.PCFG2 = 0;          //       make RB2/AN2 an analog input (the default)
  AD1CHSbits.CH0SA = 2;           //       AN2 is the positive input to the sampler
  AD1CON3bits.SAMC = 2;           //       sample for 2 Tad
  AD1CON3bits.ADCS = 2;           //       Tad = 6*Tpb
  AD1CON2bits.VCFG = 3;           //       external Vref+ and Vref- for VREFH and VREFL
  AD1CON2bits.SMPI = 7;           //       interrupt after every 8th conversion
  AD1CON2bits.BUFM = 1;           //       adc buffer is two 8-word buffers
  AD1CON1bits.FORM = 0b100;       //       unsigned 32 bit integer output
  AD1CON1bits.ASAM = 1;           //       autosampling begins after conversion
  AD1CON1bits.SSRC = 0b111;       //       conversion starts when sampling ends
  AD1CON1bits.ON = 1;             //       turn on the ADC
  IPC6bits.AD1IP = 6;             // INT step 4: IPL6, to use shadow register set
  IFS1bits.AD1IF = 0;             // INT step 5: clear ADC interrupt flag
  IEC1bits.AD1IE = 1;             // INT step 6: enable ADC interrupt
  __builtin_enable_interrupts();  // INT step 7: enable interrupts at CPU

  TMR4 = 0;                       // start timer 4 from zero
  while(storing) {
    ;                             // wait until first NUM_SAMPS samples taken
  }
  IEC1bits.AD1IE = 0;             // disable ADC interrupt

  sprintf(msg,"Values of %d analog reads\r\n",NUM_SAMPS);
  NU32_WriteUART3(msg);
  NU32_WriteUART3("Sample #   Value   Voltage   Time");

  for (i = 0; i < NUM_ISRS; ++i) {// write out NUM_SAMPS analog samples
    for (j = 0; j < 8; ++j) {
      ind = i * 8 + j;            // compute the index of the current sample
      sprintf(msg,"\r\n%5d %10d %9.3f ", ind, trace[ind], trace[ind]*3.3/1024);
      NU32_WriteUART3(msg);
    }
    tot_time = (isr_time[i] - prev_time) *0.0125; // total time elapsed, in microseconds
    sprintf(msg,"%9.4f us; %d timer counts; %6.4f us/read for last 8 reads",
        tot_time, isr_time[i]-prev_time,tot_time/8.0);
    NU32_WriteUART3(msg);
    prev_time = isr_time[i];
  }

  NU32_WriteUART3("\r\n");
  IEC1bits.AD1IE = 1;           // enable ADC interrupt. won't print the information again,
                                // but you can see the light blinking
  while(1) {
    ;
  }
  return 0;
}
```

The output should look like

```
... (earlier output snipped)
928       1019     3.284
929       1015     3.271
930       1015     3.271
931       1015     3.271
932       1015     3.271
933          4     0.013
934          4     0.013
935          4     0.013    9.0000 us; 720 timer counts; 1.1250 us/read for last
8 reads   ... (later output snipped)
```

showing the sample number, the ADC counts, and the corresponding actual voltage for samples 0 to 999. The high output voltage from OC1 is measured as approximately 3.27 V, and the low output voltage is measured as approximately 0.01 V. You can also see that the output is high for 45 consecutive samples and low for 135 consecutive samples, corresponding to the 25% duty cycle of OC1. In the snippet above, OC1's switch from high to low is measured at sample 933.

Theoretically, the time needed for one sample is 150 ns + (12 × 75 ns) = 1050 ns, but we get an extra 1 Tad (75 ns or 6 Tpb) for 1125 ns. This is 888.89 kHz sampling. Where does the extra 75 ns come from? It's not due to the extra processing time needed to enter the interrupt and read the timer: these times are constant and cancel each other when measuring the time between two interrupts. Rather, the discrepancy is from the time needed to start conversion after sampling and the time needed to start sampling after conversion. The Electrical Characteristics section of the Data Sheet lists the "Conversion Start from Sample Trigger" as being typically 1.0 Tad and the "Conversion Completion to Sample Start" as being typically 0.5 Tad. Our experiment indicates that our measured times are actually a little lower than the listed typical values, since we have only 1 Tad, not 1.5 Tad, of unexpected sample/conversion time.

10.4 Chapter Summary

- The ADC peripheral converts an analog voltage to a 10-bit digital value, where 0x000 corresponds to an input voltage at V_{REFL} (typically GND) and 0x3FF corresponds to an input voltage at V_{REFH} (typically 3.3 V). There is a single ADC on the PIC32, AD1, but it can be multiplexed to sample from any or all of the 16 pins on Port B.
- Getting an analog input is a two-step process: sampling and conversion. Sampling requires a minimum time to allow the sampling capacitor to stabilize its voltage. Once the sampling terminates, the capacitor is isolated from the input so its voltage does not change during conversion. The conversion process is performed by a Successive Approximation Register (SAR) ADC which carries out a 10-step binary search, comparing the capacitor voltage to a new reference voltage at each step.

- The ADC provides a huge array of options which are only touched on in this chapter. The sample code in this chapter provides a manual method for taking a single ADC reading in the range 0-3.3 V in just over 2 μs. For details on how to use other reference value ranges, sample and convert in the background and use interrupts to announce the end of a sequence of conversions, etc., consult the Reference Manual.

10.5 Exercises

1. Configure the ADC for manual sampling and automatic conversion. Set Tad and the sampling time as short as possible while still meeting the minimum constraints.
2. Assume that the ADC is configured for manual sampling and automatic conversion. Write a function that begins sampling from a specified ANx pin, waits for the conversion to complete, and returns the result. This function will be useful whenever you need to take an ADC reading.
3. Using the configuration code and the ADC reading function you wrote for the previous questions, write a program that prompts the user to press ENTER and then reports the voltage on AN5 (in both ADC ticks and in volts) over the UART. Test the program with a variety of voltage dividers or a potentiometer.

Further Reading

PIC32 family reference manual. Section 17: 10-Bit analog-to-digital converter (ADC). (2011). Microchip Technology Inc.

CHAPTER 11

UART

The universal asynchronous receiver/transmitter (UART) allows two devices to communicate with each other. Formerly ubiquitous as the hardware powering serial ports, the UART has been almost completely replaced by the universal serial bus (USB). Although obsolete to the average computer user, UARTs remain important in embedded systems due to their relative simplicity. A UART can be used with an external transceiver device to implement RS-232 communication, RS-485 multipoint communication, IrDA infrared wireless communication, or other types of wireless communication such as the IEEE 802.15.4 standard.

11.1 Overview

The PIC32 has six UARTs, each allowing it to communicate with one other device. Each UART uses at least two pins, one for receiving data (RX) and one for transmitting data (TX). Additionally, the devices share a common ground (GND) line. A UART can simultaneously send and receive data, a feature known as full duplex communication. For one-way communication, only one wire (in addition to GND) is required. To distinguish your PIC32 from the device with which it is communicating, which may be your computer or another PIC32, we call the other device *data terminal equipment* (DTE). The RX line for the PIC32 is the TX line for the DTE, and the TX line for the PIC32 is the RX line for the DTE.

You have used the PIC32's UARTs to communicate with your computer. FTDI driver software on your computer sends data over a USB cable, where a chip on the NU32 (the FTDI FT231X) receives the USB data and converts it into signals appropriate for one of the PIC32's UARTs. Data sent by the PIC32's UART is converted by the FTDI chip to USB signals to send to your computer, where the FTDI driver software interprets it as if received by a UART on your computer.

The important parameters for UART communication are the baud, the data length, the parity, and the number of stop bits. The two devices use the same parameters for successful communication. The baud refers to the number of bits sent per second. The PIC32's UART sends and receives data in groups of 8 or 9 bits. For data lengths of 8 bits, the PIC32 can

Embedded Computing and Mechatronics with the PIC32 Microcontroller. http://dx.doi.org/10.1016/B978-0-12-420165-1.00011-1

optionally transmit an additional parity bit as a simple transmission error-detection measure. For example, if the parity is "even," the number of bits sent that are one must be an even number; the parity bit is chosen to meet this constraint. If the receiver sees an odd number of ones in the transmission, then it knows a transmission error has occurred. Finally, the PIC32's UART can be set to one or two stop bits, which are ones sent at the end of a transmission.

The NU32 library uses a baud of 230,400, eight data bits, no parity bit, and one stop bit. Written in shorthand, this is 230,400/8N1. Parity may be odd, even, or none.

For historical reasons, common baud choices include 1200, 2400, 4800, 9600, 19,200, 38,400, 57,600, 115,200, and 230,400, but any choice is possible, as long as both devices agree. According to the Reference Manual, the PIC32's UART is theoretically capable of baud up to 20 M; however, in practice, the maximum achievable baud is much lower.

UART communication is asynchronous, meaning that there is no clock line to keep the two devices in sync. Due to differences in clock frequencies on the two devices, the baud for each device may be slightly different, and UART devices can handle slight differences by resynchronizing their baud clocks on each transmission.

Figure 11.1 shows a typical UART transmission. When not transmitting, the TX line is high. To start a transmission, the UART lowers TX for one baud period. This start bit tells the receiver that a transmission has begun so that the receiver can start its baud clock and begin receiving bits. Next, the data bits are sent. Each bit is held on the line for one baud period. The bits are sent least-significant bit first (e.g., the first bit sent for 0b11001000 will be a zero). Following the data bits, a parity bit may be optionally sent. Finally, the transmitter holds the line high, transmitting one or two stop bits. After the stop bits have been transmitted, another transmission may begin; thus using two stop bits provides the devices with extra processing time between transmissions. The start bit, parity bit, and stop bits are control bits: they do not contain data. Therefore, the baud does not directly correspond to the data rate.

As the UART receives data, hardware shifts each bit into a register. When a full byte has been received, that byte is transferred into the UART's RX first-in first-out queue (FIFO). When transmitting data, software loads bytes into the TX FIFO. The hardware then loads bytes from the FIFO into a shift register, which sends them over the wire. If either FIFO is full and

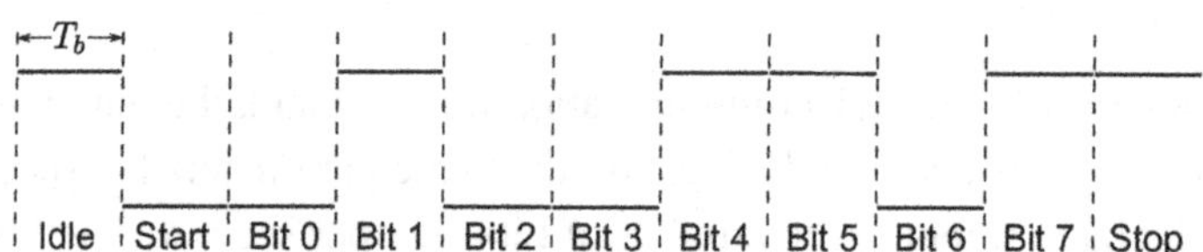

Figure 11.1
UART transmission of 0b10110010 with 8 data bits, no parity, and one stop bit.

another byte needs to be added, an overrun condition occurs and the data is lost. To prevent a TX FIFO overrun, software should not attempt to write to the UART unless the TX FIFO has space. To prevent an RX FIFO overrun, software must read the RX FIFO fast enough so that it always has space when data arrives. Hardware maintains flags indicating the status of the FIFOs and can also interrupt based on the number of items in the FIFOs.

An optional feature called hardware flow control can help software prevent overruns. Hardware flow control requires two additional wires: request to send ($\overline{\text{RTS}}$) and clear to send ($\overline{\text{CTS}}$).[1] When the RX FIFO is full, the UART hardware de-asserts (drives high) $\overline{\text{RTS}}$, which tells the DTE not to send data. When the RX FIFO has space available, the hardware asserts (drives low) $\overline{\text{RTS}}$, allowing the DTE to send data. The DTE controls $\overline{\text{CTS}}$. When the DTE de-asserts (drives high) $\overline{\text{CTS}}$, the PIC32 will not transmit data. For hardware flow control to work, both the DTE and PIC32 must respect the flow control signals. By default, when you use `make screen` or `make putty`, those terminal emulators configure your DTE to use hardware flow control.

These are the basics of UART operation. Many other options exist, far too many to cover here. The guiding principle behind all UART operation, however, remains the same: both ends of the communication must agree on all the options. When interfacing with a specific device, read its data sheet and select the appropriate options.

11.2 Details

Below is a description of the UART registers. The "x" in the SFR names stands for UART number 1 to 6. All bits default to zero except for two read-only bits in UxSTA.

UxMODE Enables or disables the UART. Determines the parity, number of data bits, number of stop bits, and flow control method.

UxMODE⟨15⟩ or UxMODEbits.ON: when set to one, enables the UART.

UxMODE⟨9:8⟩ or UxMODEbits.UEN: Determines which pins the UART uses. Common choices are

- 0b00 Only the UxTX and UxRX are used (the minimum required for UART communication).
- 0b10 UxTX, UxRX, UxCTS, and UxRTS are used. This enables hardware flow control.

UxMODE⟨3⟩ or UxMODEbits.BRGH: This is called the "high baud rate generator bit" and controls the value of a divisor M used in calculating the baud rate (see the SFR UxBRG). If this bit is 1, $M = 4$, and if it is 0, $M = 16$.

1 Some UARTs on the PIC32 do not have hardware flow control lines, and the flow control pins of one UART may coincide with the RX and TX lines of another UART. For example, using UART3 with flow control prevents the use of UART6.

UxMODE⟨2:1⟩ or UxMODEbits.PDSEL: Determines the parity and number of data bits.

0b11 9 data bits, no parity.
0b10 8 data bits, odd parity.
0b01 8 data bits, even parity.
0b00 8 data bits, no parity.

UxMODE⟨0⟩ or UxMODEbits.STSEL: The number of stop bits. 0 = 1 stop bit, 1 = 2 stop bits.

UxSTA Contains the status of the UART: error flags and busy status. Controls the conditions under which interrupts occur. Also allows the user to turn the transmitter or receiver on and off.

UxSTA⟨15:14⟩ or UxSTAbits.UTXISEL: Determines when to generate a TX interrupt. The PIC32 can hold eight bytes in its TX FIFO. Interrupts will continue to happen until the condition causing the interrupt ends.

0b10 Interrupt while TX FIFO is empty.
0b01 Interrupt after everything in the TX FIFO has been transmitted.
0b00 Interrupt whenever the TX FIFO is not full.

UxSTA⟨12⟩ or UxSTAbits.URXEN: When set, enables the UART's RX pin.

UxSTA⟨10⟩ or UxSTAbits.UTXEN: When set, enables the UART's TX pin.

UxSTA⟨9⟩ or UxSTAbits.UTXBF: When set, indicates that the transmit buffer is full. If you attempt to write to the UART when the buffer is full the data will be ignored.

UxSTA⟨8⟩ or UxSTAbits.TRMT: When clear, indicates that there is no pending transmission or data in the TX buffer.

UxSTA⟨7:6⟩ or UxSTAbits.URXISEL: Determines when UART receive interrupts are generated. The PIC32 can hold eight bytes in its RX FIFO. The interrupt will continue to happen until the condition causing the interrupt is cleared.

0b10 Interrupt whenever the RX FIFO contains six or more characters.
0b01 Interrupt whenever the RX FIFO contains four or more characters.
0b00 Interrupt whenever the RX FIFO contains at least one character.

UxSTA⟨3⟩ or UxSTAbits.PERR: Set when the parity of the received data is incorrect. For even (odd) parity the UART expects the total number of received ones (including the parity bit) to be even (odd). If not using a parity bit, then there can be no parity error, but you also lose the data integrity check that parity provides.

UxSTA⟨2⟩ or UxSTAbits.FERR: Set when a framing error occurs. A framing error happens when the UART does not detect the stop bit. This often occurs if there is a baud mismatch.

UxSTA⟨1⟩ or UxSTAbits.OERR: Set when the receive buffer is full but the UART is sent another byte. When this bit is set the UART cannot receive data; therefore, if an overrun occurs you must manually clear this bit to continue receiving data. Clearing this bit flushes the data in the receive buffer, so you may want to read the bytes in the receive buffer prior to clearing.

UxSTA⟨0⟩ or UxSTAbits.URXDA: When set, indicates that the receive buffer contains data.

UxTXREG Use this SFR to transmit data. Writing to UxTXREG places the data in an eight-byte long hardware FIFO. The transmitter removes data from the FIFO and loads it into an internal shift register, UxTSR, where the data is shifted out onto the TX line, bit by bit. Once done shifting, hardware removes the next byte and begins transmitting it.

UxRXREG Use this SFR to receive data. Hardware shifts received data bit by bit into an internal RX shift register. After receiving a full byte, hardware transfers it from the shift register into the RX FIFO. Reading from UxRXREG removes a byte from the RX FIFO. If you do not read from UxRXREG often enough, the RX FIFO may overrun. If the FIFO is full, subsequent received bytes are discarded and an overrun error status flag is set.

UxBRG Controls the baud. The value for this register should be set to achieve the desired baud B according to the following equation:

$$\text{UxBRG} = \frac{F_{PB}}{M \times B} - 1 \tag{11.1}$$

where F_{PB} is the peripheral bus frequency, and either $M = 4$ if UxMODE.BRGH = 1 or $M = 16$ if UxMODE.BRGH = 0.

Interrupt vector numbers for the UARTs are named `_UART_x_VECTOR`, where `x` is 1 to 6. The interrupt flag status bits for UART1 are IFS0bits.U1EIF (error interrupt generated by a parity error, framing error, or overrun error), IFS0bits.U1RXIF (RX interrupt), and IFS0bits.U1TXIF (TX interrupt). The interrupt enable control bits for UART1 are IEC0bits.U1EIE (error interrupt enable), IEC0bits.U1RXIE (RX interrupt enable), and IEC0bits.U1TXIE (TX interrupt enable). The priority and subpriority bits are IPC6bits.U1IP and IPC6bits.U1IS. Interrupt flag status bits, enable control bits, and priority bits for UART2 are named similarly (replacing "U1" with "U2") and are in IFS1, IEC1, and IPC8; for UART3 they are in IFS1, IEC1, and IPC7; and for UART4 to UART6 they are in IFS2, IEC2, and IPC12.

11.3 Sample Code

11.3.1 Loopback

In our first example, the PIC32 uses UART1 to talk to itself. Connect U1RX (RD2) to U1TX (RD3). The program uses the NU32 library and UART3 to prompt the user for a single byte, sends it twice from U1TX to U1RX, and reports the byte that was read on U1RX. We set the baud to an extremely low rate (100) so that you can easily see the transmission on an oscilloscope. If you set the oscilloscope into single capture mode and trigger on the falling edge, the scope will capture the signal from the beginning of the transmission, when the first start bit is sent. Sending the byte twice allows you to verify the stop bits.

Code Sample 11.1 `uart_loop.c`. **UART Code that Talks to Itself.**

```
#include "NU32.h"                  // constants, functions for startup and UART

// We will set up UART1 at a slow baud rate so you can examine the signal on a scope.
// Connect the UART1 RX and TX pins together so the UART can communicate with itself.

int main(void) {
  char msg[100] = {};
  NU32_Startup();  // cache on, interrupts on, LED/button init, UART init

  // initialize UART1:  100 baud, odd parity, 1 stop bit
  U1MODEbits.PDSEL = 0x2; // odd parity (parity bit set to make the number of 1's odd)
  U1STAbits.UTXEN = 1;     // enable transmit
  U1STAbits.URXEN = 1;     // enable receive

  // U1BRG = Fpb/(M * baud) - 1  (note U1MODEbits.BRGH = 0 by default, so M = 16)
  // setup for 100 baud.  This means 100 bits /sec or 1 bit/ 1/10ms
  U1BRG = 49999;       // 80 M/(16*100) - 1 = 49,999
  U1MODEbits.ON = 1; // turn on the uart

  // scope instructions: 10 ms/div, trigger on falling edge, single capture
  while(1) {
    unsigned char data = 0;
    NU32_WriteUART3("Enter hex byte (lowercase) to send to UART1 (i.e., 0xa1): ");
    NU32_ReadUART3(msg, sizeof(msg));
    sscanf(msg,"%2x",&data);
    sprintf(msg,"0x%02x\r\n",data);
    NU32_WriteUART3(msg); //echo back

    while(U1STAbits.UTXBF) {  // wait for UART to be ready to transmit
      ;
    }
    U1TXREG = data;            // write twice so we can see the stop bit
    U1TXREG = data;
    while(!U1STAbits.URXDA) { // poll to see if there is data to read in RX FIFO
      ;
    }
    data = U1RXREG;            // data has arrived; read the byte
    while(!U1STAbits.URXDA) { // wait until there is more data to read in RX FIFO
      ;
    }
    data = U1RXREG;            // overwriting data from previous read! could check if same
    sprintf(msg,"Read 0x%x from UART1\r\n",data);
    NU32_WriteUART3(msg);
  }
  return 0;
}
```

11.3.2 Interrupt Based

The next example demonstrates the use of interrupts. Interrupts can be generated based on the number of elements in the RX or TX buffers, or when an error has occurred. For example, you can interrupt when the RX buffer is half full or when the TX buffer is empty. The IRQs for these interrupts share the same vector; therefore, you must check within the ISR to see what event triggered it. You must also remove the condition that triggered the interrupt or it will trigger again after you exit the ISR.

The code below reads data from your terminal emulator and sends it back. It uses UART3, as does the NU32 library, but does not use the NU32 UART commands. An interrupt is triggered when the RX buffer contains at least one character, and the ISR immediately sends the data back to the terminal emulator.

Using interrupts for serial I/O allows the PIC32 to receive data from the serial port without wasting time polling for it.

Code Sample 11.2 `uart_int.c`**. UART Code that Uses Interrupts to Receive Data.**

```
#include "NU32.h"             // constants, functions for startup and UART

void __ISR(_UART_3_VECTOR, IPL1SOFT) IntUart1Handler(void) {
  if (IFS1bits.U3RXIF) {       // check if interrupt generated by a RX event
    U3TXREG = U3RXREG;         // send the received data out
    IFS1bits.U3RXIF = 0;       // clear the RX interrupt flag
  } else if(IFS1bits.U3TXIF) { // if it is a TX interrupt
  } else if(IFS1bits.U3EIF) {  // if it is an error interrupt. check U3STA for reason
  }
}

int main(void) {
  NU32_Startup();    // cache on, interrupts on, LED/button init, UART init
  NU32_LED1 = 1;
  NU32_LED2 = 1;
  __builtin_disable_interrupts();

  // set baud to 230400, to match terminal emulator; use default 8N1 of UART
  U3MODEbits.BRGH = 0;
  U3BRG = ((NU32_SYS_FREQ / 230400) / 16) - 1;

  // configure TX & RX pins
  U3STAbits.UTXEN = 1;
  U3STAbits.URXEN = 1;

  // configure using RTS and CTS
  U3MODEbits.UEN = 2;

  // configure the UART interrupts
  U3STAbits.URXISEL = 0x0; // RX interrupt when receive buffer not empty
  IFS1bits.U3RXIF = 0;     // clear the rx interrupt flag.  for
                           // tx or error interrupts you would also need to clear
                           // the respective flags
```

```
    IPC7bits.U3IP = 1;          // interrupt priority
    IEC1bits.U3RXIE = 1;        // enable the RX interrupt

    // turn on UART1
    U3MODEbits.ON = 1;
    __builtin_enable_interrupts();
    while(1) {
      :
    }
    return 0;
  }
```

11.3.3 NU32 Library

The NU32 library contains three functions that access the UART: `NU32_Setup`, `NU32_ReadUART3`, and `NU32_WriteUART3`. The setup code configures UART3 for a baud of 230400, one stop bit, 8 data bits, no parity bit, and hardware flow control. No UART interrupts are used. Notice that `NU32_ReadUART3` keeps reading from the UART until it receives a certain control character (`'\n'` or `'\r'`); thus it will wait indefinitely for input before proceeding.

The function `NU32_WriteUART3` waits for the TX FIFO to have available space before attempting to add more data to it. Also, since hardware flow control is enabled on the PIC32's UART, no data will be sent by the PIC32 unless the DTE (your computer) holds the $\overline{\text{CTS}}$ line low. The terminal emulator must have hardware flow control enabled to ensure correct operation.

Code Sample 11.3 `NU32.c`. **The NU32 Library Implementation.**

```
#include "NU32.h"

// Device Configuration Registers
// These only have an effect for standalone programs but don't harm bootloaded programs.
// the settings here are the same as those used by the bootloader
#pragma config DEBUG = OFF          // Background Debugger disabled
#pragma config FWDTEN = OFF         // WD timer: OFF
#pragma config WDTPS = PS4096       // WD period: 4.096 sec
#pragma config POSCMOD = HS         // Primary Oscillator Mode: High Speed crystal
#pragma config FNOSC = PRIPLL       // Oscillator Selection: Primary oscillator w/ PLL
#pragma config FPLLMUL = MUL_20     // PLL Multiplier: Multiply by 20
#pragma config FPLLIDIV = DIV_2     // PLL Input Divider:  Divide by 2
#pragma config FPLLODIV = DIV_1     // PLL Output Divider: Divide by 1
#pragma config FPBDIV = DIV_1       // Peripheral Bus Clock: Divide by 1
#pragma config UPLLEN = ON          // USB clock uses PLL
#pragma config UPLLIDIV = DIV_2     // Divide 8 MHz input by 2, mult by 12 for 48 MHz
#pragma config FUSBIDIO = ON        // USBID controlled by USB peripheral when it is on
#pragma config FVBUSONIO = ON       // VBUSON controlled by USB peripheral when it is on
#pragma config FSOSCEN = OFF        // Disable second osc to get pins back
#pragma config BWP = ON             // Boot flash write protect: ON
#pragma config ICESEL = ICS_PGx2    // ICE pins configured on PGx2
#pragma config FCANIO = OFF         // Use alternate CAN pins
#pragma config FMIIEN = OFF         // Use RMII (not MII) for ethernet
```

```
#pragma config FSRSSEL = PRIORITY_6 // Shadow Register Set for interrupt priority 6

#define NU32_DESIRED_BAUD 230400    // Baudrate for RS232

// Perform startup routines:
//  Make NU32_LED1 and NU32_LED2 pins outputs (NU32_USER is by default an input)
//  Initialize the serial port - UART3 (no interrupt)
//  Enable interrupts
void NU32_Startup() {
  // disable interrupts
  __builtin_disable_interrupts();

  // enable the cache
  // This command sets the CP0 CONFIG register
  // the lower 4 bits can be either 0b0011 (0x3) or 0b0010 (0x2)
  // to indicate that kseg0 is cacheable (0x3) or uncacheable (0x2)
  // see Chapter 2 "CPU for Devices with M4K Core" of the PIC32 reference manual
  // most of the other bits have prescribed values
  // microchip does not provide a _CP0_SET_CONFIG macro, so we directly use
  // the compiler built-in command _mtc0
  // to disable cache, use 0xa4210582
  __builtin_mtc0(_CP0_CONFIG, _CP0_CONFIG_SELECT, 0xa4210583);

  // set the prefectch cache wait state to 2, as per the
  // electrical characteristics data sheet
  CHECONbits.PFMWS = 0x2;

  //enable prefetch for cacheable and noncacheable memory
  CHECONbits.PREFEN = 0x3;

  // 0 data RAM access wait states
  BMXCONbits.BMXWSDRM = 0x0;

  // enable multi vector interrupts
  INTCONbits.MVEC = 0x1;

  // disable JTAG to get B10, B11, B12 and B13 back
  DDPCONbits.JTAGEN = 0;

  TRISFCLR = 0x0003;  // Make F0 and F1 outputs (LED1 and LED2)
  NU32_LED1 = 1;      // LED1 is off
  NU32_LED2 = 0;      // LED2 is on

  // turn on UART3 without an interrupt
  U3MODEbits.BRGH = 0; // set baud to NU32_DESIRED_BAUD
  U3BRG = ((NU32_SYS_FREQ / NU32_DESIRED_BAUD) / 16) - 1;

  // 8 bit, no parity bit, and 1 stop bit (8N1 setup)
  U3MODEbits.PDSEL = 0;
  U3MODEbits.STSEL = 0;

  // configure TX & RX pins as output & input pins
  U3STAbits.UTXEN = 1;
  U3STAbits.URXEN = 1;
  // configure hardware flow control using RTS and CTS
  U3MODEbits.UEN = 2;

  // enable the uart
  U3MODEbits.ON = 1;

  __builtin_enable_interrupts();
```

```
}

// Read from UART3
// block other functions until you get a '\r' or '\n'
// send the pointer to your char array and the number of elements in the array
void NU32_ReadUART3(char * message, int maxLength) {
  char data = 0;
  int complete = 0, num_bytes = 0;
  // loop until you get a '\r' or '\n'
  while (!complete) {
    if (U3STAbits.URXDA) { // if data is available
      data = U3RXREG;      // read the data
      if ((data == '\n') || (data == '\r')) {
        complete = 1;
      } else {
        message[num_bytes] = data;
        ++num_bytes;
        // roll over if the array is too small
        if (num_bytes >= maxLength) {
          num_bytes = 0;
        }
      }
    }
  }
  // end the string
  message[num_bytes] = '\0';
}

// Write a character array using UART3
void NU32_WriteUART3(const char * string) {
  while (*string != '\0') {
    while (U3STAbits.UTXBF) {
      ; // wait until tx buffer isn't full
    }
    U3TXREG = *string;
    ++string;
  }
}
```

11.3.4 Sending Data from an ISR

It is often desirable to stream data collected by the PIC32 to your computer. For example, a fixed-frequency ISR could sample data from a sensor and then send it back to your computer for plotting. The ISR may collect samples at a much higher rate than they can be sent over the UART, however. In this case, some of the data has to be discarded. The process of keeping only one piece of data per every N collected (discarding $N - 1$) is called *decimation*.[2] Decimation is common in signal processing.

In addition to decimation, another concept that enables streaming data from an ISR is that of a *circular buffer*. A circular (or ring) buffer is an implementation of a FIFO, such as the UART's

[2] The origin of this term is a disciplinary practice of the Roman Army, whereby one out of every ten soldiers who had performed disgracefully was killed.

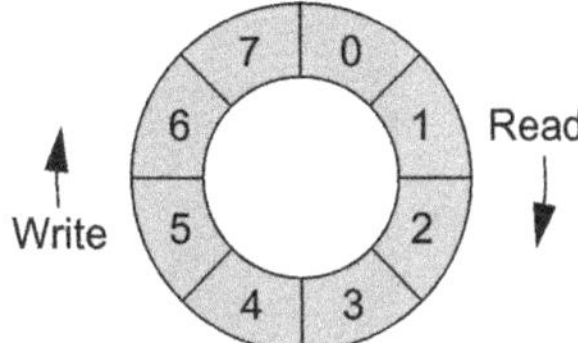

Figure 11.2
An eight-element circular buffer (FIFO), where the `write` index currently points to element 5 and the `read` index currently points to element 1.

TX and RX FIFOs. A circular buffer is implemented as an array and two index variables: `write` and `read` (see Figure 11.2). Data is added to the array at the `write` location and read from the `read` location, after which the indexes are incremented. When the indexes reach the end of the array, they wrap around to the beginning. If the `read` and `write` indexes are equal, the buffer is empty. If the `write` index is one slot behind the `read` index, the buffer is full.

Circular buffers are useful for sharing data between interrupts and mainline code. The ISR can write data to the buffer while the mainline code reads from the buffer and sends the data over the UART.

Code Sample 11.4 demonstrates the concept of decimation and Code Sample 11.5 demonstrates the concept of a circular buffer. Both programs send 5000 data samples from the PIC32 to the host computer. Code Sample 11.4 is titled `batch.c` because all decimated data is first stored in an array in RAM, then sent over the UART in one batch. Code Sample 11.5 is called `circ_buf.c` because data is streamed using a circular buffer. The use of a circular buffer (a) allows the buffer to use less RAM than the array in `batch.c` and (b) allows data to be sent immediately, not just in a batch after it is all collected.

Code Sample 11.4 `batch.c`**. Storing Data in an ISR and Sending it in a Batch Over the UART.**

```
#include "NU32.h"                // constants, functions for startup and UART

#define DECIMATE 3               // only send every 4th sample (counting starts at zero)
#define NSAMPLES 5000            // store 5000 samples

volatile int data_buf[NSAMPLES];// stores the samples
volatile int curr = 0;           // the current index into buffer

void __ISR(_TIMER_1_VECTOR, IPL5SOFT) Timer1ISR(void) {  // Timer1 ISR operates at 5 kHz
  static int count = 0;          // counter used for decimation
  static int i = 0;              // the data returned from the isr
  ++i;                           // generate the data (we just increment it for now)
  if(count == DECIMATE) {        // skip some data
    count = 0;
    if(curr < NSAMPLES) {
      data_buf[curr] = i;        // queue a number for sending over the UART
```

```
      ++curr;
    }
  }
  ++count;
  IFS0bits.T1IF = 0;              // clear interrupt flag
}

int main(void) {
  int i = 0;
  char buffer[100] = {};
  NU32_Startup();                  // cache on, interrupts on, LED/button init, UART init

  __builtin_disable_interrupts();// INT step 2: disable interrupts at CPU
  T1CONbits.TCKPS = 0b01;          // PBCLK prescaler value of 1:8
  PR1 = 1999;                      // The frequency is 80 MHz / (8 * (1999 + 1)) = 5 kHz
  TMR1 = 0;
  IPC1bits.T1IP = 5;               // interrupt priority 5
  IFS0bits.T1IF = 0;               // clear the interrupt flag
  IEC0bits.T1IE = 1;               // enable the interrupt
  T1CONbits.ON  = 1;               // turn the timer on
  __builtin_enable_interrupts(); // INT step 7: enable interrupts at CPU

  NU32_ReadUART3(buffer, sizeof(buffer)); // wait for the user to press enter
  while(curr !=NSAMPLES) { ; }            // wait for the data to be collected

  sprintf(buffer,"%d\r\n",NSAMPLES);      // send the number of samples that will be sent
  NU32_WriteUART3(buffer);

  for(i = 0; i < NSAMPLES; ++i) {
    sprintf(buffer,"%d\r\n",data_buf[i]);  // send the data to the terminal
    NU32_WriteUART3(buffer);
  }
  return 0;
}
```

Code Sample 11.5 `circ_buf.c`. **Streaming Data from an ISR Over the UART, Using a Circular Buffer.**

```
#include "NU32.h" // constants, functions for startup and UART
// uses a circular buffer to stream data from an ISR over the UART
// notice that the buffer can be much smaller than the total number of samples sent and
// that data starts streaming immediately unlike with batch.c

#define BUFLEN   1024                          // length of the buffer
#define NSAMPLES 5000                          // number of samples to collect

static volatile int data_buf[BUFLEN];          // array that stores the data
static volatile unsigned int read = 0, write = 0; // circular buf indexes
static volatile int start = 0;                 // set to start recording

int buffer_empty() {    // return true if the buffer is empty (read = write)
  return read == write;
}

int buffer_full() {     // return true if the buffer is full.
  return (write + 1) % BUFLEN == read;
}
```

```
int buffer_read() {      // reads from current buffer location; assumes buffer not empty
  int val = data_buf[read];
  ++read;                // increments read index
  if(read >= BUFLEN) {   // wraps the read index around if necessary
    read = 0;
  }
  return val;
}

void buffer_write(int data) { // add an element to the buffer.
  if(!buffer_full()) {        // if the buffer is full the data is lost
    data_buf[write] = data;
    ++write;                  // increment the write index and wrap around if necessary
    if(write >= BUFLEN) {
      write = 0;
    }
  }
}

void __ISR(_TIMER_1_VECTOR, IPL5SOFT) Timer1ISR(void) {  // timer 1 isr operates at 5 kHz
  static int i = 0;       // the data returned from the isr
  if(start) {
    buffer_write(i);      // add the data to the buffer
    ++i;                  // modify the data (here we just increment it as an example)
  }
  IFS0bits.T1IF = 0;      // clear interrupt flag
}

int main(void) {
  int sent = 0;
  char msg[100] = {};
  NU32_Startup();                     // cache on, interrupts on, LED/button init, UART init

  __builtin_disable_interrupts();   // INT step 2: disable interrupts at CPU
  T1CONbits.TCKPS = 0b01;           // PBCLK prescaler value of 1:8
  PR1 = 1999;                     // The frequency is 80 MHz / (8 * (1999 + 1)) = 5 kHz
  TMR1 = 0;
  IPC1bits.T1IP = 5;              // interrupt priority 5
  IFS0bits.T1IF = 0;              // clear the interrupt flag
  IEC0bits.T1IE = 1;              // enable the interrupt
  T1CONbits.ON  = 1;              // turn the timer on
  __builtin_enable_interrupts();    // INT step 7: enable interrupts at CPU

  NU32_ReadUART3(msg,sizeof(msg));  // wait for the user to press enter before continuing
  sprintf(msg, "%d\r\n", NSAMPLES); // tell the client how many samples to expect
  NU32_WriteUART3(msg);
  start = 1;
  for(sent = 0; sent < NSAMPLES; ++sent) { // send the samples to the client
   while(buffer_empty()) { ; }             // wait for data to be in the queue
    sprintf(msg,"%d\r\n", buffer_read());  // read from the buffer, send data over uart
    NU32_WriteUART3(msg);
  }

  while(1) {
    ;
  }
  return 0;
}
```

In circ_buf.c, if the circular buffer is full, data is lost. While circ_buf.c is written to send a fixed number of samples, it can be easily modified to stream samples indefinitely. If the UART baud is sufficiently higher than the rate at which data bits are generated in the ISR, lossless data streaming can be performed indefinitely. The circular buffer simply provides some cushion in cases where communication is temporarily delayed or disrupted.

11.3.5 Communication with MATLAB

So far, when we have used the UART to communicate with a computer, we have opened the serial port in a terminal emulator. MATLAB can also open serial ports, allowing communication with and plotting of data from the PIC32. As a first example, we will communicate with talkingPIC.c from MATLAB.

First, load talkingPIC.c onto the PIC32 (see Chapter 1 for the code). Next, open MATLAB and edit talkingPIC.m. You will need to edit the first line and set the port to be the PORT value from your Makefile.

Code Sample 11.6 talkingPIC.m. **Simple MATLAB Code to Talk to** talkingPIC **on the PIC32.**

```
port='COM3'; % Edit this with the correct name of your PORT.

% Makes sure port is closed
if ~isempty(instrfind)
    fclose(instrfind);
    delete(instrfind);
end
fprintf('Opening port %s....\n',port);

% Defining serial variable
mySerial = serial(port, 'BaudRate', 230400, 'FlowControl', 'hardware');

% Opening serial connection
fopen(mySerial);

% Writing some data to the serial port
fprintf(mySerial,'%f %d %d\n',[1.0,1,2])

% Reading the echo from the PIC32 to verify correct communication
data_read = fscanf(mySerial,'%f %d %d')

% Closing serial connection
fclose(mySerial)
```

The code talkingPIC.m opens a serial port, sends three numerical values to the PIC32, receives the values, and closes the port. Run talkingPIC.c on your PIC32, then execute talkingPIC.m in MATLAB.

We can also combine MATLAB with batch.c or circ_buf.c, allowing us to plot data received from an ISR. The example below reads the data produced by batch.c or circ_buf.c and plots it in MATLAB. Once again, change the port variable to match the serial port that your PIC32 uses.

Code Sample 11.7 uart_plot.m. **MATLAB Code to Plot Data Received from the UART.**

```
% plot streaming data in matlab
port ='/dev/ttyUSB0'

if ~isempty(instrfind)  % closes the port if it was open
  fclose(instrfind);
  delete(instrfind);
end

mySerial = serial(port, 'BaudRate', 230400, 'FlowControl','hardware');
fopen(mySerial);

fprintf(mySerial,'%s','\n'); %send a newline to tell the PIC32 to send data

len = fscanf(mySerial,'%d'); % get the length of the matrix

data = zeros(len,1);

for i = 1:len
  data(i) = fscanf(mySerial,'%d'); % read each item
end

plot(1:len,data);                  % plot the data
```

11.3.6 Communication with Python

You can also communicate with the PIC32 from the Python programming language. This freely available scripting language has many libraries available that help it be used as an alternative to MATLAB. To communicate over the serial port you need the pyserial library. For plotting, we use the libraries matplotlib and numpy. The following code reads data from the PIC32 and plots it. As with the MATLAB code, you need to specify your own port where the port variable is defined. This code will plot data generated by either batch.c or circ_buf.c.

Code Sample 11.8 uart_plot.py. **Python Code to Plot Data Received from the UART.**

```
#!/usr/bin/python
# Plot data from the PIC32 in python
# requries pyserial, matplotlib, and numpy
import serial
import matplotlib.pyplot as plt
```

```
import numpy as np

port = '/dev/ttyUSB0' # the name of the serial port

with serial.Serial(port,230400,rtscts=1) as ser:
  ser.write("\n".encode()) #tell the pic to send data. encode converts to a byte array
  line = ser.readline()
  nsamples = int(line)
  x = np.arange(0,nsamples) # x is [1,2,3,... nsamples]
  y = np.zeros(nsamples)# x is 1 x nsamples an array of zeros and will store the data

  for i in range(nsamples): # read each sample
    line = ser.readline()   # read a line from the serial port
    y[i] = int(line)        # parse the line (in this case it is just one integer)

plt.plot(x,y)
plt.show()
```

11.4 Wireless Communication with an XBee Radio

XBee radios are small, low-power radio transmitters that allow wireless communication over tens of meters, according to the IEEE 802.15.4 standard (Figure 11.3). Each of the two communicating devices connect to an XBee through a UART, and then they can communicate wirelessly as if their UARTs were wired together. For example, two PIC32s could talk to each other using their UART3s using the NU32 library.

XBee radios have numerous firmware settings that must be configured before you use them. The main setting is the wireless channel; two XBees cannot communicate unless they use the same channel. Another setting is the baud, which can be set as high as 115,200. The easiest way to configure an XBee is to purchase a development board, connect it to your computer, and use the X-CTU program provided by the manufacturer. Alternatively, you can program XBees directly using the API mode over a serial port (either from your computer or the PIC32).

After setting up the XBees to use the desired baud and communication channel, you can use them as drop-in replacements, one at each UART, for the wires that would usually connect them.[3]

[3] One caveat occurs at higher baud rates. The XBee generates its baud by dividing an internal clock signal. This clock does not actually achieve a baud of 115,200, however; when set to 115,200 the baud actually is 111,000. Such baud rate mismatches are a common issue when using UARTs. Due to the tolerances of UART timing, the XBee may work for a time but occasionally experience failures. The solution is to set your baud to 111,000 to match the actual baud of the XBee.

Figure 11.3
An XBee 802.15.4 radio. (Image courtesy of Digi International, digi.com.)

11.5 Chapter Summary

- A UART is the low-level engine underlying serial communication. Once ubiquitous, serial ports have been largely replaced by USB on consumer products.
- The NU32 board uses a UART to communicate with your PC. Software on your computer emulates a serial port, which transfers data via USB to a chip on the NU32 board. This chip then converts the USB data into a format suitable for the UART. Neither your terminal nor the PIC32 know that data is actually sent over USB, they just see a UART.
- The PIC32 maintains two eight-byte hardware FIFOs, one for receiving, one for sending. These FIFOs buffer data in hardware, allowing software to temporarily attend to other tasks while the buffers are being transmitted or filled by received data.
- Both the PIC32 and the DTE must agree upon a common communication speed (baud) and data format; otherwise data will not be interpreted properly.
- Hardware flow control provides a method to signal that your device is not ready to receive more data. Although hardware handles flow control automatically, software must ensure that the RX and TX buffers do not overrun.

11.6 Exercises

1. Plot the waveform for a UART sending the byte 0b11011001, assuming 9600 baud, no parity, 8 data bits, and one stop bit.

2. Write a program that reads characters that you type in your terminal emulator (via UART3), capitalizes them, and returns them to your computer. Rather than processing each character one line at a time, you want a result after each character is pressed; therefore, you cannot use `NU32_WriteUART3` or `NU32_ReadUART3`. For example, if you type 'x' in the terminal emulator, you should see 'X'. You can use the C standard library function `toupper`, defined in `ctype.h`, to convert characters to upper case.

Further Reading

PIC32 family reference manual. Section 21: UART. (2012). Microchip Technology Inc.
XBee/XBee-PRO RF modules (v1.xEx). (2009). Digi International.

CHAPTER 12

SPI Communication

The Serial Peripheral Interface (SPI) allows the PIC32 to communicate with other devices at high speeds. Numerous devices use SPI as their primary mode of communication, such as RAM, flash, SD cards, ADCs, DACs, and accelerometers. Unlike the UART, SPI communication is synchronous: the "master" device on an SPI bus creates a separate clock signal that dictates the timing of communication. The devices do not have to be configured in advance to share the same bit rate, and any clock frequency can be used, provided it is within the capabilities of the chips. High speeds are possible; for example, the SPI interface of the STMicroelectronics LSM303D accelerometer/magnetometer, a device considered in this chapter, supports up to 10 MHz clock signals.

12.1 Overview

SPI is a master-slave architecture, and an SPI bus has one master device and one or more slaves. A minimal SPI bus consists of three wires (in addition to GND): Master Output Slave Input (MOSI), carrying data from the master to the slave(s); Master Input Slave Output (MISO), carrying data from the slave(s) to the master; and the master's clock output, which clocks the data transfers, one bit per clock pulse. Each SPI device on the bus correspondingly has three pins: Serial Data Out (SDO), Serial Data In (SDI), and System Clock (SCK). The MOSI line is connected to the master's SDO pin and the slaves' SDI pins, and the MISO line is connected to the master's SDI pin and the slaves' SDO pins. All slaves' SCK pins are inputs, connected to the master's SCK output (Figure 12.1).

Each of the PIC32's three SPI peripherals can either be a master or a slave. We typically think of the PIC32 acting as the master.

If there is more than one slave on the bus, then the master controls which slave is active by using an active-low slave select ($\overline{\mathrm{SS}}$) line, one per slave. Only one slave-select line can have a low signal at a time, and the line which is low indicates which slave is active. Unselected slaves must let their SDO output float at high impedance, effectively disconnected from the MISO line, and they should ignore data on the MOSI line. In Figure 12.1, there are two slaves, and therefore two output slave-select lines from the master $\overline{\mathrm{SS1}}$ and $\overline{\mathrm{SS2}}$, and one slave-select input line for each slave. Thus, adding more slaves to the bus means adding more wires.

Embedded Computing and Mechatronics with the PIC32 Microcontroller. http://dx.doi.org/10.1016/B978-0-12-420165-1.00012-3

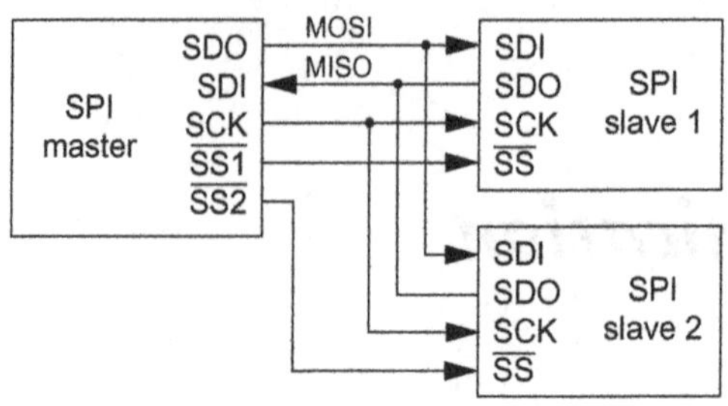

Figure 12.1
An SPI master connected to two slave devices. Arrows indicate data direction. A PIC32 SPI peripheral can either be a master or a slave. Only one slave select line can be active (low) at a time.

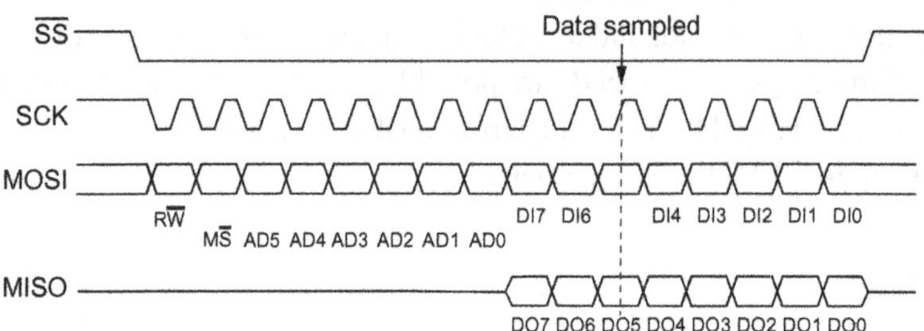

Figure 12.2
Timing for an SPI transaction with a slave LSM303D accelerometer/magnetometer. The LSM303D is selected when $\overline{SS}$ is driven low. On the falling edge of the master's SCK, both the master (controlling MOSI with its SDO pin) and the slave (controlling MISO with its SDO pin) transition to the next bit of their signal. On the rising edge of SCK, the devices read the data, the master reading MISO with its SDI pin and the slave reading MOSI with its SDI pin. In this example, the master first sends 8 bits, the first of which (R$\overline{W}$) determines whether it will be reading data from the LSM303D or writing data to it. The bits AD0 to AD5 determine the address of the LSM303D register that the master is accessing. Finally, if the operation is a read from the LSM303D, the LSM303D puts 8 bits DO0 to DO7 on the MISO line; otherwise the master sends 8 bits on the MOSI line to write to the LSM303D.

Compare this to the (typically lower speed) I^2C (Chapter 13) and CAN (Chapter 19) buses, which have a fixed number of wires regardless of the number of devices on them.

The master initiates all communication by creating a square wave on the clock line. Reading from and writing to the slave occur simultaneously, one bit per clock pulse. Transfers occur in groups of 8, 16, or 32 bits, depending on the slave. Figure 12.2, taken from the LSM303D data sheet, depicts the signals for a typical SPI transaction. The timing of the data bits relative to the clock signal are settable using SFRs to match the behavior of other devices on the SPI bus; see Section 12.2 for more information.

In addition to the basic SPI communication described above, the PIC32 SPI module has other modes of operations, which we do not cover in detail here. For example, a framing mode allows for streaming data from supported devices. By default, the SPI peripheral has only a single input and single output buffer. The PIC32 also has an enhanced buffer mode, which provides FIFOs for queuing data waiting to be transferred or processed.

12.2 Details

Each of the three SPI peripherals, SPI2 to SPI4, has four pins associated with it: SCKx, SDIx, SDOx, and $\overline{\text{SSx}}$, where x is 2 to 4. When the SPI peripheral is a slave, it can be configured so that it only receives and transmits data when the input $\overline{\text{SSx}}$ is low (e.g., when there is more than one slave on the bus). When the SPI peripheral is a master, it can be configured to drive the $\overline{\text{SSx}}$ automatically when communicating with a single slave. If there are multiple slaves on the bus, however, other digital outputs can be used to control the multiple slave select pins of the slaves.

Each SPI peripheral uses four SFRs, SPIxCON, SPIxSTAT, SPIxBUF, and SPIxBRG, x = 2 to 4. Many of the settings are related to the alternative operation modes that we do not discuss. SFRs and fields that we omit may be safely left at their default values for typical applications. All default bits are zero, except for one read-only bit in SPIxSTAT.

SPIxCON This register contains the main control options for the SPI peripheral.

- SPIxCON⟨28⟩ or SPIxCONbits.MSSEN: Master slave select enable. If set, the SPI master will assert (drive low) the slave select pin $\overline{\text{SSx}}$ prior to sending an 8-, 16-, or 32-bit transmission and de-assert (drive high) after sending the data. Some devices require you to toggle the slave select after a complete multi-word transaction; in this case, you should clear SPIxCONbits.MSSEN to zero (the default) and use any digital output as the slave select.
- SPIxCON⟨15⟩ or SPIxCONbits.ON: Set to enable the SPI peripheral.
- SPIxCON⟨11:10⟩ or SPIxCONbits.MODE32 (bit 11) and SPIxCONbits.MODE16 (bit 10): Determines the communication width.
  - 0b1X (SPIxCONbits.MODE32 = 1): 32 bits of data sent per transfer.
  - 0b01 (SPIxCONbits.MODE32 = 0 and SPIxCONbits.MODE16 = 1): 16 bits of data sent per transfer.
  - 0b00 (SPIxCONbits.MODE32 = SPIxCONbits.MODE16 = 0): 8 bits of data sent per transfer.
- SPIxCON⟨9⟩ or SPIxCONbits.SMP: determines when, relative to the clock pulses, the master samples input data. Should be set to match the slave device's specifications.
  - 1 Sample at end of a clock pulse.
  - 0 Sample in the middle of a clock pulse.

SPIxCON⟨8⟩ or SPIxCONbits.CKE: The clock signal can be configured as either active high or active low by setting SPIxCONbits.CKP (see below). This bit, SPIxCONbits.CKE, determines whether the master changes the current output bit on the edge when the clock transitions from active to idle or idle to active (see SPIxCONbits.CKP, below). You should choose this bit based on what the slave device expects.

1 The output data changes when the clock transitions from active to idle.

0 The output data changes when the clock transitions from idle to active.

SPIxCON⟨7⟩ or SPIxCONbits.SSEN: This slave select enable bit determines whether the $\overline{\text{SSx}}$ pin is used in slave mode.

1 The $\overline{\text{SSx}}$ pin must be low for this slave to be selected on the SPI bus.

0 The $\overline{\text{SSx}}$ pin is not used, and is available to be used with another peripheral.

SPIxCON⟨6⟩ or SPIxCONbits.CKP: The clock signal can be configured as being active high or active low. Chosen in conjunction with SPIxCONbits.CKE, above, this setting should match the expectations of the slave device.

1 The clock is idle when high, active when low.

0 The clock is idle when low, active when high.

SPIxCON⟨5⟩ or SPIxCONbits.MSTEN: Master enable. Usually the PIC32 operates as the master, meaning that it controls the clock and hence when and how fast data is transferred, in which case this bit should be set to 1. To use the SPI peripheral as a slave, this bit should be cleared to 0.

1 The SPI peripheral is the master.

0 The SPI peripheral is the slave.

SPIxSTAT The status of the SPI peripheral.

SPIxSTAT⟨11⟩ or SPIxSTATbits.SPIBUSY: When set, indicates that the SPI peripheral is busy transferring data. You should not access the SPI buffer SPIxBUF when the peripheral is busy.

SPIxSTAT⟨6⟩ or SPIxSTATbits.SPIROV: Set to indicate that an overflow has occurred, which happens when the receive buffer is full and another data word is received. This bit should be cleared in software. The SPI peripheral can only receive data when this bit is clear.

SPIxSTAT⟨1⟩ or SPIxSTATbits.SPITXBF: SPI transmit buffer full. Set by hardware when you write to SPIxBUF. Cleared by hardware after the data you wrote is transferred into the transmit buffer SPIxTXB, a non-memory-mapped buffer. When this bit is clear you can write to SPIxBUF.

SPIxSTAT⟨0⟩ or SPIxSTATbits.SPIRXBF: SPI receive buffer full. Set by hardware when data is received into the SPI receive buffer indicating that SPIxBUF can be read. Cleared when you read the data via SPIxBUF.

SPIxBUF Used to both read and write data over SPI. When, as the master, you write to SPIxBUF, the data is actually stored in a transmit buffer SPIxTXB, and the SPI peripheral generates a clock signal and sends the data over the SDOx pin. Meanwhile, in response to the clock signal, the slave sends data to the SDIx pin, where it is stored in a receive buffer SPIxRXB, which you do not have direct access to. To access this received data, you do a read from SPIxBUF. Therefore, perhaps unintuitively, after executing the C code

```
SPI1BUF = data1;
data2 = SPI1BUF;
```

`data1` and `data2` will not be identical! `data1` is sent data, and `data2` is received data. To avoid buffer overflow errors, every time you write to SPIxBUF you should also read from SPIxBUF, even if you do not need the data. Additionally, since the slave can only send data when it receives a clock signal, which is only generated by the master when you write data, as a master you must write to SPIxBUF before getting new data from the slave.

SPIxBRG Determines the SPI clock frequency. Only the lowest 12 bits are used. To calculate the appropriate value for SPIxBRG use the tables provided in the Reference Manual or the following formula:

$$\text{SPIxBRG} = \frac{F_{PB}}{2F_{sck}} - 1, \tag{12.1}$$

where F_{PB} is the peripheral bus clock frequency (80 MHz for the NU32 board) and F_{sck} is the desired clock frequency. SPI can operate at relatively high frequencies, in the MHz range. The master dictates the clock frequency and the slave reads the clock; therefore a slave device never needs to configure a clock frequency.

The interrupt vector for SPIx is `_SPI_x_VECTOR`, where x is 2 to 4. An interrupt can be generated by an SPI fault, SPI RX conditions, and SPI TX conditions. For SPI2, the interrupt flag status bits are IFS1bits.SPI2EIF (error), IFS1bits.SPI2RXIF (RX), and IFS1bits.SPI2TXIF (TX); the enable control bits are IEC1bits.SPI2EIE (error), IEC1bits.SPI2RXIE (RX), and IEC1bits.SPI2TXIE (TX); and the priority and subpriority bits are IPC7bits.SPI2IP and IPC7bits.SPI2IS. For SPI3 and SPI4, the bits are named similarly, replacing SPI2 with SPIx (x = 3 or 4), and with SPI3's flag status bits in IFS0, enable control bits in IEC0, and priority in IPC6; and SPI4's in IFS1, IEC1, and IPC8.

The sample code in this chapter does not include interrupts, but TX and RX interrupt conditions can be selected using SPIxCONbits.STXISEL and SPIxCONbits.SRXISEL. See the Reference Manual.

12.3 Sample Code

12.3.1 Loopback

The first example uses two SPI peripherals to allow the PIC32 to communicate with itself over SPI. Although practically useless, the code serves as vehicle for understanding the basic configuration of both an SPI master and an SPI slave. Both master (SPI4) and slave (SPI3) are configured to send 16 bits per transfer. Notice that the SPI master sets a clock frequency, whereas the slave does not. The program prompts the user to enter two 16-bit hexadecimal numbers. The first number is sent by the master and the second by the slave. The code then reads from the SPI ports and prints the results.

Remember, one bit of data is transferred in each direction per clock cycle. When the slave writes to its SPI buffer, the data is not actually sent until the master generates a clock signal. Both master and slave use the clock to send and receive the data. As sending and receiving happen simultaneously, the master should always read from SPI4BUF after writing to SPI4BUF.

Code Sample 12.1 `spi_loop.c`. **SPI Loopback Example.**

```
#include "NU32.h"  // constants, funcs for startup and UART
// Demonstrates spi by using two spi peripherals on the same PIC32,
// one is the master, the other is the slave
// SPI4 will be the master, SPI3 the slave.
// connect
// SDO4 -> SDI3 (pin F5 -> pin D2)
// SDI4 -> SDO3 (pin F4 -> pin D3)
// SCK4 -> SCK3 (pin B14 -> pin D1)

int main(void) {
  char buf[100] = {};
  // setup NU32 LED's and buttons
  NU32_Startup();

  // Master - SPI4, pins are: SDI4(F4), SDO4(F5), SCK4(B14), SS4(B8); not connected)
  // since the pic is just starting, we know that SPI is off. We rely on defaults here
  SPI4BUF;                   // clear the rx buffer by reading from it
  SPI4BRG = 0x4;             // baud rate to 8 MHz [SPI4BRG = (80000000/(2*desired))-1]
  SPI4STATbits.SPIROV = 0;  // clear the overflow bit
  SPI4CONbits.MODE32 = 0;   // use 16 bit mode
  SPI4CONbits.MODE16 = 1;
  SPI4CONbits.MSTEN = 1;    // master operation
  SPI4CONbits.ON = 1;       // turn on spi 4

  // Slave - SPI3, pins are: SDI3(D2), SDO3(D3), SCK3(D1), SS3(D9; not connected)
  SPI3BUF;                   // clear the rx buffer
  SPI3STATbits.SPIROV = 0;  // clear the overflow
  SPI3CONbits.MODE32 = 0;   // use 16 bit mode
  SPI3CONbits.MODE16 = 1;
  SPI3CONbits.MSTEN = 0;    // slave mode
  SPI3CONbits.ON = 1;       // turn spi on.  Note: in slave mode you do not set baud

  while(1) {
    unsigned short master = 0, slave = 0;
    unsigned short rmaster = 0, rslave = 0;
    NU32_WriteUART3("Enter two 16-bit hex words (lowercase) to send from ");
    NU32_WriteUART3("master and slave (i.e., 0xd13f 0xb075): \r\n");
    NU32_ReadUART3(buf, sizeof(buf));
    sscanf(buf,"%04hx %04hx",&master, &slave);
    // have the slave write its data to its SPI buffer
```

```
        // (note, the data will not be sent until the master performs a write)
        SPI3BUF = slave;
        // now the master performs a write
        SPI4BUF = master;
        // wait until the master receives the data
        while(!SPI4STATbits.SPIRBF) {
          ; // could check SPI3STAT instead; slave receives data same time as master
        }
        // receive the data
        rmaster = SPI4BUF;
        rslave = SPI3BUF;
        sprintf(buf,"Master sent 0x%04x,  Slave sent 0x%04x\r\n", master, slave);
        NU32_WriteUART3(buf);
        sprintf(buf," Slave read 0x%04x, Master read 0x%04x\r\n",rslave,rmaster);
        NU32_WriteUART3(buf);
      }
      return 0;
    }
```

12.3.2 SRAM

One use for SPI is to add external RAM to the PIC32. For example, the Microchip 23K256 256 kbit (32 KB) static random-access memory (SRAM) has an SPI interface. The data sheet describes its communication protocol. The SRAM has three modes of operation: byte, page, and sequential. Byte operation allows reading or writing a single byte of RAM. In page mode, you can access one 32-byte page of RAM at a time. Finally, sequential mode allows writing or reading sequences of bytes, ignoring page boundaries. The example code we provide uses sequential mode.

The SRAM chip requires the use of slave select (called chip select $\overline{\text{CS}}$ on the 23K256). When this signal drops low, the SRAM knows that data or commands are about to be sent, and when $\overline{\text{CS}}$ becomes high, the SRAM knows that communication is finished. We control $\overline{\text{CS}}$ using a normal digital output pin, as its state is only changed after several bytes are sent, not after every byte is sent to the device (which is what the automatic slave select enable feature of the SPI peripheral would do). After wiring the chip according to Figure 12.3 you can run the

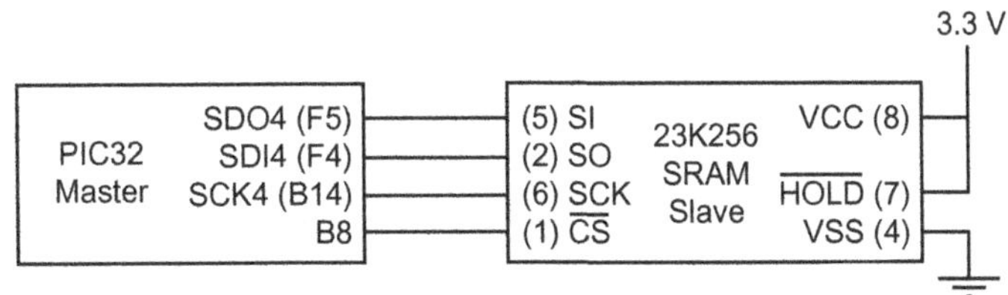

Figure 12.3
SRAM circuit diagram. SRAM pin numbers are given in parentheses.

following sample code, which writes data to the SRAM and reads it back, sending the results over UART to your computer.

The main function used by the sample code is spi_io, which writes a byte to the SPI port and reads the result. Every operation uses this command to communicate with the SRAM, since every write requires a read and vice versa. After configuring the SRAM to use sequential mode, the example reads the status of the SRAM, writes some data to it, and reads it back.

After executing this sample code, comment out the write to RAM, recompile and execute the code. Notice that, if you do not power off the SRAM, the SRAM still contains the data from the previous write, no matter how long between writes. Dynamic RAM, or DRAM, on the other hand, must periodically have its bits rewritten or it loses the data.

Code Sample 12.2 spi_ram.c. **SPI SRAM Access.**

```
#include "NU32.h"       // constants, funcs for startup and UART
// Demonstrates spi by accessing external ram
// PIC is the master, ram is the slave
// Uses microchip 23K256 ram chip (see the data sheet for protocol details)
// SDO4 -> SI (pin F5 -> pin 5)
// SDI4 -> SO (pin F4 -> pin 2)
// SCK4 -> SCK (pin B14 -> pin 6)
// SS4 -> CS (pin B8 -> pin 1)
// Additional SRAM connections
// Vss (Pin 4) -> ground
// Vcc (Pin 8) -> 3.3 V
// Hold (pin 7) -> 3.3 V (we don't use the hold function)
//
// Only uses the SRAM's sequential mode
//
#define CS LATBbits.LATB8       // chip select pin

// send a byte via spi and return the response
unsigned char spi_io(unsigned char o) {
  SPI4BUF = o;
  while(!SPI4STATbits.SPIRBF) { // wait to receive the byte
    ;
  }
  return SPI4BUF;
}

// initialize spi4 and the ram module
void ram_init() {
  // set up the chip select pin as an output
  // the chip select pin is used by the sram to indicate
  // when a command is beginning (clear CS to low) and when it
  // is ending (set CS high)
  TRISBbits.TRISB8 = 0;
  CS = 1;

  // Master - SPI4, pins are: SDI4(F4), SDO4(F5), SCK4(F13).
  // we manually control SS4 as a digital output (F12)
  // since the pic is just starting, we know that spi is off. We rely on defaults here
```

```
  // setup spi4
  SPI4CON = 0;              // turn off the spi module and reset it
  SPI4BUF;                  // clear the rx buffer by reading from it
  SPI4BRG = 0x3;            // baud rate to 10 MHz [SPI4BRG = (80000000/(2*desired))-1]
  SPI4STATbits.SPIROV = 0;  // clear the overflow bit
  SPI4CONbits.CKE = 1;      // data changes when clock goes from hi to lo (since CKP is 0)
  SPI4CONbits.MSTEN = 1;    // master operation
  SPI4CONbits.ON = 1;       // turn on spi 4

                            // send a ram set status command.
  CS = 0;                   // enable the ram
  spi_io(0x01);             // ram write status
  spi_io(0x41);             // sequential mode (mode = 0b01), hold disabled (hold = 0)
  CS = 1;                   // finish the command
}

// write len bytes to the ram, starting at the address addr
void ram_write(unsigned short addr, const char data[], int len) {
  int i = 0;
  CS = 0;                        // enable the ram by lowering the chip select line
  spi_io(0x2);                   // sequential write operation
  spi_io((addr & 0xFF00) >> 8 ); // most significant byte of address
  spi_io(addr & 0x00FF);         // the least significant address byte
  for(i = 0; i < len; ++i) {
    spi_io(data[i]);
  }
  CS = 1;                        // raise the chip select line, ending communication
}

// read len bytes from ram, starting at the address addr
void ram_read(unsigned short addr, char data[], int len) {
  int i = 0;
  CS = 0;
  spi_io(0x3);                   // ram read operation
  spi_io((addr & 0xFF00) >> 8);  // most significant address byte
  spi_io(addr & 0x00FF);         // least significant address byte
  for(i = 0; i < len; ++i) {
    data[i] = spi_io(0);         // read in the data
  }
  CS = 1;
}

int main(void) {
  unsigned short addr1 = 0x1234;                 // the address for writing the ram
  char data[] = "Help, I'm stuck in the RAM!";   // the test message
  char read[] = "***************************";   // buffer for reading from ram
  char buf[100];                                 // buffer for comm. with the user
  unsigned char status;                          // used to verify we set the status
  NU32_Startup();  // cache on, interrupts on, LED/button init, UART init
  ram_init();

  // check the ram status
  CS = 0;
  spi_io(0x5);                                    // ram read status command
  status = spi_io(0);                             // the actual status
  CS = 1;

  sprintf(buf, "Status 0x%x\r\n",status);
  NU32_WriteUART3(buf);
```

```
    sprintf(buf,"Writing \"%s\" to ram at address 0x%x\r\n", data, addr1);
    NU32_WriteUART3(buf);
                                                  // write the data to the ram
    ram_write(addr1, data, strlen(data) + 1);     // +1, to send the '\0' character
    ram_read(addr1, read, strlen(data) + 1);      // read the data back
    sprintf(buf,"Read \"%s\" from ram at address 0x%x\r\n", read, addr1);
    NU32_WriteUART3(buf);

    while(1) {
      ;
    }
    return 0;
}
```

12.3.3 LSM303D Accelerometer/Magnetometer

The STMicroelectronics LSM303D accelerometer/magnetometer, depicted in Figure 12.4, is a sensor that combines a three-axis accelerometer, a three-axis magnetometer, and a temperature sensor.[1] As a surface mount component, the accelerometer is difficult to work with; therefore, we use it with a breakout board from Pololu. The combination of an accelerometer and magnetometer is ideal for creating an electronic compass: the accelerometer gives tilt parameters relative to the earth, providing a reference frame for the magnetometer readings. The PIC32 can control this device using either SPI or I^{2}C, another communication method discussed in Chapter 13.

The sample code consists of three files: `accel.h`, `spi_accel.c`, and `accel.c`. The header file `accel.h` provides a rudimentary interface to the accelerometer. The function prototypes in `accel.h` are implemented using SPI in `spi_accel.c`. Thus you can think of `accel.h` and `spi_accel.c` together as making an LSM303D interface library. (In Chapter 13, we will make another version of the library by using I^{2}C to implement the function prototypes.)

The SPI peripheral is set for 10 MHz operation and, as in `spi_ram.c`, `spi_io` encapsulates the write/read behavior. The `main` file that uses the LSM303D library is `accel.c`. This code reads and displays the sensor values approximately once per second.

You can test the accelerometer readings by tilting the board in various directions. The accelerometer will always read 1 g of acceleration in the downward direction. If you rotate the board you should see the magnetic field readings change. You can test the temperature sensor by blowing on it: the heat from your breath will cause the reading to temporarily increase. To use this device as a magnetic compass you must perform a calibration; STMicroelectronics application note AN3192 provides a guide.

---

[1] This chip uses microelectromechanical systems (MEMS) to provide you with so many sensors in such a small package.

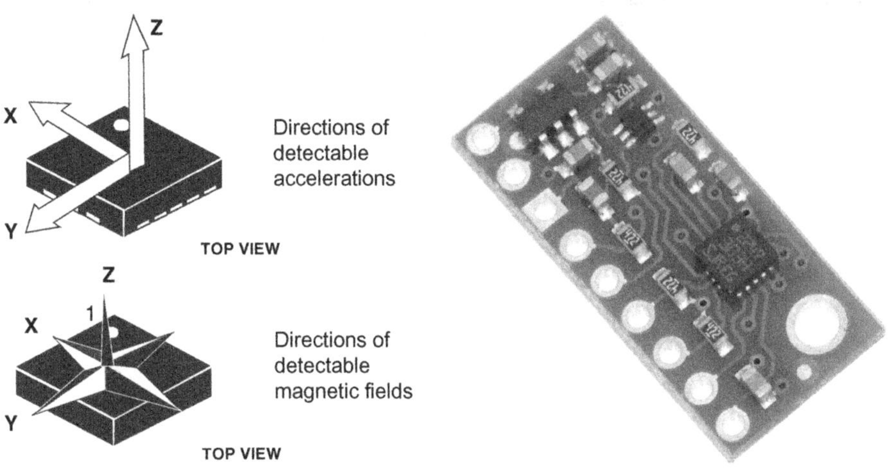

Figure 12.4
The LSM303D accelerometer on the Pololu breakout board. (Image of breakout board courtesy of Pololu Robotics and Electronics, pololu.com.)

Now that you are an expert in SPI, you can figure out the wiring yourself.

Code Sample 12.3 `accel.h`. **Header File Providing the Interface to the LSM303D Accelerometer/Magnetometer.**

```
#ifndef ACCEL__H__
#define ACCEL__H__
// Basic interface with an LSM303D accelerometer/compass.
// Used for both i2c and spi examples, but with different implementation (.c) files

                        // register addresses
#define CTRL1 0x20      // control register 1
#define CTRL5 0x24      // control register 5
#define CTRL7 0x26      // control register 7

#define OUT_X_L_A 0x28  // LSB of x-axis acceleration register.
                        // accel. registers are contiguous, this is the lowest address
#define OUT_X_L_M 0x08  // LSB of x-axis of magnetometer register

#define TEMP_OUT_L 0x05 // temperature sensor register

// read len bytes from the specified register into data[]
void acc_read_register(unsigned char reg, unsigned char data[], unsigned int len);

// write to the register
void acc_write_register(unsigned char reg, unsigned char data);

// initialize the accelerometer
void acc_setup();
#endif
```

Code Sample 12.4 accel.c. **Example Code that Reads the LSM303D and Prints the Results Over UART.**

```
#include "NU32.h" // constants, funcs for startup and UART
#include "accel.h"
// accelerometer/magnetometer example.  Prints the results from the sensor to the UART

int main() {
  char buffer[200];
  NU32_Startup(); // cache on, interrupts on, LED/button init, UART init
  acc_setup();

  short accels[3];    // accelerations for the 3 axes
  short mags[3];      // magnetometer readings for the 3 axes
  short temp;         // temperature reading
  while(1) {
    // read the accelerometer from all three axes
    // the accelerometer and the pic32 are both little endian by default
    // (the lowest address has the LSB)
    // the accelerations are 16-bit twos complement numbers, the same as a short
    acc_read_register(OUT_X_L_A, (unsigned char *)accels,6);

    // NOTE: the accelerometer is influenced by gravity,
    // meaning that, on earth, when stationary it measures gravity as a 1g acceleration
    // You could use this information to calibrate the readings into actual units
    sprintf(buffer,"x: %d y: %d z: %d\r\n",accels[0], accels[1], accels[2]);
    NU32_WriteUART3(buffer);

    // need to read all 6 bytes in one transaction to get an update.
    acc_read_register(OUT_X_L_M, (unsigned char *)mags, 6);

    sprintf(buffer, "xmag: %d ymag: %d zmag: %d \r\n",mags[0], mags[1], mags[2]);
    NU32_WriteUART3(buffer);

    // read the temperature data.  It's a right-justified 12-bit two's complement number
    acc_read_register(TEMP_OUT_L,(unsigned char *)&temp,2);
    sprintf(buffer,"temp: %d\r\n",temp);
    NU32_WriteUART3(buffer);

    //delay
   _CP0_SET_COUNT(0);
   while(_CP0_GET_COUNT() < 40000000) { ; }
  }
}
```

Code Sample 12.5 spi_accel.c. **Communicates with the LSM303D Accelerometer/Magnetometer Using SPI.**

```
#include "accel.h"
#include "NU32.h"
// interface with the LSM303D accelerometer/magnetometer using spi
// Wire GND to GND, VDD to 3.3V, Vin is disconnected (on Pololu breakout board)
// SDO4 (F5) ->  SDI (labeled SDA on Pololu board),
// SDI4 (F4) -> SDO
// SCK4 (B14) -> SCL
// RB8 -> CS
#define CS LATBbits.LATB8          // use RB8 as CS
```

```
// send a byte via spi and return the response
unsigned char spi_io(unsigned char o) {
  SPI4BUF = o;
  while(!SPI4STATbits.SPIRBF) { // wait to receive the byte
    ;
  }
  return SPI4BUF;
}

// read data from the accelerometer, given the starting register address.
// return the data in data
void acc_read_register(unsigned char reg, unsigned char data[], unsigned int len)
{
  unsigned int i;
  reg |= 0x80; // set the read bit (as per the accelerometer's protocol)
  if(len > 1) {
    reg |= 0x40; // set the address auto inc. bit (as per the accelerometer's protocol)
  }
  CS = 0;
  spi_io(reg);
  for(i = 0; i != len; ++i) {
    data[i] = spi_io(0); // read data from spi
  }
  CS = 1;
}

void acc_write_register(unsigned char reg, unsigned char data)
{
  CS = 0;               // bring CS low to activate SPI
  spi_io(reg);
  spi_io(data);
  CS = 1;               // complete the command
}

void acc_setup() {      // setup the accelerometer, using SPI 4
  TRISBbits.TRISB8 = 0;
  CS = 1;

  // Master - SPI4, pins are: SDI4(F4), SDO4(F5), SCK4(B14).
  // we manually control SS4 as a digital output (B8)
  // since the PIC is just starting, we know that spi is off. We rely on defaults here

  // setup SPI4
  SPI4CON = 0;                 // turn off the SPI module and reset it
  SPI4BUF;                     // clear the rx buffer by reading from it
  SPI4BRG = 0x3;               // baud rate to 10MHz [SPI4BRG = (80000000/(2*desired))-1]
  SPI4STATbits.SPIROV = 0;     // clear the overflow bit
  SPI4CONbits.CKE = 1;         // data changes when clock goes from active to inactive
                               //    (high to low since CKP is 0)
  SPI4CONbits.MSTEN = 1;       // master operation
  SPI4CONbits.ON = 1;          // turn on SPI 4

  // set the accelerometer data rate to 1600 Hz. Do not update until we read values
  acc_write_register(CTRL1, 0xAF);

  // 50 Hz magnetometer, high resolution, temperature sensor on
  acc_write_register(CTRL5, 0xF0);

  // enable continuous reading of the magnetometer
  acc_write_register(CTRL7, 0x0);
}
```

12.4 Chapter Summary

- An SPI device can be either a master or a slave. Usually the PIC32 is configured as the master.
- Two-way communication requires at least three wires: a clock controlled by the master, master-output slave-input (MOSI), and master-input slave-output (MISO). Some slave devices also require their slave select pin to be actively controlled, even if they are the only slave on the bus. If there is more than one slave device on the SPI bus, then there must be one slave select line for every slave. The slave whose slave select line is held low by the master controls the MISO line.
- When the master performs a write, it generates a clock signal. This clock signal also signals the active slave to send data back to the master. Therefore, every write by the master should be followed by a read, even if you do not need the data.
- Master writes to the MOSI line are initiated by writing data to SPIxBUF. Received data is obtained by reading SPIxBUF, which actually gives you access to data in the receive buffer SPIxRXB.

12.5 Exercises

1. Why must you write to the SPI bus in order to read a value from the slave?
2. Is it possible to use only two wires (plus GND) if you need to only read or only write? Why or why not?
3. Write a program that receives bytes from the terminal emulator, sends the bytes by SPI to an external chip, receives bytes back from the external chip, and sends the received bytes to the terminal emulator for display. The program should also allow the user to toggle the $\overline{SS}$ line. Such a program may prove useful when working with an unfamiliar chip. Note: you can read and write hexadecimal numbers using the `%x` format specifier with `sscanf` and `sprintf`.

Further Reading

23A256/23K256 256K SPI bus low-power serial SRAM. (2011). Microchip Technology Inc.

AN3192 using LSM303DLH for a tilt compensated electronic compass. (2010). STMicroelectronics.

LSM303D ultra compact high performance e-compass 3D accelerometer and 3D magnetometer module. (2012). STMicroelectronics.

CHAPTER 13

I^2C Communication

Each of the PIC32's four inter-integrated circuit (I^2C, pronounced *eye-squared-see*) peripherals allows it to communicate with multiple devices using only two pins. Many devices have I^2C interfaces, including RAM, accelerometers, ADCs, and OLED screens. An advantage of I^2C over SPI is that the I^2C bus has only two wires, no matter how many devices are connected, and each device can act as a master or a slave. A disadvantage is the more complicated support software and the typically lower bit rates—the standard mode is defined as 100 kbit/s and the fast mode is defined as 400 kbit/s—though some I^2C devices allow higher rates.

13.1 Overview

The I^2C bus consists of two wires, one for data (SDA) and one for a clock (SCL), in addition to the common ground reference. Multiple chips can be connected to the same two wires and communicate with each other. A chip can be a master or a slave. Multiple masters and slaves can be connected to the same bus; however, only one master can operate at one time. Masters control the clock signal and hence the speed at which data flows. Usually, I^2C devices operate either in standard 100 kHz or fast 400 kHz mode, although the PIC32 can set arbitrary frequencies.

Figure 13.1 shows a circuit diagram of a typical bus connection. Rather than driving the lines high and low, each pin on the I^2C bus switches between high impedance output (disconnected, floating, high-z) and logic low. The high-z state is equivalent to the open-drain digital output mode, where the pin, rather than being high, is effectively disconnected. Pull-up resistors on each I^2C line ensure that the line is low only when a device outputs a low signal. The PIC32's four I^2C peripherals can handle pull-up voltages between 3.3 and 5 V.

If all devices on a line are high-z, the line is pulled high, a logical 1. If any of the devices on a line are pulling it low, the line is low, a logical 0. When the bus is idle, i.e., no data is being transmitted, all device outputs are high-z, and both SDA and SCL are high.

When a device assumes the role of a master to begin a transmission, it pulls the SDA line low while leaving SCL high. This is the start bit; it tells other devices on the bus that the bus has been claimed by a master, and that they should not attempt to claim it until the transmission is

Embedded Computing and Mechatronics with the PIC32 Microcontroller. http://dx.doi.org/10.1016/B978-0-12-420165-1.00013-5

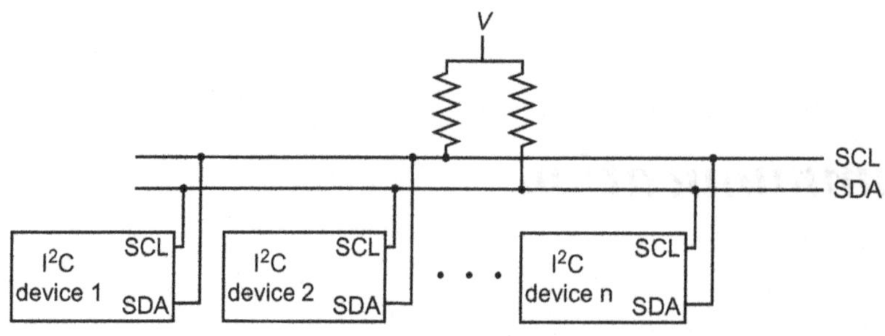

Figure 13.1
n devices on an I^2C bus. Each of the two lines is pulled up, typically to 3.3 or 5 V, by a pull-up resistor, unless one of the devices holds the line low. The PIC32 can work with either 3.3 or 5 V. A typical pull-up resistance is 2.4 kΩ, but any nearby value should be fine.

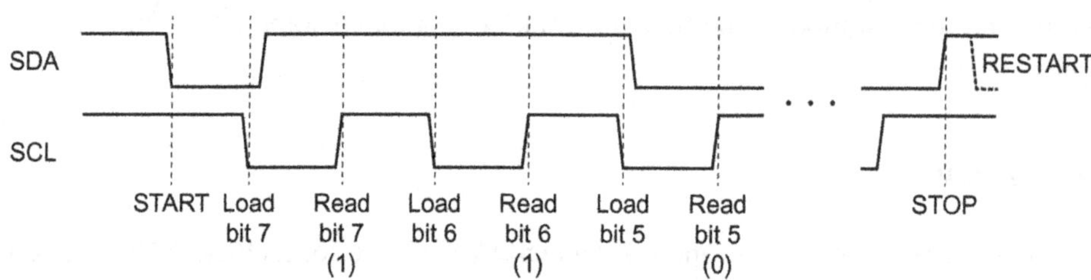

Figure 13.2
An I^2C transmission begins when the master pulls SDA low while leaving SCL high. Then the master begins driving the clock SCL. Data is loaded onto SDA at every falling edge of SCL and read from SDA on every rising edge of SCL. In this figure, the transmission begins with 0b110... (these are the highest three bits of the address of the slave being selected). Either the master or a slave can control SDA, depending on whether the master wants to send information to the slave or receive information from the slave. Transmission stops when the master releases SDA, allowing it to go high while SCL is high. A RESTART (dashed line) occurs if the master quickly pulls the SDA line low again, taking control of the bus again before another master can claim it.

finished. The master (or a slave; see below) then transmits data over the SDA line while the master controls the clock SCL. Data is loaded onto SDA when SCL drops low and read from SDA when SCL rises. Eight-bit data bytes are transferred most significant bit first. Transmission stops when the master issues the stop bit: an SDA transition from low to high while SCL is high. See the timing diagram in Figure 13.2.

Unlike UARTs and SPI, I^2C employs a handshaking protocol between master and slave. A typical data transaction consists of the following primitives, in order:

1. START: The master issues a start bit, dropping SDA from high to low while SCL stays high.

2. ADDRESS: The master transmits a byte consisting of a 7-bit address and a read-write bit, $R\overline{W}$, in the least significant bit. The 7-bit address indicates which slave the master is addressing; each device on the bus has a unique 7-bit address.[1] If $R\overline{W} = 0$, the master will write to the slave; if it is a 1, the master expects to read data from the slave.
3. ACKNOWLEDGMENT ($\overline{ACK}$/NACK): If a slave has recognized its address, it will respond with a single acknowledgment bit of 0, holding SDA low for the next clock cycle. This is called an $\overline{ACK}$. If SDA is high (no $\overline{ACK}$, also called a NACK), the master knows an error has occurred.
4. WRITE or RECEIVE: If $R\overline{W} = 0$ (write), the master sends a byte over SDA. If $R\overline{W} = 1$ (read), then the slave controls the SDA and sends a byte.
5. ACKNOWLEDGMENT ($\overline{ACK}$/NACK): If it is a write, the slave must send an $\overline{ACK}$ bit to acknowledge that it has received the data. Otherwise the master knows an error has occurred. If it is a read, the master either sends an $\overline{ACK}$ if it wants another byte or a NACK if it is done requesting bytes. If the master wishes to send another byte, or has requested another byte, return to step 4.
6. STOP or RESTART: The master issues a stop bit (SDA raised to high while SCL is high) to end the transmission. This allows other devices to claim control of the bus. If the master wants to keep control of the bus, possibly to change the communication from a write to a read or vice versa, the master can issue a RESTART (or "repeated start") instead of a STOP. This is simply a STOP followed quickly by another START, before another master can claim the bus. In this case, return to step 2.

If two or more masters attempt to take control of an idle bus at approximately the same time, an arbitration process ensures that all but one of them drop out. Each device monitors the state of the SDA line, and if it is ever a 0 (low) while the device is transmitting a 1 (high), the device knows that another master is driving the line, and therefore the device drops out. Devices that lose arbitration report a *bus collision*.

Although the master nominally drives the clock line SCL, a slave may also pull it low. For example, if the slave needs more time to process data sent to it, it can hold the SCL line low when the master tries to let SCL return to high. The master senses this and waits until the slave allows SCL to return high before resuming its normal clock rate. This is called *clock stretching*.

13.2 Details

Seven SFRs control the behavior of I^2C peripherals. Many of the fields in these SFRs initiate an "event" on the I^2C bus. All bits default to zero except I2CxCON.SCLREL. In the SFRs below, x refers to the I^2C peripheral number, 1, 3, 4, or 5.

[1] 10-bit addressing is also supported, but this chapter focuses on 7-bit addresses.

I2CxCON The I^{2}C control register. Setting bits in this register initiates primitive operations used by the I^{2}C protocol. Some bits control an I^{2}C slave's behavior.

I2CxCON⟨15⟩ or I2CxCONbits.ON: Setting this bit enables the I^{2}C module.

I2CxCON⟨12⟩ or I2CxCONbits.SCLREL: In slave mode only, this SCL release control bit is set to release the clock, telling the master it may continue to clock the communication, or cleared to hold the clock low (clock stretching). When the master detects that SCL is low, it delays sending a clock signal, giving the slave more time before responding to the master.

I2CxCON⟨6⟩ or I2CxCONbits.STREN: This SCL stretch enable bit is used in slave mode to control clock stretching. If set to one, the slave will hold SCL low prior to transmitting to and after receiving from the master. When SCL is low, the clock is stretched and the master pauses the clock. If clear, clock stretching only happens at the beginning of a slave transmission. A slave transmission begins whenever the master expects data from the slave, according to the I^{2}C protocol. The transmission does not actually occur until the slave sets I2CxCONbits.SCLREL.

I2CxCON⟨5⟩ or I2CxCONbits.ACKDT: Acknowledge data bit, used only in master mode. If set to one, the master will send a NACK when sending an acknowledgment, signaling that no more data is requested. If set to zero, the master sends an $\overline{\text{ACK}}$ during the acknowledge, signaling that it wants more data.

I2CxCON⟨4:0⟩ These bits initiate various control signals on the bus. While any of these bits is high, you should not set any of the other bits.

I2CxCON⟨4⟩ or I2CxCONbits.ACKEN: Setting this bit initiates an acknowledgment, transmitting the I2CxCON.ACKDT bit. Hardware clears this bit after the ACK ends.

I2CxCON⟨3⟩ or I2CxCONbits.RCEN: Setting this bit initiates a RECEIVE. Hardware clears this bit after the receive has finished.

I2CxCON⟨2⟩ or I2CxCONbits.PEN: Setting this bit initiates a STOP. Hardware clears the bit after the stop is sent.

I2CxCON⟨1⟩ or I2CxCONbits.RSEN: Setting this bit initiates a RESTART. Hardware clears this bit after the restart is finished.

I2CxCON⟨0⟩ or I2CxCONbits.SEN: Setting this bit initiates a START. Hardware clears this bit after the start is finished.

I2CxSTAT Contains the status of the I^{2}C peripheral and results from signals on the I^{2}C bus.

I2CxSTAT⟨15⟩ or I2CxSTATbits.ACKSTAT: If clear (0) an $\overline{\text{ACK}}$ (acknowledge) has been received; otherwise an ACK has not been received.

I2CxSTAT⟨14⟩ or I2CxSTATbits.TRSTAT: If this transmission status bit is set (1) the master is transmitting; otherwise the master is not transmitting. Only used in master mode.

I2CxSTAT⟨6⟩ or I2CxSTATbits.I2COV: Useful for debugging. If set (1), a receive overflow has occurred, meaning the receive buffer contains a byte but another byte has been received. Software must clear this bit.

I2CxSTAT⟨5⟩ or I2CxSTATbits.D_A: In slave mode, indicates whether the most recently received or transmitted information was an address (0) or data (1).

I2CxSTAT⟨2⟩ or I2CxSTATbits.R_W: In slave mode, this bit is a 1 if the master is requesting data from the slave and a 0 if the master is sending data.

I2CxSTAT⟨1⟩ or I2CxSTATbits.RBF: The receive buffer full bit, when set, indicates that a byte has been received and is ready in I2CxRCV.

I2CxSTAT⟨0⟩ or I2CxSTATbits.TBF: When set (1), indicates that the transmit buffer is full and that a transmission is occurring.

I2CxADD This register contains the I^2C peripheral's address. The address is contained in the lower ten bits (although only seven are used in 7-bit address mode). Whenever an ADDRESS is initiated on the I^2C bus, the slave will respond if the ADDRESS matches its own address. Only slaves need to set an address.

I2CxMSK Allows the slave to ignore some address bits.

I2CxBRG The lower 16 bits of this register determine the baud. Typically the baud is either 100 kHz or 400 kHz. To compute the value of I2CxBRG, use the formula

$$\text{I2CxBRG} = \left(\left(\frac{1}{2 \times F_{sck}} - T_{PGD}\right) F_{pb}\right) - 2, \tag{13.1}$$

where F_{sck} is the desired baud, F_{pb} is the peripheral bus clock frequency, and T_{PGD} is 104 ns, according to the Reference Manual. With an 80 MHz peripheral bus clock, I2CxBRG = 390 for 100 kHz and I2CxBRG = 90 for 400 kHz. Only masters need to set a baud rate.

I2CxTRN Used to transmit a byte of data. Write an address to this register when performing the ADDRESS command, or write data when sending a data byte.

I2CxRCV Used to receive data from the I^2C bus. This register contains any data received. On the master, this register only contains data after a RECEIVE request has been initiated. On the slave, this register is loaded whenever the master sends any data, including address bytes.

Interrupts can be generated for a master or the slave by any of the data transaction primitives or by errors detected in the handshaking. Interrupts can also be generated by bus collisions. See the Reference Manual for details. The interrupt vectors are `_I2C_x_VECTOR`, where x is 1, 3, 4, or 5. For I^2C peripheral 1, the flag status bits are in IFS0bits.I2C1BIF (bus collision), IFS0bits.I2C1SIF (slave event), and IFS0bits.I2C1MIF (master event); the enable bits are in IEC0bits.I2C1BIE (bus collision), IEC0bits.I2C1SIE (slave event), and IEC0bits.I2C1MIE (master event); and the priority and subpriority bits are in IPC6bits.I2C1IP and IPC6bits.I2C1IS, respectively. For I^{2}C peripherals 3 to 5, the bits are named similarly, replacing I2C1 with I2Cx, x = 3 to 5. For I2C3, the bits are also in IFS0, IEC0, and IPC6. For I2C4, the bits are in IFS1, IEC1, and IPC7, and for I2C5, they are in IFS1, IEC1, and IPC8.

13.3 Sample Code

13.3.1 Loopback

In this example, a single PIC32 communicates with itself using I^{2}C. Here we use I^{2}C 1 as the master and I^{2}C 5 as a slave.[2]

We have divided the code into three modules: the master, the slave, and the main function, allowing you to test the code on a single PIC32 and then to use two PIC32's to test inter-PIC communication.

First, the master code. The implementation of the I^{2}C master contains functions roughly corresponding to the primitives discussed earlier. Each function executes the primitive command and waits for it to complete. By calling the primitive functions in succession, you can form an I^{2}C transaction.

Code Sample 13.1 `i2c_master_noint.h`. **Header File for I^{2}C Master with No Interrupts.**

```
#ifndef I2C_MASTER_NOINT_H__
#define I2C_MASTER_NOINT_H__
// Header file for i2c_master_noint.c
// helps implement use I2C1 as a master without using interrupts

void i2c_master_setup(void);              // set up I2C 1 as a master, at 100 kHz

void i2c_master_start(void);              // send a START signal
void i2c_master_restart(void);            // send a RESTART signal
void i2c_master_send(unsigned char byte); // send a byte (either an address or data)
unsigned char i2c_master_recv(void);      // receive a byte of data
void i2c_master_ack(int val);             // send an ACK (0) or NACK (1)
void i2c_master_stop(void);               // send a stop

#endif
```

[2] Technically, a single I^{2}C peripheral can simultaneously be a master and a slave. When the master sends data to its slave address, the slave will respond!

Code Sample 13.2 `i2c_master_noint.c`. **Implementation of I^2C Master with No Interrupts.**

```
#include "NU32.h"          // constants, funcs for startup and UART
// I2C Master utilities, 100 kHz, using polling rather than interrupts
// The functions must be called in the correct order as per the I2C protocol
// Master will use I2C1 SDA1 (D9) and SCL1 (D10)
// Connect these through resistors to Vcc (3.3 V). 2.4k resistors recommended,
// but something close will do.
// Connect SDA1 to the SDA pin on the slave and SCL1 to the SCL pin on a slave

void i2c_master_setup(void) {
  I2C1BRG = 390;                    // I2CBRG = [1/(2*Fsck) - PGD]*Pblck - 2
                                    // Fsck is the freq (100 kHz here), PGD = 104 ns
  I2C1CONbits.ON = 1;               // turn on the I2C1 module
}

// Start a transmission on the I2C bus
void i2c_master_start(void) {
    I2C1CONbits.SEN = 1;            // send the start bit
    while(I2C1CONbits.SEN) { ; }    // wait for the start bit to be sent
}

void i2c_master_restart(void) {
    I2C1CONbits.RSEN = 1;           // send a restart
    while(I2C1CONbits.RSEN) { ; }   // wait for the restart to clear
}

void i2c_master_send(unsigned char byte) { // send a byte to slave
  I2C1TRN = byte;                   // if an address, bit 0 = 0 for write, 1 for read
  while(I2C1STATbits.TRSTAT) { ; }  // wait for the transmission to finish
  if(I2C1STATbits.ACKSTAT) {        // if this is high, slave has not acknowledged
    NU32_WriteUART3("I2C2 Master: failed to receive ACK\r\n");
  }
}

unsigned char i2c_master_recv(void) { // receive a byte from the slave
    I2C1CONbits.RCEN = 1;             // start receiving data
    while(!I2C1STATbits.RBF) { ; }    // wait to receive the data
    return I2C1RCV;                   // read and return the data
}

void i2c_master_ack(int val) {        // sends ACK = 0 (slave should send another byte)
                                      // or NACK = 1 (no more bytes requested from slave)
    I2C1CONbits.ACKDT = val;          // store ACK/NACK in ACKDT
    I2C1CONbits.ACKEN = 1;            // send ACKDT
    while(I2C1CONbits.ACKEN) { ; }    // wait for ACK/NACK to be sent
}

void i2c_master_stop(void) {          // send a STOP:
  I2C1CONbits.PEN = 1;                // comm is complete and master relinquishes bus
  while(I2C1CONbits.PEN) { ; }        // wait for STOP to complete
}
```

Next the slave code. The slave code uses I^2C 5, and, upon a read request, will return the last two bytes written to the slave. As the slave is interrupt driven, it will work as soon as you call `i2c_slave_setup`, providing the desired 7-bit address (the master is configured to talk to a slave on address 0x32). The slave interrupt reads the status flags so that it can discriminate between reads, writes, address bytes, and data bytes. Notice that, when the slave wishes to send data to the master, it must release SCL to be controlled by the master by setting I2C5CONbits.SCLREL to one.

Code Sample 13.3 `i2c_slave.h`. **Header File for I^{2}C Slave.**

```
#ifndef I2C_SLAVE_H__
#define I2C_SLAVE_H__
// implements a basic I2C slave

void i2c_slave_setup(unsigned char addr); // set up the slave at the given address

#endif
```

Code Sample 13.4 `i2c_slave.c`. **Implementation of an I^{2}C Slave.**

```
#include "NU32.h"   // constants, funcs for startup and UART
// Implements a I2C slave on I2C5 using pins SDA5 (F4) and SCL5 (F5)
// The slave returns the last two bytes the master writes

void __ISR(_I2C_5_VECTOR, IPL1SOFT) I2C5SlaveInterrupt(void) {
  static unsigned char bytes[2];    // store two received bytes
  static int rw = 0;                // index of the bytes read/written
  if(rw == 2) {                     // reset the data index after every two bytes
    rw = 0;
  }
  // We have to check why the interrupt occurred.  Some possible causes:
  // (1) slave received its address with RW bit = 1:  read address & send data to master
  // (2) slave received its address with RW bit = 0:  read address (data will come next)
  // (3) slave received an ACK in RW = 1 mode:        send data to master
  // (4) slave received a data byte in RW = 0 mode:   store this data sent by master

  if(I2C5STATbits.D_A) {          // received data/ACK, so Case (3) or (4)
    if(I2C5STATbits.R_W) {        // Case (3):  send data to master
      I2C5TRN = bytes[rw];        // load slave's previously received data to send to master
      I2C5CONbits.SCLREL = 1;     // release the clock, allowing master to clock in data
    } else {                      // Case (4):  we have received data from the master
      bytes[rw] = I2C5RCV;        // store the received data byte
    }
    ++rw;
  } else {                        // the byte is an address byte, so Case (1) or (2)
    I2C5RCV;                      // read to clear I2C5RCV (we don't need our own address)
    if(I2C5STATbits.R_W) {        // Case (1):  send data to master
      I2C5TRN = bytes[rw];        // load slave's previously received data to send to master
      ++rw;
      I2C5CONbits.SCLREL = 1;     // release the clock, allowing master to clock in data
    }                             // Case (2):  do nothing more, wait for data to come
  }
  IFS1bits.I2C5SIF = 0;
}

// I2C5 slave setup (disable interrupts before calling)
void i2c_slave_setup(unsigned char addr) {
  I2C5ADD = addr;                     // the address of the slave
```

```
    IPC8bits.I2C5IP  = 1;          // slave has interrupt priority 1
    IEC1bits.I2C5SIE = 1;          // slave interrupt is enabled
    IFS1bits.I2C5SIF = 0;          // clear the interrupt flag
    I2C5CONbits.ON   = 1;          // turn on i2c2
}
```

Next, the main program, in which the PIC32 communicates with itself over I^{2}C. The main program initializes the slave and master and performs some I^{2}C transactions, sending the results over the UART to your terminal. Notice how the primitive operations are assembled into a single transaction. You should connect the clock (SCL) and data (SDA) lines of I^{2}C 1 and I^{2}C 5, along with the pull-up resistors. To examine what happens when the slave does not return an ACK, disconnect the I^{2}C bus wires.

Code Sample 13.5 `i2c_loop.c`**. The Main I^{2}C Loopback Program.**

```
#include "NU32.h"          // config bits, constants, funcs for startup and UART
#include "i2c_slave.h"
#include "i2c_master_noint.h"
// Demonstrate I2C by having the I2C1 talk to I2C5 on the same PIC32
// Master will use SDA1 (D9) and SCL1 (D10).  Connect these through resistors to
// Vcc (3.3 V) (2.4k resistors recommended, but around that should be good enough)
// Slave will use SDA5 (F4) and SCL5 (F5)
// SDA5 -> SDA1
// SCL5 -> SCL1
// Two bytes will be written to the slave and then read back to the slave.
#define SLAVE_ADDR 0x32

int main() {
  char buf[100] = {};                        // buffer for sending messages to the user
  unsigned char master_write0 = 0xCD;        // first byte that master writes
  unsigned char master_write1 = 0x91;        // second byte that master writes
  unsigned char master_read0  = 0x00;        // first received byte
  unsigned char master_read1  = 0x00;        // second received byte

  NU32_Startup();                  // cache on, interrupts on, LED/button init, UART init
  __builtin_disable_interrupts();
  i2c_slave_setup(SLAVE_ADDR);               // init I2C5, which we use as a slave
                                             //  (comment out if slave is on another pic)
  i2c_master_setup();                        // init I2C2, which we use as a master
  __builtin_enable_interrupts();

  while(1) {
    NU32_WriteUART3("Master: Press Enter to begin transmission.\r\n");
    NU32_ReadUART3(buf,2);
    i2c_master_start();                      // Begin the start sequence
    i2c_master_send(SLAVE_ADDR << 1);        // send the slave address, left shifted by 1,
                                             // which clears bit 0, indicating a write
    i2c_master_send(master_write0);          // send a byte to the slave
    i2c_master_send(master_write1);          // send another byte to the slave
    i2c_master_restart();                    // send a RESTART so we can begin reading
    i2c_master_send((SLAVE_ADDR << 1) | 1); // send slave address, left shifted by 1,
                                             // and then a 1 in lsb, indicating read
    master_read0 = i2c_master_recv();        // receive a byte from the bus
    i2c_master_ack(0);                       // send ACK (0): master wants another byte!
```

```
      master_read1 = i2c_master_recv();       // receive another byte from the bus
      i2c_master_ack(1);                      // send NACK (1):  master needs no more bytes
      i2c_master_stop();                      // send STOP:  end transmission, give up bus

      sprintf(buf,"Master Wrote: 0x%x 0x%x\r\n", master_write0, master_write1);
      NU32_WriteUART3(buf);
      sprintf(buf,"Master Read: 0x%x 0x%x\r\n", master_read0, master_read1);
      NU32_WriteUART3(buf);
      ++master_write0;                        // change the data the master sends
      ++master_write1;
    }
    return 0;
  }
```

Using I^{2}C to have the PIC32 communicate with itself may not seem practical, but it helps demonstrate the peripheral without involving other chips. If you have another PIC32 available, you can compile the slave as a standalone program using `i2c_slave_loop.c` (below). By connecting the master loopback example to a slave on another chip you can witness inter-PIC communication.

To use `i2c_slave_loop.c`, compile and link it with `i2c_slave.c` and run it on another NU32. Connect the SDA and SCL pins on the two NU32s, as well as the pull-up resistors. Power the slave NU32 from the master NU32 by connecting the GND rails of the two NU32s together, and by connecting the two 6 V rails together. Plug in the master NU32 while making sure that the slave NU32 is unplugged but with its power switch on. Both NU32's will turn on and off based on the state of the master's switch.

Code Sample 13.6 `i2c_slave_loop.c`**. A Standalone I^{2}C Slave.**

```
#include "i2c_slave.h"
#include "NU32.h"

int main() {
  NU32_Startup();
  i2c_slave_setup(0x32); // enable the slave w/ address 0x32

  while(1) {             // the slave is handled in an interrupt in i2c_slave.c
    _nop();              // so we do nothing.
  }
  return 0;
}
```

13.3.2 Interrupt-Based Master

In Section 13.3.1, we created functions for each I^{2}C primitive: the functions initiate a command and wait for it to complete. In this section, we provide interrupt-based master code. The function `i2c_write_read` allows the master to initiate a write-read transaction by providing a slave address, an input array with the bytes to write, and an output array with the bytes that are read. If the length of the write or read array is zero, then that specific action will not be performed.

The interrupt removes the need to wait for each primitive operation to complete. Instead, an interrupt triggers at the end of each primitive. The ISR tracks the state of the current communication and performs the appropriate state transition. As coded, the function `i2c_write_read` waits for the whole transaction to complete before returning; however, in time-critical applications this behavior could be modified. Calling `i2c_write_read` would initiate the transaction, but not wait for it to finish. Mainline code could continue to execute, and either check for the result of the transaction at a later time or handle the transaction results from within the ISR.

Code Sample 13.7 `i2c_master_int.h`**. Header File for Interrupt-Based I^2C Master.**

```
#ifndef I2C_MASTER_INT__H__
#define I2C_MASTER_INT__H__

// buffer pointer type.  The buffer is shared by an ISR and mainline code.
// the pointer to the buffer is also shared by an ISR and mainline code.
// Hence the double volatile qualification
typedef volatile unsigned char * volatile buffer_t;

void i2c_master_setup(); //sets up I2C1 as a master using an interrupt

// Initiate an I2C write read operation at the given address.
// You can optionally only read or only write by passing 0 length for reading or writing.
// This will not return until the transaction is complete.  Returns false on error.
int i2c_write_read(unsigned int addr, const buffer_t write, unsigned int wlen,
    const buffer_t read, unsigned int rlen );

// write a single byte to the slave
int i2c_write_byte(unsigned int addr, unsigned char byte);

#endif
```

Code Sample 13.8 `i2c_master_int.c`**. Implementation of an Interrupt-Based I^2C Master.**

```
#include "NU32.h"          // constants, funcs for startup and UART
#include "i2c_master_int.h"
// I2C Master utilities, using interrupts
// Master will use I2C1 SDA1 (D9) and SCL1 (D10)
// Connect these through resistors to Vcc (3.3V). 2.4k resistors recommended, but
// something close will do.
// Connect SDA1 to the SDA pin on a slave device and SCL1 to the SCL pin on a slave.

// keeps track of the current I2C state
static volatile enum {IDLE,START,WRITE,READ,RESTART,ACK,NACK,STOP,ERROR} state = IDLE;

static buffer_t to_write = NULL;            // data to write
static buffer_t  to_read = NULL;            // data to read
```

```
static volatile unsigned char address = 0; // the 7-bit address to write to / read from
static volatile unsigned int  n_write = 0; // number of data bytes to write
static volatile unsigned int  n_read  = 0; // number of data bytes to read


void __ISR(_I2C_1_VECTOR, IPL1SOFT) I2C1MasterInterrupt(void) {
  static unsigned int write_index = 0, read_index = 0;  //indexes the read/write arrays

  switch(state) {
    case START:                        // start bit has been sent
      write_index = 0;                 // reset indices
      read_index = 0;
      if(n_write > 0) {                // there are bytes to write
        state = WRITE;                 // transition to write mode
        I2C1TRN = address << 1;        // send the address, with write mode set
      } else {
        state = ACK;                   // skip directly to reading
        I2C1TRN = (address << 1) & 1;
      }

      break;
    case WRITE:                        // a write has finished
      if(I2C1STATbits.ACKSTAT) {       // error: didn't receive an ACK from the slave
        state = ERROR;
      } else {
        if(write_index < n_write) {         // still more data to write
          I2C1TRN = to_write[write_index];  // write the data
          ++write_index;
        } else {                            // done writing data, time to read or stop
          if(n_read > 0) {                  // we want to read so issue a restart
            state = RESTART;
            I2C1CONbits.RSEN = 1;           // send the restart to begin the read
          } else {                          // no data to read, issue a stop
            state = STOP;
            I2C1CONbits.PEN = 1;
          }
        }
      }
      break;
    case RESTART: // the restart has completed
      // now we want to read, send the read address
      state = ACK;        // when interrupted in ACK mode, we will initiate reading a byte
      I2C1TRN = (address << 1) | 1; // the address is sent with the read bit sent
      break;
    case READ:
      to_read[read_index] = I2C1RCV;
      ++read_index;
      if(read_index == n_read) { // we are done reading, so send a nack
        state = NACK;
        I2C1CONbits.ACKDT = 1;
      } else {
        state = ACK;
        I2C1CONbits.ACKDT = 0;
      }
      I2C1CONbits.ACKEN = 1;
      break;
    case ACK:
            // just sent an ack meaning we want to read more bytes
      state = READ;
      I2C1CONbits.RCEN = 1;
      break;
```

```
    case NACK:
      //issue a stop
      state = STOP;
      I2C1CONbits.PEN = 1;
      break;
    case STOP:
      state = IDLE; // we have returned to idle mode, indicating that the data is ready
      break;
    default:
      // some error has occurred
      state = ERROR;
  }
  IFS0bits.I2C1MIF = 0;         //clear the interrupt flag
}

void i2c_master_setup() {
  int ie = __builtin_disable_interrupts();
  I2C1BRG = 90;                        // I2CBRG = [1/(2*Fsck) - PGD]*Pblck - 2
                                       // Fsck is the frequency (400 kHz here), PGD = 104ns
                                       // this is 400 khz mode
                                       // enable the i2c interrupts
  IPC6bits.I2C1IP  = 1;                // master has interrupt priority 1
  IEC0bits.I2C1MIE = 1;                // master interrupt is enabled
  IFS0bits.I2C1MIF = 0;                // clear the interrupt flag
  I2C1CONbits.ON = 1;                  // turn on the I2C2 module

  if(ie & 1) {
    __builtin_enable_interrupts();
  }
}

// communicate with the slave at address addr.  first write wlen bytes to the slave,
// then read rlen bytes from the slave
int i2c_write_read(unsigned int addr, const buffer_t write,
    unsigned int wlen, const buffer_t read, unsigned int rlen ) {
  n_write = wlen;
  n_read = rlen;
  to_write = write;
  to_read = read;
  address = addr;
  state = START;
  I2C1CONbits.SEN = 1;         // initialize the start
  while(state != IDLE && state != ERROR) { ; }  // initialize the sequence
  return state != ERROR;
}

// write a single byte to the slave
int i2c_write_byte(unsigned int addr, unsigned char byte) {
  return i2c_write_read(addr,&byte,1,NULL,0);
}
```

13.3.3 Accelerometer/Magnetometer

The STMicroelectronics LSM303D is a three-axis accelerometer and magnetometer that can be used as a digital compass. It also has a temperature sensor. More details about this sensor and the breakout board are discussed in Chapter 12. Here we use the same accelerometer library and example that we used in Chapter 12, except now our implementation uses I^2C rather than SPI.

When used with the Pololu breakout board, the accelerometer has an I^2C address of 0x1D. The example code continuously displays the raw accelerometer, magnetometer, and temperature sensor values. The necessary code is Code Samples 12.3 and 12.4 from Chapter 12, as well as the following code.

Code Sample 13.9 `i2c_accel.c`. **I^2C Implementation of the Basic Accelerometer Library. Requires** `i2c_master_int.h`.

```
#include "accel.h"
#include "i2c_master_int.h"
#include <stdlib.h>

#define I2C_ADDR 0x1D // the I2C slave address

// Wire GND to GND, VDD to 3.3V, SDA to SDA2 (RA3) and SCL to SCL2 (RA2)

// read data from the accelerometer, given the starting register address.
// return the data in data
void acc_read_register(unsigned char reg, unsigned char data[], unsigned int len)
{
  unsigned char write_cmd[1] = {};
  if(len > 1) { // want to read more than 1 byte and we are reading from the accelerometer
    write_cmd[0] = reg | 0x80; // make the MSB of addr 1 to enable auto increment
  }
  else {
    write_cmd[0] = reg;
  }
  i2c_write_read(I2C_ADDR,write_cmd, 1, data,len);
}

void acc_write_register(unsigned char reg, unsigned char data)
{
  unsigned char write_cmd[2];
  write_cmd[0] = reg;   // write the register
  write_cmd[1] = data;  // write the actual data
  i2c_write_read(I2C_ADDR, write_cmd, 2, NULL, 0);
}

void acc_setup() {                    // set up the accelerometer, using I2C 2
  i2c_master_setup();
  acc_write_register(CTRL1, 0xAF); // set accelerometer data rate to 1600 Hz.
                                   // Don't update until we read values
  acc_write_register(CTRL5, 0xF0); // 50 Hz magnetometer, high resolution, temp sensor on
  acc_write_register(CTRL7, 0x0);  // enable continuous reading of the magnetometer
}
```

13.3.4 OLED Screen

An organic light emitting diode (OLED) screen is a low-power, high-resolution monochrome display. In this example we use an inexpensive 128 x 64 pixel OLED screen (128 columns and 64 rows) with an onboard SSD1306 controller chip, as can be found on hobbyist websites. The PIC32 uses I^2C to communicate with the SSD1306. The OLED library we provide gives a simple interface to the OLED display; however, it does not provide comprehensive access to all of the controller's functions.

Each pixel is represented by a single bit. The PIC32 stores pixel data in a framebuffer in PIC32 RAM, a copy of the OLED controller's RAM. The functions `display_pixel_set` and `display_pixel_get` access this framebuffer rather than directly accessing the OLED controller's memory. The function `display_draw` copies the whole framebuffer to the OLED controller, which updates the screen.

The sample code consists of three files: the two files of the OLED library `i2c_display.{c,h}` and a main program `i2c_pixels.c`. The main program uses the OLED library to draw diagonal lines across the screen (Figure 13.3).

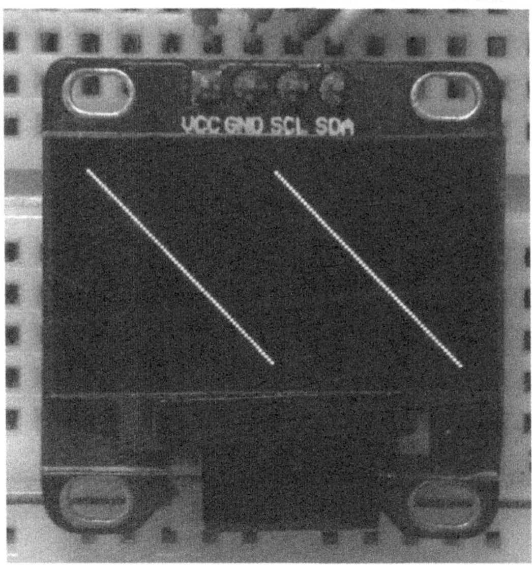

Figure 13.3
An OLED screen.

Code Sample 13.10 `i2c_display.h`. **Header File for Controlling an OLED Display.**

```
#ifndef I2C_DISPLAY_H__
#define I2C_DISPLAY_H__
// bare-bones driver for interfacing with the SSD1306 OLED display via I2C
// not fully featured, just demonstrates basic operation
// note that resetting the PIC doesn't reset the OLED display, only power cycling does

#define WIDTH 128 //display width in bits
#define HEIGHT 64 //display height, in bits

void display_init(void); // initialize I2C1

void display_command(unsigned char cmd); // issue a command to the display

void display_draw(void);                 // draw the buffer in the display

void display_clear(void);                // clear the display

void display_pixel_set(int row, int col, int val); // set pixel at given row and column

int display_pixel_get(int row, int col);  // get the pixel at the given row and column

#endif
```

Code Sample 13.11 `i2c_display.c`. **OLED Screen Interfacing Code.**

```
#include "i2c_master_int.h"
#include "i2c_display.h"
#include <stdlib.h>
// control the SSD1306 OLED display

#define DISPLAY_ADDR 0x3C

#define SIZE WIDTH*HEIGHT/8 //display size, in bytes

static unsigned char video_buffer[SIZE+1] = {0};// buffer corresponding to display pixels
                                               // for sending over I2C. The first byte
                                               // lets us to store the control character
static unsigned char * gddram = video_buffer + 1; // the video buffer start, excluding
                                                 // address byte we write these pixels
                                                 // to GDDRAM over I2C


void display_command(unsigned char cmd) {// write a command to the display
  unsigned char to_write[] = {0x00,cmd}; // 1st byte = 0 (CO = 0, DC = 0), 2nd is command
  i2c_write_read(DISPLAY_ADDR, to_write,2, NULL, 0);
}

void display_init() {
  i2c_master_setup();
                          // goes through the reset procedure
  display_command(0xAE);  // turn off display

  display_command(0xA8);     // set the multiplex ratio (how many rows are updated per
                             // oled driver clock) to the number of rows in the display
  display_command(HEIGHT-1); // the ratio set is the value sent+1, so subtract 1
```

```
                           // we will always write the full display on a single update.
  display_command(0x20); // set address mode
  display_command(0x00); // horizontal address mode
  display_command(0x21); // set column address
  display_command(0x00); // start at 0
  display_command(0xFF); // end at 127
                           // with this address mode, the address will go through all
                           // the pixels and then return to the start,
                           // hence we never need to set the address again

  display_command(0x8d); // charge pump
  display_command(0x14); // turn on charge pump to create ~7 Volts needed to light pixels
  display_command(0xAF); // turn on the display
  video_buffer[0] = 0x40;// co = 0, dc =1, allows us to send data directly from video
                           // buffer, 0x40 is the "next bytes have data" byte
}

void display_draw() {     // copies data to the gddram on the oled chip
  i2c_write_read(DISPLAY_ADDR, video_buffer, SIZE + 1, NULL, 0);
}

void display_clear() {
  memset(gddram,0,SIZE);
}

// get the position in gddram of the pixel position
static inline int pixel_pos(int row, int col) {
  return (row/8)*WIDTH + col;
}

// get a bitmask for the actual pixel position, based on row
static inline unsigned char pixel_mask(int row) {
  return 1 << (row % 8);
}

// invert the pixel at the given row and column
void display_pixel_set(int row, int col,int val) {
  if(val) {
    gddram[pixel_pos(row,col)] |= pixel_mask(row);   // set the pixel
  } else {
    gddram[pixel_pos(row,col)] &= ~pixel_mask(row);  // clear the pixel
  }
}

int display_pixel_get(int row, int col) {
  return (gddram[pixel_pos(row,col)] & pixel_mask(row)) != 0;
}
```

Code Sample 13.12 `i2c_pixels.c`**. Draw Some Lines on an OLED Screen.**

```
#include "NU32.h"          // constants, funcs for startup and UART
#include "i2c_display.h"
// Tests the OLED driver by drawing pixels

int main() {
  NU32_Startup();  // cache on, interrupts on, LED/button init, UART init
  display_init();
```

```
    int row, col;
    for(col = 0; col < WIDTH; ++col) { // draw a diagonal line
      row = col % HEIGHT;              // when we hit the last row
      display_pixel_set(row,col,1);   // start from row 0, but keep advancing
                                      // the column
      display_draw();                 // we draw every update, to display progress.
    }
    display_draw();

    return 0;
  }
```

13.3.5 Multiple Devices

To test three devices on the same I^2C bus—the PIC32, the accelerometer, and the OLED screen—and to have a little fun, we implemented the classic arcade game *Snake* (Figure 13.4). The goal of this game is to move the snake, represented by a string of pixels, to eat food without the snake's head running into the boundaries of the screen or the body of the snake itself. The snake's head moves north, south, east, or west depending on the direction the player tilts the accelerometer, and the snake's body trails along behind the head. When the snake's head passes over a food pixel, a new food pixel appears, and the body of the snake grows by one pixel, increasing the challenge.

The OLED screen and the accelerometer are used as slaves, and they have different slave addresses. The code relies on `i2c_snake.c` (below), `i2c_accel.c`, `accel.h`, `i2c_master_int.c`, and `i2c_master_int.h`.

Code Sample 13.13 `i2c_snake.c`. **The Game of Snake, on an OLED Screen.**

```
#include "NU32.h"          // cache on, interrupts on, LED/button init, UART init
#include "i2c_display.h"
#include "accel.h"

// the game of snake, on an oled display. eat those pixels!
```

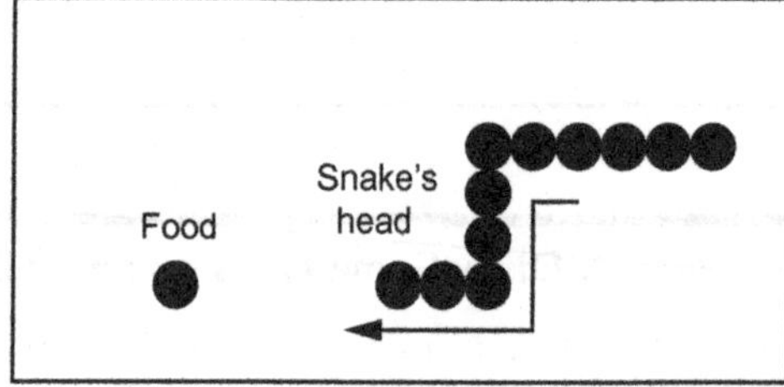

Figure 13.4
The game of *Snake*. Eat the food, and do not crash into the wall or yourself!

```
#define MAX_LEN WIDTH*HEIGHT

typedef struct {
  int head;
  int tail;
  int rows[MAX_LEN];
  int cols[MAX_LEN];
} snake_t; // hold the snake

// direction of the snake
typedef enum {NORTH = 0, EAST = 1, SOUTH = 2, WEST= 3} direction_t;

// grow the snake in the appropriate direction, returns false if snake has crashed
int snake_grow(snake_t * snake, direction_t dir) {
  int hrow = snake->rows[snake->head];
  int hcol = snake->cols[snake->head];

  ++snake->head;
  if(snake->head == MAX_LEN) {
    snake->head = 0;
  }
  switch(dir) {                  // move the snake in the appropriate direction
    case NORTH:
      snake->rows[snake->head] = hrow -1;
      snake->cols[snake->head] = hcol;
      break;
    case SOUTH:
      snake->rows[snake->head] = hrow + 1;
      snake->cols[snake->head] = hcol;
      break;
    case EAST:
      snake->rows[snake->head] = hrow;
      snake->cols[snake->head] = hcol + 1;
      break;
    case WEST:
      snake->rows[snake->head] = hrow;
      snake->cols[snake->head] = hcol -1;
      break;
   }
  // check for collisions with the wall or with itself and return 0, otherwise return 1
  if(snake->rows[snake->head] < 0 || snake->rows[snake->head] >= HEIGHT
      || snake->cols[snake->head] < 0 || snake->cols[snake->head] >= WIDTH) {
    return 0;
  } else if(display_pixel_get(snake->rows[snake->head],snake->cols[snake->head]) == 1) {
      return 0;
  } else {
    display_pixel_set(snake->rows[snake->head],snake->cols[snake->head],1);
    return 1;
  }

}

void snake_move(snake_t * snake)  { // move the snake by deleting the tail
  display_pixel_set(snake->rows[snake->tail],snake->cols[snake->tail],0);
  ++snake->tail;
  if(snake->tail == MAX_LEN) {
    snake->tail = 0;
  }
}
```

```
int main(void) {
  NU32_Startup();
  display_init();
  acc_setup();

  while(1) {
    snake_t snake = {5, 0, {20,20,20,20,20,20},{20,21,22,23,24,25}};
    int dead = 0;
    direction_t dir = EAST;
    char dir_key = 0;
    char buffer[3];
    int i;
    int crow, ccol;
    int eaten = 1;
    int grow = 0;
    short acc[2]; // x and y accleration
    short mag;
    for(i = snake.tail; i <= snake.head; ++i) { // draw the initial snake
      display_pixel_set(snake.rows[i],snake.cols[i],1);
    }
    display_draw();
    acc_read_register(OUT_X_L_M,(unsigned char *)&mag,2);
    srand(mag); // seed the random number generator with the magnetic field
                // (not the most random, but good enough for this game)
    while(!dead) {
      if(eaten) {
        crow = rand() % HEIGHT;
        ccol = rand() % WIDTH;
        display_pixel_set(crow,ccol,1);
        eaten = 0;
      }

      //determine direction based on largest magnitude accel and its direction
      acc_read_register(OUT_X_L_A,(unsigned char *)&acc,4);
      if(abs(acc[0]) > abs(acc[1])) { // move snake in direction of largest acceleration
        if(acc[0] > 0) {              // prevent snake from turning 180 degrees,
          if(dir != EAST) {           // resulting in an automatic self crash
            dir = WEST;
          }
        } else
          if( dir != WEST) {
            dir = EAST;
          }
      } else {
        if(acc[1] > 0) {
          if( dir != SOUTH) {
            dir = NORTH;
          }
        } else {
          if( dir != NORTH) {
            dir = SOUTH;
          }
        }
      }
      if(snake_grow(&snake,dir)) {
        snake_move(&snake);
      } else if(snake.rows[snake.head] == crow && snake.cols[snake.head] == ccol) {
          eaten = 1;
          grow += 15;
      } else {
        dead = 1;
```

```
            display_clear();
          }
          if(grow > 0) {
              snake_grow(&snake,dir);
              --grow;
          }
          display_draw();
        }
      }
      return 0;
    }
```

13.4 Chapter Summary

- I^2C communication requires two wires, one for the clock (SCL) and another for data (SDA). Both lines should be pulled up to 3.3 or 5 V with resistors.
- An I^2C device can be either a master or a slave. The master controls the clock and initiates all communication. Usually the PIC32 operates as the master, but it can also be a slave.
- Each slave has a unique address, and it only responds when the master issues its address, allowing multiple slaves to be connected to a single master. Multiple devices can assume the role of master, but only one master can operate at a time.
- The I^2C peripheral automatically handles the primitives of I^2C communication, like ADDRESS, WRITE, and RECEIVE. A full I^2C transaction, however, requires a sequence of these primitive commands that must be handled in software. The master can sequence the commands by polling for status flags indicating the completion of a primitive, or by generating interrupts on the completion of primitives.

13.5 Exercises

1. Why do the I^2C bus lines require pull-up resistors?
2. Write a program that reads a series of bytes (entered as hexadecimal numbers) from the terminal emulator and sends them over I^2C to an external chip. It should then display the data received over I^2C as hexadecimal numbers. If you want, you may also allow for direct control over sending various I^2C primitives such as START or RESET. Note: you can read and write hexadecimal numbers using the `\%x` format specifier with `sscanf` and `sprintf`.

Further Reading

LSM303D ultra compact high performance e-compass 3D accelerometer and 3D magnetometer module. (2012). STMicroelectronics.

PIC32 family reference manual. Section 24: Inter-integrated circuit. (2013). Microchip Technology Inc.

SSD1306 OLED/PLED segment/common driver with controller. (2008). Solomon Systech.

UM10204 I^2C*-bus specification and user manual* (v. 6). (2014). NXP Semiconductors.

CHAPTER 14

Parallel Master Port

The parallel master port (PMP) uses multiple wires to communicate multiple bits simultaneously (in parallel). This is in contrast to serial communication, where one bit is sent at a time. Parallel ports were once popular on computers and printers, but they have now been displaced by newer communication technologies, such as USB and Ethernet. Nevertheless, the PMP provides a simple interface to some devices, such as some LCD controllers.

14.1 Overview

The PIC32 has one PMP peripheral. Despite its name, the PMP can act as a master or a slave; however, we focus on using it as a master. As master, the PMP has different modes of communicating with the slave, making it suitable for interfacing with various parallel devices.

The PIC32's PMP can use up to 16 address pins (PMA0 to PMA15), 8 data pins (PMD0 to PMD7), and an assortment of control pins. The address pins are often used to indicate the slave's remote memory address to read to or write from. The peripheral can optionally increment the address after each read/write to enable reading from/writing to multiple registers on the target device. The data pins are used for sending bytes between the PMP and the device. Control pins implement hardware handshaking, a signaling mechanism that tells both the PIC32 and device when data is available on the pins and the direction of the data transfer (read or write).

The PMP supports two basic methods of hardware handshaking, master modes 1 and 2. Both methods involve two pins and pulsed signals called strobes. A strobe is a single pulse used to signal the slave device; it also refers to the pin that issues the pulse. In master mode 2, the pins are called parallel master read (PMRD, shared with RD5) and parallel master write (PMWR, shared with RD4). To read from the device, the PMP pulses the read strobe (PMRD), signaling the device to output bits to the data pins. Writing requires pulsing the write strobe (PMWR), which signals the device to read bits from the data pins. Either the PIC32 or the external device can control the data lines; when one is driving a line, the other must be in a high-impedance (input) state.

Embedded Computing and Mechatronics with the PIC32 Microcontroller. http://dx.doi.org/10.1016/B978-0-12-420165-1.00014-7

In master mode 1, the RD5 pin is called PMRD/PMWR and the RD4 pin is called PMENB, the parallel master enable strobe. The PMRD/PMWR signal determines whether the operation is a read or a write while the strobe PMENB initiates the read or write.

The timing of the data and strobes must satisfy timing requirements according to the slave's data sheet. The strobe itself must be of sufficient duration. Microchip calls this duration `WaitM`. Data setup time, before the strobe, is `WaitB`, and data should be valid on the data lines for time `WaitE` after the end of the strobe.

14.2 Details

The parallel master port uses several SFRs; however, some are not used in master mode. We omit the SFRs used only in slave mode. All bits of the SFRs below default to 0.

PMCON The main configuration register.

PMCON⟨15⟩ or PMCONbits.ON: Setting this bit enables the PMP.

PMCON⟨9⟩ or PMCONbits.PTWREN: Setting this bit enables the PMWR/PMENB pin (shared with RD4). The pin is called PMWR when using master mode 2 and PMENB when using master mode 1. Most devices require the use of PMWR/PMENB so you will usually set this bit.

PMCON⟨8⟩ or PMCONbits.PTRDEN: Setting this bit enables the PMRD/PMWR pin (shared with RD5). The pin is called PMRD when using master mode 2 and PMRD/PMWR when using master mode 1.

PMCON⟨1⟩ or PMCONbits.WRSP: Determines the polarity of either PMWR or PMENB, depending on whether you are using master mode 2 or master mode 1, respectively. When set, the strobe pulse is from low to high back to low. When clear, the strobe pulse is from high to low back to high.

PMCON⟨0⟩ or PMCONbits.RDSP: Determines the polarity of either PMRD or PMRD/PMWR, depending on whether you are using master mode 2 or master mode 1, respectively. When set, the strobe pulse is from low to high back to low. When clear, the strobe pulse is from high to low back to high.

PMMODE Controls the PMP's operating mode and tracks its status.

PMMODE⟨15⟩ or PMMODEbits.BUSY: The PMP sets this bit when the peripheral is busy. Poll this bit to determine when a read or write operation has finished.

PMMODE⟨9:8⟩ or PMMODEbits.MODE: Determines the mode of the PMP. The two master modes are

0b11 Master mode 1, uses a single strobe PMENB and a read/write direction signal PMRD/PMWR.

0b10 Master mode 2, uses a read strobe PMRD and a write strobe PMWR.

PMMODE⟨7:6⟩ or PMMODEbits.WAITB: Determines the amount of time between initiating a read/write and triggering the strobe. The time inserted will be

(1 + PMMODEbits.WAITB)T_{pb}, where T_{pb} is the peripheral bus clock period, for a minimum of one wait state (T_{pb}) and a maximum of four wait states ($4T_{pb}$).

PMMODE⟨5:2⟩ or PMMODEbits.WAITM: Determines the duration of the strobe. The duration is given by (1 + PMMODEbits.WAITM)T_{pb}, for a minimum of one wait state (T_{pb}) and a maximum of 16 wait states ($16T_{pb}$).

PMMODE⟨1:0⟩ or PMMODEbits.WAITE: Determines how long the data should be valid after the end of the strobe. For reads, the number of wait states is PMMODEbits.WAITE and for writes it is PMMODEbits.WAITE + 1. Each wait state is one T_{pb} long.

PMADDR Set this register to the desired read/write address before performing the operation. The peripheral may also be configured to automatically increment this register after each read or write, allowing you to access a series of consecutive registers on the slave device. The bits of this register that are used by the PMP are determined by PMAEN (below).

PMDIN In either master mode, use this register to read from and write to the slave. Writes to this register are sent to the slave. Reading a value from the PMP requires you to read from this SFR twice. The first read returns the value currently in the PMDIN buffer, which is old data and should be ignored. The first read also initiates the PMP read sequence, causing a strobe and new data to be latched into the PMDIN register. The second read of PMDIN returns the data actually sent by the slave.

PMAEN Determines which parallel port address pins will be used by the PMP and which pins are available to other peripherals. To claim a pin PMA0 to PMA15 for the PMP, set the corresponding bit to 1.

The PMP's interrupt vector is `_PMP_VECTOR`, its interrupt flag status bit is IFS1bits.PMPIF, its interrupt enable control bit is IEC1bits.PMPIE, and its priority and subpriority are contained in IPC7bits.PMPIP and IPC7bits.PMPIS. Interrupts can be generated on every completed read or write operation. The sample code below does not use interrupts.

14.3 Sample Code

The PMP is used to implement the LCD library for a Hitachi HD44780 or compatible LCD controller.[1] The header file may look familiar to you, as it is the one we saw in Chapter 4.

In this example, we use master mode 1, with the PMRD/PMWR direction bit (RD5) attached to the HD44780's R/$\overline{\text{W}}$ input and the PMENB strobe (RD4) attached to the HD44780's E input. Only one address pin is used, PMA10 (RB13), and in this case this single bit is used to select either the HD44780's 8-bit instruction register (PMA10 = 0) or its 8-bit data register

[1] Many controllers compatible with the HD44780 exist. We used the Samsung KS006U when testing the examples in this chapter.

(PMA10 = 1). The HD44780's data sheet describes the valid 8-bit instructions, such as "clear the display" and "move the cursor." A write to the LCD involves assigning a 0 (instruction) or 1 (data) to the PMA10 bit, then writing an 8-bit value to PMDIN, representing either the instruction (e.g., clear the screen) or data (e.g., a character to be printed).

The functions that interact directly with the PMP are `wait_pmp`, `LCD_Setup`, `LCD_Read`, and `LCD_Write`.

Code Sample 14.1 `LCD.h`. **An LCD Control Library Using the PMP.**

```
#ifndef LCD_H
#define LCD_H
// LCD control library for Hitachi HD44780-compatible LCDs.

void LCD_Setup(void);                          // Initialize the LCD
void LCD_Clear(void);                          // Clear the screen, return to position (0,0)
void LCD_Move(int line, int col);              // Move position to the given line and column
void LCD_WriteChar(char c);                    // Write a character at the current position
void LCD_WriteString(const char * string);     // Write string starting at current position
void LCD_Home(void);                           // Move to (0,0) and reset any scrolling
void LCD_Entry(int id, int s);                 // Control display motion after sending a char
void LCD_Display(int d, int c, int b);         // Turn display on/off and set cursor settings
void LCD_Shift(int sc, int rl);                // Shift the position of the display
void LCD_Function(int n, int f);               // Set number of lines (0,1) and the font size
void LCD_CustomChar(unsigned char val, const char data[7]); // Write custom char to CGRAM
void LCD_Write(int rs, unsigned char db70);   // Write a command to the LCD
void LCD_CMove(unsigned char addr);            // Move to the given address in CGRAM
unsigned char LCD_Read(int rs);                // Read a value from the LCD
#endif
```

Code Sample 14.2 `LCD.c`. **Implementation of an LCD Control Library Using the PMP.**

```
#include "LCD.h"
#include<xc.h>     // SFR definitions from the processor header file and some other macros

#define PMABIT 10 // which PMA bit number to use

// wait for the peripheral master port (PMP) to be ready
// should be called before every read and write operation
static void waitPMP(void)
{
  while(PMMODEbits.BUSY) { ; }
}

// wait for the LCD to finish its command.
// We check this by reading from the LCD
static void waitLCD() {
  volatile unsigned char val = 0x80;

  // Read from the LCD until the Busy flag (BF, 7th bit) is 0
  while (val & 0x80) {
    val = LCD_Read(0);
  }
  int i = 0;
  for(i = 0; i < 50; ++i) { // slight delay
```

```
    _nop();
  }
}

// set up the parallel master port (PMP) to control the LCD
// pins RE0 - RE7 (PMD0 - PMD7) connect to LCD pins D0 - D7
// pin RD4 (PMENB) connects to LCD pin E
// pin RD5 (PMRD/PMWR) Connects to LCD pin R/W
// pin RB13 (PMA10) Connects to RS.
// interrupts will be disabled while this function executes
void LCD_Setup() {
  int en = __builtin_disable_interrupts(); // disable interrupts, remember initial state

  IEC1bits.PMPIE = 0;     // disable PMP interrupts
  PMCON = 0;              // clear PMCON, like it is on reset
  PMCONbits.PTWREN = 1;   // PMENB strobe enabled
  PMCONbits.PTRDEN = 1;   // PMRD/PMWR enabled
  PMCONbits.WRSP = 1;     // Read/write strobe is active high
  PMCONbits.RDSP = 1;     // Read/write strobe is active high

  PMMODE = 0;             // clear PMMODE like it is on reset
  PMMODEbits.MODE = 0x3; // set master mode 1, which uses a single strobe

  // Set up wait states.  The LCD requires data to be held on its lines
  // for a minimum amount of time.
  // All wait states are given in peripheral bus clock
  // (PBCLK) cycles.  PBCLK of 80 MHz in our case
  // so one cycle is 1/80 MHz = 12.5 ns.
  // The timing controls asserting/clearing PMENB (RD4) which
  // is connected to the E pin of the LCD (we refer to the signal as E here)
  // The times required to wait can be found in the LCD controller's data sheet.
  // The cycle is started when reading from or writing to the PMDIN SFR.
  // Note that the wait states for writes start with minimum of 1 (except WAITE)
  // We add some extra wait states to make sure we meet the time and
  // account for variations in timing amongst different HD44780 compatible parts.
  // The timing we use here is for the KS066U which is faster than the HD44780.
  PMMODEbits.WAITB = 0x3;  // Tas in the LCD datasheet is 60 ns
  PMMODEbits.WAITM = 0xF;  // PWeh in the data sheet is 230 ns (we don't quite meet this)
                           // If not working for your LCD you may need to reduce PBCLK
  PMMODEbits.WAITE = 0x1;  // after E is low wait Tah (10ns)

  PMAEN |= 1 << PMABIT;   // PMA is an address line

  PMCONbits.ON = 1;        // enable the PMP peripheral
  // perform the initialization sequence
  LCD_Function(1,0);       // 2 line mode, small font
  LCD_Display(1, 0, 0);    // Display control: display on, cursor off, blinking cursor off
  LCD_Clear();             // clear the LCD
  LCD_Entry(1, 0);         // Cursor moves left to right. do not shift the display

  if(en & 0x1)             // if interrupts were enabled before, re-enable them
  {
    __builtin_enable_interrupts();
  }
}

// Clears the display and returns to the home position (0,0)
void LCD_Clear(void) {
  LCD_Write(0,0x01); //clear the whole screen
}
```

```
// Return the cursor and display to the home position (0,0)
void LCD_Home(void) {
  LCD_Write(0,0x02);
}

// Issue the LCD entry mode set command
// This tells the LCD what to do after writing a character
// id : 1 increment cursor, 0 decrement cursor
// s : 1 shift display right, 0 don't shift display
void LCD_Entry(int id, int s) {
  LCD_Write(0, 0x04 | (id << 1) | s);
}

// Issue the LCD Display command
// Changes display settings
// d : 1 display on, 0 display off
// c : 1 cursor on, 0 cursor off
// b : 1 cursor blinks, 0 cursor doesn't blink
void LCD_Display(int d, int c, int b) {
  LCD_Write(0, 0x08 | (d << 2) | (c << 1) | b);
}

// Issue the LCD display shift command
// Move the cursor or the display right or left
// sc : 0 shift cursor, 1 shift display
// rl : 0 to the left, 1 to the right
void LCD_Shift(int sc, int rl) {
  LCD_Write(0,0x1 | (sc << 3) | (rl << 2));
}

// Issue the LCD Functions set command
// This controls some LCD settings
// You may want to clear the screen after calling this
// n : 0 one line, 1 two lines
// f : 0 small font, 1 large font (only if n == 0)
void LCD_Function(int n, int f) {
  LCD_Write(0, 0x30 | (n << 3) | (f << 2));
}

// Move the cursor to the desired line and column
// Does this by issuing a DDRAM Move instruction
// line : line 0 or line 1
// col  : the desired column
void LCD_Move(int line, int col) {
  LCD_Write(0, 0x80 | (line << 6) | col);
}

// Sets the CGRAM address, used for creating custom
// characters
// addr address in the CGRAM to make current
void LCD_CMove(unsigned char addr) {
  LCD_Write(0, 0x40 | addr);
}

// Writes the character to the LCD at the current position
void LCD_WriteChar(char c) {
  LCD_Write(1, c);
}

// Write a string to the LCD starting at the current cursor
void LCD_WriteString(const char *str) {
```

```
  while(*str) {
    LCD_WriteChar(*str); // increment string pointer after char sent
    ++str;
  }
}

// Make val a custom character.  This only implements
// The small font version
// val : between 0 and 7
// data : 7 character array.  The first 5 bits of each character
//        determine whether that pixel is on or off
void LCD_CustomChar(unsigned char val, const char * data) {
  int i = 0;
  for(i = 0; i < 7; ++i) {
    LCD_CMove(((val & 7) << 2) | i);
    LCD_Write(1, data[i]);
  }
}

// Write data to the LCD and wait for it to finish by checking the busy flag.
// rs : the value of the RS signal, 0 for an instruction 1 for data
// data : the byte to send
void LCD_Write(int rs, unsigned char data) {
  waitLCD();           // wait for the LCD to be ready
  if(rs) { // 1 for data
    PMADDRSET = 1 << PMABIT;
  } else { // 0 for command
    PMADDRCLR = 1 << PMABIT;
  }
  waitPMP();           // Wait for the PMP to be ready
  PMDIN = data;        // send the data
}

// read data from the LCD.
// rs : the value of the RS signal 0 for instructions status, 1 for data
unsigned char LCD_Read(int rs) {
  volatile unsigned char val = 0; // volatile so 1st read doesn't get optimized away
  if(rs) { // 1 to read data
    PMADDRSET = 1 << PMABIT;
  } else { // 0 to read command status
    PMADDRCLR = 1 << PMABIT;
  }
  // from the PIC32 reference manual, you must read twice to actually get the data
  waitPMP();                  // wait for the PMP to be ready
  val = PMDIN;
  waitPMP();
  val = PMDIN;
  return val;
}
```

14.4 Chapter Summary

- The PMP sends multiple bits of data simultaneously; however, this requires using many more wires than serial communication protocols.
- The PMP performs hardware handshaking. The master tells the slave device when to read data and when to output data.

- Devices with parallel ports use slightly different protocols, hence the PMP is highly configurable. For more details on the options, consult the Reference Manual.

14.5 Exercises

1. Compare and contrast parallel and serial communication protocols.
2. The LCD display used in this chapter has the ability to scroll text, allowing you to display messages longer than the screen length. By consulting the data sheet and the provided `LCD` library, implement a program that displays messages entered from the terminal emulator. If the message is too long for a single line, use the LCD's scrolling features to display the whole message.

Further Reading

HD44780U (LCD-II) dot matrix liquid crystal display controller/driver. HITACHI.
KS0066U 16COM/40SEG driver and controller for dot matrix LCD. Samsung.
PIC32 family reference manual. Section 13: Parallel master port. (2012). Microchip Technology Inc.

CHAPTER 15

Input Capture

The input capture (IC) peripheral monitors an external signal and stores a timer value when that signal changes, allowing precise timing of external events. In some sense, input capture can be viewed as the opposite of output compare. Output compare changes the value of an output pin based on the value of a timer, whereas input capture stores the value of a timer based on the value of an input pin.

15.1 Overview

The PIC32 has five input capture modules, each associated with a single pin. The input capture peripheral monitors the voltage on the pin and can trigger on external events such as a rising or falling edge. When the specified event occurs, the module stores the value of a timer in a FIFO buffer and, optionally, triggers an interrupt. Thus, each event receives a timestamp from the PIC32, allowing software to know when a certain event occurred or to calculate the duration of a high or low pulse. Figure 15.1 depicts the operation of the IC module.

A simple way for one microcontroller to communicate an analog value to another microcontroller is to have the first generate a PWM signal whose duty cycle corresponds to the analog value (between 0 and 1). The second microcontroller uses input capture to measure the duty cycle. Input capture is also used in conjunction with Hall effect sensors to control the commutation of brushless motors (Chapter 29).

Prior to using the IC module, you should configure a timer. The frequency of the timer determines the precision of the times captured by the IC peripheral. The IC peripheral always synchronizes the input event with the system clock; therefore, if the timer runs at high frequency (e.g., using the 80 MHz peripheral bus clock as input with a prescaler equal to one), there may be a delay of up to three timer cycles between when the event occurs and when the timer value is recorded. Slower timers, however, do not have this issue—the data can be synchronized within one timer period.

15.2 Details

The input capture peripheral uses two SFRs, ICxCON and ICxBUF, where x = 1 to 5.

Embedded Computing and Mechatronics with the PIC32 Microcontroller. http://dx.doi.org/10.1016/B978-0-12-420165-1.00015-9

ICxCON The main IC control register. All bits default to zero.

ICxCON⟨15⟩ or ICxCONbits.ON: Set to one to enable the module. Clear to zero to disable the module and reset it.

ICxCON⟨9⟩ or ICxCONbits.FEDGE: This "First Capture Edge" bit is only relevant in interrupt capture mode 6, below (ICxCONbits.ICM = 0b110). If this bit is one, then the first edge captured is a rising edge; if it is a zero, then it is a falling edge.

ICxCON⟨8⟩ or ICxCONbits.C32: If one, the IC peripheral uses the 32-bit timer Timer23. If zero, the IC uses a 16-bit timer.

ICxCON⟨7⟩ or ICxCONbits.ICTMR: Determines which timer to use, if ICxCONbits.C32 is zero. If ICxCONbits.C32 is one, this value is ignored.

- 0 Use Timer3.
- 1 Use Timer2.

ICxCON⟨6:5⟩ or ICxCONbits.ICI: Interrupt after ICxCONbits.ICI + 1 capture events, where ICxCONbits.ICI takes values from zero to three.

ICxCON⟨4⟩ or ICxCONbits.ICOV: Input capture overrun, read-only. The IC peripheral has a FIFO buffer that can store four timer values. When this FIFO is full and another capture event occurs, hardware sets ICxCONbits.ICOV. In certain modes, ICxCONbits.ICOV will not be set even when the buffer is full. To clear ICxCONbits.ICOV you must read from the ICxBUF, removing the overflow condition.

ICxCON⟨3⟩ or ICxCONbits.ICBNE: Read-only. Set whenever event times are available in the input capture FIFO, accessed via ICxBUF.

ICxCON⟨2:0⟩ or ICxCONbits.ICM: Determines the events that cause the IC module to trigger.

- 0b111 Allows the IC pin to be used to wake the PIC32 from sleep.
- 0b110 First trigger on the edge specified by ICxCONbits.FEDGE. Afterwards, trigger on every edge.
- 0b101 Capture every sixteenth rising edge.
- 0b100 Capture every fourth rising edge.

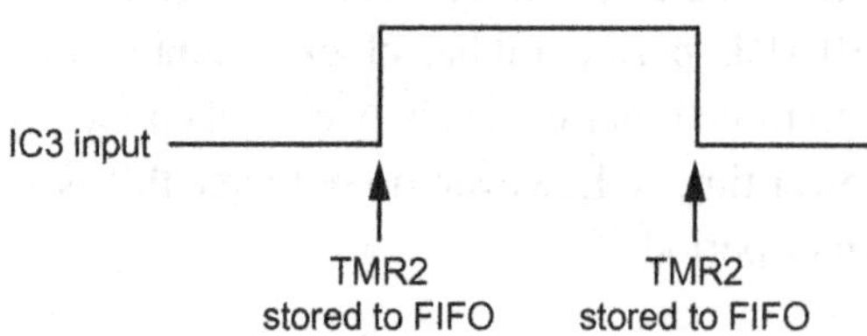

Figure 15.1
Input capture module 3 configured to store the value of Timer2 on every change of the IC3 digital input.

0b011 Capture every rising edge.
0b010 Capture every falling edge.
0b001 Capture every edge.
0b000 Input capture disabled.

ICxBUF Read-only buffer that returns the timer value captured by the IC peripheral. This SFR reads the next element from a four-value deep FIFO; thus, a maximum of four events can be captured between reads of the FIFO. When the FIFO contains data, ICxCONbits.ICBNE is set, indicating that a read from ICxBUF will contain a captured timestamp. To clear the overflow flag, ICxCONbits.ICOV, read from ICxBUF, which removes a value from the FIFO.

The interrupt vector for input capture module 1 is `_INPUT_CAPTURE_1_VECTOR`, and the flag status, enable control bits, and priority and subpriority bits are in IFS0bits.IC1IF, IEC0bits.IC1IE, IPC1bits.IC1IP, and IPC1bits.IC1IS, respectively. For ICx, x = 2 to 5, the vector names are similar, replacing `_1_` with `_x_`. The flag status and enable control bits are all in IFS0. The priority and subpriority bits are in IPC2, IPC3, IPC4, and IPC5 for IC2, IC3, IC4, and IC5, respectively.

15.3 Sample Code

The following code demonstrates how an output compare module can send a PWM signal encoding a value between 0 and 1 (the duty cycle) to an input capture module. Usually these modules would reside on different microcontrollers; in this example, we use a single PIC32.

The code configures OC1 to use Timer2 as its timer base. Timer2 uses the peripheral bus clock as input, with a prescaler of 8, meaning that TMR2 increments at 10 MHz. Timer2 rolls over at a period match value of 9999, thereby generating a PWM signal of frequency 10 MHz/(9999+1) = 1 kHz.

IC1 uses Timer3 for its timing operations. Timer3 is also configured to increment at 10 MHz. IC1 is configured to capture every rising and falling edge, beginning with a rising edge (IC1CONbits.ICM = 0b110). IC1 generates an interrupt after every four edges, and the ISR reads the four recently captured times and calculates the PWM signal's period and duty cycle.

The user is repeatedly prompted for the duty cycle of OC1's PWM by entering the number of clock cycles that the signal is high each PWM cycle. The PWM's period and duty cycle, as measured by IC1, is then printed to the user's screen. Here is example output:

```
PWM period = 10,000 ticks of 10 MHz clock = 1 ms (1 kHz PWM).
Enter the high portion in 10 MHz (100 ns) ticks, in range 5-9995.
You entered 250.
Measured period is 10000 clock cycles, high for 250 cycles,
  for a duty cycle of  2.50 percent.
```

Code Sample 15.1 `input_capture.c`**. Using Input Capture to Measure the Duty Cycle of a PWM Signal.**

```
#include "NU32.h"      // constants, funcs for startup and UART

// Use IC1 (D8) to measure the PWM duty cycle of OC1 (D0).
// Connect D8 to D0 for this to work.

#define TMR3_ROLLOVER 0xFFFF   // defines rollover count for IC1's 16-bit Timer3

static volatile int icperiod = -1;     // measured period, in Timer3 counts
static volatile int ichigh = -1;       // measured high duration, in Timer3 counts

void __ISR(_INPUT_CAPTURE_1_VECTOR, IPL3SOFT) InputCapture1() {
  int rise1, fall1, rise2, fall2;
  rise1 = IC1BUF;                 // time of first rising edge
  fall1 = IC1BUF;                 // time of first falling edge
  rise2 = IC1BUF;                 // time of second rising edge
  fall2 = IC1BUF;                 // time of second falling edge; not used below
  if (fall1 < rise1) {            // handle Timer3 rollover between rise1 and fall1
    fall1 = fall1 + TMR3_ROLLOVER+1;
    rise2 = rise2 + TMR3_ROLLOVER+1;
  }
  else if (rise2 < fall1) {       // handle Timer3 rollover between fall1 and rise2
    rise2 = rise2 + TMR3_ROLLOVER+1;
  }
  icperiod = rise2 - rise1;       // calculate period, time between rising edges
  ichigh = fall1 - rise1;         // calculate high duration, between 1st rise and 1st fall
  IFS0bits.IC1IF = 0;             // clear interrupt flag
}

int main() {
  char buffer[100] = "";
  int val = 0, pd = 0, hi = 0; // desired pwm value, period, and high duration
  int i = 0;                   // loop counter

  NU32_Startup();  // cache on, interrupts on, LED/button init, UART init
  __builtin_disable_interrupts();

  // set up PWM signal using OC1 using Timer2
  T2CONbits.TCKPS = 0x3;      // Timer2 1:8 prescaler; ticks at 10 MHz (each tick is 100ns)
  PR2 = 9999;                 // roll over after 10,000 ticks, or 1 ms (1 kHz)
  TMR2 = 0;
  OC1CONbits.OCM = 0b110;     // PWM mode without fault pin; other OC1CON are defaults
                              // (use TMR2)
  T2CONbits.ON = 1;           // turn on Timer2
  OC1CONbits.ON = 1;          // turn on OC1

  // set up IC1 to use Timer3. IC1 could also use Timer2 (sharing with OC1) in this case
  // since we set both timers to the same frequency, but we'd need to incorporate the
  // different rollover period in the ISR
  T3CONbits.TCKPS = 0x3;      // Timer3 1:8 prescaler; ticks at 10 MHz (each tick is 100ns)
  PR3 = TMR3_ROLLOVER;        // rollover value is also used in ISR to handle
                              // timer rollovers.
  TMR3 = 0;
  IC1CONbits.ICTMR = 0;       // IC1 uses Timer3
  IC1CONbits.ICM = 6;         // capture every edge
  IC1CONbits.FEDGE = 1;       // capture rising edge first
```

```
    IC1CONbits.ICI = 3;         // interrupt every 4th edge
    IFS0bits.IC1IF = 0;         // clear interrupt flag
    IPC1bits.IC1IP = 3;         // interrupt priority 3
    IEC0bits.IC1IE = 1;         // enable IC1 interrupt
    T3CONbits.ON = 1;           // turn on Timer3
    IC1CONbits.ON = 1;          // turn on IC1

    __builtin_enable_interrupts();

    while(1) {
      NU32_WriteUART3("PWM period = 10,000 ticks of 10 MHz clock = 1 ms (1 kHz PWM).\r\n");
      NU32_WriteUART3("Enter high portion in 10 MHz (100 ns) ticks, in range 5-9995.\r\n");
      NU32_ReadUART3(buffer,sizeof(buffer));
      sscanf(buffer,"%d",&val);
      if (val < 5) {
        val = 5;                 // try removing these limits and understanding the
      } else if (val > 9995) {  // behavior when val is close to 0 or 10,000
        val = 9995;
      }
      sprintf(buffer,"You entered %d.\r\n",val);
      NU32_WriteUART3(buffer);
      OC1RS = val;                      // change the PWM duty cycle
      for (i=0; i<1000000; i++) {       // a short delay as PWM updates and
        _nop();                         // IC1 measures the signal
      }
      __builtin_disable_interrupts(); // disable ints briefly to copy vars shared with ISR
      pd = icperiod;
      hi = ichigh;
      __builtin_enable_interrupts();  // re-enable the interrupts
      sprintf(buffer,"Measured period is %d clock cycles, high for %d cycles, \r\n",pd,hi);
      NU32_WriteUART3(buffer);
      sprintf(buffer,"  for a duty cycle of %5.2f percent.\r\n\n",
              100.0*((double) hi/(double) pd));
      NU32_WriteUART3(buffer);
    }
    return 0;
  }
```

15.4 Chapter Summary

- The input capture peripheral allows you to record the times of the rising and/or falling edges of a digital input.
- The precision of the time recorded depends on the frequency of the timer used. When the timer runs at high frequency (e.g., the peripheral bus 80 MHz frequency), there may be a lag of up to three timer cycles between the time the event occurs and when it is recorded.

15.5 Exercises

1. What other peripherals can you use to approximate input capture functionality?
2. What advantages does input capture provide over using other peripherals that can approximate input capture's functionality?

Further Reading

PIC32 family reference manual. Section 15: Input capture. (2010). Microchip Technology Inc.

CHAPTER 16

Comparator

A comparator compares two analog voltages, outputting a digital high signal if input V_{IN+} is greater than input V_{IN-} and a digital low signal otherwise (Figure 16.1). Each of the two inputs to a comparator can be an external input (e.g., from a sensor) or an internally generated reference. One of these internal references is a low-resolution digital-to-analog converter that can create 16 different reference voltages. This reference can also be made available at an output pin, so although the PIC32 does not have a true DAC, it does have a very simple one.

16.1 Overview

The PIC32 has two comparators, CM1 and CM2. The noninverting (+) input for CM1 can be selected from the input voltage at pin C1IN+ or CV_{REF}, the output of the four-bit internal comparator reference voltage DAC. The inverting (−) input can be selected from the input voltages at pin C1IN−, C1IN+, C2IN+, or an internal voltage IV_{REF} of 1.2 V. The comparator output can be queried in software, made available on the output pin C1OUT, or used to trigger an interrupt. CM2 behaves exactly as CM1, replacing the "1" in the names of the possible inputs and output with a "2."

Additionally, the CV_{REF} voltage can be output to an external pin (CVREFOUT/RB10), providing the functionality of a simple DAC. This output has relatively high output impedance, so if you are connecting CVREFOUT to another circuit, it is a good idea to buffer the output (see Appendix B.5). The different values that CV_{REF} can take are explained below when discussing the CVRCON SFR. The comparator and voltage reference peripherals are discussed in separate chapters in the Reference Manual.

16.2 Details

Each comparator has its own control SFR, CM1CON, and CM2CON, but they share a status SFR, CMSTAT.

CMxCON, x = 1 or 2 Control register for the comparators.

CMxCON⟨15⟩ or CMxCONbits.ON: Set to one to enable the comparator.

Embedded Computing and Mechatronics with the PIC32 Microcontroller. http://dx.doi.org/10.1016/B978-0-12-420165-1.00016-0

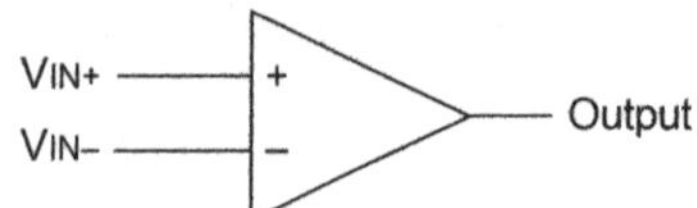

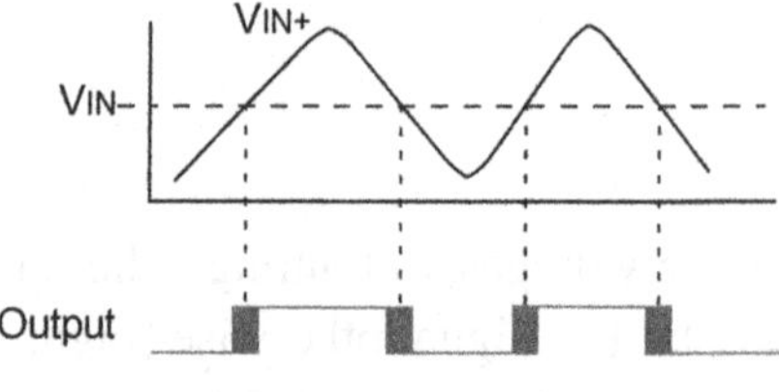

Figure 16.1
Comparator.

CMxCON⟨14⟩ or CMxCONbits.COE: Setting this comparator output enable bit to one causes the comparator's output to drive the CxOUT pin. When set, the corresponding TRIS SFR bit should be cleared to zero, making it an output (0). C1OUT corresponds to pin RB8 and C2OUT is pin RB9.

CMxCON⟨13⟩ or CMxCONbits.CPOL: Setting this comparator polarity bit to one inverts the comparator's output.

CMxCON⟨8⟩ or CMxCONbits.COUT: The output of the comparator. Can also be read from CMSTAT.

CMxCON⟨7:6⟩ or CMxCONbits.EVPOL: These interrupt event polarity bits control the conditions under which the comparator generates an interrupt.

- 0b11 Interrupt when the comparator output transitions from high to low or low to high.
- 0b10 Interrupt when the comparator output transitions from high to low.
- 0b01 Interrupt when the comparator output transitions from low to high.
- 0b00 Do not interrupt.

CMxCON⟨4⟩ or CMXCONbits.CREF: Determines what is connected to the noninverting (+) comparator input.

- 1 Internal CV_{REF}, created by the comparator reference voltage DAC (see below).
- 0 External CxIN+ pin. If you are using USB, then C1IN+ cannot be used as an input pin, so this bit should be set to 1.

CMxCON⟨1:0⟩ or CMxCONbits.CCH: Controls what is connected to the inverting (−) comparator input. Let y represent the number of the other comparator; that is, if using CM1CON (x = 1), then y = 2, and if using CM2CON (x = 2), then y = 1. Then the meaning of the CMxCONbits.CCH bit field is determined as follows:

- 0b11 Connected to IV_{REF}, the internal voltage reference, which is 1.2 V.
- 0b10 External CyIN+ pin.

0b01 External CxIN+ pin.
0b00 External CxIN− pin.

CMSTAT The comparator status register, shared by both comparators.

CMSTAT⟨1⟩ or CMSTATbits.C2OUT: The output of comparator 2.

CMSTAT⟨0⟩ or CMSTATbits.C1OUT: The output of comparator 1.

The comparator reference voltage peripheral has a single SFR, CVRCON. All bits default to zero.

CVRCON The comparator reference voltage control register. Controls the reference voltage value and whether it is output to a pin.

CVRCON⟨15⟩ or CVRCONbits.ON: Set to one to enable the comparator reference voltage.

CVRCON⟨6⟩ or CVRCONbits.CVROE: Set this output enable bit to output the reference voltage CV_{REF} on the CVREFOUT (RB10) pin. If clear, voltage will only be accessible internally.

CVRCON⟨5⟩ or CVRCONbits.CVRR: Controls the range of the CV_{REF} output voltage, according to Figure 16.2. Assuming the default setting for CVRCONbits.CVRSS (below), the 16 available output voltages are as follows:

1 From 0 V to 2.06 V, in steps of 3.3 V/24 = 0.1375 V.
0 From 0.83 V to 2.37 V, in steps of 3.3 V/32 ≈ 0.103 V.

CVRCON⟨4⟩ or CVRCONbits.CVRSS: Selects the sources for Vmax and Vmin in Figure 16.2.

1 : Uses external voltage references CV_{REF+} (RA10) and $CV_{REF}-$ (RA9).

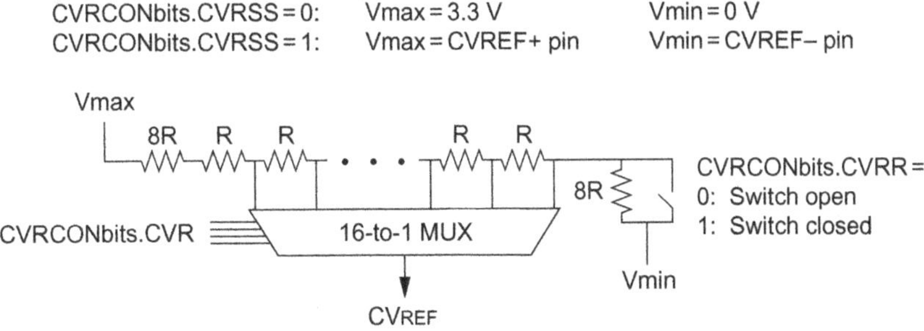

Figure 16.2

The CVREF DAC circuit. CVRCONbits.CVRSS controls whether Vmax and Vmin are 3.3 V and 0 V, respectively, or whether they are determined by inputs at the CVREF+ and CVREF− pins. CVRCONbits.CVRR determines whether the voltage steps are 1/32 of the voltage range (CVRCONbits.CVRR = 0, switch open) or 1/24 of the voltage range (1, switch closed), as well as whether the minimum voltage is 1/4 of the voltage range (0, switch open) or at the bottom of the voltage range (1, switch closed). The four bits of CVRCONbits.CVR select CVREF from the 16 available voltages in the resistor network.

0 : The default, uses the analog supply voltages AV_{dd} (3.3 V) and AV_{ss} (0 V).

CVRCON⟨3:0⟩ or CVRCONbits.CVR: These four bits, taking values 0 to 15, determine the actual voltage output, CV_{REF}. Depends on the settings of CVRCONbits.CVRSS and CVRCONbits.CVRR. When CVRCONbits.CVRR = 1,

$$CV_{REF} = \text{Vmin} + (\text{CVRCONbits.CVR}/24) \times (\text{Vmax} - \text{Vmin}), \tag{16.1}$$

and when CVRCONbits.CVRR = 0,

$$CV_{REF} = \text{Vmin} + (0.25 + \text{CVRCONbits.CVR}/32) \times (\text{Vmax} - \text{Vmin}). \tag{16.2}$$

The interrupt vector for CM1 is `_COMPARATOR_1_VECTOR`, and its interrupt flag, enable control, and priority and subpriority bits are in IFS1bits.CMP1IF, IEC1bits.CMP1IE, IPC7bits.CMP1IP and IPC7bits.CMP1IS. The vector for CM2 is `_COMPARATOR_2_VECTOR`, and the relevant bits are also in IFS1, IEC1, and IPC7, replacing "CMP1" by "CMP2."

16.3 Sample Code

16.3.1 Voltage Comparison

The following code uses CM2 to compare an external voltage to an internally generated reference signal. The internal reference voltage is IV_{REF} = 1.2 V, and the external voltage is applied to the noninverting (+) input via C2IN+ (RB3). Whenever the voltage on C2IN+ is greater than 1.2 V, both LEDs will illuminate. You can test the example by applying the output of various voltage dividers to the C2IN+ pin.

Code Sample 16.1 `comparator.c`. **Basic Comparator Example**

```
#include "NU32.h"  // constants, funcs for startup and UART

// Uses comparator 2 to interrupt on a low voltage condition.
// The + comparator terminal is connected to C2IN+ (B3).
// The - comparator terminal is connected to the internal voltage IVref (1.2 V).
// Both NU32 LEDs illuminate if the + input is > 1.2 V; otherwise, off.
// The comparator output can be viewed on C2OUT (B9) with a voltmeter or measured by ADC.

int main(void) {
  NU32_Startup();          // cache on, interrupts on, LED/button init, UART init
  CM2CONbits.COE = 1;      // comparator output is on the C2OUT pin, so you can measure it
  CM2CONbits.CCH = 0x3;    // connect - input to IVref; by default + connected to C2IN+
  TRISBbits.TRISB9 = 0;    // configure B9 as an output, which must be set to use C2OUT
  CM2CONbits.ON = 1;

  while(1) {
    // test the comparator output
    if(CMSTATbits.C2OUT) { // if output is high then the input signal > 1.2 V
      NU32_LED1 = 0;
      NU32_LED2 = 0;
```

```
    } else {
      NU32_LED1 = 1;
      NU32_LED2 = 1;
    }
  }
  return 0;
}
```

16.3.2 Analog Output

This example demonstrates how to output the internal comparator reference voltage to a pin. This reference voltage has relatively high output impedance, since it is designed to be input to a comparator; therefore, this output should be buffered in most cases. Without a buffer circuit, you can still look at the output voltage using a high-impedance input like a voltmeter or an oscilloscope. The reference voltage cycles through the 16 available values, one per second.

Code Sample 16.2 `ref_volt.c`. **Output Comparator Reference Voltage to a Pin**

```
#include "NU32.h"    // constants, funcs for startup and UART

// Use the comparator reference voltage as a cheap DAC.
// Voltage is output on CVREFOUT (RB10). You can measure it.
// You will most likely need to buffer this output to use it to
// drive a low-impedance load.

int main(void) {
  int i;
  NU32_Startup();            // cache on, interrupts on, LED/button init, UART init
  TRISBbits.TRISB10 = 0;  // make the CVrefout/RB10 pin an output
  CVRCONbits.CVROE = 1;   // output the voltage on CVrefout
  CVRCONbits.CVRR = 0;    // use the smaller output range, at higher voltages
  CVRCONbits.ON = 1;      // turn the module on
  while(1) {
    for(i = 0; i < 16; ++i) { // step through the voltages, one per second
      CVRCONbits.CVR = i;
      _CP0_SET_COUNT(0);
      while(_CP0_GET_COUNT() < 40000000) {
        ;
      }
    }
  }
  return 0;
}
```

16.4 Chapter Summary

- Comparators allow you to compare two analog voltages and determine which one is larger.
- Comparators can compare two external signals or an external signal with an internal reference voltage.
- The comparator reference voltage can be output to an external pin, serving as a simple four-bit DAC.

16.5 Exercises

1. Describe the advantages and disadvantages of using the PIC32's internal voltage reference for analog output compared to pairing a PWM signal with a low-pass filter.
2. Use the PIC32's two comparators and appropriate reference voltages to implement a 2-bit analog-to-digital converter.

Further Reading

PIC32 family reference manual. Section 19: Comparator. (2010). Microchip Technology Inc.
PIC32 family reference manual. Section 20: Comparator voltage reference. (2012). Microchip Technology Inc.

CHAPTER 17

Sleep, Idle, and the Watchdog Timer

In battery-powered applications, conserving energy is important. The PIC32 provides multiple power-saving modes which help reduce energy consumption, for example by reducing the system or peripheral bus clock frequencies. Apart from changing clock frequencies, the PIC32 offers an *idle* mode, where the CPU halts but the peripherals continue to operate, unless they are individually disabled during idle mode. To save the most power, the PIC32 can be placed in *sleep* mode, where the system clock and peripheral bus clock are shut down. Most peripherals cease to operate except for a few that do not rely on either the system clock or peripheral bus clock. For example, a changed signal on a change-notification digital input can be used to wake the PIC32 from sleep mode.

Another peripheral that can operate in sleep mode is the watchdog timer (WDT), which uses the PIC32's internal low-power RC (LPRC) oscillator to keep time. The WDT continues to operate during sleep, and the WDT rollover can be used to wake up the PIC32 after a fixed period. In addition, even when not in sleep mode, the WDT can be useful for recovering from faulty code. To use this capability, the user's code should periodically reset the WDT, before it rolls over. If the WDT expires, for example because the code is stuck in an unexpected infinite loop, the PIC32 will automatically reset. Thus the WDT provides an escape mechanism if the software ever enters an unexpected state.

This chapter describes the idle and sleep modes and the use of the WDT.

17.1 Overview

17.1.1 Power Saving

The PIC32 has several oscillator sources that can drive the system clock and peripheral clock. Lower frequencies result in less power consumption but also reduced performance. Rather than focus on the numerous oscillator sources and settings, we assume that both the system clock and peripheral bus clock are operating at 80 MHz from the primary oscillator source Posc. Thus, the power-saving methods we focus on involve disabling the CPU and peripherals: the idle and sleep modes.

Embedded Computing and Mechatronics with the PIC32 Microcontroller. http://dx.doi.org/10.1016/B978-0-12-420165-1.00017-2

In both idle and sleep modes, the CPU stops executing instructions. In idle mode, however, the system and peripheral bus clocks continue running and most peripherals function, unless selectively disabled. In sleep mode, the system clock and peripheral bus clock halt, and peripherals relying on either of these clocks cease to function. Therefore, sleep conserves more energy than idle.

To enter sleep or idle mode, you must issue the assembly instruction `wait`. When a peripheral interrupt wakes the PIC32 from sleep, code will continue executing from where the `wait` instruction was issued. The WDT, however, wakes the PIC32 from sleep by causing a reset. This is a simple software reset, not a full reset of the PIC32 as happens when you power cycle the PIC32. For example, the SFR bits do not revert to their default values, as they would with a power cycle; they keep any values that were set previously in the program.

The WDT reset causes the program to jump to the C runtime startup code installed at the reset address, which checks if the reset was caused by a WDT timeout during sleep or idle. If so, control returns to the beginning of your `main` function, where your code can check if the WDT reset the PIC32 during sleep or idle mode. You can issue an `iret` assembly instruction to resume the code from where the `wait` instruction was issued. If the C runtime startup code determines that the WDT reset did not occur during sleep or idle, the regular startup code executes, as if you had pressed the RESET button.

17.1.2 Watchdog Timer

Unlike other peripherals, the primary setup of the WDT occurs in the configuration bits, set when the device is programmed with an external programmer. For the NU32 board, these bits were set when the bootloader was installed (see Section 17.2). The configuration bits control whether the WDT is on or off and the timeout period. If enabled in the configuration bits, software cannot disable the WDT; however, software can enable it if it is not enabled in the configuration bits. Preventing software from changing WDT settings is a safety mechanism: buggy software cannot disable the error recovery mechanism that the WDT provides.

17.2 Details

Several SFRs and configuration words control the WDT and power-saving modes.

DEVCFG1 The configuration word that contains the WDT configuration. This configuration word can only be changed with a programming device. When the bootloader was loaded onto the NU32, this word was set to disable the watchdog timer and to set a rollover period of 4.096 s. The WDT can be enabled in software.

DEVCFG1⟨23⟩ or FWDTEN: Watchdog timer enable bit. If one, the WDT is enabled and cannot be disabled. The bootloader sets this bit to zero using `#pragma config FWDTEN = OFF` in the bootloader code, allowing you to enable the WDT in software or to leave it disabled.

DEVCFG1⟨20:16⟩ or WDTPS: Watchdog timer postscaler select bits. The WDT rollover period is 2^{WDTPS} ms. When the bootloader was installed, the programmer set these bits to 0b01100 = 12 using `#pragma config WDTPS = PS4096`. The LPRC timer operates at 32 kHz and provides a nominal timeout of 1 ms, so this postscaler creates a WDT period of 4.096 s.

WDTCON The WDT control register.

WDTCON⟨15⟩ or WDTCONbits.ON: Set to enable the watchdog timer. This bit is only relevant if the device configuration bit FWDTEN is zero (WDT disabled by default). If FWDTEN is one, the WDT is enabled regardless of the value of WDTCONbits.ON.

WDTCON⟨6:2⟩ or WDTCONbits.SWDTPS: Read-only bits containing the value of the WDT postscaler configuration, DEVCFG1⟨20:16⟩ (WDTPS).

WDTCON⟨0⟩ or WDTCONbits.WDTCLR: Write a 1 to this bit to clear the WDT. If the WDT is enabled and you do not write a 1 to this bit often enough, the PIC32 will reset.

RCON The reset control register. Among other functions, RCON provides information on the type of the most recent reset or whether the PIC32 just woke from a sleep or idle mode.

RCON⟨4⟩ or RCONbits.WDTO: Hardware sets this bit to 1 if the reset occurred due to a WDT timeout.

RCON⟨3⟩ or RCONbits.SLEEP: Hardware sets this bit to 1 if the device had been in sleep mode.

RCON⟨2⟩ or RCONbits.IDLE: Hardware sets this bit to 1 if the device had been in idle mode.

OSCCON Allows you to change some oscillator settings.

OSCCON⟨4⟩ or OSCCONbits.SLPEN: When set, issuing the `wait` instruction causes the PIC32 to enter sleep mode. When clear, the `wait` instruction causes the PIC32 to enter idle mode.

SIDL bit This is not an SFR, but rather the "Stop in Idle Mode" bit in the control register of many peripherals. For example, Timer1's control register has the bit T1CONbits.SIDL. This bit controls whether the peripheral will stop in idle mode. If this bit is set to one, the peripheral will stop functioning (saving power) when the PIC32 is in idle mode. Otherwise the peripheral remains on.

17.3 Sample Code

The following code demonstrates sleep mode and the use of the WDT. The WDT acts as both (1) a guard against faulty code where the WDT is not reset and (2) a way to wake the PIC32 from sleep.

First, the code checks the reason for the last reset. If the WDT caused the PIC32 to reset while in sleep mode, an `iret` instruction is issued, allowing the code to resume where it left off.

Otherwise, the WDT reset happened due to the software getting unexpectedly stuck in an infinite loop, so the program starts from the beginning.

Next, the WDT is enabled and the PIC32 is set to enter sleep mode when a `wait` instruction is issued. The program then enters a loop, printing the alphabet to the serial terminal at approximately one letter per second. After each letter, the software writes to WDTCONbits.WDTCLR, resetting the watchdog timer.

Pressing the NU32's USER button prior to the letter "j" being printed causes the PIC32 to enter sleep mode. While the PIC32 sleeps the WDT continues to tick but WDTCONbits.WDTCLR is no longer set; thus the WDT will time out and reset the PIC32. The code detects that the PIC32 has awoken, and, rather than restarting the program, resumes from where it left off.

Pressing the USER button after the letter "j" causes the PIC32 to enter an infinite loop that does nothing, simulating faulty code. As the code no longer sets WDTCONbits.WDTCLR, the WDT will eventually time out, resetting the PIC32. Upon resetting, the program will start printing the letters from the beginning. When the PIC32 resets in this manner it typically indicates a bug in the code.

Code Sample 17.1 `wdt.c`. **Sleep Mode and WDT Demonstration.**

```
#include "NU32.h"          // constants, funcs for startup and UART

int main(void) {

  if(RCONbits.WDTO && RCONbits.SLEEP) { // reset due to WDT waking PIC32 from sleep?
      __asm__ __volatile__ ("iret");    // if so, resume where we left off.
  }

  NU32_Startup();       // cache on, interrupts on, LED/button init, UART init

  if(RCONbits.WDTO) { // WDT caused the reset, but it was not from sleep mode
    RCONbits.WDTO = 0;// clear WDT reset.  Subsequent resets due to the reset button
                      // won't be interpreted as a WDT reset
    NU32_WriteUART3("\r\nReset after a WDT timeout in infinite loop.\r\n");
  }

  char letters[2] = "a";  // second char is the string terminator
  int pressed = 0;        // true if button pressed during the delay

  OSCCONbits.SLPEN = 1;   // enters sleep (not idle) when 'wait' instruction issued

  // print instructions
  NU32_WriteUART3("Press USER button before 'j' to go to sleep; after to enter a\r\n");
  NU32_WriteUART3("faulty infinite loop. If sleep, the WDT will wake the PIC32.\r\n");
  NU32_WriteUART3("If infinite loop, the WDT will reset the PIC32 and start over.\r\n");

  WDTCONbits.ON = 1;            // turn on the WDT (rollover of 4.096s in DEVCFG1)
```

```
  while(1) {
    if(!NU32_USER || pressed) { // USER button pushed (NU32_USER is low if USER pressed)
      if(letters[0] < 'j') {    // if button pushed early in program, go to sleep
        NU32_WriteUART3(" Going to sleep ... ");
        __asm__ __volatile__ ("wait"); // until the WDT rolls over and wakes from sleep
        NU32_WriteUART3(" Waking up. ");
      } else {                  // if button pushed late, get stuck in infinite loop
        NU32_WriteUART3(" Getting stuck in infinite loop.\r\n");
        while(1) {
          _nop();               // fortunately the WDT will reset the PIC32
        }
      }
    }
    NU32_WriteUART3(letters);
    ++letters[0];
    pressed = 0;
    _CP0_SET_COUNT(0);
    while(_CP0_GET_COUNT() < 40000000) {
      pressed |= !NU32_USER;    // delay for ~1 second, still poll the user button
    }
    WDTCONbits.WDTCLR = 1;      // clear the watchdog timer
  }
  return 0;
}
```

17.4 Chapter Summary

- Lower oscillator frequencies result in lower power consumption but also reduced performance.
- In idle mode, the CPU stops executing instructions but most peripherals can continue to operate. Peripherals, including the WDT, can wake the PIC32 from idle.
- In sleep mode, the CPU stops executing instructions and only peripherals that do not rely on the system clock or peripheral bus clock can operate. Some peripherals, like change notification or the WDT, can wake the PIC32 from sleep.
- You must periodically write to WDTCONbits.WDTCLR to reset an enabled WDT; otherwise the PIC32 will reset.
- Important WDT settings must be set in the configuration bits.
- Software can check RCON at reset to determine the reason for the reset.

17.5 Exercises

1. Describe a situation in which saving power is important. In such a situation would you prefer to be in sleep mode or idle mode more often? Why?
2. Pretend that, to protect against the possibility of an inadvertent infinite loop, you have enabled the watchdog timer. Your code works most of the time; however, in some situations it resets due to the watchdog timer expiring. Does this behavior indicate an error in your code?

3. In the scenario of the previous Exercise, your friend suggests that you should simply disable the watchdog timer to fix the problem. Do you agree? Why or why not?

Further Reading

PIC32 family reference manual. Section 10: Power-saving modes. (2011). Microchip Technology Inc.

PIC32 family reference manual. Section 07: Resets. (2013). Microchip Technology Inc.

PIC32 family reference manual. Section 09: Watchdog timer and power-up timer. (2013). Microchip Technology Inc.

CHAPTER 18

Flash Memory

Flash memory retains its contents even when powered off. So far, we have used this nonvolatile memory (NVM) to store your programs and the bootloader; however, we have not explicitly accessed it from software. By writing to the appropriate SFRs, software can write data to flash memory, allowing you to save data even when the PIC32 loses power. The process of writing to flash via software, outlined here, is often referred to as run-time self programming (RTSP). The bootloader uses RTSP to store your programs on the PIC32.

Flash can also be written using an external programmer such as the PICkit 3. This chapter does not discuss how external programmers work; their operation is described in the PIC32 Flash Programming Specification. The bootloader was originally stored on your PIC32 using an external programmer.

18.1 Overview

Our PIC32 contains 12 kB of boot flash and 512 kB of flash program memory. Program memory is divided into 128 *pages*, each 4 kB (4096 bytes). Each page is further divided into eight *rows*, each containing 128 four-byte words.

As per the PIC32 memory model (see Chapter 2), each byte of flash resides at a unique physical address. The CPU (or the prefetch cache module) can directly read instructions or data from flash memory during the execution of your program. If you have ever declared any initialized `const` global variables, the linker stored these in flash, unlike other data stored in RAM. This behavior is specific to the XC32 compiler and is not a rule for C generally.

Unlike reading from flash, which is straightforward, writing to flash is a bit peculiar, for the following reasons:

- If you are not careful which physical address you write to, you could accidentally overwrite program instructions!
- To prevent accidental corruption of program instructions and data in flash, your code must implement a specific unlocking sequence before writing to flash.
- "Erasing" a region of flash memory means setting all bits to one. A "write" operation only changes some of the ones to zeros; it cannot change zeros to ones. Therefore any write

Embedded Computing and Mechatronics with the PIC32 Microcontroller. http://dx.doi.org/10.1016/B978-0-12-420165-1.00018-4

operation must be preceded by an erase operation. For example, if we erase a four-byte word at a particular physical address (so it now contains 0xFFFFFFFF), then write the four-byte word 0xFFFFFF00 to that address, the address will hold the proper value. But if we then attempt to write the four-byte word 0xFF00FFFF to the same address, without first erasing, the address will now hold the value 0xFF00FF00—the second write was unable to convert the zero bits back to ones.

- The smallest region you can erase is an entire 4 kB page. The only write operations available are a write of a four-byte word or an entire row (128 four-byte words).
- Flash memory fails after too many erase-write cycles, so writing to flash should only be done infrequently. According to the Electrical Characteristics section of the Data Sheet, only 1000 erase/write cycles are guaranteed before flash cells may degrade.

18.2 Details

Writes to flash memory are controlled via the flash SFRs, described in more detail in the Flash Programming section of the Reference Manual. All SFR bits default to zero. While certain pages in program flash memory can be made off-limits to RTSP by setting write-protection bits in the device configuration register DEVCFG0, no pages of program flash were write-protected when DEVCFG0 was configured when the bootloader was installed on the NU32. The boot flash was write-protected, however.

NVMCON The main control register for flash memory.

NVMCON⟨15⟩ or NVMCONbits.WR: Set this write control bit to one to start the flash operation stored in NVMCONbits.NVMOP, below. When the operation completes, the PIC32 clears this bit. This bit can only be set if the flash unlock sequence has been performed and NVMCONbits.WREN = 1.

NVMCON⟨14⟩ or NVMCONbits.WREN: Write enable bit. When set, NVMCONbits.WR can be written, as long as the unlock sequence has been performed. You should set this bit prior to writing to flash and clear it when you are finished.

NVMOP⟨3:0⟩ or NVMCONbits.NVMOP: Determines the operation that is performed when NVMCONbits.WR is set.

0b0101 Erase all program memory pages (mostly useful for bootloaders).

0b0100 Erase a single page, selected by the address NVMADDR.

0b0011 Write a row to the row chosen by NVMADDR. Data to be written is stored at the address in NVMSRCADDR.

0b0001 Write a word to the address stored in NVMADDR. The word to write is stored in NVMDATA.

NVMKEY Used to unlock the NVMCON register to enable erasing and writing. Issue the following commands in order, with interrupts disabled and NVMCONbits.WREN already set to one, to unlock NVMCON and perform the flash operation stored in NVMCONbits.NVMOP:

```
NVMKEY = 0xAA996655;  // unlock step 1
NVMKEY = 0x556699AA;  // unlock step 2; the unlock sequence is complete
NVMCONSET = 0x8000;   // while unlocked, set write control bit to start
                         operation
```

These steps should be performed consecutively to prevent the unlock sequence from timing out.

NVMADDR Stores the physical address of flash memory of the page that will be erased, the row that will be written, or the word that will be written, depending on NVMCONbits.NVMOP.

NVMDATA Stores the data that will be written to flash when a single word is written.

NVMSRCADDR The word-aligned physical address of data to be written when a whole row is programmed. "Word-aligned" means that the physical memory address must be divisible by four (since there are four bytes in a flash word).

18.3 Sample Code

The `flash.{c,h}` library below allocates one page of flash memory and provides functions for erasing the page, writing a single word, or reading a single word.

Examine `flash.c`. Notice the buffer declaration

```
static const unsigned int buffer[PAGE_WORDS] __attribute__ ((__aligned__(PAGE_
SIZE))) = {0};
```

This declaration looks similar to that of a normal global array, except for a few extra keywords. Because the array is declared as `const` and is initialized using `= {0}`, the XC32 linker allocates `buffer` to flash memory rather than RAM.[1] (When you compile a program using this flash library, you can examine the map file to see where this read-only data is allocated in flash.) This data is loaded to flash when you load your program onto the PIC32; it is not re-initialized every time you run your program, as an initialized array in RAM would be.

The `__attribute__ ((__aligned__(PAGE_SIZE)))` code ensures that the linker places `buffer` on a page boundary; that is, the address of `buffer` must be the start of a page in flash memory. In other words, `buffer` (equivalently `&buffer[0]`) must be evenly divisible by the page size (4096).

[1] There is no brief C syntax to initialize the buffer to have all bits equal to one, which would have amounted to an erase of the buffer.

Although declaring a `const` global array is the easiest way to allocate space in flash program memory to hold your persistent data, you can also allocate flash memory directly in the linker script.

The function `flash_op` handles unlocking the flash and executing the appropriate operation. The only operations we have implemented are writing a single word in the buffer and erasing the buffer.

Code Sample 18.1 `flash.h`**. Flash Memory Header File.**

```
#ifndef FLASH__H__
#define FLASH__H__
// the flash module allocates a page of flash and provides read/write accesss

#define PAGE_SIZE 4096                                // size of a page, in bytes
#define PAGE_WORDS (PAGE_SIZE/4)                      // size of a page, in 4-byte words

// erases the flash page by setting all bits to 1's
void flash_erase(void);

// writes the 0's of a 4-byte word
void flash_write_word(unsigned int index, unsigned int data);

// reads a word from flash
unsigned int flash_read_word(unsigned int index);

#endif
```

Code Sample 18.2 `flash.c`**. Flash Memory Implementation.**

```
#include "flash.h"
#include <xc.h>
#include <sys/kmem.h> // macros for converting between physical and virtual addresses

#define OP_ERASE_PAGE 4 // erase page operation, per NVMCONbits.NVMOP specification
#define OP_WRITE_WORD 1 // write word operation, per NVMCONbits.NVMOP specification

// Making the array const and initializing it to 0 ensures that the linker will
// store it in flash memory.
// Since one page is erased at a time, array must be a multiple of PAGE_SIZE bytes long.
// The aligned attribute ensures that the page falls at an address divisible by 4096.

static const unsigned int buf[PAGE_WORDS] __attribute__ ((__aligned__(PAGE_SIZE))) = {0};

static void flash_op(unsigned char op) { // perform a flash operation (op is NVMOP)
  int ie = __builtin_disable_interrupts();

  NVMCONbits.NVMOP = op;  // store the operation
  NVMCONbits.WREN = 1;    // enable writes to the WR bit

  NVMKEY = 0xAA996655;    // unlock step 1
  NVMKEY = 0x556699AA;    // unlock step 2
  NVMCONSET = 0x8000;     // set the WR bit to begin the operation
```

```
  while (NVMCONbits.WR) { // wait for the operation to finish
    ;
  }
  NVMCONbits.WREN = 0;     // disables writes to the WR bit

  if (ie & 0x1) {          // re-enable interrupts if they had been disabled
    __builtin_enable_interrupts();
  }
}

void flash_erase() {          // erase the flash buffer. resets the memory to ones
  NVMADDR = KVA_TO_PA(buf);  // use the physical address of the buffer
  flash_op(OP_ERASE_PAGE);
}

void flash_write_word(unsigned int index, unsigned int data) { // writes a word to flash
  NVMADDR = KVA_TO_PA(buf + index);   // physical address of flash to write to
  NVMDATA = data;                     // data to write
  flash_op(OP_WRITE_WORD);
}

unsigned int flash_read_word(unsigned int index) { // read a word from flash
  return buf[index];
}
```

The following code, `flashbasic.c`, uses the `flash` library to store up to 1023 `unsigned int` words. It first checks whether a valid flash buffer has already been created, perhaps prior to the most recent power-up of the PIC32. If no buffer exists, then it creates the buffer by erasing the page and writing the four-byte hex "password" 0xDEADBEEF at the last element of the buffer. The presence of this password tells the program that a valid buffer exists, and the page should not be erased.

Each time through the infinite loop in `main`, the program tells the user how many words have already been stored in flash and asks whether the user would like to add a new word to the buffer (by typing "a"), show the words that have already been stored ("s"), erase the page and make a new flash buffer ("m"), or generate an error ("e"). The error overwrites the 0xDEADBEEF password, and the next time through the loop the code sees that there is no buffer with a valid password and therefore erases the flash page and starts over. This is the same effect as making a new page.

To add a word to the buffer, the code looks for the first unoccupied element of the buffer to store it in. This is the first element with the "erased" value of 0xFFFFFFFF. Since 0xFFFFFFFF is interpreted as an erased element, the user cannot store this value. Thus only $2^{32} - 1$ different values can be stored by the user, not 2^{32} different values.

The buffer is considered full when the first unoccupied element is the last element of the array, the element holding the password.

After entering some data into the buffer, if you power-cycle the PIC32, you will see that the data is still there. Output from a sample run is shown below.

```
No flash memory allocated currently; making a page.
Currently 0 words stored in flash.
(a)dd word, erase & (m)ake new page, (s)how words, (e)rror? // user enters a
Enter an unsigned int to store at location 0.              // enters 11223344
Adding 0x11223344 at location 0.

Currently 1 words stored in flash.
(a)dd word, erase & (m)ake new page, (s)how words, (e)rror? // user enters a
Enter an unsigned int to store at location 1.              // enters abcdef01
Adding 0xabcdef01 at location 1.

Currently 2 words stored in flash.
(a)dd word, erase & (m)ake new page, (s)how words, (e)rror? // user enters s
at index    0: 0x11223344                                   // data printed
at index    1: 0xabcdef01

Currently 2 words stored in flash.
(a)dd word, erase & (m)ake new page, (s)how words, (e)rror? // user enters a
Enter an unsigned int to store at location 2.              // enters ffffffff,
Adding 0xffffffff at location 2.                           // unstorable word

Currently 2 words stored in flash.                         // still 2 words
(a)dd word, erase & (m)ake new page, (s)how words, (e)rror? // user enters s
at index    0: 0x11223344                                   // ffffffff is
at index    1: 0xabcdef01                                   // an erased element
```

Code Sample 18.3 `flash_basic.c`**. Basic Flash Memory Demonstration.**

```
#include "NU32.h"          // constants, funcs for startup and UART
#include "flash.h"         // allocates buffer of PAGE_WORDS (1024) unsigned ints

// Uses flash.{c,h} library to allocate a page in flash and then write 32-bit
// words there, consecutively starting from the zeroth index in the array.
// LIMITATION:  Words that are all 1's (0xFFFFFFFF) cannot be saved; array
// elements holding this value are considered to be empty (not written since erase).

#define PAGE_IN_USE_PWD     0xdeadbeef // password meaning a valid flash buffer is present
#define PAGE_IN_USE_INDEX  (PAGE_WORDS-1) // the password is at the last index of buffer

int page_exists() {  // valid flash page is allocated if password is at the right index
  return (flash_read_word(PAGE_IN_USE_INDEX) == PAGE_IN_USE_PWD);
}

// returns the index where the next word should be written
unsigned int next_page_index() {
  unsigned int count = 0;

  while ((flash_read_word(count) != 0xFFFFFFFF)
          && (count < PAGE_IN_USE_INDEX)) {
    count++;
  }
  return count;
}
```

```
// page is full if the next word would be written at the password index
int page_full() {
  return (next_page_index() == PAGE_IN_USE_INDEX);
}

// erase page and write the password indicating a flash page is available
void make_page() {
  flash_erase();
  flash_write_word(PAGE_IN_USE_INDEX,PAGE_IN_USE_PWD);
}

void add_new_word() { // add a four-byte word at the next_page_index if page is not full
  char msg[100];
  unsigned int ind = next_page_index();  // where to write the word in the flash page
  unsigned int val;

  if (page_full()) {
    NU32_WriteUART3("Flash page full; no more data will be stored.\r\n");
  }
  else {
    sprintf(msg,"Enter an unsigned int to store at location %d.\r\n",ind);
    NU32_WriteUART3(msg);
    NU32_ReadUART3(msg,sizeof(msg)); // enter word using 8 hex characters, like f01dable
    sscanf(msg,"%x",&val);
    flash_write_word(ind,val);
    sprintf(msg,"Adding 0x%x at location %d.\r\n",val,ind);
    NU32_WriteUART3(msg);
  }
}

void show_words() {  // print out the currently saved four-byte words in hex
  char msg[100];
  unsigned int i;
  for (i = 0; i < next_page_index(); i++) {
    sprintf(msg,"at index %4d: 0x%x \r\n",i,flash_read_word(i));
    NU32_WriteUART3(msg);
  }
}

int main(void) {
  char msg[100]="";

  NU32_Startup(); // cache on, interrupts on, LED/button init, UART init
  while(1) {
    if (!page_exists()) {  // initialize a flash page if the password is not present
      NU32_WriteUART3("\r\nNo flash memory allocated currently; making a page.");
      make_page();
    }
    sprintf(msg,"\r\nCurrently %d words stored in flash.\r\n",next_page_index());
    NU32_WriteUART3(msg);
    NU32_WriteUART3("(a)dd word, erase & (m)ake new page, (s)how words, (e)rror?\r\n");
    NU32_ReadUART3(msg,sizeof(msg));
    switch (msg[0]) {       // check the first character entered by user
      case 'a':
        add_new_word();
        break;
      case 'm':
        make_page();
        break;
      case 's':
        show_words();
```

```
            break;
        case 'e':// shows that if we obliterate the password, then no valid flash page exists
            flash_write_word(PAGE_IN_USE_INDEX,0);
            break;
        default:
            break;
        }
    }
    return 0;
}
```

The code above attempts to minimize page erases, and therefore maximize the flash lifetime, by filling the page with up to 1023 saved data words before having to erase again. One limitation is the inability to distinguish the data 0xFFFFFFFF from an erased element. Another is the slow method for finding the next index where data should be stored—we had to step through the buffer looking for the first unoccupied (erased) element.

More sophisticated methods for managing flash could use some of the buffer elements as *control* bits that represent which array elements are occupied by data. When you write data to an array cell, you write a zero to the corresponding control bit. This scheme overcomes both limitations above, at the cost of having fewer array elements available for data. To lessen the wear on any single page, more than one page could be allocated to hold data, up to the flash memory capacity of the PIC32.

18.4 Chapter Summary

- Flash memory can be used to store data that you want to retain across power cycling of the PIC32.
- Program flash memory is divided into 128 pages of 4 kB each. Each page is divided into eight rows, each consisting of 128 four-byte words.
- Flash can only be erased a page at a time, by setting all bits to ones. Writes can only flip ones to zeros, never zeros to ones. Writes can only be done to a single four-byte word or an entire row at once.
- Flash has a limited lifespan so you should minimize writes and erasures.

18.5 Exercises

1. To minimize flash erasure, pretend that the 4096 words of a page are divided into control and data words. Each bit of a control word corresponds to one or more data words, called a block; a one indicates that the block is empty and a zero indicates that the block is full. Assuming four data words per block, how many data words can a page of flash store?

2. Assume that you want to save a single four-word block at a time, and that each page of flash can be erased 1000 times.
   a. How many times could you store a block of data, assuming that you need to erase the page before each write?
   b. How many times could you write a block of data using the control/data scheme described in Exercise 1?
3. Implement and test the control/data word scheme described in Exercise 1.

Further Reading

PIC32 family reference manual. Section 05: Flash programming. (2012). Microchip Technology Inc.
PIC32 flash programming specification. (2014). Microchip Technology Inc.

CHAPTER 19

Controller Area Network (CAN)

The high-speed controller area network (CAN) is an asynchronous protocol used extensively in industrial automation and for communication between many microprocessors in modern automobiles. CAN allows many devices to communicate over only two wires in electrically noisy environments. Speeds up to 1 Mbps are possible over distances up to tens of meters, with lower bit rates over longer distances. The PIC32 provides two CAN controller peripherals.

19.1 Overview

A CAN bus consists of two wires, CANH and CANL, terminated by 120 Ω resistors at either end. The PIC32, like many devices, cannot directly create the required CAN bus voltages, and therefore connects to the bus through a separate *CAN transceiver*, such as the Microchip MCP2562. Figure 19.1 shows a CAN bus with n devices attached to it, including the PIC32's CAN1 module connected to the bus through an MCP2562. In the discussion below, *devices* have CAN controller peripherals, and they interact with the bus through transceivers.

The DEVCFG3 configuration bit FCANIO on the NU32's PIC32 was cleared to zero when the bootloader was installed, meaning that the CAN1 module uses the alternate pins AC1TX (RF4) and AC1RX (RF5) instead of the default pins C1TX (RF0) and C1RX (RF1). This is because RF0 and RF1 are used as digital outputs to control LED1 and LED2 on the NU32.

To send a message over the bus using CAN1, the PIC32 sends bits from its AC1TX pin to the MCP2562's TXD pin, and the MCP2562 controls the output voltages V_{CANH} and V_{CANL} to send the information over the bus. The MCP2562 also converts bits on the bus to the PIC32's logic levels (0 and 3.3 V) and sends them from its RXD pin to the PIC32's AC1RX pin.

To send a logic low signal (0), the MCP2562 drives V_{CANH} to a high voltage and V_{CANL} to a low voltage. This state of the bus is called *dominant*. The exact value of $V_{\text{CANH}} - V_{\text{CANL}}$ is not critical, provided it exceeds some minimum threshold that allows transceivers on the bus to recognize the dominant (logic 0) state. For example, the MCP2562 requires $V_{\text{CANH}} - V_{\text{CANL}} > 0.9$ V to ensure that the bus is measured as dominant.

To send a logic high signal (1), the MCP2562 CANH and CANL outputs are left floating (high impedance). Because the bus is a closed loop (terminated by resistors), this means that

Embedded Computing and Mechatronics with the PIC32 Microcontroller. http://dx.doi.org/10.1016/B978-0-12-420165-1.00019-6

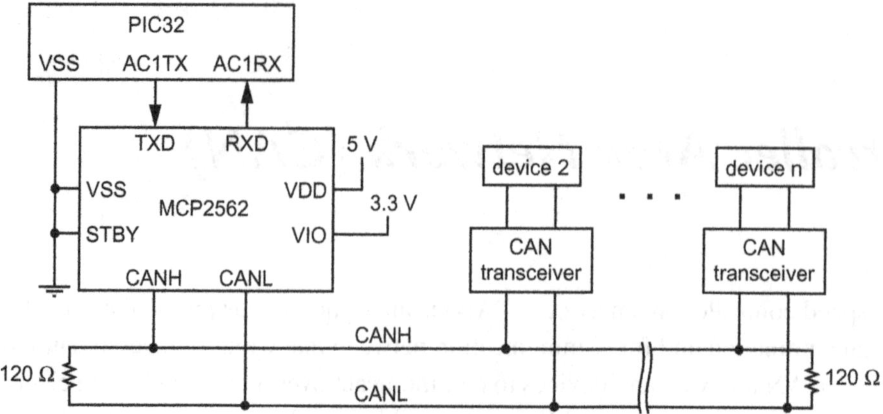

Figure 19.1
The PIC32 and other devices connected to the CAN bus through CAN transceivers. In this figure the PIC32 is connected using CAN module 1; it could also (or instead) be connected using CAN module 2 with pins C2TX and C2RX.

$V_{CANH} \approx V_{CANL}$, assuming that no other transceiver on the bus is making it dominant. This state of the bus is called *recessive*. The MCP2562 recognizes the recessive (logic 1) state if $V_{CANH} - V_{CANL} < 0.5$ V.

The dominant (0) and recessive (1) states of the bus are given these names because the bus is only recessive if all devices are transmitting ones. If any device transmits a zero, its transceiver puts the bus in the dominant (0) state, since the high-impedance outputs of the recessive transceivers cannot affect the bus voltage.

Each device on the bus should be configured with the same nominal bit rate (since, as with the UART, there is no clock signal). If no device is communicating, all transceivers have high-impedance outputs and the bus is in the recessive (logic 1) state. When a device begins to send a data frame, it first sends a logic 0 (drives the bus to dominant) for the duration of one bit. This is called the start-of-frame (SOF) bit. The next 11 bits in the frame carry the standard identifier (SID) of the message, indicating the type of information in the message. The next bit is the remote transmission request (RTR), which is a 0 if the device is sending data and a 1 if it is requesting data from another device. The next bit is the identifier extension bit (IDE), which is 0 for standard CAN frames.[1] The next bit is a reserved bit RB0 which is 0 by convention. The next 4 bits are the data length code (DLC), which indicates the number of data bytes in the transmission, which must be 0-8 by the CAN standard. The next 0-64 bits are the 0-8 bytes of data (DATA) as dictated by the DLC. The next 16 bits are the cyclic redundancy check

[1] The IDE bit is a 1 for *extended* CAN frames. We do not discuss extended CAN data frames, which differ primarily by allowing 29 bits for the message identifier, rather than 11 bits.

(CRC), an error-detection code which receiving devices can use to check if the data was received correctly. The next two bits are the acknowledgment field (ACK); the transmitter sends two ones (recessive), but any receiver that received the data correctly should send a zero (dominant) during the first bit to indicate that the message was received. If there is no acknowledgment, the transmitter realizes that there has been a transmission error and should resend the frame. Finally, the frame is concluded by seven consecutive recessive (1) bits (EOF). The standard CAN frame is illustrated in the following table.

field	SOF	SID	RTR	IDE	RB0	DLC	DATA	CRC	ACK	EOF
# bits	1	11	1	1	1	4	0-64	16	2	7

A standard CAN frame.

Because the timing of bus transitions (dominant to recessive and vice versa) are used to help synchronize devices on the bus (see the description of the CiCFG SFR in Section 19.2), the CAN protocol requires a bit transition at least once every six bits. If the frame has five consecutive bits of the same sign, a bit of the opposite sign is inserted by the transmitter. This is called *bit stuffing*. Thus, depending on the SID and DATA, the length of the actual CAN frame may be extended by stuffed bits. Since each receiving device understands the bit stuffing rule, it discards the stuffed bits. If a receiving device senses six consecutive bits of the same sign, an error has occurred, and the device can transmit six consecutive dominant bits to declare the error. (This chapter does not discuss CAN error handling.) Stuffbits are not added during the ACK or EOF fields, and the seven consecutive recessive bits in the EOF do not signal an error.

After a frame, at least three consecutive recessive (1) bits are inserted as *interframe spacing*. Any recessive-to-dominant transition after this is considered a SOF.

If the bus has been quiet and then one device transmits a SOF bit, other devices wait until the frame is complete before attempting to send a message. If two or more devices attempt to send a message simultaneously, the device with the message with the lower SID wins the arbitration and is able to send its message, while other devices must wait. Arbitration can be achieved because each transmitting device is also measuring the bus voltage, and if the bus is ever dominant (logic 0) when device x is sending a recessive (logic 1), device x knows another device has won the arbitration, and device x stops attempting to transmit. Since the SID is the first bit string sent in the frame, and since the most significant bits of the SID are sent first, the device sending the lowest SID has the most zeros at the beginning of its frame and therefore wins the arbitration.

Messages broadcast over the bus are not necessarily relevant to all devices on the bus, so devices only process messages with SIDs they care about. The PIC32 uses user-specified SID masks and filters to look for relevant messages, and only stores relevant messages in RAM.

19.2 Details

The PIC32 handles CAN frames by using first-in first-out queues (FIFOs). These FIFOs are stored in RAM that you allocate. The CAN peripheral can use up to 32 separate FIFOs, each configurable as either a receive (RX) or transmit (TX) FIFO. Each FIFO consists of a specified number of message buffers, up to 32, to hold messages that are ready to go out (TX FIFO) or have been received (RX FIFO).

Each message buffer is four 32-bit words (16 bytes total), containing all the data needed to construct a full data frame.[2] The four four-byte words are

- CMSGSID: Contains the SID of the message. If it is a received message, then it also contains timestamp information and the ID of the filter that accepted the message.
- CMSGEID: Contains the RTR, IDE, RB0, and DLC bits, as well as other bits if using extended CAN frames.
- CMSGDATA0: Contains bytes 0-3 of the DATA.
- CMSGDATA1: Contains bytes 4-7 of the DATA.

For example, you could have one TX FIFO queue able to hold up to 12 messages, therefore occupying $1 \times 12 \times 16 = 192$ bytes of RAM, and four RX FIFOs each holding up to 32 received messages, therefore occupying $4 \times 32 \times 16 = 2048$ bytes of RAM.

Receive (RX) masks and filters are used to ignore irrelevant messages and to automatically sort relevant messages into the appropriate RX FIFO. A MASK is an 11-bit number, and a message is accepted if the bitwise AND

MASK & SID

matches a user-programmed filter. An accepted message is stored in a FIFO associated with the filter. The user can define up to 32 filters and four masks.

The CAN peripheral relies on a large number of SFRs. Some apply to the CAN peripheral as a whole, while others only apply to individual filters or FIFOs. We provide the information you need to establish basic CAN communication; the Reference Manual section provides a full description. Unlike other peripherals, you must turn on the CAN peripheral and put it in a mode that allows it to be configured. After setting the configuration, you change modes so that CAN can send and receive messages.

The PIC32 has two CAN modules, and in the SFRs below, i is either 1 or 2, indicating CAN1 or CAN2.

---

[2] For an RX FIFO, it is possible to use message buffers with only two 32-bit words, containing only CMSGDATA0 and CMSGDATA1, ignoring the SID, timestamp, DLC, etc. In this chapter we focus on RX messages with four four-byte words.

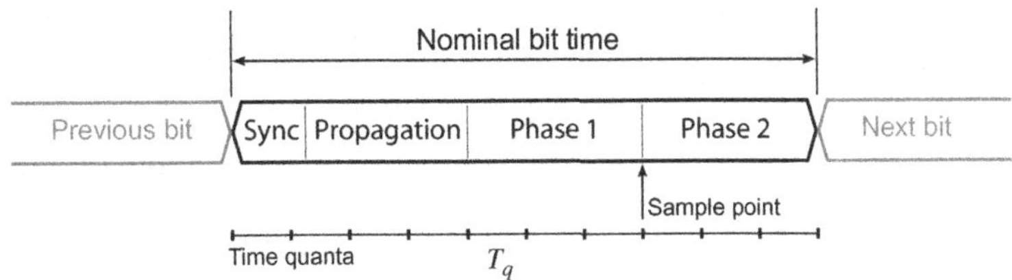

Figure 19.2
Timing for a single bit on the CAN bus, showing the four segments within a bit duration.

CiCON The main configuration register. Allows you to set the operating mode.

CiCON⟨26:24⟩ or CiCONbits.REQOP: Request a change in the operating mode. After requesting a mode change you must wait for the mode to actually change by polling CiCONbits.OPMOD. Switching between certain modes may require a condition on the bus to be met; thus, errors in the wiring of the bus may prevent some mode switches. The main modes are:

0b100 Configuration mode, which allows all CAN SFRs to be modified.

0b010 Loopback mode. The CAN is internally wired to itself, allowing you to test sending and receiving messages. When in loopback mode, the CAN is not actually connected to the bus, so no transceiver is required.

0b000 The CAN peripheral operates normally.

CiCON⟨23:21⟩ or CiCONbits.OPMOD: The current operating mode. Uses the same values as CiCONbits.REQOP. You should poll this value after requesting an operating mode transition to wait for CAN bus to actually enter the desired mode. The CAN peripheral cannot change modes until certain bus conditions are met; therefore, if your code hangs when polling OPMOD you may have a wiring error.

CiCON⟨15⟩ or CiCONbits.ON: Set to 1 to turn the CAN module on. You cannot switch modes unless this bit is set, so, unlike other peripherals, enabling the peripheral is not the last step in its configuration.

CiCFG Configures the bit rate for the CAN peripheral as well as parameters controlling synchronization to other devices on the bus. According to the CAN protocol, each bit duration consists of an integral number of *time quanta* of duration T_q. According to the Reference Manual, the CAN module should be configured so that the duration of a single bit is eight to 25 T_q. For example, at a 1 Mbps bit rate, the duration of each bit is 1 μs, and if there are ten T_q per bit, then $T_q = 100$ ns.

Also according to the CAN protocol, the bit duration is broken into four segments: synchronization, propagation, phase 1, and phase 2 (Figure 19.2). The synchronization phase lasts one T_q, but the others generally last for multiple T_q. Data on the bus is actually sampled at the transition between phase 1 and phase 2.

These segments are used to help CAN devices resynchronize with each other, accounting for oscillator frequency and phase differences and propagation delays. On the SOF recessive-to-dominant transition at the beginning of every data frame, the CAN module restarts its bit timer. In addition, within a data frame, if a transition is measured outside of the synchronization segment, the CAN module resynchronizes by lengthening or shortening phase 1 or phase 2 by an integral number of T_q, depending on the timing error.

The CiCFG SFRs are used to configure the time quantum T_q; the maximum number of T_q that phase 1 and phase 2 can be lengthened or shortened to achieve resynchronization; and the number of T_q in the propagation, phase 1, and phase 2 segments. The overall bit duration is the sum of the durations of these three segments and the synchronization segment (which lasts one T_q). For example, for $T_q = 100$ ns and propagation, phase 1, and phase 2 segments lasting $3T_q$, the total bit duration is $10T_q = 1$ μs and the bit rate is 1 Mbps.

CiCFG⟨13:11⟩ or CiCFGbits.SEG1PH: The duration of phase 1 is (CiCFGbits.SEG1PH + 1)T_q. In the default configuration, CiCFGbits.SEG2PH for phase 2 will be set automatically to CiCFGbits.SEG1PH, making phase 1 and phase 2 equal duration.

CiCFG⟨10:8⟩ or CiCFGbits.PRSEG: The duration of the propagation segment is (CiCFGbits.PRSEG + 1)T_q. Should be at least twice the time it takes signals to propagate down the bus.

CiCFG⟨7:6⟩ or CiCFGbits.SJW: The maximum amount that the phase 1 or phase 2 segments can be adjusted is (CiCFGbits.SJW + 1)T_q. SJW must not be greater than SEG2PH.

CiCFG⟨5:0⟩ or CiCFGbits.BRP: This baud rate prescaler determines T_q, where

$$T_q = 2 \times (\text{CiCFG.BRP} + 1)/F_{sys},$$

and F_{sys} is the frequency of SYSCLK. For example, for an 80 MHz SYSCLK, CiCFGbits.BRP = 0b000011 = 3 means that $T_q = 2 \times (3 + 1)/80\text{ MHz} = 100$ ns.

Setting the bit timing for CAN becomes critical when attempting to communicate quickly or over long distances. Microchip's Application Note AN754 provides more details about choosing the proper settings for your purposes.

CiTREC Tracks the number of receive and transmit errors. This register is especially useful when debugging or in situations where managing failures is critical.

CiFLTCONr, r = 0 to 7 There are eight filter control SFRs. A filter directs data to various RX FIFOs based on the frame's SID. Each register contains configuration bits for four of the 32 available filters. The fields for each filter are

CiFLTCONrbits.FLTENx, x = 0 to 31: Set to enable filter x.

CiFLTCONrbits.MSELx, x = 0 to 31: These two bits, taking values 0b00 to 0b11 (0 to 3), select the mask register for filter x. Mask registers determine which SID bits to ignore when hardware attempts to match an SID to a filter.

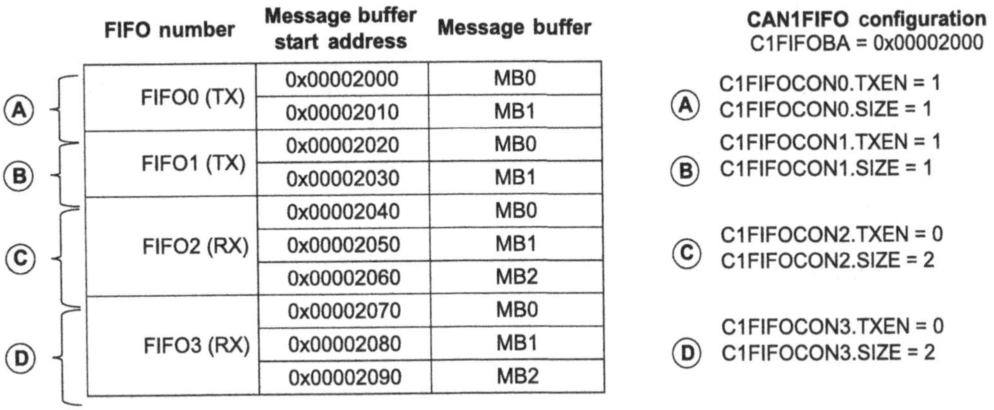

	FIFO number	Message buffer start address	Message buffer
Ⓐ	FIFO0 (TX)	0x00002000	MB0
		0x00002010	MB1
Ⓑ	FIFO1 (TX)	0x00002020	MB0
		0x00002030	MB1
Ⓒ	FIFO2 (RX)	0x00002040	MB0
		0x00002050	MB1
		0x00002060	MB2
Ⓓ	FIFO3 (RX)	0x00002070	MB0
		0x00002080	MB1
		0x00002090	MB2

CAN1FIFO configuration
C1FIFOBA = 0x00002000

Ⓐ C1FIFOCON0.TXEN = 1
C1FIFOCON0.SIZE = 1

Ⓑ C1FIFOCON1.TXEN = 1
C1FIFOCON1.SIZE = 1

Ⓒ C1FIFOCON2.TXEN = 0
C1FIFOCON2.SIZE = 2

Ⓓ C1FIFOCON3.TXEN = 0
C1FIFOCON3.SIZE = 2

Figure 19.3
Memory layout for CAN FIFOs. In this example, four FIFOs are used: two two-message TX FIFOs (FIFO0 and FIFO1) and two three-message RX FIFOs (FIFO2 and FIFO3). Each message buffer consists of four four-byte words, hence the spacing of 0x10 = 16 addresses between message buffers.

CiFLTCONrbits.FSELx, x = 0 to 31: These five bits, taking values 0b00000-0b11111 (0 to 31), associate a FIFO (numbered 0-31) with filter x. When the filter matches, the data will be stored in the specified FIFO.

CiRXFn, n = 0 to 31 One SFR per filter. Specifies the SID that the filter matches. Only operates if enabled by setting the appropriate bits in CiFLTCONr. When using standard CAN frames, only 11 bits are relevant:

CiRXFn⟨31:21⟩ or CiRXFnbits.SID: The SID to match.

CiRXMr, r = 1 to 4 There are four mask registers. Masks determine which bits of an SID to ignore. Each filter uses one mask register.

CiRXMr⟨31:21⟩ or CiRXMrbits.SID: Bits that are zero in this field are the bits of the message SID that will be ignored when matching a filter. For example, if you use the mask C1RXM0bits.SID = 0b101, only bits 0 and 2 of the message's SID matter when matching against a filter.

CiFIFOBA Stores the base physical address of the contiguous region in RAM where the CAN FIFOs are located. This region must be large enough to store all of the FIFOs, up to 32, each having up to 32 message buffers, as prescribed by the user. Each message buffer consists of four 4-byte words. If you use *n* FIFOs, to avoid wasting RAM, you should use consecutive FIFOs from FIFO0 to FIFO(n − 1) (Figure 19.3).

CiFIFOCONn, n = 0 to 31 The control register for FIFOn. Determines the size of the FIFO and whether it is for transmitting or receiving.

CiFIFOCONn⟨20:16⟩ or CiFIFOCONnbits.FSIZE: The number of messages in the FIFO is CiFIFOCONnbits.FSIZE + 1.

CiFIFOCONn⟨14⟩ or FRESET: Set this bit to reset the FIFO. Hardware clears this bit after the reset process finishes. According to Microchip's PIC32MX795 Family Silicon Errata sheet, this bit can only be set using CiFIFOCONnSET = 0x5000. (You cannot set the bit using CiFIFIOCONnbits.FRESET.)

CiFIFOCONn⟨13⟩ or UINC: Set this bit after adding a message to a TX FIFO for sending, or after reading a message from an RX FIFO. This will cause the FIFO pointer CiFIFOUAn (below) to be incremented to the next message buffer in the FIFO. For a TX FIFO, the next message added to the FIFO for sending out on the bus should be placed at this new address, and for an RX FIFO, the next message read from the bus will be placed at this new address. The pointer rolls over to the beginning of the FIFO when it reaches the end of the FIFO. According to Microchip's PIC32MX795 Family Silicon Errata sheet, this bit can only be set using CiFIFOCONnSET = 0x2000. (You cannot set the bit using CiFIFOCONnbits.UINC.)

CiFIFOCONn⟨12⟩ or CiFIFOCONnbits.DONLY: For an RX FIFO, if this bit is set, only the data bytes of received messages will be stored in the FIFO. By default this bit is cleared and received messages consist of four 32-bit words. We use the default in this chapter.

CiFIFOCONn⟨7⟩ or CiFIFOCONnbits.TXEN: When set (1), the FIFO is a TX FIFO. When clear (0), the FIFO is an RX FIFO.

CiFIFOCONn⟨4⟩ or CiFIFOCONnbits.TXERR: This bit is set when an error occurred during transmission. Cleared when read. Useful for debugging.

CiFIFOCONn⟨3⟩ or CiFIFOCONnbits.TXREQ: For TX FIFOs, requests that the data in the FIFO be sent out on the CAN bus. Cleared after messages are successfully sent.

CiFIFOINTn, n = 0 to 31 Contains the interrupt enable (IE) bits and interrupt flag (IF) bits, which indicate which interrupt conditions have been triggered by FIFOn. Interrupt flags indicate the state of the FIFO and may be polled even if the particular interrupt is disabled. Some important status flags are:

CiFIFOINTn⟨10⟩ or CiFIFOINTnbits.TXNFULLIF: Read only. Set to one when a TX FIFO is not full, cleared to zero when a TX FIFO is full.

CiFIFOINTn⟨0⟩ or CiFIFOINTnbits.RXNEMPTYIF: Read only. Set to one when an RX FIFO is not empty and therefore has at least one message, cleared to zero when an RX FIFO is empty.

CiFIFOUAn, n = 0 to 31 This user address stores the physical address of the current position in FIFOn. For TX FIFOs, this address is where you place your next message to go out on the bus. For RX FIFOs, this address is where the next message received from the bus is placed. These addresses are maintained automatically once the user specifies the physical base address of the FIFOs, CiFIFOBA.

The interrupt vectors for CAN modules 1 and 2 are `_CAN_1_VECTOR` (46) and `_CAN_2_VECTOR` (47), respectively. The interrupt flag status bit for CAN1 is IFS1bits.CAN1IF, the interrupt enable control bit is IEC1bits.CAN1IE, the priority is IPC11bits.CAN1IP, and the subpriority is IPC11bits.CAN1IS. Similarly, for CAN2 the relevant bits are IFS1bits.CAN2IF, IEC1bits.CAN2IE, IPC11bits.CAN2IP, and IPC11bits.CAN2IS. Many events can generate a CAN module interrupt, including FIFO events and events described in the CiINT SFR (not covered here). To determine which event caused an interrupt, examine the flag bits in the relevant CAN SFRs. The sample code below does not use interrupts; see the Reference Manual section for more details.

19.2.1 Addresses

The CAN bus manages all of its FIFOs in RAM using physical addresses. Your program, however, uses virtual addresses (see Chapter 3 for more details about the memory map). The header file `<sys/kmem.h>` provides macros for converting between physical and virtual addresses. To convert virtual addresses to physical addresses, such as when you calculate the physical address CiFIFOBA from a pointer (virtual address) to a block of RAM allocated for the FIFOs, use `KVA_TO_PA(virtual_address)`. To convert physical addresses to virtual addresses, such as when you write to a FIFO virtual address based on the CAN's physical address FIFO pointer CiFIFOUAn, use `PA_TO_KVA1(physical_address)`.

19.2.2 Transmitting a Message

To transmit a message, you must load it into a TX FIFO at the current physical address held by CiFIFOUAn. A TX message consists of four four-byte words, CMSGSID, CMSGEID, CMSGDATA0, and CMSGDATA1, which can be viewed as an array of four unsigned integers stored in RAM consecutively. CMSGSID is at the lowest address `addr`, given by `unsigned int * addr = PA_TO_KVA1(CiFIFOUAn)`. The bit fields are illustrated in Figure 19.4, where SRR, EID, and RB1 are only used in an extended CAN frame and are cleared to zero for a standard CAN frame.

The 11-bit SID can take any value between 0 and $2^{11} - 1$. To set the SID in CMSGSID to 27, for example, we can simply write `addr[0] = 27`. The next word, called CMSGEID and stored at `addr[1]`, contains the fields IDE, RTR, RB0, and DLC. IDE is zero for a standard CAN frame, RTR is zero for a normal data transmission, and RB0 is zero by the CAN protocol, so we only need to choose the data length code (DLC). Since DLC occupies the lowest four bits, we can simply write `addr[1] = length`, where `length` is the number of data bytes, 0 to 8. The last two words, CMSGDATA0 at `addr[2]` and CMSGDATA1 at `addr[3]` are the data bytes. If you are sending unsigned integers, you can simply assign the values to `addr[2]` and `addr[3]`. If you are sending other data types, like `chars`, `floats`, or `signed ints`, Exercise 14 in Appendix A gives an idea of how to use `union` to put data of different types into a single array.

Name		Bit 31/2315/7	Bit 30/22/14/6	Bit 29/21/13/5	Bit 28/20/12/4	Bit 27/19/11/3	Bit 26/18/10/2	Bit 25/17/9/1	Bit 24/16/8/0
CMSGSID	31:24	—	—	—	—	—	—	—	—
	23:16	—	—	—	—	—	—	—	—
	15:8	—	—	—	—	—	SID<10:8>		
	7:0	SID<7:0>							
CMSGEID	31:24	—	—	SRR	IDE	EID<17:14>			
	23:16	EID<13:6>							
	15:8	EID<5:0>						RTR	RB1
	7:0	—	—	—	RB0	DLC<3:0>			
CMSGDATA0	31:24	Transmit buffer data byte 3							
	23:16	Transmit buffer data byte 2							
	15:8	Transmit buffer data byte 1							
	7:0	Transmit buffer data byte 0							
CMSGDATA1	31:24	Transmit buffer data byte 7							
	23:16	Transmit buffer data byte 6							
	15:8	Transmit buffer data byte 5							
	7:0	Transmit buffer data byte 4							

Figure 19.4

Four 32-bit (four-byte) words are used to generate a CAN message for transmission. Bits 0-7 of CMSGSID are at the lowest address, while bits 0-7 of CMSGEID, CMSGDATA0, and CMSGDATA1 are at addresses 0x10 (16), 0x20 (32), and 0x30 (48) higher, respectively. For a standard CAN frame, the bit fields SRR, EID, and RB1 are cleared to zero.

A typical process for transmitting a CAN frame is the following:

1. Ensure that the TX FIFO is not full by checking to make sure that CiFIFOINTnbits.TXNFULLIF = 1.
2. Get the address where the next message to be sent on the bus should be stored by reading CiFIFOUAn and converting it to a virtual address using `PA_TO_KVA1`.
3. Store the desired message in the FIFO at the virtual address calculated above.
4. Set the FIFO's UINC bit, using `CiFIFOCONnSET = 0x2000`, to notify the CAN peripheral to increment CiFIFOUAn by one message buffer size (16 bytes).
5. Set CiFIFOCONnbits.TXREQ to one, requesting transmission. This bit will be cleared by hardware when the transmission completes successfully.

19.2.3 Receiving a Message

To receive a message, you must configure an RX FIFO and a filter that matches the desired message. Configuring a filter n requires enabling it and assigning it an SID and mask using CiFLTCONr (r = 0 to 7), CiRXFn (n = 0 to 31), and CiRXMr (r = 1 to 4). When the CAN

Name		Bit 31/2315/7	Bit 30/22/14/6	Bit 29/21/13/5	Bit 28/20/12/4	Bit 27/19/11/3	Bit 26/18/10/2	Bit 25/17/9/1	Bit 24/16/8/0
CMSGSID	31:24	CMSGTS<15:8>							
	23:16	CMSGTS<7:0>							
	15:8	FILHIT<4:0>					SID<10:8>		
	7:0	SID<7:0>							
CMSGEID	31:24	—	—	SRR	IDE	EID<17:14>			
	23:16	EID<13:6>							
	15:8	EID<5:0>						RTR	RB1
	7:0	—	—	—	RB0	DLC<3:0>			
CMSGDATA0	31:24	Receive buffer data byte 3							
	23:16	Receive buffer data byte 2							
	15:8	Receive buffer data byte 1							
	7:0	Receive buffer data byte 0							
CMSGDATA1	31:24	Receive buffer data byte 7							
	23:16	Receive buffer data byte 6							
	15:8	Receive buffer data byte 5							
	7:0	Receive buffer data byte 4							

Figure 19.5
Layout used by the CAN peripheral for storing a received CAN frame.

peripheral matches the SID on the bus to a mask and SID in a filter, it stores the message in the desired FIFO. For example, if a filter is configured with a mask of 0x7FF (all 11 bits are 1), then all bits of the SID on the bus must match all bits of the filter's SID. If there is a match, the message will be stored in the FIFO specified by CiFLTCONr.

Like a transmit message, a receive message is also four four-byte words long. Figure 19.5 shows the full layout of an RX message. As with a transmit message, we assume that the current RX FIFO position is stored in `addr`, a pointer to an array of `unsigned ints`. Then CMSGSID is at `addr[0]`, CMSGEID is at `addr[1]`, CMSGDATA0 is at `addr[2]`, and CMSGDATA1 is at `addr[3]`. Only the first DLC bytes in the data words are valid. The five bits in CMSGSID⟨15:11⟩ contain the number of the filter that put the message here. For more information on how the timestamp data in CMSGSID⟨31:16⟩ is generated from the SYSCLK and the CiTMR SFR, see the Reference Manual.

To read a message, you should either enable message receive interrupts or poll, waiting for CiFIFOINTnbits.RXNEMPTYIF to be set. To read the message, perform the following steps.

1. Ensure that RX filters have been set and enabled.
2. The RX FIFO should be non-empty, so check to make sure that CiFIFOINTnbits.RXNEMPTYIF = 1.

3. Get the address of the next RX message by reading CiFIFOUAn and converting it to a virtual address using `PA_TO_KVA1`.
4. Process the message.
5. Set the FIFO's UINC bit, using `CiFIFOCONnSET = 0x2000`, to notify the CAN peripheral to move CiFIFOUAn to the next message buffer.

19.3 Sample Code

19.3.1 Loopback

The first example uses the CAN's loopback mode, allowing the CAN module to send data to itself. Loopback mode allows testing the CAN module without worrying about physical layer issues such as transceivers, propagation delay, or bus impedance. The code creates a single filter to respond to only one specific SID, and it uses one TX FIFO and one RX FIFO. It prompts the user for data, sends and receives that data using the CAN peripheral, and reports the result to the user.

Code Sample 19.1 `can_loop.c`. **Basic CAN Loopback Functionality.**

```
#include "NU32.h"          // constants, funcs for startup and UART
#include <sys/kmem.h>      // used to convert between physical and virtual addresses
// Basic CAN example using loopback mode, so this functions with no external hardware.
// Prompts user to enter numbers to send via CAN.
// Sends the numbers and receives them via loopback, printing the results.

#define MY_SID 0x146  // the sid that this module responds to

#define FIFO_0_SIZE 4 // size of FIFO 0 (RX), in number of message buffers
#define FIFO_1_SIZE 2 // size of FIFO 1 (TX), in number of message buffers
#define MB_SIZE 4     // number of 4-byte integers in a message buffer

// buffer for CAN FIFOs
static volatile unsigned int fifos[(FIFO_0_SIZE + FIFO_1_SIZE) * MB_SIZE];

int main() {
  char buffer[100];
  int to_send = 0;
  unsigned int * addr;              // used for storing fifo addresses

  NU32_Startup();                   // cache on, interrupts on, LED/button init, UART init

  C1CONbits.ON = 1;                         // turn on the CAN module
  C1CONbits.REQOP = 4;                      // request configure mode
  while(C1CONbits.OPMOD != 4) { ; }         // wait to enter config mode

  C1FIFOCON0bits.FSIZE = FIFO_0_SIZE-1;// set fifo 0 size.  Actual size is 1 + FSIZE
  C1FIFOCON0bits.TXEN = 0;                  // fifo 0 is an RX fifo

  C1FIFOCON1bits.FSIZE = FIFO_1_SIZE-1;// set fifo 1 size. Actual size is 1 + FSIZE
  C1FIFOCON1bits.TXEN = 1;                  // fifo 1 is a TX fifo
  C1FIFOBA = KVA_TO_PA(fifos);              // tell CAN where the fifos are

  C1RXM0bits.SID = 0x7FF;                   // mask 0 requires all SID bits to match
```

```
    C1FLTCON0bits.FSEL0 = 0;                // filter 0 is for FIFO 0
    C1FLTCON0bits.MSEL0 = 0;                // filter 0 uses mask 0
    C1RXF0bits.SID = MY_SID;                // filter 0 matches against SID
    C1FLTCON0bits.FLTEN0 = 1;               // enable filter 0

                                            // skipping baud settings since loopback only
    C1CONbits.REQOP = 2;                    // request loopback mode
    while(C1CONbits.OPMOD != 2) { ; }       // wait for loopback mode

    while(1) {
      NU32_WriteUART3("Enter number to send via CAN:\r\n");
      NU32_ReadUART3(buffer,100);
      sscanf(buffer,"%d", &to_send);
      sprintf(buffer,"Sending: %d\r\n",to_send);
      NU32_WriteUART3(buffer);

      addr = PA_TO_KVA1(C1FIFOUA1);         // get FIFO 1 (the TX fifo) current message address
      addr[0] = MY_SID;                     // only the sid must be set for this example
      addr[1] = sizeof(to_send);            // only DLC field must be set, we indicate 4 bytes
      addr[2] = to_send;                    // 4 bytes of actual data
      C1FIFOCON1SET = 0x2000;               // setting UINC bit tells fifo to increment pointer
      C1FIFOCON1bits.TXREQ = 1;             // request that data from the queue be sent

      while(!C1FIFOINT0bits.RXNEMPTYIF) { ; } // wait to receive data
      addr = PA_TO_KVA1(C1FIFOUA0);         // get the VA of current pointer to the RX FIFO
      sprintf(buffer,"Received %d with SID = 0x%x\r\n",addr[2], addr[0] & 0x7FF);
      NU32_WriteUART3(buffer);
      C1FIFOCON0SET = 0x2000;               // setting UINC bit tells fifo to increment pointer
    }
    return 0;
  }
```

19.3.2 Light Control

The next example requires at least two PIC32s to be connected to the CAN bus through transceivers. One PIC32 (on an NU32 board) is a virtual traffic control police officer, sending messages to the other PIC32 to change a traffic light to red, yellow, or green. The second PIC32 (on an NU32 board) simulates a traffic light by illuminating two LEDs if the light should be green, one LED if it should be yellow, and no LEDs if it should be red. This example is easily extensible to any number of traffic lights on the CAN bus by using different SIDs to address different traffic lights.

The traffic cop PIC32 communicates with your computer via UART, allowing you to tell the cop what color you would like the light. The cop then relays the message to the other PIC32 via CAN. The traffic cop PIC32 runs the program `can_cop.c`, while the traffic light PIC32 runs the program `can_light.c`. Note that `can_cop.c` performs some rudimentary error checking and prints some diagnostic information. If you see such information, you should confirm that you have wired the bus correctly. If you run `can_cop.c` without any other devices on the CAN bus, you will eventually trigger the error condition, as CAN requires at least two devices on the bus for acknowledgment generation.

The bit timing for this example uses a time quantum of $T_q = 34/F_{sys} = 34 \times 12.5\text{ ns} = 425\text{ ns}$ and segment durations of $4T_q$ for propagation, $4T_q$ for phase 1, and $4T_q$ for phase 2. Adding the T_q for the synchronization segment, this makes a bit duration of $T_b = 13T_q = 5.525\ \mu\text{s}$ and a bit rate of $1/T_b \approx 180,995$ Hz.

Code Sample 19.2 `can_cop.c`. **Control the Lights on Another PIC32 via CAN.**

```
#include "NU32.h"          // constants, funcs for startup and UART
#include <sys/kmem.h>      // used to convert between physical and virtual addresses
#include <ctype.h>         // function "tolower" makes uppercase chars into lowercase
// The CAN cop is the "police officer" that controls the traffic light
// connect AC1TX to TXD on the transceiver, AC1RX to RXD on the transceiver
// The CANH and CANL transceiver pins should be connected to the same wires as
// the PIC running can_light.c.
//
// if your transceiver chip is the Microchip MCP2562 then connect
// Vss and STBY to GND, VDD to 5V VIO to 3.3V
#define FIFO_0_SIZE 4      // size of FIFO 0, in number of message buffers
#define FIFO_1_SIZE 4
#define MB_SIZE 4          // number of 4-byte integers in a message buffer

#define LIGHT1_SID 1       // SID of the first traffic light

volatile unsigned int fifos[(FIFO_0_SIZE + FIFO_1_SIZE)* MB_SIZE]; // buffer for CAN FIFOs

int main(void) {
  char buffer[100] = "";
  char cmd = '\0';
  unsigned int * addr = NULL;     // used to store fifo address

  NU32_Startup();                 // cache on, interrupts on, LED/button init, UART init

  C1CONbits.ON = 1;                       // turn on the CAN module
  C1CONbits.REQOP = 4;                    // request configure mode
  while(C1CONbits.OPMOD != 4) { ; }       // wait to enter config mode

  C1FIFOCON0bits.FSIZE = FIFO_0_SIZE-1;// set fifo 0 size in message buffers
  C1FIFOCON0bits.TXEN = 0;                // fifo 0 is an RX fifo

  C1FIFOCON1bits.FSIZE = FIFO_1_SIZE-1;// set fifo 1 size
  C1FIFOCON1bits.TXEN = 1;                // fifo 1 is a TX fifo
  C1FIFOBA = KVA_TO_PA(fifos);            // tell CAN where the fifos are

  C1CFGbits.BRP = 16;    // Tq = (2 x (BRP + 1)) x 12.5 ns = 425 ns
  C1CFGbits.PRSEG =  3; // 4Tq in propagation segment
  C1CFGbits.SEG1PH = 3; // 4Tq in phase 1. Phase 2 is set automatically to be the same.
  // bit duration = 1Tq(sync) + 4Tq(prop) + 4Tq(phase 1) + 4Tq(phase 2) = 13Tq = 5.525 us
  // so baud is 1/5.525 us = 180,995 Hz

  C1CFGbits.SJW = 0;    // up to (SJW+1)*Tq adjustment possible in phase 1 or 2 to resync

  C1CONbits.REQOP = 0;                  // request normal mode
  while(C1CONbits.OPMOD != 0) { ; }     // wait for normal mode
  NU32_LED1 = 0;
  while(1) {
    NU32_WriteUART3("(R)ed, (Y)ellow, or (G)reen?\r\n");
    NU32_ReadUART3(buffer,100);
```

```
    sscanf(buffer,"%c", &cmd);
    sprintf(buffer,"Setting %c\r\n", cmd);
    NU32_WriteUART3(buffer);

    if(C1TRECbits.TXWARN) {          // many bad transmissions have occurred,
                                     // print info to help debug bus (no ACKs?)
      sprintf(buffer,"Error: C1TREC 0x%08x\r\n",C1TREC);
      NU32_WriteUART3(buffer);
    }

    addr = PA_TO_KVA1(C1FIFOUA1);    // get VA of FIFO 1 (TX) current message buffer
    addr[0] = LIGHT1_SID;            // specify SID in word 0
    addr[1] = sizeof(cmd);           // specify DLC in word 1 (one byte being sent)
    addr[2] = tolower(cmd);          // the data (make uppercase chars lowercase)
                                     // since only 1 byte, no addr[3] given
    C1FIFOCON1SET = 0x2000;          // setting the UINC bit icrements fifo pointer
    C1FIFOCON1bits.TXREQ = 1;        // request that fifo data be sent on the bus
  }
  return 0;
}
```

Code Sample 19.3 `can_light.c`**. The** `can_cop.c` **Program can Control the LED Status via CAN.**

```
#include "NU32.h"          // config bits, constants, funcs for startup and UART
#include <sys/kmem.h>      // used to convert between physical and virtual addresses
// simulates a traffic light that can be controlled via can_cop
// LED1 on,  LED2 on  = GREEN
// LED1 on,  LED2 off = YELLOW
// LED1 off, LED2 off = RED

// connect AC1TX to TXD on the transceiver, AC1RX to RXD on the transceiver
// The CANH and CANL transceiver pins should be connected to the same wires as
// the PIC running can_cop.c
//
// if your transceiver chip is the Microchip MCP2562 then connect
// Vss and STBY to GND, VDD to 5V VIO to 3.3V
#define FIFO_0_SIZE 4 // size of FIFO 0, in number of message buffers
#define FIFO_1_SIZE 4
#define MB_SIZE 4     // number of 4-byte integers in a message buffer

#define LIGHT1_SID 1  // SID of the first traffic light

volatile unsigned int fifos[(FIFO_0_SIZE + FIFO_1_SIZE)* MB_SIZE]; // buffer for CAN FIFOs

int main(void) {
  unsigned int * addr = NULL;     // used to store fifo address

  NU32_Startup();                 // cache on, interrupts on, LED/button init, UART init

  C1CONbits.ON = 1;                   // turn on the CAN module
  C1CONbits.REQOP = 4;                // request configure mode
  while(C1CONbits.OPMOD != 4) { ; }   // wait to enter config mode

  C1FIFOCON0bits.FSIZE = FIFO_0_SIZE-1;// set fifo 0 size
  C1FIFOCON0bits.TXEN = 0;            // fifo 0 is an RX fifo

  C1FIFOCON1bits.FSIZE = FIFO_1_SIZE-1;// set fifo 1 size
  C1FIFOCON1bits.TXEN = 1;            // fifo 1 is a TX fifo
```

```
  C1FIFOBA = KVA_TO_PA(fifos);            // tell CAN where the fifos are

  C1RXM0bits.SID = 0x7FF;                 // mask 0 requires all SID bits to match

  C1FLTCON0bits.FSEL0 = 0;                // filter 0 uses FIFO 0
  C1FLTCON0bits.MSEL0 = 0;                // filter 0 uses mask 0
  C1RXF0bits.SID = LIGHT1_SID;            // filter 0 matches against SID
  C1FLTCON0bits.FLTEN0 = 1;               // enable filter 0

  C1CFGbits.BRP = 16;                     // copy the bit timing info for can_cop.c;
  C1CFGbits.PRSEG =  3;                   // see comments in can_cop.c
  C1CFGbits.SEG1PH = 3;
  C1CFGbits.SJW = 0;

  C1CONbits.REQOP = 0;                    // request normal mode
  while(C1CONbits.OPMOD != 0) { ; }       // wait for normal mode
  NU32_LED1 = 1;                          // turn both LEDs off
  NU32_LED2 = 1;
  while(1) {
    if(C1FIFOINT0bits.RXNEMPTYIF) {       // we have received data
      addr = PA_TO_KVA1(C1FIFOUA0);       // get VA of the RX fifo 0 current message
      switch(addr[2]) {
        case 'r':                         // switch to red
          NU32_LED1 = 1;
          NU32_LED2 = 1;
          break;
      case 'y':                           // switch to yellow
          NU32_LED1 = 0;
          NU32_LED2 = 1;
          break;
      case 'g':                           // switch to green
          NU32_LED1 = 0;
          NU32_LED2 = 0;
          break;
        default:
          ;                               // error! do something here
      }
      C1FIFOCON0SET = 0x2000;             // setting the UINC bit increments RX FIFO pointer
    }
  }
  return 0;
}
```

19.4 Chapter Summary

- CAN is an asynchronous protocol, developed for and used extensively in the automotive industry. It is also used in industrial control systems.
- Devices with a CAN controller, like the PIC32, typically connect to the two-wire CAN bus through a CAN transceiver, which translates between logic-level bits on the device and CAN bus voltages V_{CANH} and V_{CANL}.
- The CAN peripheral has several operating modes. You must request a mode and then wait for it to enter the appropriate mode before continuing.
- The CAN peripheral operates on FIFOs stored in RAM. A FIFO can be either an RX or TX FIFO.

- All CAN devices on the same bus receive and acknowledge all messages. Hardware filters determine whether received messages are stored in an RX FIFO, depending on the message's SID. Messages with different SIDs can be stored in different FIFOs.
- Choosing the bit timing for the CAN bus depends on the physical properties of the bus. Proper bit timing is crucial for CAN to operate at long distances and high speeds.

19.5 Exercises

1. Assume that both the accelerator and brake pedal on a car both send their input values over a CAN network. Which message, the brake or accelerator, should have the higher SID? Explain.
2. Pretend you are designing a cruise control system for a golf cart. One microcontroller controls the motor's PWM signal, one measures the wheel speed, one reads the speed input from the user, and another implements the controller. Design the CAN messages that should be sent between these four microcontrollers to implement the cruise control system.

Further Reading

AN754 understanding Microchip's CAN module bit timing. (2001). Microchip Technology Inc.
CAN specification (2.0 ed.). (1991). Robert Bosch GmbH.
Corrigan, S. (2008). *Introduction to the controller area network CAN* (Tech. Rep.).
PIC32 family reference manual. Section 34: Controller area network (CAN). (2012). Microchip Technology Inc.

CHAPTER 20

Harmony and Its Application to USB

Programming by directly manipulating SFRs, as we have done so far, requires detailed knowledge. Microchip's Harmony framework presents a different programming model, one that hides hardware intricacies behind functions, macros, and automatic code generation intended to make it easier to develop code that is portable across different PIC32 models.

Rather than being a comprehensive reference, this chapter provides an introduction to Harmony. Using a progression of examples, we demonstrate how Harmony's abstractions interact with each other and the SFRs. The chapter concludes with an example using the PIC32's USB peripheral.

As with all previous code in this book, code in this chapter assumes the bootloader is installed, which configures certain configuration bits, enables the prefetch cache module, and enables multi-vector interrupts (Chapter 3.6). Just as with the programs we have been writing until now, programs written using the Harmony framework include `p32mx795f512h.h` and link with the correct `processor.o` file, as described in Chapter 3. This allows you to use the SFR-manipulation code given earlier in the book within Harmony applications. We avoid that in this chapter, however, and adopt "the Harmony way" to interact with SFRs and peripherals.

The code in this chapter is significantly more complex than the sample code until now. This chapter can be skipped if you will not be exploring the Harmony software distribution. Since Harmony is relatively new, you should be aware that updates to Harmony may result in changes to specific function names or function behavior that will not be reflected in this chapter.

20.1 Overview

The Harmony framework attempts to accomplish three goals: abstraction, code portability, and code generation.

Abstraction hides implementation details behind a higher-level application programming interface (API) consisting of functions, data types, and macros. You have already used abstraction extensively; for example, the definitions provided by `pic32mx795f512h.h` (included

Embedded Computing and Mechatronics with the PIC32 Microcontroller. http://dx.doi.org/10.1016/B978-0-12-420165-1.00020-2

via xc.h), allow us to access SFRs by name (e.g., PORTB) rather than needing to directly enter a virtual address (VA). Another example of abstraction is NU32_WriteUART3, which allows you to send text over UART3 without needing to know how UARTs operate. The Harmony API provides an interface not just for manipulating peripherals but also for more complex tasks such as reading files from a USB flash drive.

Portable code is code that works across multiple microcontrollers with minimal modification. Abstraction aids portability. The PORTB SFR, for example, has different VAs on different PIC32MX models; however, by using PORTB structure rather than directly entering its VA, the same code works across multiple microcontrollers. Based on the -mprocessor=<proc> compiler argument, xc.h includes the appropriate pic32<proc>.h file and links against the appropriate processor.o file, which provides the appropriate VA for PORTB (see Chapter 3 for more details). Harmony, at its lowest level, uses the -mprocessor compiler flag, xc.h, and processor.o to achieve portability.

The aforementioned -mprocessor approach to portability is often insufficient because different microcontrollers have different peripherals and pin layouts. Harmony addresses this issue by imposing a structure on your program that separates the hardware-specific code from the more general logic. All code that depends on specific hardware is placed in its own configuration directory. At compile time, you select one configuration to use, and only that hardware-dependent code is included. Although useful for portable code development, we ignore Harmony's suggested directory structure for most examples in this chapter because its flexibility adds complexity.

Harmony's code generation facilities, integrated into the MPLAB X IDE, allow you to graphically configure peripherals; the tool generates the necessary source code for you. Additionally, the code generation tools automatically add Harmony dependencies to your project (a major benefit, as you will soon see). Although we do not discuss MPLAB X or these code generation tools, you may want to explore them on your own. The foundation provided in this chapter will allow you to not only use but also understand the output of the code generation tools.

20.2 The Framework

If you have not already installed Harmony, do so now. Refer to Chapter 1 for details. Throughout this chapter we refer to the Harmony installation directory as <harmonyDir> and the Harmony version as <harmonyVer>. So if Harmony is installed in /opt/microchip/harmony/v1_06, then <harmonyDir> refers to /opt/microchip/harmony and <harmonyVer> is v1_06. An important Harmony subdirectory, which contains all of the Harmony source code, is <harmonyDir>/<harmonyVer>/framework; we refer to it as

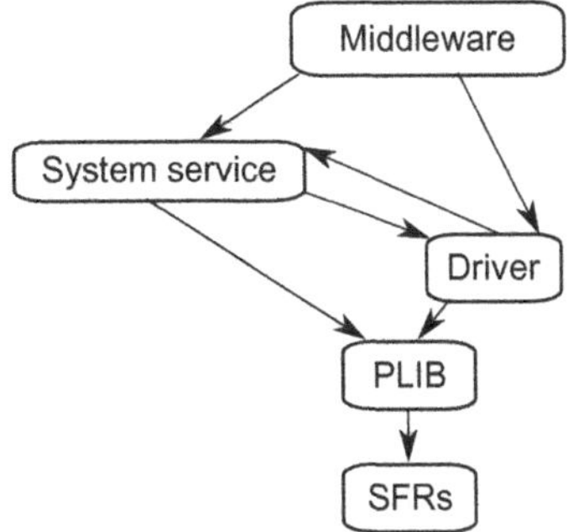

Figure 20.1
Hierarchy of Harmony module types. If a module type depends on another module type, there is an arrow from the second to the first.

`<framework>`. The Harmony documentation is installed in `<harmonyDir>/doc` and you may wish to refer to it throughout this chapter.

Conceptually, the Harmony API is divided into distinct layers, organized into a hierarchy (see Figure 20.1). Higher layers generally indicate more abstraction. The SFR layer is the lowest, corresponding to directly manipulating SFRs using definitions from `p32mx795f512h.h`.

The next layer, PLIB (short for peripheral library), contains functions for directly manipulating SFRs. Rather than setting an SFR to a value directly (e.g., `LATB = 0xFF`) you would use a function call (e.g., `PLIB_PORTS_Write(PORTS_ID_0,PORTS_CHANNEL_B,0xFF)`).

Usually you do not use the PLIB layer directly when using Harmony. The driver layer builds upon PLIB and provides easier access to peripherals and other common functions. Drivers have a common interface; for example, all drivers have a `DRV_<drivername>_Initialize` function to configure the driver.

The system services layer manages drivers and other system resources. For example, a system service might manage access to a timer driver, allowing you to run multiple tasks at different frequencies using the same underlying timer. Sometimes, drivers depend on system services; for example, the UART driver requires the interrupt system service to help it manage UART interrupts. Unlike drivers, system services do not always adhere to a consistent interface.

The highest layer we discuss, middleware, provides higher-level functionality such as implementing the USB protocol (discussed in Section 20.8). There are other aspects of Harmony (e.g., support of real-time operating systems) that we do not discuss.

A good way to understand the Harmony programming model is to work through some examples. We begin with the PLIB layer and work our way up, allowing you to see the relationship between the various API layers.

20.2.1 Setup

Just as you needed to configure your development environment prior to programming the PIC32, Harmony also requires configuration. Rather than relying on the MPLAB X IDE, we have provided you with a `Makefile` for building Harmony applications. It is available on the book's website. Just as you needed to edit your original `Makefile`, you also need to edit the Harmony `Makefile` so it knows where to find your development tools. The variables used in the Harmony `Makefile` are the same as those used in the original `Makefile` and should be entered according to the instructions in Chapter 1. You may notice an additional variable assignment at the top of the file, `CONFIG=pic32_NU32`, which controls the target platform; you will learn about this feature later.

Although the Harmony `Makefile` is different from the original `Makefile`, you use it in the same manner. It compiles all `.c` files in the current directory into a single executable. The file also has additional capabilities, allowing it to compile projects that adhere to Harmony's recommended directory structure. We recommend putting the Harmony `Makefile` in a skeleton directory, perhaps named `harmony_skel`. This directory should also include the linker script `NU32bootloaded.ld`. It need not include `NU32.h` and `NU32.c` as Harmony projects do not need these files.[1] You will copy this Harmony skeleton directory to create new projects, just as you had been doing with `skeleton`.

20.3 PLIB

The PLIB layer requires the smallest leap from SFR-based programming—nearly every line you have used to control an SFR can be implemented with a PLIB function. Code Sample 20.1 `demo_plib.c`, below, demonstrates using PLIB to toggle LEDs at 5 Hz, closely resembling `TMR_5Hz.c` from Chapter 8. Every PLIB function call corresponds to accessing the appropriate SFR. For example, instead of setting B5 high using `LATBSET = 0x20`, you use `PLIB_PORTS_PinSet`. More details about individual functions can be found in the Harmony documentation.

Notice the naming scheme employed: the PLIB functions all begin with `PLIB_`. The next part of the name indicates which category of SFRs they manipulate; for example, `PLIB_INT` relates to interrupt SFRs and `PLIB_TMR` controls timer SFRs. The first argument to every PLIB function is an ID. All ports use `PORTS_ID_0` and all interrupts use `INT_ID_0`. For timers, the ID corresponds to the timer number: `TMR_ID_x` indicates Timerx, where x is 1 to 5.

---

[1] You can still use the functions in the NU32 library in a Harmony project if you want. Just do not attempt to separately configure UART3 in your Harmony project; it is already claimed by `NU32`! Also, you must be careful about using `NU32`'s blocking functions, as you will see in Section 20.4.

demo_plib.c does not include NU32.h. Instead it includes sys/attribs.h in the XC32 distribution (Chapter 3) so that we can use the __ISR macro. (NU32.h had included this file for us.) It also includes <framework>/peripheral/peripheral.h, which provides functions from the Harmony PLIB library.

Code Sample 20.1 demo_plib.c**. Toggle LEDs at 5 Hz, Using Harmony PLIB.**

```
#include <peripheral/peripheral.h>    // harmony peripheral library
#include <sys/attribs.h>              // defines the __ISR macro
// Almost a direct translation of TMR_5Hz.c to use the harmony peripheral library.
// The main difference is that it flashes two LEDs out of phase,
// instead of just flashing LED2.

void __ISR(_TIMER_1_VECTOR, IPL5SOFT) Timer1ISR(void) {
  // toggle LATF0 (LED1) and LATF1 (LED2)
  PLIB_PORTS_PinToggle(PORTS_ID_0, PORT_CHANNEL_F, PORTS_BIT_POS_0);
  PLIB_PORTS_PinToggle(PORTS_ID_0, PORT_CHANNEL_F, PORTS_BIT_POS_1);
  // clear the interrupt flag
  PLIB_INT_SourceFlagClear(INT_ID_0, INT_SOURCE_TIMER_1);
}

int main(void) {
  // F0 and F1 are  is output (LED1 and LED2)
  PLIB_PORTS_PinDirectionOutputSet(PORTS_ID_0, PORT_CHANNEL_F, PORTS_BIT_POS_0);
  PLIB_PORTS_PinDirectionOutputSet(PORTS_ID_0, PORT_CHANNEL_F, PORTS_BIT_POS_1);

  // turn LED1 on by clearing F0 to zero
  PLIB_PORTS_PinClear(PORTS_ID_0, PORT_CHANNEL_F, PORTS_BIT_POS_0);

  // turn off LED2 by setting F1 to one
  PLIB_PORTS_PinSet(PORTS_ID_0, PORT_CHANNEL_F, PORTS_BIT_POS_1);

  PLIB_TMR_Period16BitSet(TMR_ID_1, 62499);                     // set up PR1: PR1 = 62499
  PLIB_TMR_Counter16BitSet(TMR_ID_1,0);                         // set up TMR1: TMR1 = 0
  PLIB_TMR_PrescaleSelect(TMR_ID_1, TMR_PRESCALE_VALUE_256);    // 1:256 prescaler

  // set up the timer interrupts
  // clear the interrupt flag
  PLIB_INT_SourceFlagClear(INT_ID_0, INT_SOURCE_TIMER_1);

  // set the interrupt priority
  PLIB_INT_VectorPrioritySet(INT_ID_0, INT_VECTOR_T1, INT_PRIORITY_LEVEL5);

  // enable the timer interrupt
  PLIB_INT_SourceEnable(INT_ID_0, INT_SOURCE_TIMER_1);

  // start the timer
  PLIB_TMR_Start(TMR_ID_1);
  // enable interrupts
  PLIB_INT_Enable(INT_ID_0);
  while (1) {
      ;       // infinite loop
  }
  return 0;
}
```

To build demo_plib.c, copy harmony_skel to a new directory and add demo_plib.c. Your directory now has three files: the Harmony Makefile, NU32bootloaded.ld, and demo_plib.c. Type make to compile the program. Notice that make issues the commands

```
> xc32-gcc -g -O1 -x c -c -mprocessor=32MX795F512H -I<framework> -I./
  -o demo_plib.o demo_plib.c
> xc32-gcc -mprocessor=32MX795F512H -o out.elf demo_plib.o
    -l:<harmonyDir>/<harmonyVer>/bin/framework/peripheral/PIC32MX795F512H_
     peripherals.a
    -Wl,--script="NU32bootloaded.ld",-Map=out.map
```

These commands are similar to what you have seen before to compile and link non-Harmony programs (see Chapter 3). We have added a few additional options here. The compiler is invoked with options to set the include path, which tells the compiler where to find header files: -I<framework> and -I./. The first path specifies the Harmony framework directory. When you include files in your project, all paths to Harmony headers are specified relative to this directory. For example, peripheral.h, which we included as peripheral/peripheral.h, is located at <framework>/peripheral/peripheral.h. The -I./ tells the compiler to add the current directory to the include path; this is needed because, in later programs, Harmony files include files that you write.

The linking step also has an additional option:
-l:<harmonyDir>/<harmonyVer>/bin/framework/peripheral/PIC32MX795F512H_peripherals.a.
This causes the linker to link against the Harmony pre-compiled library PIC32MX795F512H_peripherals.a. Every supported PIC32 model has its own peripheral library that implements the PLIB functions, and you must link against the library appropriate to your microcontroller.[2]

20.4 Harmony Concepts

To move beyond the PLIB layer and use Harmony effectively, it helps to understand a few programming concepts that we have not used until now: finite state machines (FSMs), tasks, non-blocking functions, and callback functions.

An FSM consists of two types of objects: states and transitions. For example, a simple drink vending machine has four states: wait-for-money, has-money, refund-money, and vend. When in the wait-for-money state, any drink button pressed by the user is ignored; the FSM waits for sufficient money to be inserted. Once sufficient money has been inserted, the FSM transitions to the has-money state. In this state, the FSM waits for a drink button to be pressed. When a button is pressed, the FSM transitions to the vend state, dispensing the drink and any change.

[2] Usually, due to compiler optimizations, the functions in the .a library are unnecessary as they are present as inlined functions in the peripheral header files.

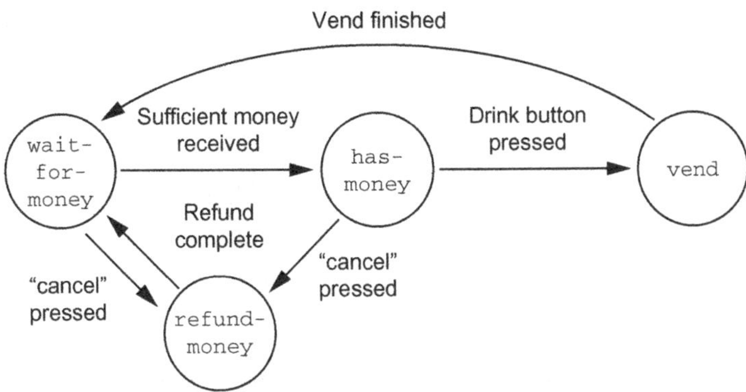

Figure 20.2
The vending machine finite state machine.

When finished vending, the FSM returns to the `wait-for-money` state. If the user presses the "cancel" button in either the `wait-for-money` or `has-money` state, the FSM transitions to the `refund-money` state. After any inserted money is returned to the user, the FSM transitions again to the `wait-for-money` state.

An FSM can be viewed as a graph: the states are nodes and the transitions are arrows between the nodes. The vending machine FSM is illustrated in Figure 20.2.

When you write code in Harmony, you may find it useful to implement an FSM. You define a finite set of states, like with the vending machine, and for each state you define (1) the actions to perform while in that state and (2) the conditions that cause transitions to other states.

More importantly, many capabilities in Harmony are implemented as FSMs. For example, the Harmony UART driver implements FSMs to control sending and receiving bytes. Each FSM is called a *task*. Your code is likely to use multiple Harmony tasks, and conceptually these tasks run simultaneously. To keep these tasks running properly:

1. Your code should regularly call Harmony functions associated with each task. Each function performs the actions for the current FSM state and evaluates whether conditions have been satisfied to transition to a new state. Typically, your code enters an infinite loop and calls the Harmony task update functions at the end of the loop.[3]
2. Many Harmony function calls associated with a task return quickly. Such a function is called *non-blocking*, meaning that it does not occupy the CPU for a long time and block the execution of subsequent code. Non-blocking function calls are critical; otherwise they may delay another FSM too long, preventing it from performing its functions or checking

[3] Some Harmony tasks can manage their FSMs using interrupts, in which case you call the Harmony task update function from the appropriate ISR.

conditions for state transitions. Similarly, functions you write (perhaps to implement your own FSM) should be non-blocking, to allow the Harmony task functions to be called regularly.

Many Harmony modules allow you to specify a function to be executed when a certain state transition occurs in the module's FSM. This function is called a *callback function*. For example, a Harmony timer driver module could be configured to call your callback function when the driver recognizes a timer rollover. Thus your callback function acts similarly to a fixed frequency ISR, but without using an interrupt.

What we have learned about FSMs and tasks, non-blocking functions, and callback functions will be used repeatedly in the following examples.

20.5 Drivers

In Harmony, drivers are used to control specific devices or peripherals. Unlike PLIB functions, drivers impose several additional requirements on your project. All drivers follow the same basic usage patterns. Although they ultimately provide abstract interfaces to the peripherals, drivers require more setup than using PLIB functions or SFRs alone. You not only need to include the appropriate headers, but also must add specific `.c` and `.h` files to your project, define certain macros, and execute code in a specific manner.

All the files needed for drivers exist in subdirectories `<framework>/driver/<drvname>`. For example, the I^{2}C driver is located at `<framework>/driver/i2c`. Within the driver directory is the header file `drv_<drvname>.h` (e.g., `drv_i2c.h`), which you need to include in your projects using a `#include` directive. There also exists a `<framework>/driver/<drvname>/src` directory which may contain `.c` files, `.h` files, and subdirectories with additional source code. To use a driver, you need to include certain files from within the `src` directory or its subdirectories in your project's compilation; however, the specific files depend on which features of the driver you wish to use. For example, some drivers have both static and dynamic modes, where the static mode typically has several additional options that can be set at compile time. The Harmony documentation specifies which files you need based on which features you desire. To use the driver files in your project, copy them into your project directory.[4]

In addition to copying the driver files, you also need to create two header files: `system_config.h` and `system_definitions.h`. These two files are present in nearly every Harmony project. `system_config.h` should define macros needed by Harmony code. Without these, drivers may not compile. `system_definitions.h` should `#include` Harmony and other header files needed by the program.

[4] Harmony's code generation tool within MPLAB automatically adds Harmony dependencies to your project.

20.5.1 UART

In Code Sample 20.2 demo_uart.c, below, we configure and use the UART driver.[5] The result will be two functions that roughly duplicate the functionality of NU32_ReadUART3 and NU32_WriteUART3. Refer to Chapter 11 for details about UARTs. This program behaves like Code Sample 1.2 (talkingPIC.c): it reads text sent from a terminal emulator and displays it to the user.

To build this project, you need the following in your project directory:

- demo_uart.c: Your source code.
- drv_usart.c: Copied from <framework>/driver/usart/src/dynamic/. The main Harmony UART driver file.
- drv_usart_read_write.c: Copied from <framework>/driver/usart/src/dynamic/. Provides access to the file I/O mode of the driver.
- system_config.h: Defines various macros needed by the Harmony code used by the program.
- system_definitions.h: #includes Harmony and other standard header files needed by the program.

With the five files above as well as the Harmony Makefile and NU32bootloaded.ld in your project directory, you can use make as usual to build the project.

demo_uart.c follows a pattern typical of Harmony programs. First it includes system_config.h and system_definitions.h. Next it provides values for an initialization structure that determines how the UART driver will be used. All drivers have an initialization struct named DRV_<drvname>_INIT, where <drvname> is the driver name (in this case USART).

The main function has two parts: the initialization code and the main loop. All drivers must be initialized using a function named DRV_<drvname>_Initialize (in this case DRV_USART_Initialize), which uses the initialization struct to configure the driver.

Next, the driver must be opened using DRV_<drvname>_Open (in this case DRV_USART_Open). As you will see in later examples, some drivers must be opened from the main loop. The DRV_<drvname>_Open function must be called after the driver is initialized. DRV_<drvname>_Open provides a DRV_HANDLE which can be used to access the driver. Each access through a given DRV_HANDLE is considered a client. Some drivers support multiple clients to provide concurrent access to the driver from multiple tasks while others do not. In our examples, we always use one client per driver.

The main loop implements the program's logic. In demo_uart.c, the logic is an FSM with three states: APP_STATE_QUERY (ask the user to enter text), APP_STATE_RECEIVE (wait for text

[5] Harmony refers to the driver as a USART, but we refer to it as a UART wherever possible.

from the user), and APP_STATE_ECHO (write the user's text back to the terminal emulator). The function WriteUart is used by the APP_STATE_QUERY and APP_STATE_ECHO states, and the function ReadUart is used by the APP_STATE_RECEIVE state. ReadUart and WriteUart are similar to their NU32 counterparts NU32_ReadUART3 and NU32_WriteUART3, with two major differences: they can be used with any UART (not just UART3) and they are non-blocking, meaning that they return quickly even if they have not fully completed their task. These functions return zero if they have not yet finished with their task (to be resumed the next time they are called) or one if the process has finished. It is worth examining ReadUart and WriteUart to see how to write a non-blocking function that returns quickly and that may be called multiple times before it completes its task.

ReadUart and WriteUart are non-blocking because the main loop must also regularly update the FSMs of the Harmony modules that you use, by calling function(s) named DRV_<drvname>_Tasks<subtask>, where <subtask> is an additional part of the function name and is sometimes omitted, depending on the specific driver. In demo_uart.c, the main loop updates three FSMs handling different aspects of the UART: receive, transmit, and receive/transmit errors.

Code Sample 20.2 demo_uart.c. **Demonstrates the Harmony UART Driver. The User Types in the Terminal and the PIC32 Responds.**

```
// Demonstrates the harmony UART driver.
// Implements a program similar to talkingPIC.c.

#include "system_config.h"       // macros needed for this program
#include "system_definitions.h"  // includes header files needed by the program

// UART_init, below, is of type DRV_USART_INIT, a struct.  Here we initialize uart_init.
// The fields in DRV_USART_INIT, according to framework/driver/usart/drv_usart.h, are
// .moduleInit, .usartID, .mode, etc.  Syntax below doesn't give the field names, so the
// values are assigned to fields in the order they appear in the definition of the
// DRV_USART_INIT struct.

const static DRV_USART_INIT uart_init = {             // initialize struct with driver options
  .moduleInit = {SYS_MODULE_POWER_RUN_FULL},          // no power saving
  .usartID = USART_ID_3,                              // use UART 3
  .mode = DRV_USART_OPERATION_MODE_NORMAL,            // use normal UART mode
  .modeData = 0,                                      // not used in normal mode
  .flags = 0,                                         // no flags needed
  .brgClock = APP_PBCLK_FREQUENCY,                    // peripheral bus clock frequency
  .lineControl = DRV_USART_LINE_CONTROL_8NONE1,       // 8 data bits, no parity, 1 stop bit
  .baud = 230400,                                     // baud
  .handshake = DRV_USART_HANDSHAKE_FLOWCONTROL,       // use flow control
                                                      // remaining fields are not needed here
};

// Write a string to the UART. Does not block.  Returns true when finished writing.
int WriteUart(DRV_HANDLE handle, const char * msg);

// Read a string from the UART.  The string is ended with a '\r' or '\n'.
// If more than maxlen characters are read, data wraps around to the beginning.
```

```
// Does not block, returns true when '\r' or '\n' is encountered.
int ReadUart(DRV_HANDLE handle, char * msg, int maxlen);

#define BUF_SIZE 100

// states for our FSM
typedef enum {APP_STATE_QUERY, APP_STATE_RECEIVE, APP_STATE_ECHO} APP_STATE;

int main(void) {
  char buffer[BUF_SIZE];
  APP_STATE state = APP_STATE_QUERY;  // initial state of our FSM will ask user for text
  SYS_MODULE_OBJ uart_module;
  DRV_HANDLE uart_handle;

  // Initialize the UART.
  uart_module = DRV_USART_Initialize(DRV_USART_INDEX_0,(SYS_MODULE_INIT*)&uart_init);

  // Open the UART for non-blocking read/write operations.
  uart_handle = DRV_USART_Open(
      uart_module, DRV_IO_INTENT_READWRITE | DRV_IO_INTENT_NONBLOCKING);

  while (1) {
    switch(state) {
      case APP_STATE_QUERY:
        if(WriteUart(uart_handle,"\r\nWhat do you want? ")) {
                                     // Start/continue writing to UART.
                                     // If we get here, the message has been completed.
          state = APP_STATE_RECEIVE;   // Switch to receive message state.
        }
        break;
      case APP_STATE_RECEIVE:
        if(ReadUart(uart_handle,buffer,BUF_SIZE)) {
                                     // Start/continue reading msg from user.
                                     // If we get here, the user's message is concluded.
          state = APP_STATE_ECHO;      // Switch to echo state.
        }
        break;
      case APP_STATE_ECHO:
        if( WriteUart(uart_handle,buffer)) {
                                     // Start/continue echoing message to UART.
                                     // If we get here, we're finished echoing.
          state = APP_STATE_QUERY;     // Switch to user query state.
        }
        break;
      default:
        ;// logic error, impossible state!
    }

    // Update the UART FSMs.  Since we are not using UART interrupts, the FSM
    // updating must be done in mainline code, and it should be done often.
    // Typically done at the end of the main loop, and there should be no
    // blocking functions in the main loop.
    DRV_USART_TasksReceive(uart_module);
    DRV_USART_TasksTransmit(uart_module);
    DRV_USART_TasksError(uart_module);
  }
  return 0;
}

// WriteUart keeps track of the number of characters already sent in the most recent
// message send request.  Once it realizes the last character has been sent, it returns
```

```
// TRUE (1), indicating it is finished.  Otherwise it tries to send another byte of
// the msg.  In any case, it returns quickly (non-blocking).

int WriteUart(DRV_HANDLE handle, const char * msg) {
  static int sent = 0;     // number of characters sent (static so saved between calls)
  if(msg[sent] == '\0') { // we are at the last string character
    sent = 0;              // reset the "sent" count for the next time
    return 1;              // finished sending message
  } else {
    // DRV_USART_Write(handle,str,numbytes) tries to add numbytes from str to the UART
    // send buffer, returning the number of bytes that were placed in the buffer,
    // so we can keep track. Note that DRV_USART_Write takes a void *, hence the cast
    sent += DRV_USART_Write(handle,(char*)(msg + sent),1);
    return 0;
  }
}

// ReadUart reads bytes into msg.  It keeps track of the number of characters received.
// If the number exceeds maxlen, then wraps around and begins to write to msg at
// beginning.  Returns TRUE (1) if the entire user message has been received, or FALSE (0)
// if the end of the message has not been reached.  Regardless, it returns quickly
// (non-blocking).

int ReadUart(DRV_HANDLE handle, char * msg, int maxlen) {
  static int recv = 0;  // number of characters received
  int nread = 0;        // number of bytes read
  // DRV_USART_Readh(handle,str,numbytes) tries to read one byte from uart receive buffer.
  // Returns the number of bytes that were actually placed into str.  If no bytes
  // are available, then recv is unchanged.
  nread = DRV_USART_Read(handle,msg + recv, 1);
  if(nread) {           // if we have read one byte
    if(msg[recv] == '\r' || msg[recv] == '\n') {  // check for newline / carriage return
      msg[recv] = '\0'; // insert the null character
      recv = 0;         // prepare to receive another string
      return 1;         // indicate that the string is ready
    } else {
      recv += nread;
      if(recv >= maxlen) { // wrap around to the beginning
        recv = 0;
      }
      return 0;
    }
  }
  return 0;
}
```

We now discuss `demo_uart.c` in more depth. The variable `uart_init` is the driver initialization structure. The first element in this `struct` controls the driver's behavior with respect to power saving modes. We always use `SYS_MODULE_POWER_RUN_FULL` (no power saving); for more details see the Harmony documentation. The next element of the initialization `struct` determines which UART peripheral to use, in this case UART3. The other fields control settings related to UARTs, such as the baud; refer to the Harmony documentation and Chapter 11 for details.

The main function first initializes the UART driver by calling DRV_USART_Initialize, which requires an index to the driver and the driver-specific initialization struct. The driver initializer index determines which instance of the driver to initialize. You use multiple driver instances to manage multiple peripherals of the same type. For example, if you wanted to use both UART3 and UART2 you would need to use two UART driver instances. The initializer function creates an instance of the driver and returns a SYS_MODULE_OBJ that allows you to access the driver later.

After initializing a UART driver instance, we open the driver using DRV_USART_Open, which requires a handle to the initialized driver and a parameter describing how you will use it. We open the driver with read and write access and in non-blocking mode. Non-blocking mode means that function calls to the driver will return even if the hardware has not completed the commanded task. We need non-blocking operations so that we can continue to update Harmony drivers in the main loop while waiting for input. The non-blocking operations are what allow us to write ReadUart and WriteUart as non-blocking functions.

The non-blocking nature of ReadUart and WriteUart necessitates the FSM architecture. Each call to these functions may not actually complete the sending or receiving of the entire string. If the function returns before its task completes, it returns zero. Subsequent calls then attempt to complete the task. When the task completes, the function returns one.

The FSM facilitates calling these functions multiple times, only moving to the next state after the desired task completes. If you wanted to use ReadUart or WriteUart in a blocking manner you could loop until they finished. For example

```
while(!ReadUart(uart_handle,buffer, BUF_SIZE)) { ; }
```

would loop until an entire line was read.

The functions ReadUart and WriteUart are implemented using the DRV_USART_Read and DRV_USART_Write functions from the UART driver (see the comments in the demo_uart.c listing). Both DRV_USART_Read and DRV_USART_Write are part of the UART driver's read/write mode. Other modes include byte mode and buffer queue mode. The different modes are enabled by setting certain macros in system_config.h (discussed subsequently) and by including the appropriate .c files in your project. To use read/write mode, for example, we included drv_usart_read_write.c in the project.

Now let us take a look at the two header files we included at the beginning of the demo_uart.c, starting with system_definitions.h:

Code Sample 20.3 `system_definitions.h`**. System Definitions for the UART Example. This File Should** `#include` **Any Harmony or Other Standard Header Files Your Project Needs.**

```
#ifndef SYSTEM_DEFINITIONS_H
#define SYSTEM_DEFINITIONS_H

#include "driver/usart/drv_usart.h"  // only one header file needed

#endif
```

The purpose of `system_definitions.h` is to `#include` Harmony and other headers needed by the project. For this project, only one header file is included, `<framework>/driver/usart/drv_usart.h`, giving access to macros and function prototypes associated with the UART.

Let us now examine `system_config.h`. Every Harmony project that uses drivers must include a file named `system_config.h`, which contains macros that determine configuration options for various Harmony components. Reminder: in `system_config.h` you must only define macros using `#define`; do not include other files or declare data types. The meaning of each macro is explained in the comments.

Code Sample 20.4 `system_config.h`**. System Configuration for the UART Example.**

```
#ifndef SYSTEM_CONFIG_H__
#define SYSTEM_CONFIG_H__

// Suppresses warnings from parts of Harmony that are not yet fully
// implemented by Microchip.
#define _PLIB_UNSUPPORTED

// The number of UART driver instances needed by the program.  If you wanted
// to use UART1, UART2, and UART3, for example, this value should be 3.
#define DRV_USART_INSTANCES_NUMBER 1

// Multiple clients could concurrently use the driver.  The function
// DRV_USART_Open creates a client that code uses to access the driver.
// To allow the driver to manage concurrent perhipheral access from multiple
// tasks (e.g., mainline code and a timer ISR), you should have each task
// create its own client.  Not all drivers support concurrent access;
// consult the Harmony documentation.  Usually sufficient to set this to 1.
#define DRV_USART_CLIENTS_NUMBER 1

// Can be true or false.  If true, the driver FSMs are updated in interrupts,
// meaning that the various DRV_USART_Tasks should be called from ISRs.  Our
// program does not use interrupts.
#define DRV_USART_INTERRUPT_MODE false

// Use the read/write UART model.
#define DRV_USART_READ_WRITE_MODEL_SUPPORT true

// This is not used by Harmony, so we use APP_ as the prefix (our application).
// This is used in the initialization of the UART so the driver can generate
```

```
// the proper baud rate.
#define APP_PBCLK_FREQUENCY 80000000L

#endif
```

20.5.2 Timers

In this example, we use the Harmony timer (TMR) driver to toggle LED1 and LED2 at 5 Hz. The timer driver depends on several other modules, specifically the clock (CLK) system service, the device control (DEVCON) system service, and the interrupt (INT) system service. We explain system services later; for now, think of them as drivers. In addition to the Harmony `Makefile` and the `NU32bootloaded.ld` linker script, you need the following files in your project directory:

`demo_tmr.c`: The main source code for the example.
`system_config.h`: Defines various macros needed by the Harmony code used by the program.
`system_definitions.h`: `#include`s Harmony and other standard header files needed by the program.
`drv_tmr.c`: Copied from `<framework>/driver/tmr/src/dynamic/`. The TMR driver implementation.
`sys_clk.c`: Copied from `<framework>/system/clk/src/`. Processor-independent part of the CLK system service implementation.
`sys_clk_pic32mx.c`: Copied from `<framework>/system/clk/src/`. PIC32MX-specific part of the CLK system service implementation.
`sys_devcon.c`: Copied from `<framework>/system/devcon/src/`. Processor-independent part of the DEVCON system service.
`sys_devcon_pic32mx.c`: Copied from `<framework>/system/devcon/src/`. PIC32MX-specific DEVCON implementation.
`sys_devcon_local.h`: Copied from `<framework>/system/devcon/src/`. A necessary DEVCON header that is not otherwise on the include path.
`sys_int_pic32.c`: Copied from `<framework>/system/int/src/`. The INT system service implementation.

The `system_definitions.h` file is given below:

Code Sample 20.5 `system_definitions.h`. **System Definitions Required to Use the Timer Driver.**

```
#ifndef SYSTEM_DEFINITIONS_H
#define SYSTEM_DEFINITIONS_H

#include <stddef.h>          // Standard C header defining NULL and types used by Harmony
```

```
#include <stdbool.h>        // Standard C header defining type bool (true/false)
#include "peripheral/peripheral.h"  // The Harmony PLIB library
#include "system/devcon/sys_devcon.h" // DEVCON handles cache, other config tasks.
#include "system/clk/sys_clk.h"// Clock header. Control and query oscillator properties.
#include "system/common/sys_module.h"//Basic system module, used by most Harmony projects
#include "driver/tmr/drv_tmr.h"      // The timer driver
#include <sys/attribs.h>    // defines the __ISR macro, needed when we use interrupts
#endif
```

`system_config.h` contains macros for driver and system service settings. We also use it to define macros for pins controlling the LEDs on the NU32 board. These macros begin with `NU32_`.

Code Sample 20.6 `system_config.h`**. System Configuration for the Timer Driver.**

```
#ifndef SYSTEM_CONFIG_H__
#define SYSTEM_CONFIG_H__

// Suppresses warnings from parts of Harmony that are not yet fully
// implemented by Microchip.
#define _PLIB_UNSUPPORTED

// system clock settings

// The NU32 system clock oscillator frequency, 8 MHz.  (This is the oscillator frequency,
// not the final SYSCLK frequency.)
#define  SYS_CLK_CONFIG_PRIMARY_XTAL 8000000L

// The secondary oscillator frequency.  There is no secondary oscillator on the NU32.
#define SYS_CLK_CONFIG_SECONDARY_XTAL 0

// If we were asking Harmony to automatically determine clock multipliers and divisors
// to achieve our 80 MHz SYSCLK from the 8 MHz external oscillator, this tolerance
// would be the largest acceptable error.
#define SYS_CLK_CONFIG_FREQ_ERROR_LIMIT 10

// timer driver settings

// The PIC32 has five hardware timers.  This example uses only one, so we could set
// this number to one to save some RAM (driver instances are stored in a statically
// allocated array, so the more instances, the more RAM used).
#define DRV_TMR_INSTANCES_NUMBER 5

// Set this to true to enable interrupts.  We do not use interrupts in this example.
#define DRV_TMR_INTERRUPT_MODE false

// Some definitions for the NU32 board.
#define NU32_LED_CHANNEL PORT_CHANNEL_F  // port channel for the NU32 LEDs
#define NU32_LED1_POS    PORTS_BIT_POS_0
#define NU32_LED2_POS    PORTS_BIT_POS_1

#endif
```

Now that we have explained all of the Harmony infrastructure, we can examine the file that implements the timer demonstration. We define an initialization structure for the timer driver and use it to initialize the timer. We also use PLIB functions to initialize the output pins needed to control the LEDs.

After initializing and opening the timer, we register an *alarm* with the timer driver. The timer driver's alarm calls a callback function at a specified frequency, in this case 5 Hz. In this example, the callback function inverts the LEDs. Only one alarm can be registered at any given time. We then start the timer and enter the main loop.

The main loop, in this case, only updates Harmony module FSMs. We do not implement our own FSM logic because there is only one state (flashing the LEDs).

Code Sample 20.7 `demo_tmr.c`. **Harmony Timer Demonstration Using Polling.**

```
// demonstrates the timer driver

#include "system_config.h"
#include "system_definitions.h"

void invert_leds_callback(uintptr_t context, uint32_t alarmCount) {
  // context is data passed by the user that can then be used in the callback.
  // alarmCount tracks how many times the callback has occurred.
  PLIB_PORTS_PinToggle(PORTS_ID_0, NU32_LED_CHANNEL, NU32_LED1_POS); // toggle led 1
  PLIB_PORTS_PinToggle(PORTS_ID_0, NU32_LED_CHANNEL, NU32_LED2_POS); // toggle led 2
}

const static DRV_TMR_INIT init =   // used to configure timer; const so stored in flash
{
  .moduleInit = {SYS_MODULE_POWER_RUN_FULL},       // no power saving
  .tmrId = TMR_ID_1,                               // use timer 1
  .clockSource = DRV_TMR_CLKSOURCE_INTERNAL,       // use pbclk
  .prescale = TMR_PRESCALE_VALUE_256,              // use a 1:256 prescaler value
  .interruptSource = INT_SOURCE_TIMER_1,           // ignored since system_config has set
                                                   // interrupt mode to false
  .mode = DRV_TMR_OPERATION_MODE_16_BIT,           // use 16 bit mode
  .asyncWriteEnable = false                        // no asynchronous write
};

int main(void) {
  SYS_MODULE_OBJ timer_handle;       // handle to the timer driver
  SYS_MODULE_OBJ devcon_handle;      // device configuration handle
  DRV_HANDLE timer1;                 // handle to the timer

  SYS_CLK_Initialize(NULL);          // initialize the clock, but tell it to use
                                     // configuration bit
                                     // settings that were set with the bootloader

  // initialize the DEVCON system service, default init settings are fine.
  devcon_handle = SYS_DEVCON_Initialize(SYS_DEVCON_INDEX_0, NULL);

  // initialize the LED pins as outputs
  PLIB_PORTS_PinDirectionOutputSet(PORTS_ID_0, NU32_LED_CHANNEL, NU32_LED1_POS);
  PLIB_PORTS_PinDirectionOutputSet(PORTS_ID_0, NU32_LED_CHANNEL, NU32_LED2_POS);
  PLIB_PORTS_PinClear(PORTS_ID_0, NU32_LED_CHANNEL, NU32_LED1_POS); // turn on LED1
  PLIB_PORTS_PinSet(PORTS_ID_0, NU32_LED_CHANNEL, NU32_LED2_POS);   // turn off LED2

  // initialize the timer driver
```

```
    timer_handle = DRV_TMR_Initialize(DRV_TMR_INDEX_0, (SYS_MODULE_INIT*)&init);

    // open the timer, this is the only client (only place where DRV_TMR_Open is called)
    timer1 = DRV_TMR_Open(DRV_TMR_INDEX_0, DRV_IO_INTENT_EXCLUSIVE);

    // the timer driver will call invert_leds_callback at 5 Hz. This is a "periodic alarm."
    DRV_TMR_Alarm16BitRegister(timer1, 62499, true, 0, invert_leds_callback);

    // start the timer
    DRV_TMR_Start(timer1);

    while (1) {
      // update device configuration. Does nothing in Harmony v1.06 but may in the future
      SYS_DEVCON_Tasks(devcon_handle);
      // update the timer driver state machine
      DRV_TMR_Tasks(timer_handle);
    }
    return 0;
  }
```

Just as in the UART example, we define an initialization `struct` (`DRV_TMR_INIT init`) to set parameters such as which timer peripheral to use (Timer1), the prescaler, and the clock source. In this example, the timer is configured with a 1:256 prescaler, so each tick occurs at 80,000,000 Hz/256 = 312,500 Hz and takes 3.2 μs. The period count is 62,499, so the timer rolls over every $3.2 \times (62{,}499 + 1)$ μs = 200 ms (5 Hz) (see Chapter 8 for details).

In `main`, we first initialize the CLK and DEVCON system services, upon which the TMR driver depends. We also set the LED pins as outputs and set the initial LED states. As with all drivers we must initialize the driver and open a client. Here we open the client with `DRV_IO_INTENT_EXCLUSIVE` to indicate that only one client will be used; the TMR driver does not support multiple clients. Finally, we must register the callback function that is executed when the timer rolls over. We pass `DRV_TMR_Alarm16BitRegister` a driver handle, a period count, a Boolean indicating whether the alarm repeats or occurs once, a context value that gets passed to the callback, and finally a pointer to a function that should be called when the alarm expires.[6]

Note that the callback function, `invert_leds_callback`, is just an ordinary function, not an ISR. Unlike many of our previous timer examples, the example does not use an ISR even though it has code that should be executed periodically. Instead, the TMR driver invokes the callback function by using *polling*. In polling mode, the driver constantly checks (i.e., polls) whether the timer's period has expired, and if so, it calls the callback function (in this case `invert_leds_callback`). The polling occurs in the timer driver's state machine, which is updated by the call to `DRV_TMR_Tasks` in the main loop. If you were to add code to the main loop that delayed the call to `DRV_TMR_Tasks` for too long, then the timer callback would be delayed.

[6] In C, pointers can point to functions. The address of a function is that function's name without the parentheses.

After registering the callback, we start the timer by calling `DRV_TMR_Start` and enter the main loop. The main loop updates two Harmony state machines, one belonging to the DEVCON system service and the other belonging to the TMR driver. The call to `DRV_TMR_Tasks` is where the timer polling occurs. If `DRV_TMR_Tasks` detects that the timer has rolled over it will call the alarm function registered in `DRV_TMR_Alarm16BitRegister`.

20.5.3 Timers with Interrupts

Instead of polling, and perhaps being delayed to call the callback function if the computations in the main loop take too long, we can employ an interrupt-based approach, while still using the Harmony TMR driver.

Using interrupts with the TMR driver requires only a few small changes. First, edit `system_config.h`, changing the value of `DRV_TMR_INTERRUPT_MODE` to `true`. Next, you must set the interrupt priority (and optionally the sub-priority), anywhere prior to the call to `DRV_TMR_Start`:

```
PLIB_INT_VectorPrioritySet(INT_ID_0, INT_VECTOR_T1, INT_PRIORITY_LEVEL5);
```

In `demo_tmr.c`, immediately before entering the main loop, enable interrupts:

```
PLIB_INT_Enable(INT_ID_0); // enable interrupts
```

Finally, make `timer_handle` a global variable, not local to `main`, and implement the timer ISR:

```
void __ISR(_TIMER_1_VECTOR, IPL5SOFT) Timer1ISR(void) {
  DRV_TMR_Tasks_ISR(timer_handle); // update the timer state machine
}
```

Note that the ISR does not perform any action other then calling `DRV_TMR_Tasks_ISR`, which updates the TMR driver state, dispatching the alarm callback function as appropriate. You need not clear any interrupt flags as the TMR driver handles these details for you. As the TMR driver state is now updated in the ISR, you should remove the call to `DRV_TMR_Tasks` from the main loop.

20.6 System Services

In Harmony, system services are similar to drivers: you add `.c` and `.h` files to your project, include the appropriate header files in `system_definitions.h`, and define necessary compile-time parameters in `system_config.h`. Files for system services reside under `<framework>/system`. Unlike drivers, however, system services lack a consistent interface (they do not all have initialize and open functions, for example, although many do) and vary widely in purpose. Some system services depend on drivers, and some drivers depend on system services. In the previous section we used the device configuration (DEVCON), clock (CLK), and interrupt (INT) system services. Both the DEVCON and CLK system services provide low-level hardware functionality: CLK handles oscillators while DEVCON handles settings such as the cache. The INT system service mostly duplicates functionality available in the interrupt PLIB so we did not use it explicitly; however, the TMR driver will not compile without the INT system service. Another system service that mostly duplicates PLIB functionality is the PORT system service; therefore, when manipulating ports we have used PLIB functions directly.

Although the aforementioned system services provide low-level functionality, other services work on a higher level. For example, the messaging (MSG) system service provides an interface for sophisticated inter-task communication. The timer (TMR) system service that we introduce here abstracts TMR drivers, allowing you to specify timer periods in ms rather than ticks. You use the TMR service to create one or more logical (not hardware) timers at different frequencies, all of which are derived from a single hardware timer. The TMR system service uses a TMR driver to manage the underlying hardware timer. The Harmony documentation refers to logical timers derived from the system service as clients, since they are created by calling functions provided by the TMR system service module.

To understand how the TMR system service works, imagine that you want one task to occur every 10 ms and another to occur every 55 ms. You could configure two separate hardware timers at those respective frequencies. If you wanted to use only one timer, however, you could configure it to interrupt every 5 ms. The timer ISR would then perform the first task every other interrupt and the second task every 11 interrupts. The TMR system service handles this logic for you, and allows you to use timers either in an interrupt or polled mode.

To build this project, you need all the files from the TMR driver program in Section 20.5.2 (because the TMR system service requires a TMR driver) plus `<framework>/system/tmr/src/sys_tmr.c`. You should modify the `system_definitions.h` file from Code Sample 20.5, adding `#include "system/tmr/sys_tmr.h"` to include the TMR system service header; the rest of the file remains the same.

The `system_config.h` header requires the same macros to be defined as in Section 20.5.2. For this example, use polled rather than interrupt mode by making sure `DRV_TMR_INTERRUPT_MODE` is set to `false`. The TMR system service also requires you to define some additional macros in `system_config.h` (see the comments below):

```
// The maximum number of timers that can be created using the system service.
// We will actually create two timers.
#define SYS_TMR_MAX_CLIENT_OBJECTS 5

// The frequency, in Hz, used for timing calculations.  The higher the value, the
// better the resolution of the timer, but the shorter its longest possible
   duration.
```

```
// This is used in determining the integer number of clock ticks for each client
   timer
// rollover.  (No hardware timers actually operate at this speed.)  We choose
   10 kHz.
#define SYS_TMR_UNIT_RESOLUTION 10000

// The hardware timer's base frequency is derived from a clock frequency
   (PBCLK here).
// Due to limited resolution of the timer, not all frequencies are exactly
   available.
// If the closest available frequency differs from the requested frequency by more
// than this amount (in percent), a run-time error occurs.  We choose ten percent.
#define SYS_TMR_FREQUENCY_TOLERANCE 10

// Just as the hardware timer's base frequency is derived from the peripheral bus
// frequency, the client timer frequency is derived from the base TMR
// system service frequency.  The requested client frequency may not be exactly
// available.  If the closest available client frequency differs from the requested
// frequency by more than this amount (in percent), a run-time error occurs.  We
// choose ten percent.
#define SYS_TMR_CLIENT_TOLERANCE 10
```

Now let us study code that uses the TMR system service to toggle one LED at 5 Hz and the other at 1 Hz. As in Code Sample 20.7 we define a callback, initialization structures, and initialize a timer driver. We also must initialize the system service. Unlike the timer driver demo, we implement an FSM in this demonstration. The program has two states: initialization (`APP_STATE_INIT`) and running (`APP_STATE_RUN`). These states are necessary because the TMR system service FSM must be updated a few times before logical timers can be started.

Code Sample 20.8 `demo_service.c`. **Demonstration of the TMR System Service.**

```
// Demonstrates the timer service.
// The timer service allows us to use one hardware timer to run tasks
// at different frequencies.

#include "system_config.h"
#include "system_definitions.h"

void invert_leds_callback(uintptr_t context, uint32_t alarmCount) {
  // context is data passed by the user that can then be used in the callback.
  // Here the context is NU32_LED1_POS or NU32_LED2_POS to tell us which LED to toggle.
  PLIB_PORTS_PinToggle(PORTS_ID_0, NU32_LED_CHANNEL, context);
}

const static DRV_TMR_INIT init =   // used to configure timer; const so stored in flash
{
  .moduleInit = {SYS_MODULE_POWER_RUN_FULL}, // no power saving
  .tmrId = TMR_ID_1,                         // use Timer1
  .clockSource = DRV_TMR_CLKSOURCE_INTERNAL, // use pbclk
  .prescale = TMR_PRESCALE_VALUE_256,        // use a 1:256 prescaler value
  .interruptSource = INT_SOURCE_TIMER_1,     // ignored because system_config.h
                                             // has set interrupt mode to false
  .mode = DRV_TMR_OPERATION_MODE_16_BIT,     // use 16-bit mode
  .asyncWriteEnable = false                  // no asynchronous write
};

const static SYS_TMR_INIT sys_init =
```

```
{
  .moduleInit = {SYS_MODULE_POWER_RUN_FULL}, // no power saving
  .drvIndex = DRV_TMR_INDEX_0,               // use timer driver 0
  .tmrFreq = 1000                    // base frequency of the system service (Hz)
};

// holds the state of the application
typedef enum {APP_STATE_INIT, APP_STATE_RUN} AppState;

int main(void) {
  SYS_MODULE_OBJ timer_handle;    // handle to the timer driver
  SYS_MODULE_OBJ devcon_handle;   // device configuration handle
  SYS_TMR_HANDLE sys_tmr;
  SYS_CLK_Initialize(NULL);       // initialize the clock,
                                  // but tell it to use configuration bit settings
                                  // that were set with the bootloader

  // initialize the device, default init settings are fine
  devcon_handle = SYS_DEVCON_Initialize(SYS_DEVCON_INDEX_0, NULL);

  // initialize the pins for LEDs
  PLIB_PORTS_PinDirectionOutputSet(PORTS_ID_0, NU32_LED_CHANNEL, NU32_LED1_POS);
  PLIB_PORTS_PinDirectionOutputSet(PORTS_ID_0, NU32_LED_CHANNEL, NU32_LED2_POS);
  PLIB_PORTS_PinClear(PORTS_ID_0, NU32_LED_CHANNEL, NU32_LED1_POS);
  PLIB_PORTS_PinSet(PORTS_ID_0, NU32_LED_CHANNEL, NU32_LED2_POS);

  // initialize the timer driver
  timer_handle = DRV_TMR_Initialize(DRV_TMR_INDEX_0, (SYS_MODULE_INIT*)&init);

  // initialize the timer system service
  sys_tmr = SYS_TMR_Initialize(SYS_TMR_INDEX_0,(SYS_MODULE_INIT*)&sys_init);

  AppState state = APP_STATE_INIT; // initialize the application state

  while (1) {
    // based on the application state, we may need to initialize timer callbacks
    switch(state) {
      case APP_STATE_INIT:
        if(SYS_STATUS_READY == SYS_TMR_Status(sys_tmr)) {
          // If the timer is ready:
          // Register the timer callbacks to invert LED1 at 5 Hz (200 ms period)
          // & LED2 at 1 Hz (1000 ms period). Both tasks use the same callback function
          // and use the context to determine which LED to invert; however, we could
          // have registered different callback functions.  Note that the context type
          // could also be a pointer, if you want more information passed to the callback
          SYS_TMR_CallbackPeriodic(200,NU32_LED1_POS,invert_leds_callback);
          SYS_TMR_CallbackPeriodic(1000,NU32_LED2_POS,invert_leds_callback);
          state = APP_STATE_RUN;
        } else {
          // the timer is not ready, so do nothing and let the state machines update
        }
        break;
      case APP_STATE_RUN:
        break; // we are just running
    }

    //update the device configuration
    SYS_DEVCON_Tasks(devcon_handle);
    SYS_TMR_Tasks(sys_tmr);
    // update the timer driver state machine
    DRV_TMR_Tasks(timer_handle);
```

```
        }
        return 0;
    }
```

The first implementation detail we examine is the timer callback `invert_leds_callback`. Notice that we use the first argument, `context`, to determine which LED to invert. We use the system service to create two logical timers that use the same callback but pass a different value as the `context`.

We also must initialize the TMR system service and the timer driver upon which it depends. The `sys_init struct` defines which TMR driver the system service uses and what base frequency (in Hz) the underlying TMR driver uses. Notice that we never explicitly open a TMR driver; rather, a system service opens the TMR driver using the frequency we specified in the initialization data. However, we still must initialize the TMR driver and update its state machine in the main loop.

We cannot create client timers until the TMR system service is ready. We therefore enter the main loop and use a state machine to determine whether some initialization must be performed or if the application is ready to run. When in the initialization state, we query the TMR system service's status using `SYS_TMR_Status` and check if it is `SYS_STATUS_READY`. Once the TMR system service is ready, we create two client timers, one with a period of 200 ms (5 Hz) and another with a period of 1000 ms (1 Hz). Each of these clients calls `invert_leds_callback` at their respective frequency; however, one timer passes `NU32_LED1_POS` and the other passes `NU32_LED2_POS` as the context parameter to the callback function. Thus, the 5 Hz timer inverts LED1 while the 1 Hz timer inverts LED2. After we create the client timers we enter the run state; otherwise, we would be creating new timers on every loop iteration! After handling the application tasks we call the appropriate `Tasks` function for each Harmony component that we use, updating their state machines.

20.7 Program Structure

Earlier we mentioned that Harmony suggests that your program have a certain structure. We have already seen the unavoidable elements of that structure in the form of `system_config.h` and `system_definitions.h`, but we still have placed all our code in a single directory. Harmony, however, is designed to allow code re-use across multiple hardware configurations. To achieve this goal, it suggests creating a logical separation between your application code and the code that is specific to a single processor. To accomplish this goal, Harmony suggests the following file and directory structure:

`app.c`: The main application logic.
`app.h`: Header for the application.

- `main.c`: Boilerplate code that ties everything together.
- `system_config`: The system configuration directory. Contains one subdirectory for every supported platform. For example:
    - `pic32_NU32`: A configuration directory for code designed to run on the NU32 board. Generally, subdirectories of `system_config` should describe the hardware they support, but they can be whatever you want. We will assume that you use `pic32_NU32` in this example as, by default, the `Makefile` looks for this directory.
    - `pic32_Standalone`: This hypothetical directory could contain configuration information for a program that does not use the bootloader.
    - `pic32_picMZ`: This hypothetical directory could contain configuration information for a program running on a PIC32MZ processor.

Within each `system_config` subdirectory (e.g., `system_config/pic32_NU32`) should be the following files:

- `system_config.h`: The system configuration header.
- `system_definitions.h`: Data type definitions and `#include` directives needed by other files.
- `system_init.c`: Performs the system initialization.
- `system_tasks.c`: Updates the Harmony modules' state machines.
- `system_interrupt.c`: Implements the ISRs.
- `framework`: Subdirectory for holding the files the project needs that are provided by Harmony. Generally mirrors the structure of `<framework>`.
    - `drivers`: Harmony driver files, organized into subdirectories for each driver used (e.g., `tmr` for the timer driver).
    - `system`: Harmony system service files, organized into subdirectories for each system service used (e.g., `tmr` for the TMR system service).

The automatic Harmony code configuration tools provided by Microchip and installed with MPLAB create a structure similar to the one above. Optionally, however, you can direct the IDE to include the files at their home locations in `<framework>` during the build rather than copying them to your project directory.

The Harmony `Makefile` that we have provided will compile and link the files in your project directory with files in `system_config/pic32_NU32` and the mentioned subdirectories. The variable `CONFIG` in the `Makefile` determines which `system_config` subdirectory is used when you compile. You can use this variable to compile your project for different platforms either by editing the `Makefile` or by overriding the variable at the command line (i.e., issuing `make CONFIG=dir`). Finally, if you remove the linker script `NU32bootloaded.ld` from the base directory, you can place a different linker script in each of the `system_config` subdirectories (e.g., system_config/pic32_NU32), allowing you to use different linker scripts for different configurations.

Now that you have a general sense of the organization of a full Harmony project, we demonstrate the use of this structure by revisiting the interrupt-based timer driver example of Section 20.5.3. By examining the code you should gain a better understanding about how this structure allows you to isolate your main program logic from hardware-specific considerations, making it more portable.

We begin with the files that remain the same, regardless of hardware configuration. All Harmony applications that follow the recommended structure have a very simple main function, defined in main.c. This file delegates initialization tasks to SYS_Initialize in system_init.c and the state machine logic to SYS_Tasks, implemented in system_tasks.c.

Code Sample 20.9 main.c. **Main File for All Canonical Harmony Programs.**

```
#include <stddef.h>                    // defines NULL
#include "system/common/sys_module.h" // SYS_Initialize and SYS_Tasks prototypes

int main(void) {

  SYS_Initialize(NULL); // initializes the system

  while(1) {
    SYS_Tasks();          // updates the state machines of polled harmony modules
  }

  return 0;
}
```

The header file app.h defines any data types needed by the application and also provides prototypes for the two primary application functions, APP_Initialize and APP_Tasks. These functions will be called by SYS_Initialize and SYS_Tasks, respectively.

Code Sample 20.10 app.h. **Header File for the Application.**

```
#ifndef APP__H__
#define APP__H__

// The application states.  APP_STATE_INIT is the initial state, used to perform
// application-specific setup.  Then, during program operation, we enter
// APP_STATE_WAIT as the timer takes over.
typedef enum { APP_STATE_INIT, APP_STATE_WAIT} APP_STATES;

// Harmony structure suggests that you place your application-specific data in a struct.
typedef struct {
  APP_STATES state;
  DRV_HANDLE handleTmr;
```

```
} APP_DATA;

void APP_Initialize(void);
void APP_Tasks(void);

#endif
```

The actual application logic is implemented in `app.c`, and is driven by an FSM.

Code Sample 20.11 `app.c`**. The Application Implementation.**

```
#include "system_config.h"
#include "system_definitions.h"
#include "app.h"

APP_DATA appdata;

void invert_led_callback(uintptr_t context, uint32_t alarmCount) {
  // context is data passed by the user that can then be used in the callback.
  // alarm count tracks how many times the callback has occured
  PLIB_PORTS_PinToggle(PORTS_ID_0, NU32_LED_CHANNEL, NU32_LED1_POS); // toggle led 1
  PLIB_PORTS_PinToggle(PORTS_ID_0, NU32_LED_CHANNEL, NU32_LED2_POS); // toggle led 2
}

// initialize the application state
void APP_Initialize(void) {
  appdata.state = APP_STATE_INIT;
}

void APP_Tasks(void) {
  switch(appdata.state) {
    case APP_STATE_INIT:
      // turn on LED1 by clearing A5
      PLIB_PORTS_PinClear(PORTS_ID_0, NU32_LED_CHANNEL, NU32_LED1_POS);
      // turn off LED2 by setting A5
      PLIB_PORTS_PinSet(PORTS_ID_0, NU32_LED_CHANNEL, NU32_LED2_POS);

      // only one client at a time, open the timer
      appdata.handleTmr = DRV_TMR_Open(DRV_TMR_INDEX_0, DRV_IO_INTENT_EXCLUSIVE);

      // timer driver calls invert_led_callback at 5 Hz.  Register this "periodic alarm."
      DRV_TMR_Alarm16BitRegister(appdata.handleTmr, 62499, true, 0, invert_led_callback);
      DRV_TMR_Start(appdata.handleTmr);
      appdata.state = APP_STATE_WAIT;
      break;
    case APP_STATE_WAIT:
      break; // we need not do anything here
  }
}
```

Next, we visit the files that have hardware-dependent implementations and are therefore located under the `system_config/<hardware>` directory, where `<hardware>` is a platform-specific name: for our purposes we use `<hardware> = pic32_NU32`.

The `system_config.h` file is almost the same as Code Sample 20.6 except, since this example uses timer interrupts, `DRV_TMR_INTERRUPT_MODE` is defined as `true`.

The `system_definitions.h` file contains an extra data structure, used to hold handles to the Harmony modules that the program uses. The handles are stored in a single global variable, made accessible to all modules via the `extern` keyword, as per the Harmony documentation's recommendation.

Code Sample 20.12 `system_definitions.h`**. System Definitions for a Harmony Application.**

```
#ifndef SYSTEM_DEFINITIONS_H
#define SYSTEM_DEFINITIONS_H

#include <stddef.h>          // some standard C headers with types needed by harmony
                             // defines integer types with fixed sizes, for example
                             // uint32_t is guaranteed to be a 32-bit unsigned integer
                             // uintptr_t is an integer that can be treated as a pointer
                             // (i.e., an unsigned int large enough to hold an address)
#include <stdbool.h>
#include "peripheral/peripheral.h"     // all the peripheral (PLIB) libraries
#include "system/devcon/sys_devcon.h"  // device configuration system service
#include "system/clk/sys_clk.h"        // clock system service
#include "system/common/sys_module.h"  // basic system module
#include "driver/tmr/drv_tmr.h"        // the timer driver

// system object handles
typedef struct {
  SYS_MODULE_OBJ sysDevcon;  // device configuration object
  SYS_MODULE_OBJ drvTmr;     // the timer driver object
} SYSTEM_OBJECTS;

// Declares a global variable that holds the system objects so that all files including
// system_definitions.h can access the handles. Is actually defined and initialized in
// system_init.c.
extern SYSTEM_OBJECTS sysObj;

#endif
```

The `system_init.c` file performs the initialization. It also calls the application's initialization function. For standalone applications, you would include the necessary configuration bit setup (`#pragma config`) in this file; however, we do not need to set configuration bits as the bootloader has already done that for us.

Code Sample 20.13 `system_init.c`**. Platform-Specific Hardware Initialization.**

```
#include "system_config.h"
#include "system_definitions.h"
#include "app.h"

// Code to initialize the system.
// For standalone projects (those without a bootloader)
```

```
// you would define the configuration values here, e.g.,
// #pragma config FWDTEN = OFF
// See NU32.h for configuration bits.  These are set by the bootloader in this example.

const static DRV_TMR_INIT init = // used to configure the timer; const so stored in flash
{
  .moduleInit = SYS_MODULE_POWER_RUN_FULL,        // no power saving
  .tmrId = TMR_ID_1,                              // use timer 1
  .clockSource = DRV_TMR_CLKSOURCE_INTERNAL,      // use pbclk
  .prescale = TMR_PRESCALE_VALUE_256,             // use a 1:256 prescaler value
  .interruptSource = INT_SOURCE_TIMER_1,          // use timer one interrupts
  .mode = DRV_TMR_OPERATION_MODE_16_BIT,          // use 16-bit mode
  .asyncWriteEnable = false
};

SYSTEM_OBJECTS sysObj; // handles to harmony modules. Defined in system_definitions.h.

// called from the beginning of main to initialize the harmony components etc.
void SYS_Initialize(void * data) {
  SYS_CLK_Initialize(NULL);  // initialize the clock, but use configuration bit settings
                             // that were set with the bootloader

                           // Initialize the device, default init settings are fine.
                           // As of harmony 1.06 this call is not needed for our purposes,
                           // but this may change in future versions so we include it.
                           // It is necessary if you want to set the prefetch cache
                           // and wait states using SYS_DEVCON_PerformanceConfig.
                           // However, we need not configure the cache and wait states
                           // the bootloader has already done this for us.
  sysObj.sysDevcon = SYS_DEVCON_Initialize(SYS_DEVCON_INDEX_0, NULL);

  // initialize the pins for the LEDs
  PLIB_PORTS_PinDirectionOutputSet(PORTS_ID_0, NU32_LED_CHANNEL, NU32_LED1_POS);
  PLIB_PORTS_PinDirectionOutputSet(PORTS_ID_0, NU32_LED_CHANNEL, NU32_LED2_POS);

  // initialize the timer driver
  sysObj.drvTmr = DRV_TMR_Initialize(DRV_TMR_INDEX_0, (SYS_MODULE_INIT*)&init);

  PLIB_INT_MultiVectorSelect(INT_ID_0); // enable multi-vector interrupt mode

  // set timer int priority
  PLIB_INT_VectorPrioritySet(INT_ID_0, INT_VECTOR_T1, INT_PRIORITY_LEVEL5);
  PLIB_INT_Enable(INT_ID_0);               // enable interrupts

  // initialize the apps
  APP_Initialize();
}
```

The `system_interrupt.c` file implements the timer ISR, and, in general, all ISRs.

Code Sample 20.14 `system_interrupt.c`**. ISR Definitions for a Harmony Application.**

```
#include "system_config.h"
#include "system_definitions.h"
#include <sys/attribs.h>

// the timer interrupt.  note that this just updates the state of the timer
void __ISR(_TIMER_1_VECTOR, IPL5SOFT) Timer1ISR(void) {
  DRV_TMR_Tasks_ISR(sysObj.drvTmr); // update the timer state machine
}
```

The system_tasks.c file updates state machines for Harmony modules that operate in a polled mode (in this case it is only the DEVCON system service). It also calls the application code, allowing the application to update its state machine.

Code Sample 20.15 system_tasks.c. **Update Harmony Modules and the Application's State Machines.**

```
#include "system_config.h"
#include "system_definitions.h"
#include "app.h"

// update the state machines
void SYS_Tasks(void) {
  SYS_DEVCON_Tasks(sysObj.sysDevcon);
  // timer tasks are interrupt driven, not polled.
  // for a polled application uncomment below
  // DRV_TMR_Tasks(sysObj.drvTmr);
  APP_Tasks(); // application specific tasks
}
```

The Harmony files required for this example are the same as those needed in the Timer example in Section 20.5.2. You should copy these files into the system_config/pic32_NU32/framework directory and its subdirectories. For example, copy <framework>/system/int/src/sys_int_pic32.c into system_config/pic32_NU32/framework/system/int.

With your Makefile and NU32bootloaded.ld in the top-level directory, you should be able to make the project and see the LEDs toggle.

20.8 USB

20.8.1 USB Basics

If you have used a computer in the past 20 years, you probably have used universal serial bus (USB) devices. An unimaginable number of USB devices have been created over the years. To support such a profusion of devices, the USB protocol is relatively complicated; therefore, we describe only the minimum required to use Harmony's USB middleware to create simple USB devices.

With USB, the bus is controlled by a single master called the *host*, typically your computer or a smart phone (although the PIC32 can act as a host too). Each host can control up to 127 *devices* through a hub. The physical connectors for hosts and devices are incompatible; this prevents two devices or two hosts from being connected to each other. Apart from the mini-B USB connector you use for programming the NU32, the board has a micro-B connector, which makes it a device by default. A USB On-The-Go cable can be used to convert the micro-B connector to an A connector, allowing the NU32 to act as a host. Such a cable is used by smart phones (which typically use a micro-B port) to allow you to connect devices such as USB flash drives to them.

USB uses four wires: 5 V power (V+), which allows the host to optionally provide power to devices; ground (V−); and two data lines, D+ and D−. The PIC32MX795F512H has a single USB port, and the micro-B connector's D+ and D− lines are directly connected to the PIC32's D+ and D− pins.

USB uses a complicated protocol for communication. Multiple versions of the USB protocol exist, with newer hosts typically being backwards compatible with older protocols. The PIC32 supports USB 2.0 in low speed (1.5 Mb/s) and full speed (12 Mb/s) modes; all our examples will use USB full speed. The USB protocol includes support for many types of devices, known as device classes. Device classes include human interface devices (HID) such as keyboards and mice; mass storage devices (MSD) such as external hard drives; communications device classes (CDC) such as a virtual serial port; and a generic device class used to implement a vendor-specific protocol. Harmony provides support for all of these device classes; however, we focus on HID because it is the simplest. Despite its name, HID devices provide a flexible interface for transferring data bidirectionally between host and device.

Harmony handles the USB protocol details for us; however, a familiarity with two basic USB concepts, endpoints and descriptors, will be helpful. USB devices transfer data between themselves and the host through "endpoints." An endpoint has a type, which determines properties of the data transfer such as latency and the host-centric direction, either in or out. In-endpoints are used for transferring data into the host (the device sends the data) and out-endpoints are used for transferring data out of the host (the device receives the data). Descriptors provide information to the host about the device's capabilities and endpoints. For example, a descriptor tells the host what device class the device uses. There are numerous types of device descriptors; we will encounter several when implementing a device and discuss the relevant details when necessary.

20.8.2 Powering the NU32 by USB

When acting as a device, you can power the PIC32 from the external power supply, as usual, or from the 5 V provided by the host's USB port. To power the PIC32 from the host, ensure that NU32's power jack is unplugged and connect the VBUS pin to the 5 V pin. Be especially careful when powering the PIC32 from the USB port of your computer: incorrect circuits could damage your PIC32, the USB port on your computer, or even your computer's motherboard! Note that USB devices must specify a current request from the host; our code always requests 100 mA.

It is also possible to use the NU32 as a USB host; see the book's website for details.

20.8.3 USB HID Device

Our USB example creates a generic HID device that can be used to transfer data between the PIC32 and a host (either your computer or a smartphone).

Unlike the virtual UART over USB approach used earlier in this book, we need to write a C program that runs on the host to interact with our new USB device. The C program uses the HID API library, a freely available, cross-platform library for communicating with USB HID devices. We leave the details of the installation and use of this library to the book's website and only provide an overview here; the details vary by operating system and may change. Overall, installation involves compiling the HID API library into a binary format. To use the library, you must specify some flags at the compiler command line: `-I<include path>` to tell the compiler where to find the `hidapi.h` header, `-L<library path>` to tell it where to find the compiled HID library, and `-l<library name>` to indicate that your code should be linked against the HID API library (the name varies by platform and the options used when compiling the library).

Prior to discussing the PIC32 code, we first examine the client code, `client.c`, to get a sense of the goals of the example. The code opens the device based on some hardware identifiers. It then prints some information about the device. Next it enters an infinite loop where it prompts the user for a string, sends that string to the PIC32, and then prints the PIC32's reply. Overall, the behavior is similar to Code Sample 1.2 (`talkingPIC.c`). Here is sample output:

```
Opened HID device.
Manufacturer String: Microchip Technology Inc.
Product String: Talking HID
Say something to PIC (blank to exit):  I'm talking via usb.
PIC Replies:  I'M TALKING VIA USB.
Say something to PIC (blank to exit):
```

Code Sample 20.16 `client.c`. **C Client for Our Custom HID Device.**

```
#include <stdio.h>
#include <stddef.h>
#include "hidapi.h"

// Client that talks to the HID device
// using hidapi, from www.signal11.us.
// hidapi allows you to directly communicate with hid devices
// without needing a special driver.
// To use hidapi you first must compile the library.
// You need its header to be on your path and you need to
// link against the library.  The procedure for doing this
// varies by platform.
// Note: error checking code omitted for clarity
#define REPORT_LEN 65 // 64 bytes per report plus report ID
#define MAX_STR 255   // max length for a descriptor string

int main(void) {
  char outbuf[REPORT_LEN] = "";
  char inbuf[REPORT_LEN] = "";
  wchar_t wstr[MAX_STR] = L""; // use 2-character "wide chars" for USB string descriptors
  hid_device *handle = NULL;

  // open the hid device using the VID and PID
  handle = hid_open(0x4d8, 0x1769, NULL);
  printf("Opened HID device.\n");

  // use blocking mode so hid_read will wait for data before returning
  hid_set_nonblocking(handle, 0);

  // get the manufacturer string
  hid_get_manufacturer_string(handle, wstr, MAX_STR);
  printf("Manufacturer String: %ls\n", wstr); // the ls is to print a wide string

  // get the product string
  hid_get_product_string(handle, wstr, MAX_STR);
  printf("Product String: %ls\n", wstr);

  while(1) {
    printf("Say something to PIC (blank to exit): ");
    // get string of max length REPORT_LEN-1 from user
    // first byte is the report id (always 0)
    fgets(outbuf + 1, REPORT_LEN - 1, stdin);
    if(outbuf[1] == '\n') { // if blank line, exit
      break;
    }
    hid_write(handle, (unsigned char *)outbuf, REPORT_LEN);// send report to the device

    // read the pic's reply, wait for bytes to actually be read
    while(hid_read(handle, (unsigned char *)inbuf, REPORT_LEN) == 0) {
      ; // (on some platforms hid_read returns 0 even in blocking mode, hence the loop)
    }
    printf("\nPIC Replies: %s\n", inbuf);
  }
  return 0;
}
```

Notice that the code includes `hidapi.h`, allowing us to use functions from the HID API library. The first interesting function call is to `hid_open`, which provides access to a given HID device based on its vendor identifier (VID) and product identifier (PID). All USB devices have a VID and a PID that should uniquely identify the device. For a hefty fee you can obtain your own VID from the USB Implementers Forum (USB-IF); if you wanted to manufacture many USB devices, your company would purchase a VID. Here we just use Microchip's VID (`0x4d8`). We choose a PID that does not conflict with any Microchip products, `0x1769`. These values are provided to the host by a descriptor that we implement on the device.

After we open the device we read the manufacturer and product strings, which provide human-readable information about the device.

In the infinite loop we use two HID API functions, `hid_write` and `hid_read`, to write data to, and read data from, the device. Data is sent and received from HID devices in structures called reports. Reports are described by the USB HID standard, and provide a flexible but complicated method for structuring data. Here we simply send and receive 64-byte-long packets. Notice that we use the packets to store strings, but technically they could store any type of data.

Prior to continuing, you should compile `client.c`. (Make sure it is not in your project directory, as it is not meant for the PIC32!) This step will ensure that you have installed the HID API successfully. Next, we will examine the PIC32 code needed to make the device work.

In this project we will dispense with the subdirectory structure of the previous example in Section 20.7 and simply put all code in the same directory. In addition to the usual Harmony `Makefile` and `NU32bootloaded.ld`, you need

- `talkingHID.c`: The main project code.
- `hid.c`: A simple HID library we created.
- `hid.h`: The header file for the HID library.
- `system_config.h`
- `system_definitions.h`
- `sys_int_pic32.c`: Copied from `<framework>/system/int/src/`. The system interrupt system service, which is needed by other Harmony modules.
- `drv_usbfs.c`: Copied from `<framework>/driver/usb/usbfs/src/dynamic/`. General USB device driver for full-speed mode. This USB layer contains code common to all Harmony USB drivers (including host implementations).
- `drv_usbfs_device.c`: Copied from `<framework>/driver/usb/usbfs/src/dynamic/`. The USB driver for all full-speed devices.

usb_device.c: Copied from <framework>/usb/src/dynamic/. The Harmony middleware that contains code common to all USB devices.

usb_device_hid.c: Copied from <framework>/usb/src/dynamic/. USB HID middleware used to implement a HID device.

After copying the necessary Harmony files into the project directory, take a look at the main file, talkingHID.c, to learn about the overall program logic. This file implements main and offloads most of the USB details to a HID library we have created, hid.{c,h}. We first define a HID report descriptor, which describes the format of the data sent and received between the device and the host. In this case each report contains 64 data bytes. Following a pattern similar to a Harmony driver, the HID library has a function for initialization (hid_setup), a function for opening the device hid_open, and a function for updating an internal FSM (hid_update). The main loop implements an FSM to appropriately receive data and send a response. We use our HID library functions hid_send and hid_receive to communicate with the host. Much like the UART example, these functions are non-blocking.

Code Sample 20.17 talkingHID.c. **The Main Talking HID File.**

```
#include "hid.h"
#include "system_config.h"
#include "system_definitions.h"
#include <sys/attribs.h>
#include <ctype.h> //for toupper
#define REPORT_LEN 0x40 // reports have 64 bytes in them

// the HID report descriptor (see Universal Serial Bus HID Usage Tables document).
// This example is from the harmony hid_basic example.
// This descriptor contains input and output reports that are 64 bytes long.  The
// data can be anything. Borrowed from the microchip generic hid example
const uint8_t HID_REPORT[NU32_REPORT_SIZE] = {
  0x06, 0x00, 0xFF,     // Usage Page = 0xFF00 (Vendor Defined Page 1)
  0x09, 0x01,           // Usage (Vendor Usage 1)
  0xA1, 0x01,           // Collection (Application)
  0x19, 0x01,           // Usage Minimum
  0x29, REPORT_LEN,     // Usage Maximum  64 input usages total (0x01 to 0x40)
  0x15, 0x01,           // Logical Minimum (data bytes in the report have min value = 0x00)
  0x25, 0x40,           // Logical Maximum (data bytes in the report have max value = 0xFF)
  0x75, 0x08,           // Report Size: 8-bit field size
  0x95, REPORT_LEN,     // Report Count: 64 8-bit fields
                        //    (for next "Input", "Output", or "Feature" item)
  0x81, 0x00,           // Input (Data, Array, Abs): input packet fields
                        // ased on the above report size, count, logical min/max, and usage
  0x19, 0x01,           // Usage Minimum
  0x29, REPORT_LEN,     // Usage Maximum  64 output usages total (0x01 to 0x40)
  0x91, 0x00,           // Output (Data, Array, Abs): Instantiates output packet fields.
                        // Uses same report size and count as "Input" fields, since nothing
                        // new/different was specified to the parser since the "Input" item.
  0xC0                  // End Collection
};

// states for the application
```

```
typedef enum {APP_STATE_INIT, APP_STATE_RECEIVE, APP_STATE_SEND} APP_STATE;

int main (void) {
  char report[REPORT_LEN]="";// message report buffer 64 bytes per hid report descriptor
  APP_STATE state = APP_STATE_INIT;
  hid_setup();               // initialize the hid usb helper module

  // enable interrupts
  PLIB_INT_Enable(INT_ID_0);

  while (1) {
    switch(state) {
      case APP_STATE_INIT:
        if(hid_open()) {          // wait for the hid device to open
          state = APP_STATE_RECEIVE;
        }
        break;
      case APP_STATE_RECEIVE:
        if(hid_receive((unsigned char *)report, REPORT_LEN)) {
          // we are finished receiving the message
          char * curr = report;
          while(*curr) { // convert to upper case
            *curr = toupper(*curr);
            ++curr;
          }
          state = APP_STATE_SEND; // send data to client
        }
        break;
      case APP_STATE_SEND:
        if(hid_send((unsigned char *)report, REPORT_LEN) ) { // finished sending
          state = APP_STATE_RECEIVE;                          // receive data again
        }
        break;
      default:
        ;// logic error, impossible state!
    }

    // update the usb hid state
    hid_update();
  }
  return 0;
}
```

Note that the report descriptor is a byte array that describes the format of the HID report. The format and interpretation of a HID report descriptor is flexible but rather complicated. It is defined in the Device Class Definition for Human Interface Devices supplement to the USB Standard.

The next file to examine is `hid.h`. We have created `hid.h` to implement some common HID functionality for you. This eliminates much code duplication when creating different types of basic HID devices.

Following the basic Harmony structure, the NU32 HID library requires you to define several macros in `system_config.h`. These macros control several aspects of a HID device that you may wish to change across projects:

- NU32_PID: The product identifier for the device. Note that we always use Microchip's vendor ID.
- NU32_REPORT_SIZE: The size, in bytes, of the HID report descriptor. Note that this is not the same as the number of bytes sent, rather it is the length of the array that describes the communication protocol between the PIC32 and the host. The actual array, HID_REPORT, was defined in Code Sample 20.17 talkingHID.c.
- NU32_DEVICE_NAME: The name of the device, in USB string descriptor format. The first byte of the string is the string descriptor's length, the second byte is the string descriptor ID (3), and then each character is represented by two bytes.
- NU32_HID_SUBCLASS: The HID subclass, which tells the host what type of HID device to expect. We use a generic subclass for this example, but devices such as keyboards and mice have their own subclasses, as defined by the USB standard.
- NU32_HID_PROTOCOL: Again, we use a generic protocol here. The protocol, for certain subclasses, provides information about the format of the HID report.

In addition to the expected macros, the hid.{c,h} library assumes you have defined the array HID_REPORT somewhere in your code (here, in Code Sample 20.17 talkingHID.c). This array contains the HID report descriptor.

Code Sample 20.18 hid.h. **The NU32 HID Library.**

```
#ifndef HID__H__
#define HID__H__
// code common to all hid examples

#include <stdbool.h>        // bool type with true, false
#include <stdint.h>         // uint8_t
#include "system_config.h"

// following harmony's lead you must define the following variables and macros
// macros in system_config.h
// #define NU32_PID - the product ID for the device
// #define NU32_REPORT_SIZE  the size of the hid report
// #define NU32_DEVICE_NAME  the name of the device in usb string descriptor format
// #define NU32_HID_SUBCLASS the hid subclass
// #define NU32_HID_PROTOCOL the hid protocol
// in one of your .c files you must also define the HID_REPORT
const extern uint8_t HID_REPORT[NU32_REPORT_SIZE]; // the hid report

// initialize the hid device
void hid_setup(void);

// attempt to open the hid device, return true when
// the keyboard is successfully opened
bool hid_open(void);

// request a hid report from the host. when the report is available, return true
// returning false indicates that the current request is pending
bool hid_receive(uint8_t report[], int length);
```

```
// send a hid report.  return true when finished sending
// returning false indicates that the current request is pending
bool hid_send(uint8_t report[], int length);

// update the necessary harmony state machines, call from the main loop
void hid_update(void);

// return true if the idle time has expired, indicating that the host expects a report
bool hid_idle_expired(void);

// get the time, in ms, based on the usb clock
uint32_t hid_time(void);

#endif
```

We saw many of the functions declared in `hid.h` used in Code Sample 20.17 `talkingHID.c`. We provide a complete description below:

- `hid_setup`: Initializes the Harmony USB middleware and prepares the PIC32 to use the USB peripheral.
- `hid_open`: Opens the Harmony HID middleware.
- `hid_receive`: Requests a report from the host. While the request is pending, returns false. The first call after a request completes returns true and the next call issues a new request.
- `hid_send`: Sends a report to the host. While the request is pending, it returns false. The first call after a request completes returns true and the next call issues a new send request.
- `hid_update`: Updates the Harmony HID FSM. Should be called frequently from the main loop.
- `hid_idle_expired`: Hosts may set an idle rate, which is the minimum rate at which the device must send data. If the device (PIC32) has not sent data for too long, `hid_idle_expired` will return true, indicating that the PIC32 should send a report.
- `hid_time`: The Harmony USB code allows us to access a "timer" with a 1 ms period; this function provides that access.

Prior to examining the implementation of the HID library, we first look at `system_config.h` and `system_definitions.h`.

We need to set up many macros for the Harmony USB driver and middleware layers. The Harmony documentation describes the details, but what is important to understand is that

1. The PIC32 is configured as a device, not a host.
2. It has two endpoints (in addition to the mandatory control endpoint, endpoint 0). One is used to send data to the host (an in-endpoint), and the other is used to receive data from the host (an out-endpoint).
3. The macros prefixed with `DRV_USB` are for the USB driver, whereas those prefixed with `USB_DEVICE` are for the Harmony USB middleware.

Code Sample 20.19 `system_config.h`. **System Configuration Header for the USB HID Project.**

```
#ifndef SYSTEM_CONFIG_H
#define SYSTEM_CONFIG_H

// avoid superfluous warnings when building harmony
#define _PLIB_UNSUPPORTED

// USB driver configuration

// work as a USB device not as a host
#define DRV_USB_DEVICE_SUPPORT      true
#define DRV_USB_HOST_SUPPORT        false

// use only one instance of the usb driver
#define DRV_USB_INSTANCES_NUMBER    1


// operate using usb interrupts
#define DRV_USB_INTERRUPT_MODE      true

// there are 2 usb endpoints
#define DRV_USB_ENDPOINTS_NUMBER    2

// USB device configuration
// use only one device layer instance
#define USB_DEVICE_INSTANCES_NUMBER     1

// size of the endpoint 0 buffer, in bytes
#define USB_DEVICE_EP0_BUFFER_SIZE      8

// enable the USB start of frame event. it happens at 1 ms intervals
#define USB_DEVICE_SOF_EVENT_ENABLE

// USB HID configuration
// use only one instance of the hid driver
#define USB_DEVICE_HID_INSTANCES_NUMBER     1

// total size of the hid read and write queues
#define USB_DEVICE_HID_QUEUE_DEPTH_COMBINED 2

// ports used by NU32 LEDs and USER button
#define NU32_LED_CHANNEL PORT_CHANNEL_F
#define NU32_USER_CHANNEL PORT_CHANNEL_D

// positions of the LEDs and user buttons
#define NU32_LED1_POS PORTS_BIT_POS_0
#define NU32_LED2_POS PORTS_BIT_POS_1
#define NU32_USER_POS PORTS_BIT_POS_7

// macros used by hid.c
#define NU32_PID 0x1769      // usb product id

#define NU32_REPORT_SIZE 28 // hid report is 28 bytes long

// name of the device. first byte is the length, next byte is the string descriptor id
// (always 3), then the following characters are two bytes each, spelling "Talking HID"
```

```
#define NU32_DEVICE_NAME "\x18\x03T\0a\01\0k\0i\0n\0g\0 \0H\0I\0D\0"
#define NU32_HID_SUBCLASS USB_HID_SUBCLASS_CODE_NO_SUBCLASS
#define NU32_HID_PROTOCOL USB_HID_PROTOCOL_CODE_NONE

#endif
```

You may notice the strange definition for `NU32_DEVICE_NAME`. This macro defines a USB string descriptor. This string descriptor begins with a one-byte length (`\x18`, which means `0x18`), a one-byte string descriptor ID (3), and then each subsequent character is two bytes long. The `\0`, which appears every other byte, inserts a 0 to conform to the two-byte character format.

We also specify, using `NU32_HID_SUBCLASS` and `NU32_HID_PROTOCOL`, that we are not using a particular HID subclass or protocol, because we are using HID to transfer raw data that the host operating system need not interpret. If we were making, for example, a USB keyboard, we would specify the keyboard subclass.

The `system_definitions.h` file includes the necessary Harmony headers:

Code Sample 20.20 `system_definitions.h`**. System Definitions for a Generic HID Implementation.**

```
#ifndef SYSTEM_DEFINITIONS_H
#define SYSTEM_DEFINITIONS_H

#include <stddef.h>
#include "system/common/sys_common.h"
#include "system/common/sys_module.h"
#include "usb/usb_device.h"
#include "usb/usb_device_hid.h"
#include "peripheral/peripheral.h"

#endif
```

We now examine the HID library's implementation. Much of `hid.c` is devoted to configuring various USB descriptors, according to the USB standard. After defining the descriptors, they are placed into structures defined by Harmony so that Harmony functions can use them. We also define initialization structures for the Harmony USB middleware. The middleware relies on the USB driver layer to detect USB events and dispatch them to two callback functions. We use the callback functions to maintain the state of the current USB transaction; for example, the state tracks whether we are currently sending or receiving data.

Code Sample 20.21 `hid.c`. **NU32 HID Library Implementation.**

```
#include "hid.h"
#include "system_config.h"
#include "system_definitions.h"
#include <sys/attribs.h>

// the USB device descriptor, part of the usb standard.
const USB_DEVICE_DESCRIPTOR device_descriptor = {
    0x12,                      // the descriptor size, in bytes
    USB_DESCRIPTOR_DEVICE,     // 0x01, indicating that this is a device descriptor
    0x0200,                    // usb version 2.0, BCD format of AABC == version AA.B.C
    0x00,                      // Class code 0, class will be in configuration descriptor
    0x00,                      // subclass 0, subclass will be in configuration descriptor
    0x00,                      // protocol unused, it is in the configuration descriptor
    USB_DEVICE_EP0_BUFFER_SIZE,// max size for packets to control endpoint (endpoint 0)
    0x04d8,                    // Microchip's vendor id, assigned by usb-if
    NU32_PID,                  // product id (do not conflict with existing pid's)
    0x0000,                    // device release number
    0x01,                   // string descriptor index; string describes the manufacturer
    0x02,                   // product name string index
    0x00,                   // serial number string index, 0 to indicate not used
    0x01                    // only one possible configuration
};

// Configuration descriptor, from the USB standard.
// All configuration descriptors are stored contiguously in memory in
// this byte array.  Remember, the pic32's CPU is little endian.
const uint8_t configuration_descriptor[] = {
    // configuration descriptor header
    0x09,                          // descriptor is 9 bytes long
    USB_DESCRIPTOR_CONFIGURATION,  // 0x02, this is a configuration descriptor
    41,0, // total length of all descriptors is 41 bytes (remember, little endian)
    1,                            // configuration has only 1 interface
    1,                            // configuration value (host uses this to select config)
    0,                            // configuration string index, 0 indicates not used
    USB_ATTRIBUTE_DEFAULT | USB_ATTRIBUTE_SELF_POWERED,  // device is self-powered
    50,                                        // max power needed 100 mA (2 mA units)

    // interface descriptor
    0x09,                          // descriptor is 9 bytes long
    USB_DESCRIPTOR_INTERFACE,      // 0x04, this is an interface descriptor
    0,                             // interface number 0
    0,                             // interface 0 in the alternate configuration
    2,                             // 2 endpoints (not including endpoint 0)
    USB_HID_CLASS_CODE,            // uses the hid class
    NU32_HID_SUBCLASS,             // hid boot interface subclass, in system_config.h
    NU32_HID_PROTOCOL,             // hid protoocol, defined in system_config.h
    0,                             // no string for this interface

    // the hid class descriptor
    0x09,                          // descriptor is 9 bytes long
    USB_HID_DESCRIPTOR_TYPES_HID,  // 0x21 indicating that this is a HID descriptor
    0x11, 0x01,                    // use HID version 1.11, (BCD format, little endian)
    0x00,                          // no country code
    0x1,                           // Number of class descriptors, including this one
    // as part of the hid descriptor, class descriptors follow (only one for this example)
    USB_HID_DESCRIPTOR_TYPES_REPORT,// this is a report descriptor
```

```
    sizeof(HID_REPORT),0x00,        // size of the report descriptor

    // in endpoint descriptors
    0x07,                           // the descriptor is 7 bytes long
    USB_DESCRIPTOR_ENDPOINT,        // 0x05, endpoint descriptor type
    0x1 | USB_EP_DIRECTION_IN,      // in to host direction, address 1
    USB_TRANSFER_TYPE_INTERRUPT,    // use interrupt transfers
    0x40, 0x00,                     // maximum packet size, 64 bytes
    0x01,                           // sampling interval 1 frame count

    // out endpoint descriptor
    0x07,                           // the descriptor is 7 bytes long
    USB_DESCRIPTOR_ENDPOINT,        // 0x05, endpoint descriptor type
    0x1 | USB_EP_DIRECTION_OUT,     // in to host direction, address 1
    USB_TRANSFER_TYPE_INTERRUPT,    // use interrupt transfers
    0x40, 0x00,                     // maximum packet size, 64 bytes
    0x01                            // sampling interval 1 frame count
};

// String descriptor table.  String descriptors provide human readable information
// to the hosts.
// The syntax \xRR inserts a byte with value 0xRR into the string.
// As per the USB standard, the first byte is the total length of the descriptor
// the next byte is the descriptor type, (0x03 for string descriptor). The following bytes
// are the string itself.  Since each character is two bytes, we insert a \0 after
// every character.  The descriptors are placed into a table for use with harmony.
const USB_DEVICE_STRING_DESCRIPTORS_TABLE string_descriptors[] = {
  // 1st byte:  length of string (0x04 = 4 bytes)
  // 2nd byte:  string descriptor (3)
  // 3rd and 4th byte:  language code, 0x0409 for English (remember, little endian)
  "\x04\x03\x09\x04",
  // manufacturer string: Microchip Technology Inc.
  "\x34\x03M\0i\0c\0r\0o\0c\0h\0i\0p\0 \0T\0e\0c\0h\0n\0o\0l\0o\0g\0y\0 \0I\0n\0c\0.\0",
 // name of the device, defined in system_config.h
  NU32_DEVICE_NAME
};

//  512-byte-aligned table needed by the harmony device layer
static uint8_t __attribute__((aligned(512)))
  endpoint_table[USB_DEVICE_ENDPOINT_TABLE_SIZE];

// harmony structure for storing the configuration descriptors.
// a device can have multiple configurations but only one can be active at one time
// we have only one configuration
const USB_DEVICE_CONFIGURATION_DESCRIPTORS_TABLE configuration_table[]= {
  configuration_descriptor };

// table of descriptors used by the harmony USB device layer
const USB_DEVICE_MASTER_DESCRIPTOR master_descriptor = {
    &device_descriptor,       // Full speed descriptor
    1,                        // Total number of full speed configurations available
    configuration_table,      // Pointer to array of full speed configurations descriptors
    NULL, 0, NULL,            // usb high speed info, high speed not supported on PIC32MX
    3,                        // Total number of string descriptors available
    string_descriptors,       // Pointer to array of string descriptors
    NULL, NULL, NULL          // unsupported features, should be NULL
};

// harmony HID initialization structure
const USB_DEVICE_HID_INIT hid_init = {
  sizeof(HID_REPORT), // size of the hid report descriptor
```

```
    &HID_REPORT,           // the hid report descriptor
    1,1                    // send and receive queues of 1 byte each
  };

  // register hid functions with the Harmony device layer
  const USB_DEVICE_FUNCTION_REGISTRATION_TABLE function_table[] = {
    {
      USB_SPEED_FULL,                         // full speed mode
      1,                                      // use configuration number 1
      0,                                      // use interface 0 of configuration number 1
      1,                                      // only one interface is used
      0,                                      // use instance 0 of the usb function driver
      (void*)&hid_init,                       // the initialization for the driver
      (void*)USB_DEVICE_HID_FUNCTION_DRIVER, // use the HID function layer
    }
  };

  // used to initialize the device layer
  const USB_DEVICE_INIT usb_device_init = {
    {SYS_MODULE_POWER_RUN_FULL}, // power state
    USB_ID_1,                    // use usb module 1 (PLIB USB_ID to use)
    false,  false,               // don't stop in idle or suspend in sleep modes
    INT_SOURCE_USB_1, 0,         // use usb 1 interrupt, not using dma so set source to 0
    endpoint_table,              // the endpoint table
    1,                           // only one function driver is registered
    (USB_DEVICE_FUNCTION_REGISTRATION_TABLE*)function_table, // function drivers for HID
    (USB_DEVICE_MASTER_DESCRIPTOR*)&master_descriptor,       // all of the descriptors
    USB_SPEED_FULL,              // use usb full speed mode
    //1,1                        // endpoint read/write queues of 1 byte each
  };

  volatile SYS_MODULE_OBJ usb;  // handle to the usb device middleware

  // maintains the status of the usb system, based on the callback events responses
  typedef struct {
    bool configured;          // true if the device is configured
    bool sent;                // true if the device report has been sent
    uint16_t idle_rate;       // how often a report should be sent, in 4 ms units
    uint32_t time;            // time in ms, based on usb clock
    unsigned int idle_count;  // the idle count, in 1 ms ticks
    bool received;            // true if a report has been received
    USB_DEVICE_HANDLE device; // harmony device handle
  } usb_status;

  // the initial status of the device
  static usb_status status = {false,false,0,0,0,false,USB_DEVICE_HANDLE_INVALID};

  // prototypes for usb event handling functions
  static void usb_device_handler(
      USB_DEVICE_EVENT event, void * eventData, uintptr_t context);

  static void usb_hid_handler(
      USB_DEVICE_HID_INDEX hidInstance, USB_DEVICE_HID_EVENT event,
      void * eventData, uintptr_t userData);

  void __ISR(_USB_1_VECTOR, IPL4SOFT) USB1_Interrupt(void)
  {
    // update the USB state machine
      USB_DEVICE_Tasks_ISR(usb);
  }
```

```
void hid_setup(void) {
    // set the USB ISR priority
    SYS_INT_VectorPrioritySet(INT_VECTOR_USB1, INT_PRIORITY_LEVEL4);

    // initialize the usb device middleware
    usb = USB_DEVICE_Initialize(USB_DEVICE_INDEX_0, (SYS_MODULE_INIT*)&usb_device_init);
}

bool hid_open(void) {
  // attempt to open the usb device
  status.device = USB_DEVICE_Open(USB_DEVICE_INDEX_0, DRV_IO_INTENT_READWRITE);

  // if the device is successfully opened
  if(status.device != USB_DEVICE_HANDLE_INVALID) {
    // register a callback for USB device events
    USB_DEVICE_EventHandlerSet(status.device,  usb_device_handler, 0);
    return true;
  } // otherwise opening failed, but this is not usually an error,
    // we just need to wait more iterations until the USB system is ready
  return false;
}

// update the usb state machine, should be called from the main loop
void hid_update() {
  USB_DEVICE_Tasks(usb);
}

bool hid_receive(uint8_t report[], int length) {
  USB_DEVICE_HID_TRANSFER_HANDLE handle;
  static bool requested = false; // true if we have requested a report
  if(status.configured) {        // the device is configured and plugged in
    if(!requested) {             // have not already requested a report
      requested = true;          // request the report
      status.received = false;   // not received the report yet
                                 // the next line issues the recieve request. When it
                                 // completes,  usb_hid_handler will be called with
                                 // event = USB_DEVICE_HID_EVENT_REPORT_RECEIVED
      USB_DEVICE_HID_ReportReceive(USB_DEVICE_HID_INDEX_0, &handle,report,length);
    }
    if(status.received) {        // requested report has been received
      requested  = false;        // ready for a new receive request
      return true;               // indicate that the report is ready
    }
  }
  return false;                  // requested report is not ready
}

// send a hid report, if we are not busy sending, otherwise return false
bool hid_send(uint8_t report[], int length) {
  USB_DEVICE_HID_TRANSFER_HANDLE handle;
  static bool requested = false;
  if(status.configured) {      // the device is configured and plugged in
    if(!requested) {           // have not requested a hid report to be sent
      requested = true;        // issue the hid report send request
      status.sent = false;     // request has not been sent
      status.idle_count = 0;   // sending a report so reset the idle count
                               // the next line issues the send request. When it
                               // completes, usb_hid_handler will be called with
                               // event = USB_DEVICE_HID_EVENT_REPORT_SENT
      USB_DEVICE_HID_ReportSend(USB_DEVICE_HID_INDEX_0, &handle, report, length);
    }
```

```
    if(status.sent) {           // finished a send request
      requested = false;        // ready for a new send request
      return true;              // indicate that the report has been sent
    }
  }
  return false;                 // send request is not finished
}

uint32_t hid_time(void) {
  // get a time count in ms ticks from the usb subsystem
  return status.time;
}

bool hid_idle_expired(void) {
 return (status.idle_rate > 0 && status.idle_count*4 >= status.idle_rate);
}

// handles HID events, reported by the Harmony HID layer
static void usb_hid_handler(
    USB_DEVICE_HID_INDEX hidInstance,
    USB_DEVICE_HID_EVENT event, void * eventData, uintptr_t context)
{
  static uint16_t protocol = 0; // store the protocol
  static uint8_t blank_report[sizeof(HID_REPORT)] = ""; // a blank report to return
                                                         // if requested
  switch(event)
  {
    case USB_DEVICE_HID_EVENT_REPORT_SENT:
      // we have finished sending a report to the host
      status.sent = true;
      break;
    case USB_DEVICE_HID_EVENT_REPORT_RECEIVED:
      // the host has sent a report to us.  Ignore zero length reports
      status.received = true;
      break;
    case USB_DEVICE_HID_EVENT_GET_REPORT:
      // send blank report when requested. Per HID spec, we must send a report when asked
      USB_DEVICE_ControlSend(status.device,blank_report,sizeof(blank_report));
      break;
    case USB_DEVICE_HID_EVENT_SET_IDLE:
      // acknowledge the receipt of the set idle request
      USB_DEVICE_ControlStatus(status.device, USB_DEVICE_CONTROL_STATUS_OK);
      // set new idle rate, in units of 4 ms. report must be sent before period expires
      status.idle_rate = ((USB_DEVICE_HID_EVENT_DATA_SET_IDLE*)eventData)->duration;
      break;
    case USB_DEVICE_HID_EVENT_GET_IDLE:
        // send the idle rate to the host
        USB_DEVICE_ControlSend(status.device, &status.idle_rate,1);
        break;
    case USB_DEVICE_HID_EVENT_SET_PROTOCOL:
      // all usb hid devices that support the boot protocol must implement SET_PROTOCOL &
      // GET_PROTOCOL which allows the host to select between the boot protocol and the
      // report descriptor we made. our operation remains the same regardless so we just
      // store the request and return it when asked
      protocol = ((USB_DEVICE_HID_EVENT_DATA_SET_PROTOCOL*)eventData)->protocolCode;
      USB_DEVICE_ControlStatus(status.device, USB_DEVICE_CONTROL_STATUS_OK);
      break;
    case  USB_DEVICE_HID_EVENT_GET_PROTOCOL:
      // return the currently selected protocol to the host
      USB_DEVICE_ControlSend(status.device, &protocol,1);
      break;
```

```
    default:
      break; // many other events we simply don't handle
  }
}

// handles USB device events, reported by the Harmony device layer
static void usb_device_handler(
    USB_DEVICE_EVENT event, void * eventData, uintptr_t context) {
  switch(event) {
    case USB_DEVICE_EVENT_SOF:
      // this event occurs at the USB start of frame, every 1 ms per the usb spec
      // the event is enabled by defining USB_DEVICE_EVENT_SOF_ENABLE in system_config.h
      ++status.idle_count; // keep track of how long device has not sent reports
      ++status.time;       // also keep a running time, in ms
      break;
    case USB_DEVICE_EVENT_RESET:
      // usb bus was reset
      status.configured = false;
      break;
    case USB_DEVICE_EVENT_DECONFIGURED:
      // device was deconfigured
      status.configured = false;
      break;
    case USB_DEVICE_EVENT_CONFIGURED:
      // we have been configured.  eventData holds the selected configuration,
      // but this device has only have one configuration.
      // we can now register a hid event handler
      USB_DEVICE_HID_EventHandlerSet(USB_DEVICE_HID_INDEX_0, usb_hid_handler, 0);
      status.configured = true;
      break;
    case USB_DEVICE_EVENT_POWER_DETECTED:
      // Vbus is detected meaning the device is attached to a host
      USB_DEVICE_Attach(status.device);
      break;
    case USB_DEVICE_EVENT_POWER_REMOVED:
      // the device was removed from a host
      USB_DEVICE_Detach(status.device);
      break;
    default:
      break; // there are other events that we do not handle
  }
}

// The USB device layer, when it initializes the driver layer,
// attempts to call this function, but Harmony does not implement it as of
// v1.06.  Therefore we place it here
void DRV_USB_Tasks_ISR_DMA(SYS_MODULE_OBJ o)
{
  DRV_USB_Tasks_ISR(o);
}
```

We define the following descriptors:

`device_descriptor`: Defines basic USB information such as the VID, the PID, and the USB version.

`configuration_descriptors`: Actually a collection of several descriptors, stored in a byte array. These descriptors describe the device's configuration and are used, for example, to

inform the host that the device is a HID device with one in-endpoint and one out-endpoint.
`string_descriptors`: A Harmony data type containing a table of string descriptors. This table contains, for example, the manufacturer string and the device name string (which is determined based on what you define `NU32_DEVICE_NAME` to be). Remember, string descriptors use a different format than C strings: they start with one byte for the length, a byte with the value 3 (indicating that this is a string descriptor), and then the subsequent characters are two bytes long.

To be used with Harmony, the USB descriptors must be placed into Harmony-specific variables: `endpoint_table`, `configuration_table`, and `master_descriptor`. We also must define initialization structures for the Harmony HID layer (`hid_init`) and the USB Device layer (`usb_device_init`). The `function_table` array is used to inform the Harmony device layer that it should use Harmony's HID functions. We then define a variable to store a handle to the Harmony USB middleware.

As stated earlier, we must implement two callbacks: one for the device layer (`usb_device_handler`) and another for the HID layer (`usb_hid_handler`). As the program runs, various USB events will occur and these callbacks will be called appropriately. Both of these callbacks modify the state of the system, which we track using the `usb_status struct`. The device callback is registered with Harmony when the device is opened, whereas the HID callback is registered in the device callback, in response to the event that occurs when the host configures the device.

The event handlers that we implement for both the HID layer and generic USB device layer consist mainly of a `switch` statement determining the reason for the callback. Our callbacks only handle a small subset of the possible events, but this is enough for our purposes. The most important HID events (handled in `usb_hid_handler`) are

- `USB_DEVICE_HID_EVENT_REPORT_SENT`: A HID report has been sent to the host.
- `USB_DEVICE_HID_EVENT_REPORT_RECEIVED`: A HID report has been received from the host.

The important events handled by the generic USB device layer are

- `USB_DEVICE_EVENT_CONFIGURED`: This event occurs when the host has configured the device. When configured, we register the HID event handler.
- `USB_DEVICE_EVENT_POWER_DETECTED`: The device has detected power from the host. We must call `USB_DEVICE_Attach` to tell Harmony to configure the device and receive subsequent events.

To help update the USB device module's state, we must call `USB_DEVICE_Tasks` frequently and implement the USB ISR, which calls `USB_DEVICE_Tasks_ISR`. The need to call `USB_DEVICE_Tasks` motivates the `hid_update` function.

The initialization function (hid_setup) and opening function (hid_open) must be called before the device starts receiving events. The functions hid_receive, hid_send, hid_time, and hid_idle_expired all depend upon usb_status being updated in the Harmony USB callbacks. For example, to receive or send a HID report (i.e., the data), the USB device must be configured and another report must not be pending. If these conditions are not met, the functions return false, allowing the main loop to continue and the state of the Harmony USB state machine to update.

20.9 Chapter Summary

- Harmony is a comprehensive software framework for all PIC32 microcontrollers that aims to promote modular and portable code. Its documentation comprises over 4000 pages, or more than 8 reams of paper if you wish to print it!
- Harmony is divided into different layers that fulfill different roles. This chapter discusses various PLIB, driver, and system service modules. There are also modules providing support for real-time operating systems.
- Microchip suggests that programs using Harmony adhere to a specific directory structure. For large projects it may help keep your files organized, but for smaller projects it may seem overwhelming. The Makefile we provide can compile programs either using Microchip's suggested structure or just a flat directory structure.
- Harmony can be used to implement a USB device that communicates with a host.

20.10 Exercises

1. Add an additional logical timer to the system timer example. At a frequency of 0.25 Hz, switch which LED blinks faster and which blinks slower.
2. Describe a situation when using the TMR system service would be beneficial compared to using the timer driver directly.
3. What happens to the USB example code if talkingHID.c wants to communicate a string exceeding 64 bytes to/from the host? Modify talkingHID.c and client.c so that strings longer than 64 bytes can be handled.

Further Reading

Axelson, J. (2015). *USB complete: The developer's guide* (5th ed.). Madison, WI: Lakeview Research LLC.

Device class definition for human interface devices (HID) (Version 1.11). (2001). USB Implementers' Forum.

HID API for Linux, Mac OS X, and Windows. (2015). http://signal11.org/oss/hidapi. (Accessed: May 20, 2015)

MPLAB Harmony help (v1.06). (2015). Microchip Technology Inc.

Universal serial bus specification (Revision 2.0). (2000). Compaq, Hewlett-Packard, Intel, Lucent, Microsoft, NEC, and Philips.

PART IV

Mechatronics

CHAPTER 21

Sensors

Your PIC32 interacts with the outside world through sensors and actuators. "Mechatronics" is the design of microprocessor-controlled electromechanical systems incorporating sensors and actuators. There is no clear distinction between mechatronics and robotics, but we typically think of robots as higher-level, more complex and general purpose devices, often with more sophisticated sensing and artificial intelligence than we associate with mechatronics. Robots often integrate multiple mechatronic subsystems.

In this chapter we focus on sensors. Sensors transduce physical properties of interest to a signal that a microcontroller can understand. Some sensors produce a simple digital or analog voltage signal, which may need *signal conditioning* circuitry before sending to the PIC32. Examples of signal conditioning include switch debouncing, voltage amplification, and low-pass filtering (see Appendix B). Other sensors, like rotary encoders, encode their signals in digital pulse trains, to be decoded either by the PIC32 itself or by an external circuit. Finally, some sensors incorporate their own microprocessor or ASIC (application-specific integrated circuit) and can communicate by one of the peripherals discussed earlier (e.g., UART, CAN, I^2C, or SPI).

An overview of sensors could be organized by the transduction principle involved (e.g., a voltage proportional to a magnetic field's strength due to the Hall effect or a current proportional to light intensity by the photoelectric effect). These transduction principles can be applied to measure many other quantities of interest; for example, sensors can be constructed to measure the rotation angle of a joint using either the Hall effect or photodiodes. To the mechatronics designer, the specific transduction principle employed is typically secondary to the sensor's capability of sensing the quantity of interest (e.g., the joint angle). Therefore, in this chapter, we roughly organize the presentation around typical quantities of interest: angle, angular velocity, acceleration, force, etc. We do not go into details of the physics of the transduction principles, but provide a broad overview of relatively inexpensive sensors common in mechatronics and how they can be interfaced to the PIC32.

Many of the sensors in this chapter can be purchased from vendors like Digikey or Mouser, or on convenient breakout boards from vendors like SparkFun, Pololu, or Adafruit.

Embedded Computing and Mechatronics with the PIC32 Microcontroller. http://dx.doi.org/10.1016/B978-0-12-420165-1.00021-4

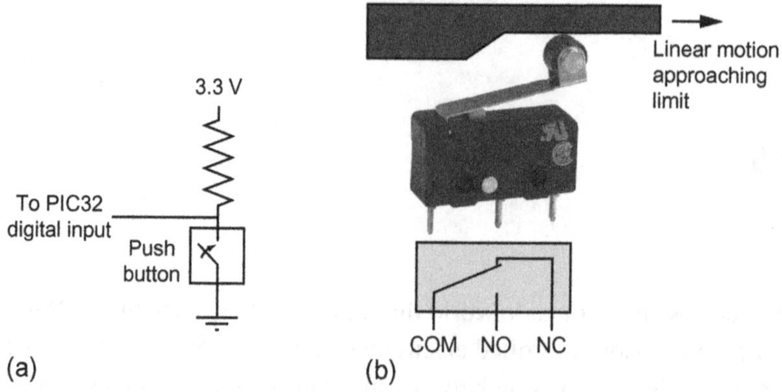

Figure 21.1
(a) A button interfacing to the PIC32. (b) A hinged roller-lever limit switch, and its use in a linear joint to detect the end of travel. (Image courtesy of Digi-Key Electronics, digikey.com.)

21.1 Contact: Buttons and Switches

Perhaps the simplest sensors are buttons and switches. Buttons and switches can be used to get information from a user (e.g., a keyboard or the USER button on the NU32) or to sense when a robot joint has reached the limit of its allowable travel (a *limit switch*).

A simple button interface to the PIC32 is shown in Figure 21.1(a). The button has two connections, one to a pull-up resistor and one to ground. When the button is unpressed, the internal switch is open circuit, and the digital input to the PIC32 is high (3.3 V). When the button is pressed, the switch is closed, pulling the digital input low (ground). This kind of button is called *normally open* (or NO for short). There are also buttons that are *normally closed* (NC), requiring the button to be pressed to open the internal switch.

One common application for a switch is as a limit switch. When a robot linear or rotary joint reaches its limit of travel, the limit switch depresses, sending a signal to the controller to stop driving the joint. The limit switch in Figure 21.1(b) has both an NO and an NC output.

The switch interface in Figure 21.1(a) shows an external pull-up resistor. The digital inputs on the PIC32 supporting Change Notification have internal pull-up resistors that can be used instead, eliminating the external resistor (see Chapter 7).

A common problem with mechanical switches such as those in Figure 21.1 is *bounce*—many on-off transitions in a brief period of time as the switch establishes or breaks contact. If bounce is a problem for the particular application, the designer should decide the shortest amount of time allowed between "real" transitions (as opposed to mechanical bounces), then design either a circuit or a software routine to *debounce* the switch. See, for example, the debouncing circuit in Appendix B.2.1.

A switch is often characterized by the number of *poles* and *throws*. The number of poles is the number of internal moving levers, and the number of throws is the number of different connections each lever can make contact with. Thus the switch in Figure 21.1(a) is a single-pole single-throw switch (or SPST for short), and the switch in Figure 21.1(b) is a single-pole double-throw (SPDT) switch. Other common configurations are DPST (two internal switches of the type in Figure 21.1(a), meaning four external connections) and DPDT (two internal switches of the type in Figure 21.1(b), meaning six external connections). Each of the two poles in DPST and DPDT switches is activated by the same external button or lever.

Mechanical switches are distinguished by their current rating. Switches with higher current ratings have larger contact surfaces between the throws and the poles.

Switches can be used in many different ways. For example, attaching a stiff wire to the end of a limit switch as in Figure 21.1(b) could allow you to use the wire as a binary "whisker" sensor.

21.2 Light

21.2.1 Types of Light Sensors

Light sensors include photocells (also called photoresistors), photodiodes, and phototransistors. Photodiodes and phototransistors are used not only to sense light levels directly, but as building blocks in many other types of sensors.

Photocell

A photocell is a resistor that changes resistance depending on the amount of light incident on it. A photocell operates on semiconductor photoconductivity: the energy of photons hitting the semiconductor frees electrons to flow, decreasing the resistance.

An example photocell is the Advanced Photonix PDV-P5002, shown in Figure 21.2. In the dark, this photocell has a resistance of approximately 500 kΩ, and in bright light the resistance drops to approximately 10 kΩ. The PDV-P5002 is sensitive to light in the wavelengths 400-700 nm, approximately the same wavelengths the human eye is responsive to. Figure 21.2 shows a simple circuit illustrating how it can be used as an ambient light sensor feeding either a digital or an analog input to the PIC32.

Photodiode

Photocells are easy to use, but their resistance changes relatively slowly. For example, the PDV-P5002 may take tens of milliseconds to fully change resistance in response to ambient light change. A much faster response can be obtained with a photodiode. As with a photocell, a photodiode operates by photons "kicking up" electrons that allow current to flow, but unlike a photocell, current can flow even without an externally imposed voltage due to the electric

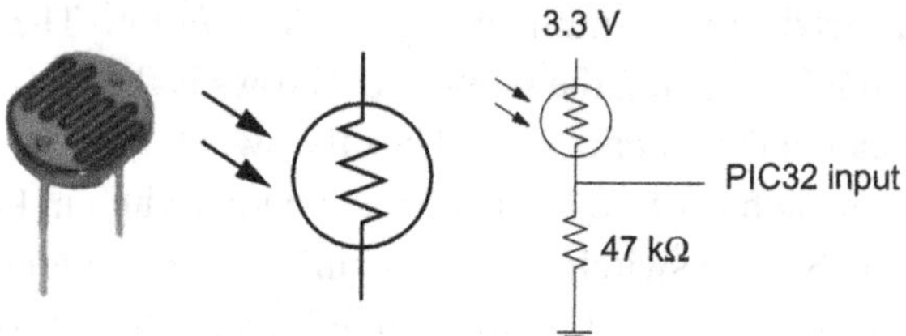

Figure 21.2
(Left) The PDV P5002 photocell. (Image courtesy of Advanced Photonix, Inc., advancedphotonix.com.) (Middle) Circuit symbol for a photocell. (Right) A simple light-level-detection circuit. In bright light, the photocell's resistance is around 10 kΩ, making an output of about 2.7 V. In darkness, the photocell's resistance is around 500 kΩ, making an output of about 0.3 V. The sensor output could go to a PIC32 digital or analog input.

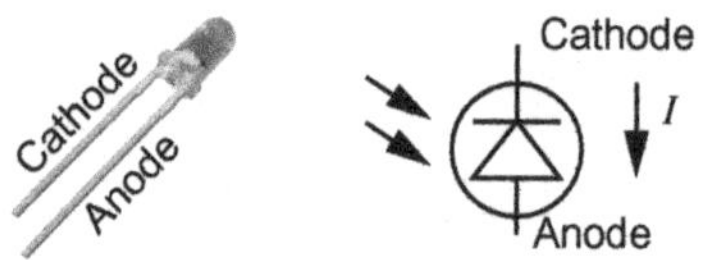

Figure 21.3
(Left) A photodiode. The cathode of a diode is the shorter leg, and the anode is the longer leg. (Right) The circuit symbol for a photodiode, and the direction that photocurrent flows when light hits the photodiode.

field in the diode. In response to a rapidly changing light source, this photocurrent can turn on and off in just a few nanoseconds, depending on the design of the circuit the photodiode is used in.

When light hits the photodiode, reverse photocurrent flows, from the cathode to the anode (Figure 21.3). This current is quite small; for the OPTEK Technology OP906, for example, the maximum current is on the order of tens of microamps. While it may be possible to simply pass this current through a large resistance to generate a measurable voltage, it is common to use an op amp or instrumentation amp circuit to create a sensor with a low-impedance output, a sufficient gain from light levels to voltage, and a fast switching time. It is also common to put a reverse bias voltage across the photodiode to reduce the diode's capacitance, allowing faster current switches. A drawback of the reverse bias voltage is the creation of a reverse *dark current*, in addition to the photocurrent. Thus the reverse bias voltage decreases switching time but reduces the sensitivity of the circuit.

When a photodiode is used without an imposed reverse bias, for maximum sensitivity, it is used in *photovoltaic* mode. When a photodiode is used with an imposed reverse bias, for maximum switching speed, it is used in *photoconductive* mode. This chapter does not cover amplifier circuit designs for these cases; see the references at the end of this chapter.

Some photodiodes come with light filters to adjust their sensitivity to different light wavelengths. The OP906 has no filter, and responds to light at wavelengths between approximately 500 nm and 1100 nm, with a peak response at 880 nm (infrared, invisible). The OP906 can be paired with a Fairchild QED123 LED, which emits IR light at 880 nm, for applications like photointerrupters and reflective object sensors (below).

Photodiodes also come with lenses to direct the incoming light, and the lens on the OP906 ensures that there is little response to light arriving at an angle more than 20 degrees off the central axis of the sensor. Other photodiodes have wider or even narrower viewing angles. Which is best for you depends on your application.

Phototransistor

A phototransistor is a type of bipolar junction transistor including a photodiode junction. An NPN phototransistor has a photodiode at its base-collector junction, and the photocurrent generated there acts as the base current I_B. Below saturation, the phototransistor implements the equations $I_C = \beta I_B$, where I_C is the collector current and β is the transistor's gain, and $I_E = I_C + I_B$, where I_E is the emitter current. (See Appendix B.3 for more on bipolar junction transistors.) Since a typical β is 100, a phototransistor has a higher gain from light to current than a photodiode.

For example, the OSRAM SFH 310 NPN phototransistor creates emitter currents of up to a few milliamps, as compared to the microamps of a photodiode. This higher current makes phototransistors much easier to interface to than photodiodes. See the example circuit in Figure 21.4. A drawback compared to a photodiode is the longer rise and fall times of the current, on the order of 10 μs for the SFH 310.

Like photodiodes, phototransistors may have filters to alter their sensitivity spectrum and lenses to control their viewing angle. The SFH 310 has a viewing angle of up to about 25 degrees off the central axis, and it is sensitive to light of wavelengths 450 nm to 1100 nm, which includes much of the visible spectrum (about 390 nm to 700 nm). Peak sensitivity of the SFH 310 is at 880 nm. The SFH 310 can be paired with the IR LED QED123, mentioned above, with its 880 nm wavelength. If a visible LED is preferred, you could use the Kingbright WP7113SRC/DU red LED at 640 nm (Figure 21.4). While this wavelength is below the SFH 310's 880 nm peak sensitivity, the response is still about 60% of peak.

21.2.2 Basic Applications

Photodiodes and phototransistors are often paired with LEDs to make a variety of different types of sensors. Two of the simplest are photointerrupters and reflective object sensors.

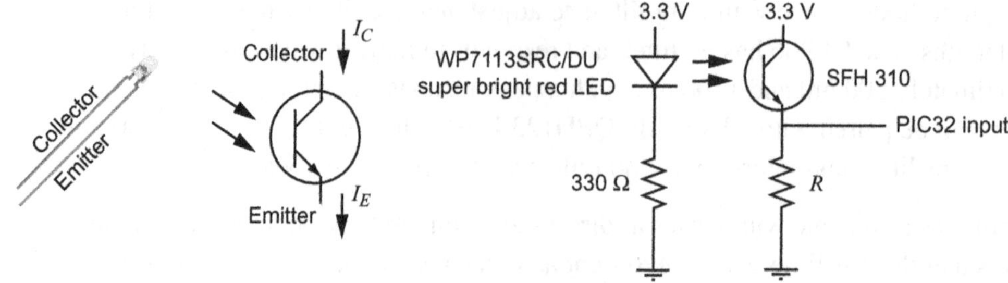

Figure 21.4

(Left) The SFH 310 NPN phototransistor. The shorter leg is the collector and the longer leg is the emitter. (Image courtesy of Digi-Key Electronics, digikey.com.) (Middle) The circuit symbol for an NPN phototransistor. (Right) A circuit with a WP7113SRC/DU red LED illuminating an SFH 310 phototransistor. The resistance *R* should be chosen to get the right voltage range at the input to the PIC32, which could be an analog or digital input, depending on the application. For a sufficiently large resistance *R*, the sensor's output voltage ranges from close to 0 V (no light on the phototransistor) to close to 3.15 V (transistor saturated, with 0.15 V drop from collector to emitter).

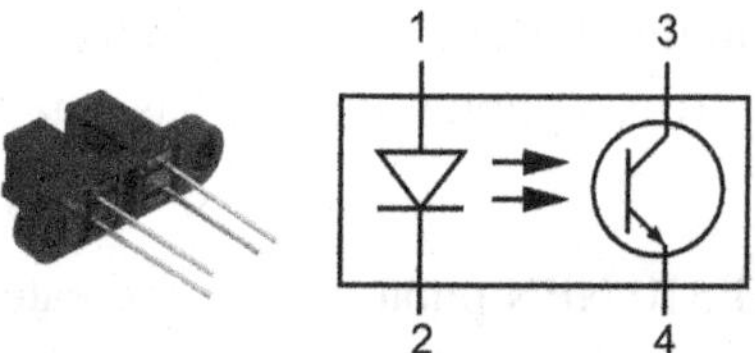

Figure 21.5

(Left) The OPB370T51 photointerrupter. (Image courtesy of Digi-Key Electronics, digikey.com.) (Right) The connections of the package's pins 1 to 4 to the LED and phototransistor.

Photointerrupter

A photointerrupter, or slotted optical switch, contains an LED and a phototransistor or photodiode in a single package. The two are aimed at each other across a small gap, as with the OPTEK Technology OPB370T51, which uses an infrared LED and a phototransistor (Figure 21.5). The LED is always powered, so if the gap is clear, current flows through the phototransistor. If the gap is blocked by an opaque object, blocking the LED light, current through the phototransistor drops. A complete circuit is similar to that illustrated in Figure 21.4(right). An optointerrupter can be used as a type of a limit switch or as a building block for an optical encoder (Section 21.3.2).

To obtain clean digital pulses to a PIC32 input, rather than slowly varying analog voltages as the gap transitions from unblocked to blocked and back, the phototransistor output in Figure 21.4(right) can be passed through a Schmitt trigger (Appendix B.2, Figure B.6).

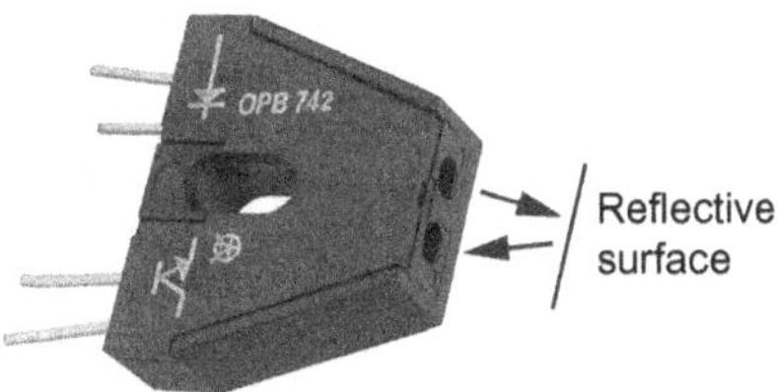

Figure 21.6
The OPB742 reflective object sensor uses an LED and a phototransistor to detect the presence of nearby reflective surfaces. (Image courtesy of Digi-Key Electronics, digikey.com.)

Reflective object sensor

A reflective object sensor is very similar to a photointerrupter, except instead of directly facing each other, the LED and phototransistor are pointed nearly parallel to each other, with a slight inward focus; see the OPTEK Technology OPB742 in Figure 21.6. The OPB742 is designed to detect reflective surfaces at distances between about 0.2 cm and 0.8 cm. When there is no reflective surface nearby, little current flows through the phototransistor; when there is a reflective surface within range, significant current flows through the phototransistor. As with the photointerrupter, a complete circuit is similar to that illustrated in Figure 21.4(right).

21.3 Angle of a Revolute Joint

There are many ways to measure the angle or angular velocity of a joint; here we mention a few of the most common.

21.3.1 Potentiometer

A potentiometer (Appendix B.2), or pot for short, is a variable resistor, typically with a rotating knob that determines the variable resistance (Figure 21.7). A pot has three output terminals: two at either end of the internal resistor, with a fixed resistance between them, and one at the *wiper*. As the knob rotates, the wiper slides over the resistive element, and the resistance between one end of the resistor and the wiper increases smoothly from zero to the max resistance of the pot, while the resistance between the wiper and the other end of the resistor drops smoothly from max resistance to zero. Thus, by putting a voltage across the two ends of the internal resistor, the pot's wiper provides a voltage proportional to the angle of the knob, which can be read by the PIC32's analog input (Figure 21.7(c)).

Pots come in many different styles, distinguished by total resistance across the resistive element; the type of knob or other attachment (e.g., the hollow-shaft pot in Figure 21.7(d)); the number of turns that the pot allows (from less than a single turn to multi-turn); the *taper* of the resistive element, which dictates how the resistance changes with rotation of the knob

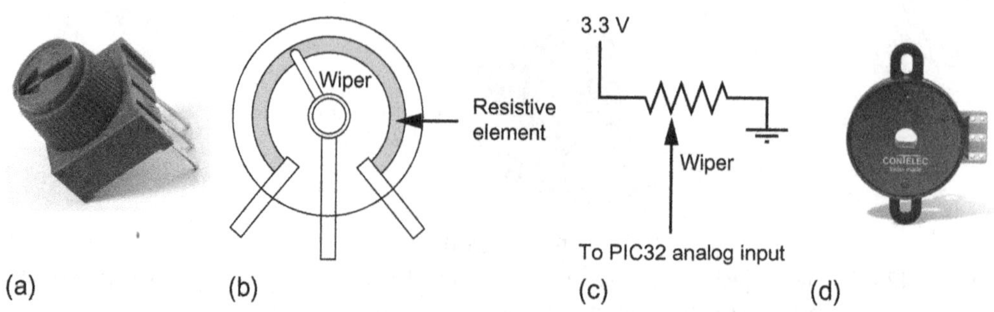

Figure 21.7

(a) A breadboardable 10 kΩ pot. (b) A representation of the wiper sliding over the resistive element as the knob is rotated. (c) The circuit symbol for a pot, and its use in a simple circuit that measures pot rotation as an analog value between 0 and 3.3 V. (d) The Contelec WAL305 hollow-shaft pot. The interior rotating element of this pot can be press-fitted on a motor shaft. (Image courtesy of Contelec AG, www.contelec.ch/en.)

(typically linear or logarithmic, where the latter is often used in audio applications); and the amount of power the resistive element can dissipate without damage. Since a typical pot has a sliding contact between the wiper and the resistive element, pots can only endure a limited number of cycles. More expensive pots have a longer lifetime and a more precise relationship between the rotation of the knob and the variable resistance.

Pots are relatively inexpensive and easy to use, but since they transmit their angle readings as analog voltages, their readings are subject to electrical noise. For applications where electrical noise is an issue, or where more precise angle estimates are needed, encoders, with their digital outputs, are more common choices.

21.3.2 Encoder

There are two major types of encoders: *incremental* and *absolute*.

Incremental encoder

An incremental encoder creates two pulse trains, A and B, as the encoder shaft rotates a *codewheel*. These pulse trains can be created by magnetic field sensors (Hall effect sensors) or light sensors (LEDs and phototransistors or photodiodes). The latter technique, used in *optical encoders*, is illustrated in Figure 21.8. The codewheel could be an opaque material with slots or a transparent material (glass or plastic) with opaque lines.

The relative phase of the A and B pulses determines whether the encoder is rotating clockwise or counterclockwise. A rising edge on B after a rising edge on A means the encoder is rotating one way, and a rising edge on B after a falling edge on A means the encoder is rotating the

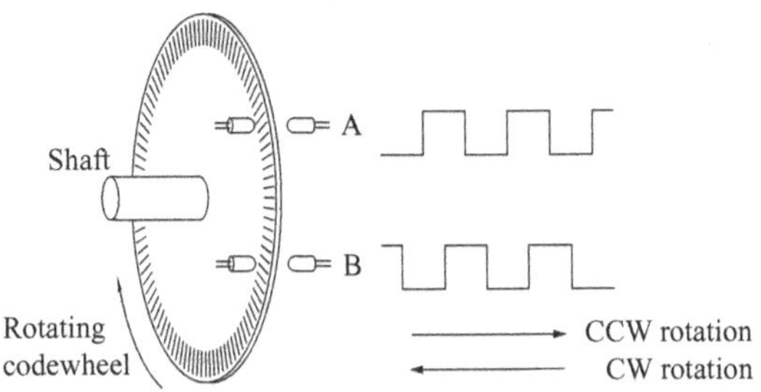

Figure 21.8
A rotating optical encoder creates 90 degree out-of-phase pulse trains on A and B using LEDs and phototransistors. Although two LEDs are shown in the image, it is common to use one LED and a mask to create the two light streams.

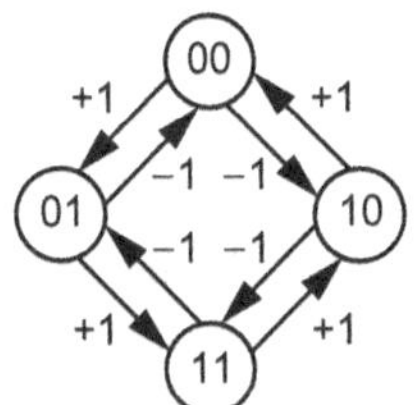

Figure 21.9
4x decoding of A/B quadrature encoder channels. Each node of the state machine shows the digital A/B signals as a two-bit number AB. When the signals change, the encoder count is either incremented or decremented according to the specific transition.

opposite direction. A rising edge on B followed by a falling edge on B (with no change in A) means that the encoder has undergone no net motion. The out-of-phase A and B pulse trains are known as *quadrature* signals.

In addition to determining the rotation direction, the pulses can be counted to determine how far the encoder has rotated. The encoder signals can be "decoded" at 1x, 2x, or 4x resolution, where 1x resolution means that a single count is generated for each full cycle of A and B (e.g., on the rising edge of A), 2x resolution means that two counts are generated for each full cycle (e.g., on the rising and falling edges of A), and 4x means that a count is generated for every rising and falling edge of A and B (four counts per cycle, Figure 21.9). Thus an encoder with "100 lines" or "100 pulses per revolution" can be used to generate up to 400 counts per revolution of the encoder. If the encoder is attached to a motor shaft, and the motor shaft is also attached to a 20:1 speed-reducing gearhead, then the encoder generates $400 \times 20 = 8000$ counts per revolution of the gearhead output shaft.

Some encoders offer a third output channel called the *index* channel, usually labeled I or Z. The index channel creates one pulse per revolution of the joint and can be used to determine when the joint is at a "home" position. Some encoders also offer differential outputs $\bar{A}$, $\bar{B}$, and $\bar{Z}$, which are always opposite A, B, and Z, respectively. This is for noise immunity in electrically noisy environments. For example, if a transient magnetic field induces a voltage change on all of the encoder lines, a *single-ended* reading (e.g., channel A only) may incorrectly interpret the voltage change as movement of the encoder. A differential reading of $A - \bar{A}$ would reject this noise, since it is common to both A and $\bar{A}$.

Some microcontrollers, but not the PIC32, are equipped with a "quad encoder interface" (QEI) peripheral that accepts A and B inputs directly and maintains the encoder count on an internal counter. On the PIC32, the A and B channels can be used with a change notification ISR that implements the state machine of Figure 21.9, provided the A and B lines do not change too quickly. A better solution is to use external encoder decoder circuitry to maintain the count, then query the count using SPI, I^2C, or parallel communication.

Absolute encoder

An incremental encoder can only tell you how far the joint has moved since the encoder was turned on. An absolute encoder can tell you where the joint is at any time, regardless of the position of the joint at power on. To provide absolute position information, an absolute encoder uses many more LED/phototransistor pairs, and each one provides a single bit of information on the joint's position. For example, an absolute encoder with 17 channels, like the Avago Technologies AEAT-9000-1GSH1, can distinguish the absolute orientation of a joint up to a resolution of $360°/(2^{17}) = 0.0027°$ (131,072 unique positions).

As the codewheel rotates, the binary count represented by the 17 channels increments according to *Gray code*, not the typical binary code, so that at each increment, only one of the 17 channels changes signal.[1] This removes the need for the infinite manufacturing precision needed to make two signals switch at exactly the same angle. Compare the following two three-bit sequences, for example (Figure 21.10):

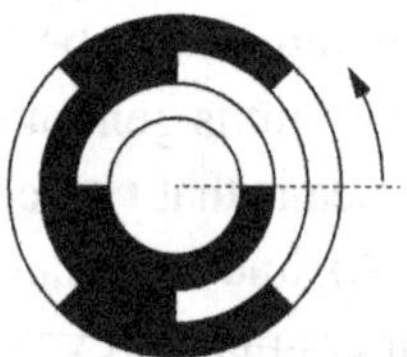

Figure 21.10
A 3-bit Gray code codewheel for an optical absolute encoder. If the innermost ring corresponds to the most significant bit, then as we proceed counterclockwise around the codewheel, the count is 000, 001, 011, 010, 110, 111, 101, and 100.

1 The AEAT-9000-1GSH1 actually uses a 12-bit Gray code codewheel. The other five bits are obtained by advanced methods not discussed in this chapter.

State	0	1	2	3	4	5	6	7
Binary code	000	001	010	011	100	101	110	111
Gray code	000	001	011	010	110	111	101	100

Absolute encoders typically employ some type of serial communication to send their readings.

21.3.3 Magnetic Encoders

The angle of a rotational joint can be measured by the orientation of the magnetic field due to a magnet attached to the joint. An example is the Avago magnetic encoder sensor illustrated in Figure 21.16 and described in Section 21.6.

21.3.4 Resolver

A resolver consists of three coils: an input excitation coil on the rotor and "sine" and "cosine" measurement coils on the stator. The rotor coil is driven by a sinusoidal reference excitation voltage, $V_r(t) = V_{r,max} \sin \omega t$, where the precise voltage $V_{r,max}$ and frequency ω are not critical, but the frequency $\omega/2\pi$ is typically multiple kHz. The current through the excitation coil generates a magnetic field, which induces a current and therefore a voltage across the stator measurement coils. The sine and cosine coils are offset by 90 degrees relative to each other, so that the voltages induced across the measurement coils are given by

$$V_{\sin}(t) = V \sin \omega t \sin \theta,$$
$$V_{\cos}(t) = V \sin \omega t \cos \theta,$$

where θ is the angle of the rotor coil (see Figure 21.11).

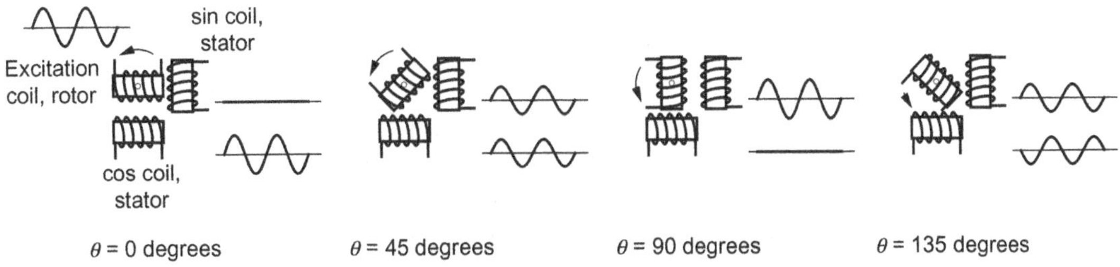

Figure 21.11

A resolver consists of an excitation coil on the rotor and sine and cosine pickup coils on the stator. The excitation coil is driven by a sinusoidal voltage as a function of time. The excitation coil induces sinusoidal voltages across the pickup coils. The amplitude and phase (zero or 180 degrees) of the pickup voltage sinusoids is a function of the angle of the rotor.

The angle of the resolver's rotor is decoded by a resolver-to-digital converter (RDC) chip, such as the Analog Devices AD2S90. The two pickup coil voltages are sent to the RDC, which decodes the angle and provides three kinds of outputs: (a) serial output simulating a 12-bit absolute encoder; (b) A and B outputs simulating a 1024-line incremental encoder (1024 A and B pulses per revolution, allowing 4x decoding for a resolution of 4096 counts per revolution); and (c) an analog voltage proportional to the angular velocity.

21.3.5 Tachometer

A tachometer refers to any device that produces a signal proportional to the speed of rotation of a joint. There are many different types of tachometers, some based on measuring the frequency of, or the time between, pulses generated by the rotating shaft. Any angle-measuring device can be used to simulate a tachometer by numerical differencing, taking angle measurements at times t and $t + \delta t$ and calculating $\dot{\theta}(t + \delta t) \approx (\theta(t + \delta t) - \theta(t))/\delta t$.

21.4 Position of a Prismatic Joint

Linear or *prismatic* joints are the second-most common type of joint, after rotary joints. Often prismatic joints are driven by rotary motors with a transmission that converts rotational motion to linear motion, such as a ball screw or a rack and pinion (Chapter 26). In this case, the linear motion can be indirectly measured by a rotational sensor (e.g., a pot or encoder) on the motor.

In other cases, it is necessary or desirable to measure the linear displacement directly. For these cases, potentiometers and encoders have direct linear analogs. Linear potentiometers are sometimes called *slide pots*, and are common on analog audio equipment. Linear incremental and absolute encoders simply take the codewheel and straighten it out into a line.

Just as a resolver employs an AC excitation signal and induction to determine the angle of a revolute joint, a *linear variable differential transformer* (LVDT) employs induction to determine the position of a prismatic joint (Figure 21.12). A stationary excitation coil, driven by a sinusoidal voltage in the kHz frequency range, is coupled to two stationary pickup coils by a ferromagnetic core that moves with the linear joint. The coupling between the excitation coil and each pickup coil changes with the position of the core. The voltages across the two pickup coils are differenced (hence the "D" in LVDT), and this differenced signal is used to determine the position of the core. When the core is centered, the difference between the two pickup voltages is zero. As the core moves away from the center position, the amplitude of the difference increases, and the phase of the differenced sinusoid, relative to the excitation sinusoid, is zero as the core moves in one direction and 180 degrees as the core moves in the other direction.

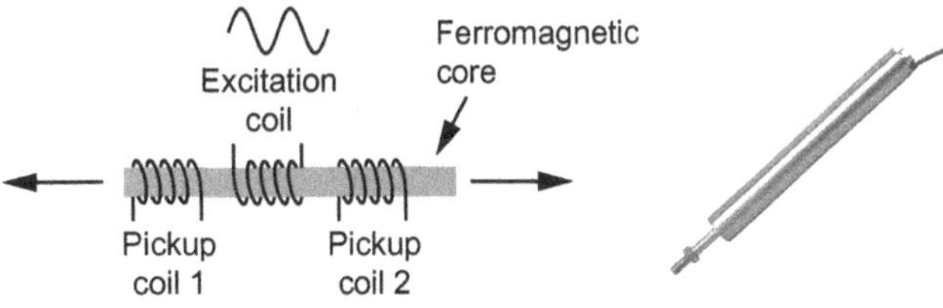

Figure 21.12

(Left) An LVDT consists of a movable core, a stationary excitation coil, and two stationary pickup coils. (Right) An Omega LD320 LVDT. (Image courtesy of Omega Engineering, Inc., omega.com.)

Figure 21.12 shows an Omega LD320 LVDT. An LVDT is typically interfaced to a microcontroller using an LVDT signal conditioning chip such as the Analog Devices AD698, which generates the excitation voltage and turns the differential pickup voltage into an analog voltage output proportional to the position of the LVDT core.

21.5 Acceleration and Angular Velocity: Gyros, Accelerometers, and IMUs

Gyroscopes (gyros) and accelerometers use sensing of inertial forces to measure the angular velocity of a body about one, two, or three axes (gyros) or linear acceleration of a body along one, two , or three axes (accelerometers). A three-axis gyro and a three-axis accelerometer can be used together to make an *inertial measurement unit* (IMU) that attempts to track the motion of a rigid body, based solely on inertial forces. The ability to do this is fundamentally limited by the fact that linear velocity relative to a "fixed" frame cannot be directly measured based on inertial forces—any reference frame translating at a constant velocity is an inertial frame, indistinguishable from other inertial frames.

Gyros and accelerometers have existed for many years, but the advent of microelectromechanical systems (MEMS) has brought these formerly expensive devices within reach of low-cost consumer applications. In this section we focus on MEMS gyros and accelerometers.

21.5.1 MEMS Accelerometer

A MEMS accelerometer measures acceleration (which includes the gravitational acceleration g) by measuring the deflection of a tiny mass m suspended on springs. For example, for a one-axis accelerometer that measures acceleration in a single direction, let $x = 0$ be the rest position of the mass when the acceleration is zero (Figure 21.13). Then the acceleration a of the accelerometer can be calculated from the deflection x using the equation $kx + ma = 0$, where k is the stiffness of the springs. The deflection is typically sensed by the change in capacitance between plates fixed to the mass and plates fixed to the body of the accelerometer.

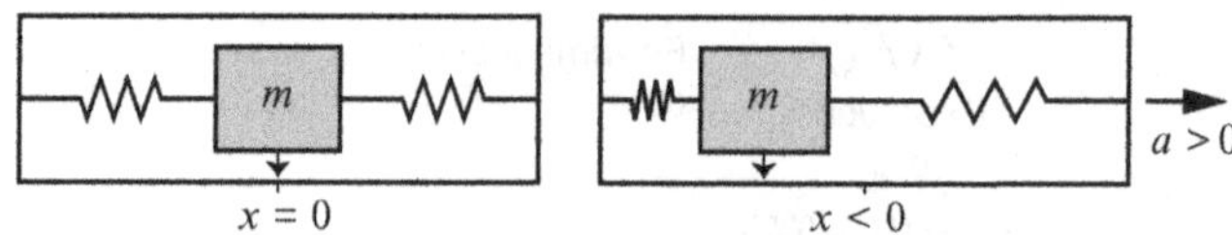

Figure 21.13
(Left) A MEMS accelerometer mass m at its home position $x = 0$ relative to the body of the accelerometer. (Right) An acceleration $a > 0$ leads to a deflection of the mass $x = -ma/k < 0$, where k is the total stiffness of the springs.

Accelerometers come in one-, two-, and three-axis (x, y, z) devices; different ranges of detectable accelerations; and different output types, including I^2C, SPI, and analog output. The Analog Devices ADXL362 is a three-axis accelerometer capable of measuring x-y-z accelerations in the range $\pm 2g$, $\pm 4g$, or $\pm 8g$, as selected by the user, and provides both analog and 12-bit resolution SPI output.

The STMicroelectronics LSM303D accelerometer, discussed in Chapters 12 (SPI) and 13 (I^2C), includes a three-axis accelerometer as well as a three-axis magnetometer. The magnetometer can be used to sense the Earth's magnetic field, and combined with the accelerometer to measure the gravity direction, allows the implementation of a tilt-compensated compass.

21.5.2 MEMS Gyro

Like an accelerometer, a MEMS gyro uses masses supported by springs and capacitive deflection sensors. Unlike an accelerometer, the masses in a gyro are forced to constantly vibrate. Because of this motion, when the gyro is rotated, the masses experience "Coriolis forces" that deflect the nominal vibration relative to a gyro-fixed frame (see Figure 21.14 for an explanation of Coriolis forces). These Coriolis effects are proportional to the rotation rate. The masses and capacitance sensors are physically configured so that the differential capacitance change is zero if deflection is caused by a linear acceleration of the gyro, ensuring that only deflections due to angular velocity are measured (Figure 21.15).

The STMicroelectronics L3GD20H is a three-axis gyro with a user-selectable full-scale range of up to ± 2000 degrees/s in each of the three axes. Data is transmitted by I^2C or SPI.

21.5.3 MEMS IMU

An IMU combines a gyro and an accelerometer, and optionally a magnetometer and/or a barometric pressure sensor. The barometer allows approximate altitude readings while the magnetometer allows sensing the orientation of the Earth's magnetic field.

The primary goal of an IMU is to track the 3D position and orientation of a rigid body in time without using any external references. For example, if the body begins at rest, the

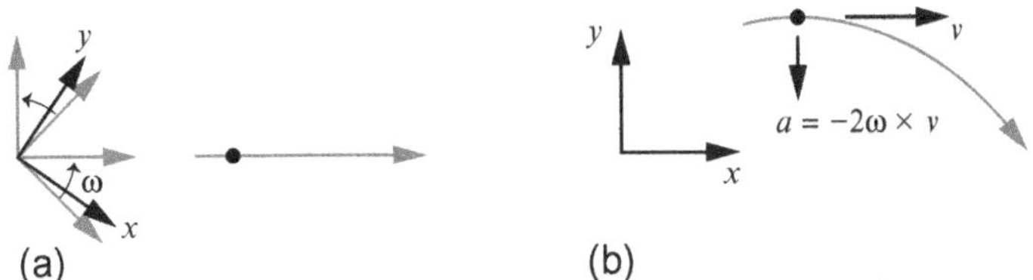

Figure 21.14

The Coriolis effect, illustrated by an *xyz* frame rotating with positive angular velocity about the *z*-axis out of the page. The constant angular velocity is written in vector form as $\omega = (0, 0, \omega_z)$. (a) Viewed in a stationary frame fixed to the page, a point mass *m* moves at a constant velocity to the right while the gray frame rotates. The frame's angle and the mass are shown in black at a specific instant in time. (b) Viewed by an observer in the non-inertial rotating frame, the velocity of the point mass does not appear constant. Instead, the mass follows a spiral trajectory. The mass is shown in black at the same instant as in (a). In the coordinates of the rotating frame, the acceleration of the point mass is $a = -2\omega \times v$, where v is the velocity vector of the mass as viewed from the rotating frame, and ω in the non-inertial frame is the same as in the inertial frame. The "Coriolis force" is simply $ma = -2m\omega \times v$. This is not a true force on the mass; the apparent acceleration of the mass is a result of observing its motion from a rotating frame.

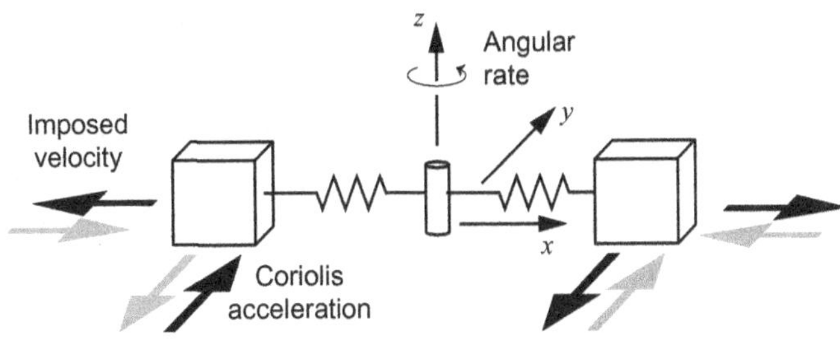

Figure 21.15

A schematic of a one-axis MEMS gyro. The two masses are oscillated opposite each other in the *x*-direction. When one is moving with $\dot{x} > 0$, the other is moving with $\dot{x} < 0$. If the gyro is rotated with a positive angular velocity about the *z*-axis, the masses experience equal and opposite "Coriolis forces" in the $\pm y$ direction. If the two masses are currently moving outward (black arrows), the masses experience accelerations as indicated by the corresponding black acceleration arrows; if the two masses are currently moving inward (gray arrows), the masses experience accelerations in the directions of the corresponding gray acceleration arrows.

accelerometer can use the gravitational field to determine the orientation of the body. There is no way for an IMU to determine the initial *x*-*y*-*z* position of the body,[2] so only relative motion can be estimated. When the body begins to move, it must experience either linear accelerations, angular velocities, or both. Angular velocity can be numerically integrated by

[2] The *z* position can be approximately measured by a barometer.

microcontroller software to estimate the orientation of the body, and linear accelerations can be integrated to get linear velocities and again to get the net position change from the initial position. Sophisticated software integrators exist; they typically employ extended Kalman filters.

Because of sensor errors and errors due to numerical integration, estimates of linear velocity and position tend to accumulate error over time. These errors can be mitigated by occasionally referencing an external reference such as GPS. IMUs allow systems that rely on GPS to continue functioning when GPS is briefly unavailable.

An IMU can consist of a PCB with separate accelerometer and gyro ICs; a single IC incorporating both an accelerometer and a gyro; or a complete integrated solution, including onboard estimation software. An example IMU is the STMicroelectronics ASM330LXH, which features a three-axis accelerometer with a user-selectable full scale between $\pm 2g$ and $\pm 16g$ and a three-axis gyro with a user-selectable full scale between ± 125 and ± 2000 degrees/s. It communicates by I^2C or SPI.

21.6 Magnetic Field Sensing: Hall Effect Sensors

The *Hall effect* is a manifestation of the *Lorentz force law*, which states that a moving charge carrier in a magnetic field experiences a force if the magnetic field flux lines are not aligned with the direction of motion. (The Lorentz force law is discussed in more detail in Chapter 25, as the basis for converting electrical energy to mechanical energy in DC motors.) Current flowing through a flat, stationary semiconductor plate is deflected by this force until there is a charge buildup at the edges of the plate that balances the effect of the Lorentz force. The charge buildup can be measured as a voltage across the plate, transverse to the direction of the current flow. The magnitude of this *Hall voltage* is a function of the strength of the magnetic field and its alignment relative to the current direction.

The Hall effect is used in a great variety of sensors, including 3D magnetic-field-sensing magnetometers (e.g., for a digital compass or for measuring orientation in a known, artificially created magnetic field), current sensors, rotary encoders, and angular position sensors for brushless DC motors (Chapter 29), and sensors that detect proximity to a magnet or ferromagnetic material. To sense whether a piece of metal is nearby, for example, a sensor can be designed consisting of a Hall sensor with a magnet fixed closed by. If the sensor comes close to a piece of metal, the magnet's magnetic field changes, changing the voltage read by the Hall sensor.

The Toshiba TCS20DPR is a digital Hall effect switch IC with three pins, for power (e.g., 3.3 V), GND, and a digital output that reads low when the magnetic flux density exceeds a limit B_{on} and high when the flux density drops below B_{off}, where $B_{on} > B_{off}$. If the flux is between B_{off} and B_{on}, the sensor reading does not change from its previous value. This *hysteresis* assures that the output only changes if the magnetic field has changed significantly.

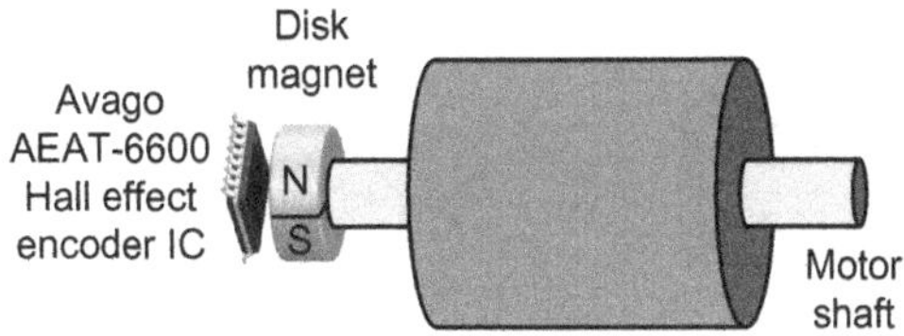

Figure 21.16
Many motors have a shaft extending on both sides of the motor: one side for a sensor element and the other side to connect to a gearhead or the load. Here, the orientation of a disk magnet mounted on one end of the shaft is sensed by the Avago AEAT-6600 magnetic encoder IC. (Image of AEAT-6600 courtesy of Avago Technologies, avagotech.com.)

The values of B_{off} and B_{on} are a few milliTeslas (mT). For reference, the strength of the Earth's magnetic field at the Earth's surface is a bit less than 0.1 mT, while a typical small magnet at a distance of a centimeter might have a field strength on the order of 100 mT. Thus the TCS20DPR could be used to test for the presence of a nearby magnet, perhaps on a moving joint.

Another application of the Hall effect is embodied in the Avago Technologies AEAT-6600 angular magnetic encoder IC (Figure 21.16). A diametrically magnetized two-pole disk magnet is mounted on a rotating shaft directly above the IC. The AEAT-6600 is capable of sensing the orientation of the magnet with 16-bit resolution, or 65,536 unique angles. The orientation can be communicated in several ways: as incremental encoder A/B/I signals; as three Hall sensor signals for brushless DC motors (Chapter 29); as a value encoded in the duty cycle of a PWM signal; or as a 16-bit position using asynchronous serial communication.

21.7 Distance

Inexpensive ultrasonic and infrared ranging sensors can sense the distance to nearby surfaces, from distances of a few centimeters to a few meters. The HC-SR04 ultrasonic sensor and the Sharp GP2Y0A60SZ are two such sensors (Figure 21.17).

The HC-SR04 has four pins: Vcc (powered by 5 V), GND, Trigger input, and Echo output. When the HC-SR04 receives a 10 μs high pulse on the Trigger input, its ultrasonic transmitter emits a brief burst of 40 kHz ultrasonic pulses. The time until the receiver hears the echo, together with the speed of sound (approximately 340 m/s in dry air at room temperature at sea level), is used to determine the distance to the nearby surface. A digital high pulse is created on the Echo output for a time proportional to the distance to the surface, where the distance in centimeters equals approximately the duration of the pulse in microseconds divided by 58.[3] The sensor works best when the reflecting surface is approximately parallel with, and directly in front of, the face of the HC-SR04, and large enough to generate a good reflection. The

[3] A pulse of over 30 ms indicates that no reflection was detected.

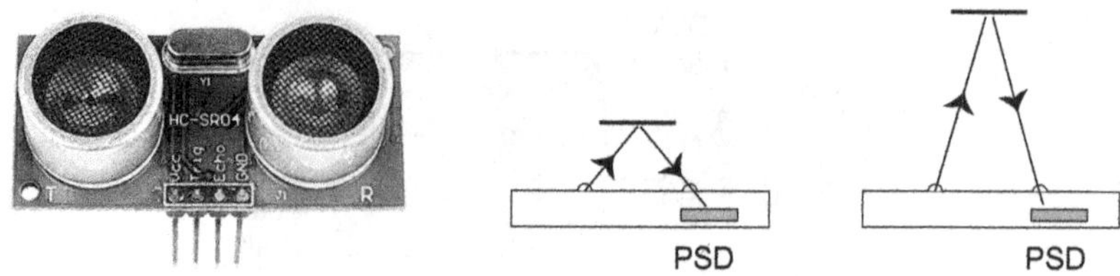

Figure 21.17
(Left) The HC-SR04 ultrasonic distance sensor. (Right) The Sharp GP2Y0A60SZ infrared distance sensor detects distance using a position-sensitive detector to triangulate the distance to the reflective surface.

HC-SR04 may also detect surfaces not parallel to the face of the HC-SR04, and up to 20 degrees off the central axis of the emitter/receiver. Under ideal conditions, the HC-SR04's distance readings can be accurate to up to about 1 cm.

The Sharp GP2Y0A60SZ consists of an infrared beam emitter and a position-sensitive detector (PSD). The emitted IR beam is reflected by the sensed surface and detected by the PSD. The PSD uses the photovoltaic effect, also used by photodiodes and phototransistors, to measure the linear position of the reflected spot of light (see Figure 21.17). The PSD output is returned as an analog voltage, updated at approximately 60 Hz, and distance is calculated by triangulation. The GP2Y0A60SZ is sensitive to ranges between 10 and 150 cm.

21.8 Force

A force sensor is used to sense the force transmitted through a body. For example, a robot arm may have a force-torque sensor between the arm and its gripper, to sense the mass of the object being grasped. Such a sensor actually senses the force in three axes (x, y, and z) as well as the torque about the three axes. Digital scales also employ force sensors.

Forces are commonly measured using *strain gauges*. A strain gauge is a resistor whose resistance changes as it is stretched or compressed. For example, consider a resistive, slightly flexible rod, with current flowing from one end to the other. As the rod is compressed, it becomes shorter and fatter, and resistance decreases. As it is stretched, it becomes longer and thinner, and resistance increases (Figure 21.18).

A typical strain gauge, such as the Vishay strain gauge in Figure 21.18, consists of a metallic foil resistor mounted on a flexible insulating backing. When glued to a (typically metallic) substrate, the change in resistance measures the *strain* (compression or stretching) of the underlying substrate. The stiffness properties of the substrate are then used to estimate the forces that the substrate is experiencing. The substrate is typically quite stiff, as flexibility is generally an undesired property in a force sensor, so strains tend to be quite small. To sense

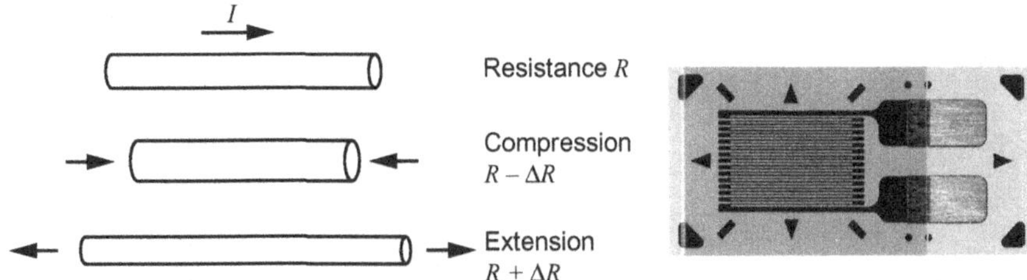

Figure 21.18

(Left) The principle behind a strain gauge, modeled here as a compressible/extensible rod: compression reduces the resistance and extension increases the resistance. (Right) A closeup of a Vishay metallic foil strain gauge. The long and thin resistor is patterned to accentuate the resistance change due to small strains. The large solder pads on the right are the ends of the resistor. (Image courtesy of Micro-Measurements, a brand of Vishay Precision Group, vpgsensors.com.)

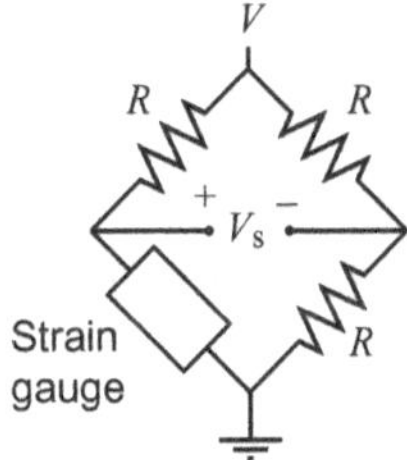

Figure 21.19

A Wheatstone bridge to sense the change in resistance of a strain gauge, whose nominal resistance is R. The small sensed voltage V_S is usually amplified by an instrumentation amplifier.

the small resistance change, a Wheatstone bridge (Figure 21.19) is often used in conjunction with an instrumentation amplifier (Appendix B.4). Because of the differential nature of the sensed voltage V_S, the reading is relatively immune to variations in supply voltage and effects of resistance changes due to temperature.

Working with strain gauges is a bit of an art. For example, forces in different directions will affect the reading of a single strain gauge. For this reason, the design of the shape of the substrate, and therefore its stiffness in different directions, is quite important. It is also common to use multiple strain gauges, for example two strain gauges at right angles to each other, to better identify the direction of the force. Finally, the output of a set of strain gauges must be carefully calibrated by applying known loads to the force sensor and fitting a mapping between sensor readings and applied forces.

Rather than gluing your own strain gauges, you are more likely to buy an integrated *load cell* including the substrate, strain gauge(s), and Wheatstone bridge(s). Load cells come in single-axis and multi-axis versions, up to a full six axes (forces and torques about three

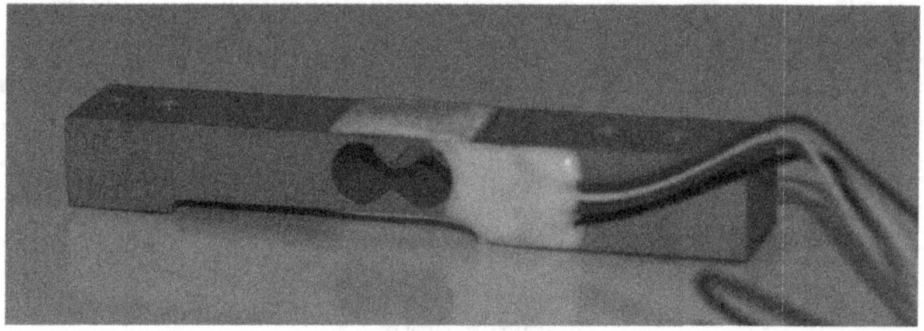

Figure 21.20
A one-axis load cell. The metal substrate is shaped to achieve the desired force sensitivity.

orthogonal axes). Professional load cells can cost several thousand dollars, but inexpensive one- and two-axis load cells are available from manufacturers of load cells for consumer digital scales (Figure 21.20). These load cells typically integrate the Wheatstone bridge but require you to supply your own instrumentation amplifier. Vendors include seeedstudio.com and elane.net.

A much more flexible force-sensitive resistor is the Flex Sensor from Spectra Symbol. Flex Sensors are flexible strips with the resistor printed on one side in conductive ink. As the strip is bent in the other direction, by up to 180 degrees, the resistance increases up to twice its original value.

21.9 Temperature

Thermistors are resistors whose resistances vary significantly with temperature, and they come in two types: NTC (negative temperature coefficient) and PTC (positive temperature coefficient). An example NTC thermistor is the TDK B57164K103J (Figure 21.21) which has a nominal resistance of 10 kΩ at 25° C, rising to 35.6 kΩ at 0° C and dropping to 549 Ω at 100° C.

While the resistance exhibited by the B57164K103J is nonlinear with respect to temperature in Celsius, the Analog Devices TMP37 is designed to produce an analog voltage proportional to the temperature in Celsius. The three pins of the TMP37 in Figure 21.21 are for power (e.g., 3.3 V), GND, and the output voltage, 20 mV/degree Celsius for the range 5-100° C, with a typical accuracy of ±1° C.

Figure 21.21
(Left) The TDK B57164K103J thermistor. (Right) The Analog Devices TMP37 temperature sensor. (Images courtesy of Digi-Key Electronics, digikey.com.)

The B57164K103J and the TMP37 each cost approximately 1 USD. To measure temperatures down to −200° C and up to 1000° C and higher, a more expensive thermocouple and thermocouple amplifier can be used.

21.10 Current

Two common methods for measuring the current flowing through a wire are to (1) use a Hall effect sensor and (2) put a low-resistance resistor in series with the wire and measure the voltage across the resistor. We consider the latter case first.

21.10.1 Current-Sense Resistor and Amplifier

To measure current, a current-sensing resistor can be placed in series with it. Current flowing through this resistor creates a voltage drop across it, which is then measured. To have minimum effect on the current, the sensing resistance should be small. For good accuracy, the resistor should have a tight tolerance on its resistance, and the resistor's power rating should be high enough to allow it to survive the largest current that can flow through it. For example, a 15 mΩ resistor used on a wire that may carry up to 5 A should be rated for at least $(5\text{ A})^2\ 0.015\ \Omega = 0.375$ W to ensure that it will not burn up.

The voltage across a current-sense resistor is intended to be small, e.g., $5\text{ A} \times 0.015\ \Omega = 0.075$ V for 5 A through a 15 mΩ resistor. A specialized current-sense amplifier chip can be used to turn this small signal into a signal usable by a microcontroller. One such chip is the Maxim Integrated MAX9918 current-sense amplifier (Figure 21.22). The voltage across the current-sense resistor is registered by pins RS+ and RS−. The analog output voltage OUT is given by

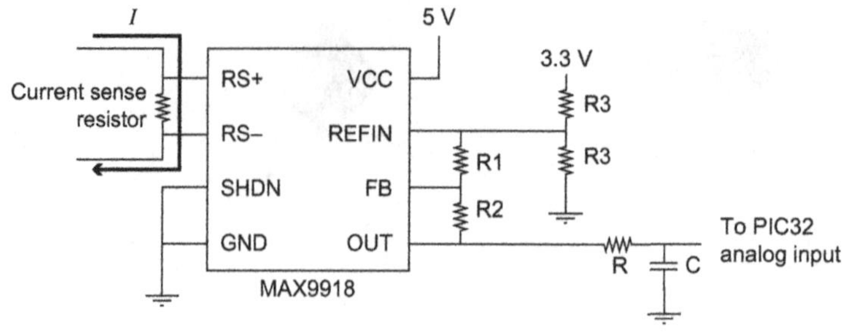

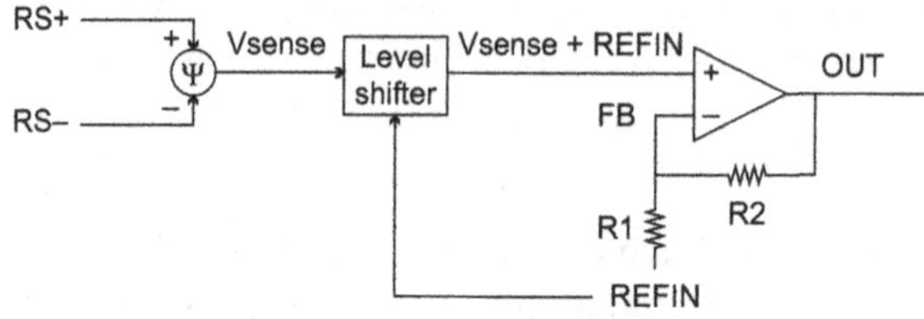

Figure 21.22

Top: Wiring the MAX9918 current-sense amplifier. Bottom: Effective internal circuit, showing how the R1 and R2 resistors are used to set the gain of a noninverting amplifier.

$$\mathrm{OUT} = \mathrm{G} * (\mathrm{RS+} - \mathrm{RS-}) + \mathrm{REFIN}$$

where the gain G is set by the external feedback resistors R1 and R2 as G = 1 + (R2/R1). To implement this equation, the chip uses a noninverting instrumentation amplifier along with a level-shifting circuit (Figure 21.22).

The circuit shows REFIN as 1.65 V, so that zero current through the sense resistor reads as 1.65 V at OUT. This offset voltage allows OUT voltages less than 1.65 V to represent negative currents. The R3 voltage divider resistors feeding the reference voltage REFIN should be relatively small, perhaps a few hundred ohms, to prevent small currents in the feedback resistor network from affecting REFIN.[4] Lowering R3 further would waste power unnecessarily.

The gain G should be chosen so that the maximum expected current gives the maximum voltage at the output. For example, if the maximum expected current is 2 A, then the maximum expected voltage across the 15 mΩ resistor is ±0.03 V. To use the full resolution of the PIC32's ADC, this should map to ±1.65 V, meaning G = 1.65 V/0.03 V = 55. Then a current of 2 A reads as 3.3 V at OUT and a current of −2 A reads as 0 V. The feedback

[4] Ideally the voltage reference to REFIN would be from a lower-impedance source, like a buffered output, but here we are trying to keep the component count down.

resistors R1 and R2 should be relatively high resistance, so as not to load the REFIN voltage divider. See, for example, Exercise 13 of Chapter 27.

One common application of a current sensor is to sense the current flowing through a motor. Since motors are typically driven by a rapidly switching pulse-width modulated voltage (Chapter 27), the current through the motor may also be rapidly changing. We are less interested in this fast variation at the PWM frequency (typically tens of kHz), and more interested in the time-averaged current over several PWM cycles. One way to approximate this time-averaged current is by low-pass RC filtering the sensor output (Appendix B.2.2).[5] A good choice for the RC time constant would give a filter cutoff frequency $f_c = 1/(2\pi RC)$ of a few hundred Hz, approximately 100 times less than a typical PWM frequency. The filter, therefore, attenuates most of the variation due to the PWM without making current sensing overly sluggish.

21.10.2 Hall Effect Current Sensor

The Allegro ACS711 is a Hall effect-based current sensor. The current is passed into the IP+ pins of the IC and back out the IP− pins. The internal conduction path generates a magnetic field which is sensed by the internal Hall effect sensor. At 0 A current, the ACS711 output is half of the supply voltage (e.g., 3.3 V/2 = 1.65 V). The output voltage increases (decreases) proportionally with the positive (negative) current flowing into the ACS711 with a constant of proportionality of 45, 55, 90, or 110 mV/A, depending on the specific ACS711 model number. Like the output of the MAX9918, this analog output can be sent to a PIC32 analog input.

21.11 GPS

For well under 100 USD, you can buy a GPS receiver that listens to satellites in Medium Earth Orbit at an altitude of approximately 20,200 km. GPS satellites carry precise atomic clocks that are regularly synchronized with Earth-based clocks. GPS satellites continuously broadcast their time and their 3D location, as determined in communication with Earth-based stations. If a GPS receiver is able to receive transmissions from at least four satellites, then the difference between the satellites' transmission times and their relative arrival times, along with knowledge of the speed of light, can be used to calculate the receiver's latitude, longitude, and altitude.

Many GPS receivers are on the market. It is common to communicate with GPS receivers via UART or USB.

[5] An even better solution would be a high-impedance input active filter, using an op-amp.

21.12 Exercises

For any of the sensors described in this chapter, or any other sensor, get the data sheet for the sensor and understand the key sensor specifications; design and build a circuit to interface it to your PIC32; create a library consisting of a header file and a C file that gives the user access to the main sensor functions; write a demo program that uses that library; and calibrate or test your sensor under different conditions and report your results to show that the sensor works as expected.

Further Reading

ACS711 hall effect linear current sensor with overcurrent fault output for <100 V isolation appliations data sheet revision 3. (2015). Allegro Microsystems, LLC.

AD2S90 low cost, complete 12-bit resolver-to-digital converter data sheet revision D. (1999). Analog Devices.

ADXL362 micropower, 3-axis, ±2g/±4g/±8g digital output MEMS accelerometer data sheet. (2012). Analog Devices.

AEAT-6600-T16 10 to 16-bit programmable angular magnetic encoder IC data sheet. (2013). Avago Technologies.

AEAT-9000-1GSH1 ultra-precision 17-bit absolute single turn encoder data sheet. (2014). Avago Technologies.

ASM330LXH automotive inertial module: 3D accelerometer and 3D gyroscope data sheet. (2015). STMicroelectronics.

Flex sensor FS data sheet. (2014). Spectra Symbol.

GP2Y0A60SZ0F/GP2Y0A60SZLF distance measuring sensor unit data sheet. Sharp.

Global positioning system standard positioning service performance standard (4th ed.). (2008). Department of Defense, United States of America.

Hambley, A. R. (2000). *Electronics* (2nd ed.). Upper Saddle River, NJ: Prentice Hall.

Horowitz, P., & Hill, W. (2015). *The art of electronics* (3rd ed.). New York, NY: Cambridge University Press.

Kingbright WP7113SRC/DU data sheet. (2015). Kingbright.

L3GD20H MEMS motion sensor: Three-axis digital output gyroscope data sheet. (2013). STMicroelectronics.

LSM303D ultra compact high performance e-compass 3D accelerometer and 3D magnetometer module. (2012). STMicroelectronics.

MAX9918/MAX9919/MAX9920 −20V to +75v input range, precision uni-/bidirectional, current-sense amplifiers. (2015). Maxim Integrated.

NTC thermistors for temperature measurement series B57164K data sheet. (2013). TDK.

Orozco, L. (2014). *Technical article MS-2624: Optimizing precision photodiode sensor circuit design.* Analog Devices.

PDV-P5002 CdS photoconductive photocells data sheet. (2006). Advanced Photonix, Inc.

PIN silicon photodiode type OP906 product bulletin. (1996). OPTEK Technology, Inc.

QED121, QED122, QED123 plastic infrared light emitting diode data sheet. (2008). Fairchild Semiconductor Corporation.

Reflective object sensor, OPB 740 series data sheet. (2011). OPTEK Technology, Inc.

Silicon NPN phototransistor SFH 310, SFH 310 FA data sheet. (2014). OSRAM Opto Semiconductors.

Slotted optical switch, series OPB370 data sheet. (2012). OPTEK Technology, Inc.

TCS20DPR digital output magnetic sensor data sheet. (2014). Toshiba.

Technical article SBOA061: Designing photodiode amplifier circuits with OPA128. (1994). Texas Instruments.

Technical article TA0343: Everything about STMicroelectronics' 3-axis digital MEMS gyroscopes. (2011). STMicroelectronics.

TMP35/TMP36/TMP37 low voltage temperature sensors data sheet revision F. (2010). Analog Devices.

CHAPTER 22

Digital Signal Processing

We have already used RC filters to low-pass-filter high-frequency PWM signals, creating analog output signals. Filters have many other uses: for example, suppressing high-frequency or 60 Hz electrical noise, extracting high-frequency components from change-sensitive sensors, and integrating or differentiating a signal. If the signal is an analog voltage, these filters can be implemented with resistors, capacitors, and op amps (Appendix B).

Filters can also be implemented in software. In this case, the signal is first converted to digital form, perhaps using an analog-to-digital converter to sample an analog signal at fixed time increments. Once in this form, a *digital filter* can be used to difference or integrate the signal, or to suppress, enhance, or extract different frequency components in the signal. Digital filters offer advantages over their analog electronic counterparts:

- No need for extra external components, such as resistors, capacitors, and op amps.
- Tremendous flexibility in the filter design. Filters with excellent properties can be implemented very easily in software.
- The ability to operate on signals that do not originate from analog voltages.

Digital filtering is one example of *digital signal processing* (DSP). We start this chapter by providing some background on sampled signal representation. We then provide an introduction to the *fast Fourier transform* (FFT), which can be used to decompose a digital signal into its frequency components. The FFT is among the most important and heavily used algorithms in video, audio, and many other signal processing and control applications. We then discuss a class of digital filters called *finite impulse response* (FIR) filters, which calculate their output values as weighted sums of their past input samples. Next we briefly describe *infinite impulse response* (IIR) filters, which calculate their output as weighted sums of their past inputs and outputs. We also discuss FFT-based filters. Finally we conclude with sample code for DSP on a PIC32.

This chapter provides a brief introduction and some practical hints on how to use FFTs, FIR filters, and IIR filters. We skip most of the mathematical underpinnings, which are covered in books and courses focusing solely on signal processing.

Embedded Computing and Mechatronics with the PIC32 Microcontroller. http://dx.doi.org/10.1016/B978-0-12-420165-1.00022-6

22.1 Sampled Signals and Aliasing

Let $x(t)$ be a periodic signal that varies as a function of (continuous) time with period T ($x(t) = x(t+T)$), and therefore frequency $f = 1/T$. A periodic signal $x(T)$ can be written as the sum of a DC (constant) component and an infinite sequence of sinusoids at frequencies $f, 2f, 3f$, etc.:

$$x(t) = A_0 + \sum_{m=1}^{\infty} A_m \sin(2\pi mft + \phi_m). \tag{22.1}$$

Thus the T-periodic signal $x(t)$ can be uniquely represented by the amplitudes $A_0, A_1, \ldots$ and the phases $\phi_1, \phi_2, \ldots$ of the component sinusoids. These amplitudes and phases form the *frequency domain* representation of $x(t)$.

An example is a square wave signal that swings between $+1$ and -1 at frequency f and 50% duty cycle. The Fourier series that creates this signal is given by $A_m = 0$ for even m and $A_m = 4/(m\pi)$ for odd m, with all phases $\phi_m = 0$. Figure 22.1 illustrates the first four components of the square wave.

The first step in analyzing an analog signal using DSP is to sample the continuous-time signal $x(t)$ at time intervals T_s (sampling frequency $f_s = 1/T_s$), yielding N samples $x(n) \equiv x(nT_s) = x(t)$ for $n = 0, 1, 2, \ldots, N-1$, as shown in Figure 22.2. The sampling process also quantizes the signal; for example, the PIC32's 10-bit ADC module converts a continuous voltage to one of 1024 levels. While quantization is an important consideration in DSP, in this chapter we ignore quantization effects and assume that $x(n)$ can take arbitrary real values.

Suppose the original analog input signal is a sinusoid

$$x(t) = A\sin(2\pi ft + \phi),$$

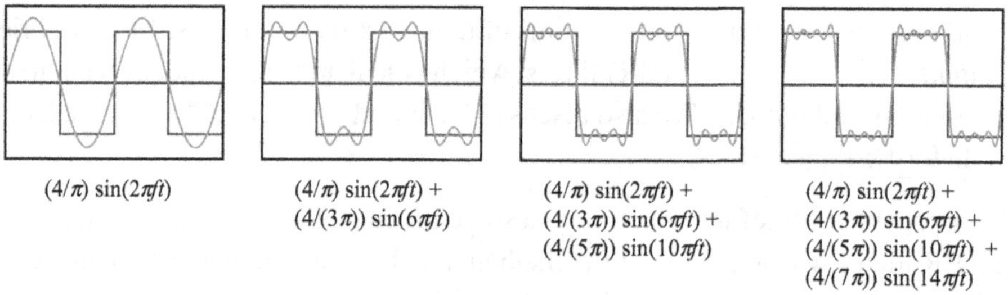

$(4/\pi)\sin(2\pi ft)$

$(4/\pi)\sin(2\pi ft) +$
$(4/(3\pi))\sin(6\pi ft)$

$(4/\pi)\sin(2\pi ft) +$
$(4/(3\pi))\sin(6\pi ft) +$
$(4/(5\pi))\sin(10\pi ft)$

$(4/\pi)\sin(2\pi ft) +$
$(4/(3\pi))\sin(6\pi ft) +$
$(4/(5\pi))\sin(10\pi ft) +$
$(4/(7\pi))\sin(14\pi ft)$

Figure 22.1
An illustration of the sum of the first four nonzero frequency components of the Fourier series of a square wave. The sum converges to the square wave as higher frequency components are included in the sum.

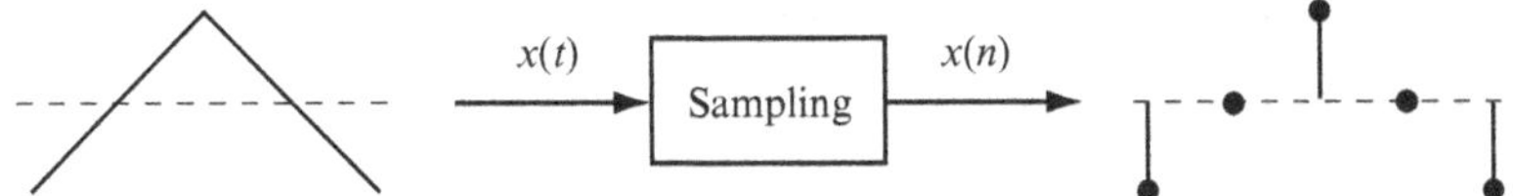

Figure 22.2
The sampling module converts the continuous-time signal $x(t)$ to a discrete-time signal $x(n)$.

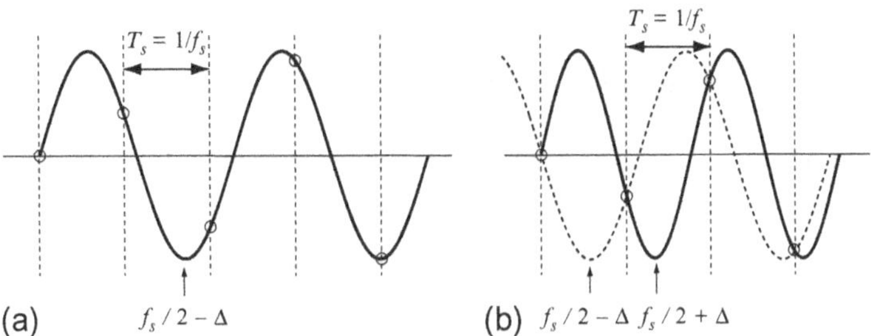

Figure 22.3
(a) The underlying sinusoid $x(t)$ with frequency $f = f_s/2 - \Delta, \Delta > 0$, can be reconstructed from its samples $x(n)$, shown as circles. (b) An input sinusoid of frequency $f = f_s/2 + \Delta$, however, appears to be a signal of frequency $f = f_s/2 - \Delta$ with a different phase.

where f is the frequency, $T = 1/f$ is the period, A is the amplitude, and ϕ is the phase. Given samples $x(n)$ taken at the sampling frequency f_s, and knowing the input is a sinusoid, it is possible to use the samples to uniquely determine A, f, and ϕ of the underlying signal, provided f is less than $f_s/2$. As we increase the signal frequency f beyond $f_s/2$, however, something interesting happens, as illustrated in Figure 22.3. The samples of a signal with frequency $f_1 = f_s/2 + \Delta, \Delta > 0$, with phase ϕ_1, are indistinguishable from the samples of a signal with lower frequency $f_2 = f_s/2 - \Delta$ with a different phase ϕ_2. For example, for $\Delta = f_s/2$, an input signal of frequency $f_1 = f_s/2 + \Delta = f_s$ looks the same as a constant (DC) input signal ($f_2 = f_s/2 - \Delta = 0$), because our once-per-cycle sampling returns the same value each time.

The phenomenon of signals of frequency greater than $f_s/2$ "posing" as signals of frequency between 0 and $f_s/2$ is called *aliasing*. The frequency $f_s/2$, the highest frequency we can uniquely represent with a discrete-time signal, is known as the *Nyquist frequency* f_N. Because we cannot distinguish higher-frequency signals from lower-frequency signals, we assume that all input frequencies are in the range $[0, f_N]$. If the sampled signal is obtained from an analog voltage, it is common to put an analog electronic low-pass *anti-aliasing* filter before the sampler to remove frequency components greater than f_N.

Aliasing is familiar from watching a low-frame-rate video of a spinning wheel. Your eyes track a mark on the wheel as it speeds up at a constant rate, and initially you see the wheel spinning forward faster and faster. In other words, the wheel appears to have an increasingly positive rotational frequency. As the wheel continues to accelerate, just after the point where the video camera captures only two images per revolution, the wheel begins to appear to be rotating backwards at a high speed (rotating with a large negative rotational frequency). As its actual forward speed increases further, the apparent negative speed begins to slow (the negative rotational frequency grows toward zero), until eventually the wheel appears to be at rest again (zero frequency) when the camera takes exactly one image per revolution.

22.2 The Discrete Fourier Transform

To design digital filters, it is important to understand the frequency domain representation of a digital signal. This representation allows us to see the amount of signal present at different frequencies, and to assess the performance of filters designed to suppress signals or noise at unwanted frequencies.

To find the frequency domain representation of an N-sample signal $x(n)$, $n = 0, \ldots, N-1$, we use the *discrete Fourier transform* (DFT) of x. As we will see, the DFT allows us to calculate the frequency domain representation of the N-periodic signal that repeats the same N samples infinitely. Assuming N is even, this representation is given by $N/2$ sinusoid phases ϕ_m, where $m = 1 \ldots N/2$, and $N/2+1$ amplitudes: the DC amplitude A_0 and the sinusoidal amplitudes A_m. The frequencies of the sinusoids are $mf = mf_s/N$ in (22.1). The spacing between frequencies represented by the A_m and ϕ_m is $f_s/N = 1/(NT_s)$—the more samples N, the higher the frequency resolution.

The DFT $X(k), k = 0, \ldots, N-1$, of $x(n), n = 0, \ldots, N-1$, is given by

$$X(k) = \sum_{n=0}^{N-1} x(n) e^{-j2\pi kn/N}, \quad k = 0, \ldots, N-1. \tag{22.2}$$

Considering that $e^{-j2\pi kn/N} = \cos(2\pi kn/N) - j\sin(2\pi kn/N)$, generally the $X(k)$ are complex numbers. Additionally, the form of (22.2) shows that $X(N/2+\Delta)$, where $\Delta \in \{1, 2, \ldots, N/2-1\}$, is the complex conjugate of $X(N/2-\Delta)$. In particular, this means that their magnitudes are equal, $|X(N/2-\Delta)| = |X(N/2+\Delta)|$.

Without going into details, the normalized "frequency" k/N associated with the sinusoids in $X(k)$ represents the actual frequency kf_s/N. Thus $X(N/2)$ is associated with the Nyquist frequency $f_N = f_s/2$. Higher frequencies ($k > N/2$) are equivalent to negative frequencies $(k-N)f_s/N$, as described in the spinning wheel aliasing analogy.

The $X(k)$ contain all the information we need to find the A_m and ϕ_m frequency domain representation of the sampled signal x. For a given $X(k) = a + bj$, the magnitude

$|X(k)| = \sqrt{a^2 + b^2}$ corresponds to N times the magnitude of the frequency component at $f_s k/N$, and the phase is given by the angle of $X(k)$ in the complex plane, i.e., the two-argument arctangent atan2(b, a). In this chapter we focus only on the amplitudes of the frequency components, ignoring the phase, since the phase is essentially random when the amplitude $\sqrt{a^2 + b^2}$ is near zero.

Because the $|X(k)|$ represent the magnitudes as a component at DC ($|X(0)|$), a component at the Nyquist frequency f_N ($|X(N/2)|$), and equal-magnitude complex conjugate pairs $X(k)$ for all other k, the A_m can be calculated as

$$A_0 = |X(0)|/N, \tag{22.3}$$

$$A_{N/2} = |X(N/2)|/N, \tag{22.4}$$

$$A_m = 2|X(m)|/N, \quad \text{for all } m = 1, \ldots, N/2 - 1, \tag{22.5}$$

where the frequency corresponding to A_m is mf_s/N.

22.2.1 The Fast Fourier Transform

The *Fast Fourier Transform* (FFT) refers to one of several methods for efficiently calculating the DFT. Many implementations of the FFT require that N be a power of two. If we have a number of samples that is not a power of two, we can simply "pad" the signal with "virtual" samples of value zero at the end. This process is called "zero padding." For example, if we have 1000 samples, we can pad the signal with 24 zeros to reach $2^{10} = 1024$ samples.

For example, assume an underlying analog signal

$$x(t) = 0.5 + \sin(2\pi(20\text{ Hz})t + \pi/4) + 0.5\sin(2\pi(200\text{ Hz})t + \pi/2)$$

with components at DC, 20 Hz, and 200 Hz. We collect 1000 samples at $f_s = 1$ kHz (0.001 s intervals) and zero pad to get $N = 1024$. The signal and its FFT magnitude plot is shown in Figure 22.4. The magnitude components are spaced at frequency intervals of f_s/N, or 0.9766 Hz. The DC, 20 Hz ($0.04 f_N$), and 200 Hz ($0.4 f_N$) components are clearly visible, though the numerical procedure has spread the components over several nearby frequencies since the actual signal frequencies are not represented exactly as mf_s/N for any integer m.

The mathematics of the DFT (and FFT) implicitly assume that the signal repeats every N samples. Thus the DFT may have a significant component at the lowest nonzero frequency, f_s/N, even if this frequency is not present in the original signal. This component suggests the following:

- If the original signal is known to be periodic with frequency f, the N samples should contain several complete cycles of the signal (as opposed to one or less than one cycle).

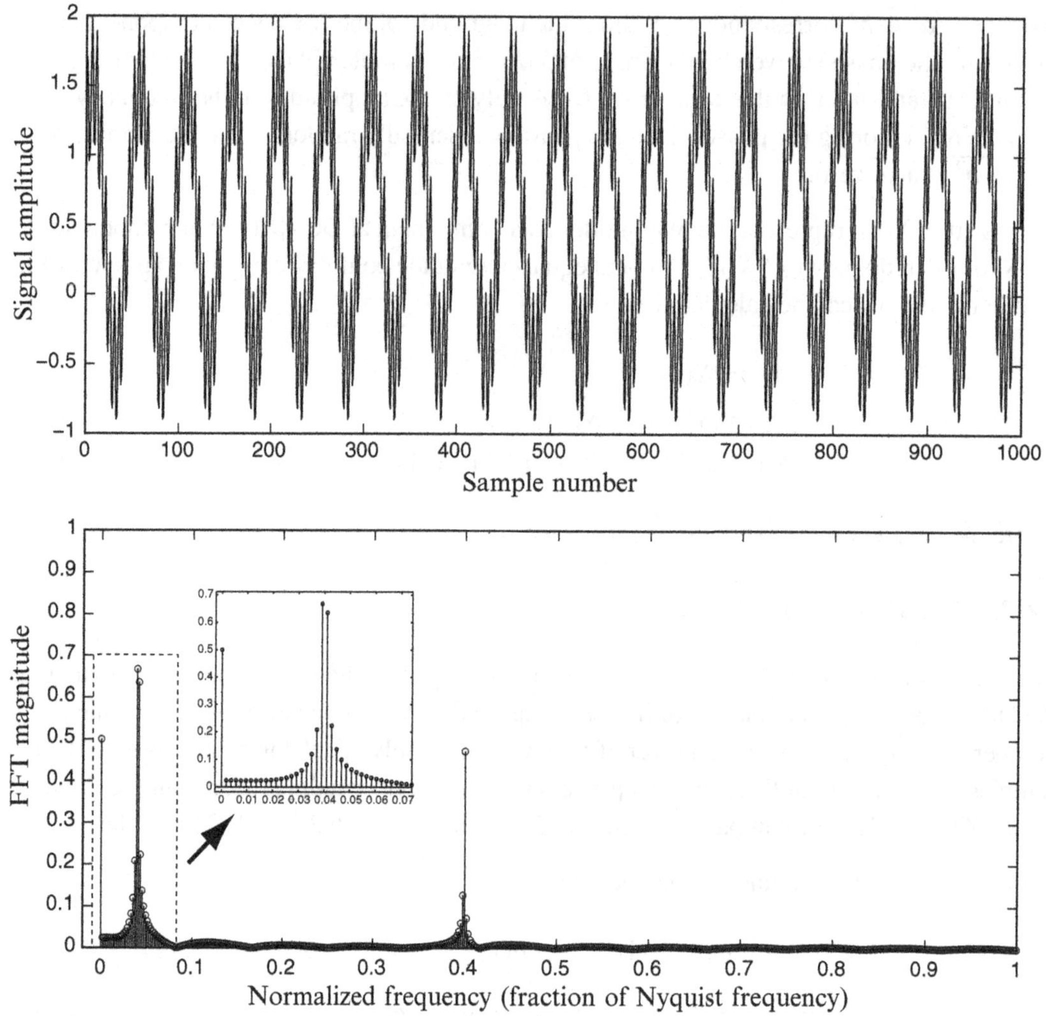

Figure 22.4
(Top) The original sampled signal. (Bottom) The FFT magnitude plot, with a portion of it magnified.

Having many cycles means that the smallest nonzero frequency represented, f_s/N, is much smaller than f, isolating the non-DC signals we care about (at f and higher) from the lower frequencies that appear due to the finite sampling.

- If the original signal is known to be periodic with frequency f, then, if possible, the samples should contain an integer number of cycles. In this case, there should be very little magnitude at the lowest nonzero frequency f_s/N.
- If the original signal is not periodic, zero padding can be used to isolate the lowest nonzero frequency component of the repeated signal from f_s/N.

22.2.2 The FFT in MATLAB

Given an even number of samples N in a row vector `x = [x(1) ... x(N)]` in MATLAB (note the index starts at 1 in MATLAB), the command

```
X = fft(x);
```

returns an N-vector `X = [X(1) ... X(N)]` of complex numbers corresponding to the amplitude and phase at different frequencies. Let us try an FFT of $N = 200$ samples of a 50% duty cycle square wave, where each period consists of 10 samples equal to 2 and 10 samples equal to 0 (i.e., the square wave of Figure 22.1 plus a DC offset of 1). The frequency of the square wave is $f_s/20$, and our entire sampled signal consists of 10 full cycles. According to Figure 22.1, the continuous-time square wave consists of frequency components at $f_s/20$, $3f_s/20$, $5f_s/20$, $7f_s/20$, etc. Thus we expect the frequency domain magnitude representation of the sampled square wave to consist of the DC component and nonzero components at these frequencies.

Let us build the signal and plot it (Figure 22.5(a)):

```
x=0;   % clear any array that might already be in x
x(1:10) = 2;
x(11:20)= 0;
x = [x x x x x x x x x x];
N = length(x);
plot(x,'Marker','o');
axis([-5 205 -0.1 2.1]);
```

Now let us plot the FFT amplitude plot, using the procedure described in Section 22.2:

```
plotFFT(x);
```

This code uses our custom MATLAB function `plotFFT`:

Code Sample 22.1 `plotFFT.m`**. Plotting the Single-Sided FFT Magnitude Plot of a Sampled Signal** `x` **with an Even Number of Samples.**

```
function plotFFT(x)

if mod(length(x),2) == 1          % x should have an even number of samples
  x = [x 0];                      % if not, pad with a zero
end
N = length(x);
X = fft(x);
mag(1) = abs(X(1))/N;             % DC component
mag(N/2+1) = abs(X(N/2+1))/N;     % Nyquist frequency component
mag(2:N/2) = 2*abs(X(2:N/2))/N;   % all other frequency components
freqs = linspace(0, 1, N/2+1);    % make x-axis as fraction of Nyquist freq
stem(freqs, mag);                 % plot the FFT magnitude plot
axis([-0.05 1.05 -0.1*max(mag) 1.1*max(mag)]);
xlabel('Frequency (as fraction of Nyquist frequency)');
```

```
ylabel('Magnitude');
title('Single-Sided FFT Magnitude');
set(gca,'FontSize',18);
```

Figure 22.5 shows the original signal and the single-sided FFT magnitude plot. Notice that the FFT very clearly picks out the frequency components at DC, $0.1f_N, 0.3f_N, 0.5f_N, 0.7f_N,$ and $0.9f_N$.

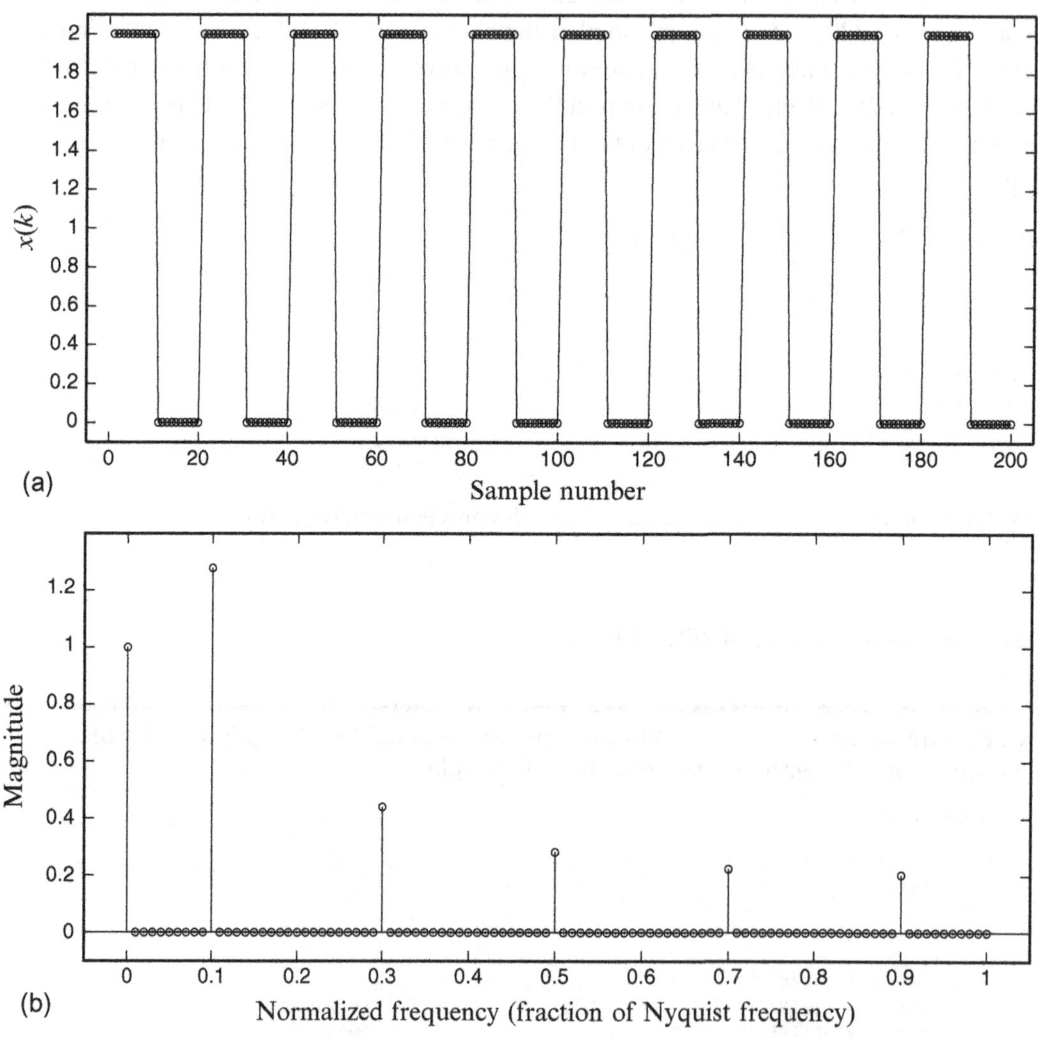

Figure 22.5
(a) The original sampled signal. (b) The single-sided FFT magnitude plot with frequencies expressed as a fraction of the Nyquist frequency.

FFT with $N = 2^n$

For efficiency reasons, the MIPS PIC32 DSP code only performs FFTs on sampled signals that have a power-of-2 length. Let us increase the number of samples with MATLAB from 200 to the next highest power of 2, $2^8 = 256$. We can either pad the original $x(k)$ with 56 zeros at the end, or we can take more samples.

Let us try the zero-padding option first:

```
x = 0;
x(1:10) = 2;
x(11:20)= 0;
x = [x x x x x x x x x x];         % get the signal samples
N = 2^nextpow2(length(x));         % compute the number of zeros to pad
xpad = [x zeros(1,N-length(x))]; % add the zero padding
plotFFT(xpad);
```

And now if the signal were sampled 256 times in the first place:

```
x = 0;
x(1:10) = 2;
x(11:20)= 0;
x = [x x x x x x x x x x x x 2*ones(1,10) zeros(1,6)];
plotFFT(x);
```

The results are plotted in Figure 22.6. The frequency components are still visible, though the results are not as clear as in Figure 22.5. A major reason for the lower quality plot is that the signal frequencies $0.1f_N, 0.3f_N$, etc., are not exactly represented in the FFT, as they were before. The frequency intervals are now $f_s/256$, not $f_s/200$. As a result, the FFT spreads the original frequency components across nearby frequencies rather than concentrating them in spikes at exact frequencies. This kind of spread is typical in most applications, as it is unlikely that the original signal will have component frequencies exactly at frequencies of the form mf_s/N.

22.2.3 The Inverse Fast Fourier Transform

Given the frequency domain representation $X(k)$ obtained from `X = fft(x)`, the inverse FFT uses the FFT algorithm to recover the original time-domain signal $x(n)$. In MATLAB, this is the procedure:

```
N = length(x);
X = fft(x);
xrecovered = fft(conj(X)/N);
plot(real(xrecovered));
```

The inverse FFT is accomplished by applying `fft` to the complex conjugate of the frequency representation `X` (the imaginary components of all entries are multiplied by -1), scaled by `1/N`.

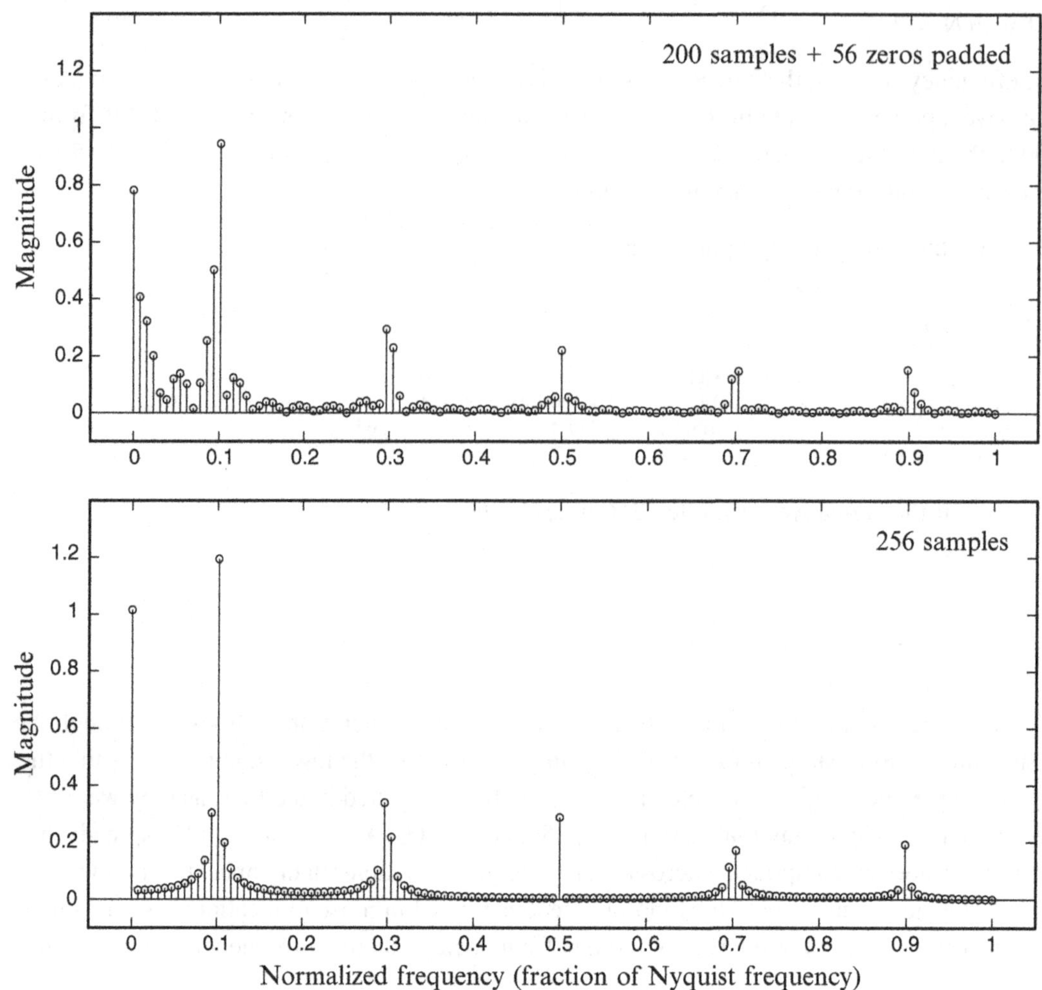

Figure 22.6
(Top) The FFT of the 200-sample square wave signal with 56 zeros padded. (Bottom) The FFT of the 256-sample square wave signal.

The vector `xrecovered` is equal to the original `x` up to numerical errors, so its entries have essentially zero imaginary components. The `real` operation returns only the real components, ensuring that the imaginary components are exactly zero.

22.3 Finite Impulse Response (FIR) Digital Filters

Now that we understand frequency domain representations of sampled signals, we turn our attention to filtering those signals (Figure 22.7). A finite impulse response (FIR) filter

Figure 22.7
A digital filter produces filtered output $z(n)$ based on the inputs $x(n)$.

produces a filtered signal $z(n)$ by multiplying the $P+1$ current and past inputs $x(n-j), j=0\ldots P$, by *filter coefficients* b_j and adding:

$$z(n)=\sum_{j=0}^{P} b_j x(n-j).$$

Such filters can be used for several operations, such as differencing a signal or suppressing low-frequency or high-frequency components. For example, if $P=1$ and we choose $b_0=1$ and $b_1=-1$, then

$$z(n)=x(n)-x(n-1),$$

i.e., the output of the filter at time n is the difference between the input at time n and the input at time $n-1$. This differencing filter in discrete time is similar to a derivative filter in continuous time.

Since FIR filtering is a linear operation on the samples, filters in series can be performed in any order. For example, a differencing filter followed by a low-pass filter gives equivalent output to the low-pass filter followed by the differencing filter.

An FIR filter has $P+1$ coefficients, and P is called the *order* of the filter. The filter coefficients are directly evident in the *impulse response*, which is the response $z(n)$ to a unit impulse $\delta(n)$, where

$$\delta(n)=\begin{cases}1 & \text{for } n=0\\ 0 & \text{otherwise.}\end{cases}$$

The output is simply $z(0)=b_0, z(1)=b_1, z(2)=b_2$, etc. The impulse response is typically written as $h(k)$.

Any input signal x can be represented as the sum of scaled and time-shifted impulses. For example,

$$x=3\delta(n)-2\delta(n-2)$$

is a signal that has value 3 at time $n=0$ and -2 at time $n=2$ (Figure 22.8(left)). Because an FIR filter is linear, the output is simply the sum of the scaled and time-shifted impulse responses,

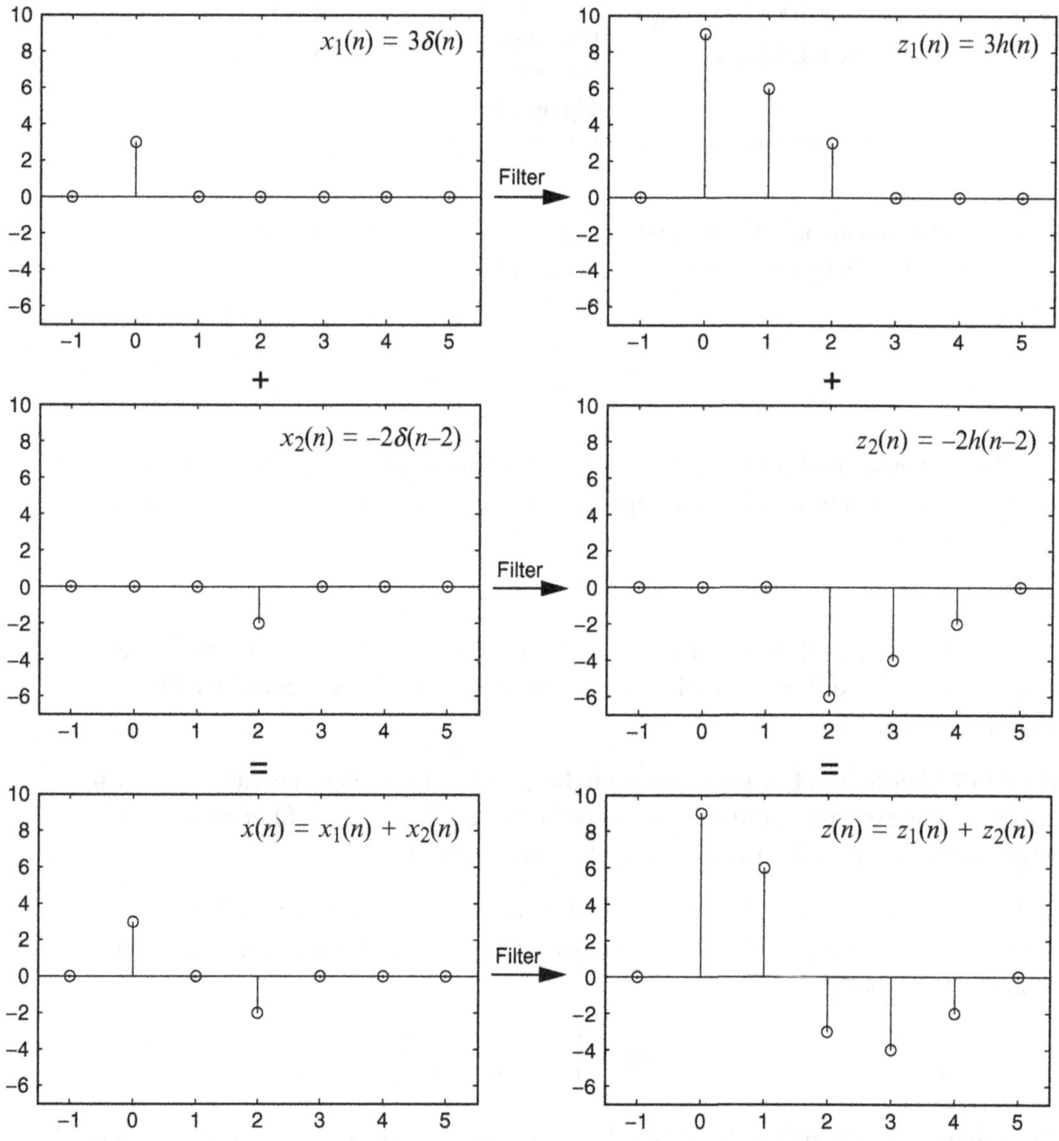

Figure 22.8
(Left) The two scaled and time-shifted impulses sum to give the signal x in time. (Left to right) The filter with coefficients $b_0 = 3, b_1 = 2, b_2 = 1$ applied to each of the individual and composite signals. (Right) The response z to the signal x can be obtained by summing the responses z_1 and z_2 to the individual components of x.

$$z = 3h(n) - 2h(n-2).$$

For the second-order filter with coefficients $b_0 = 3, b_1 = 2, b_2 = 1$, for example, the response to the input x is shown in Figure 22.8(right). When reading these signals, be aware that the leftmost samples are oldest; for example, the output $z(0)$ happens three timesteps before the output $z(3)$.

The output z is called the *convolution* of the input x and the filter's impulse response h, commonly written $z = x * h$. The convolution is obtained by simply summing the scaled and time-shifted impulse responses corresponding to each sample $x(n)$, as illustrated in Figure 22.8.

An FIR filter response can be calculated using MATLAB's `conv` command. We collect the filter coefficients into the impulse response vector `h = b = [b0 b1 b2] = [3 2 1]` and the input into the vector `x = [3 0 -2]`, and then

```
z = conv(h,x)
```

produces `z = [9 6 -3 -4 -2]`.

"Finite Impulse Response" filters get their name from the fact that the impulse response goes to zero in finite time (i.e., there is a finite number of filter coefficients). As a result, for any input that goes to zero eventually, the response goes to zero eventually. The output of an "Infinite Impulse Response" filter (Section 22.4) may *never* go to zero.

A filter is fully characterized by its impulse response. Often it is more convenient to look at the filter's *frequency response*, however. Because the filter is linear, a sinusoidal input will produce a sinusoidal output, and the filter's frequency response consists of its frequency-dependent gain (the ratio of the filter's output sinusoid amplitude to the input amplitude) and phase (the shift in the phase of the output sinusoid relative to the input sinusoid).[1] To begin to understand the discrete-time frequency response, we start with the simplest of FIR filters: the moving average filter.

22.3.1 Moving Average Filter

Suppose we have a sensor signal $x(n)$ that has been corrupted by high-frequency noise (Figure 22.9). We would like to find the low-frequency signal underneath the noise.

The simplest filter to try is a *moving average filter* (MAF). A moving average filter calculates the output $z(n)$ as a running average of the input signals $x(n)$,

$$z(n) = \frac{1}{P+1} \sum_{j=0}^{P} x(n-j), \tag{22.6}$$

[1] See Appendix B.2.2 for related information on frequency response for analog circuits.

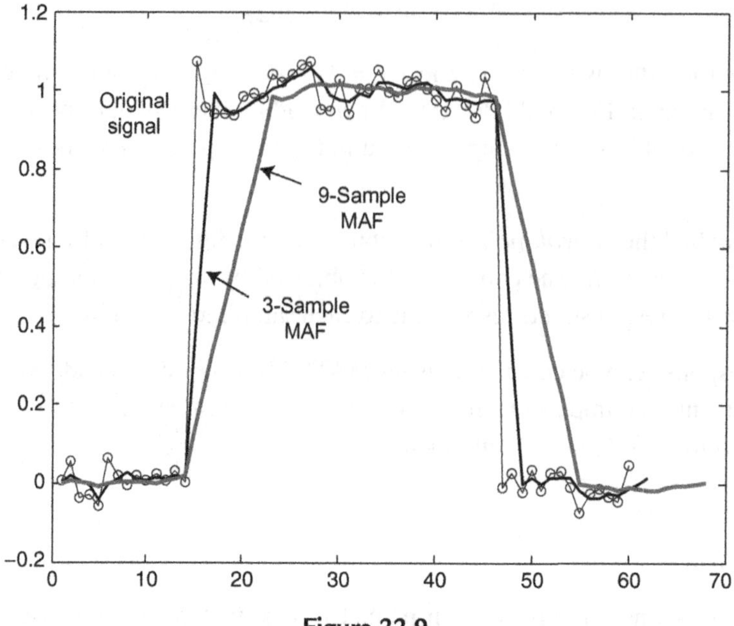

Figure 22.9
The original noisy signal with samples $x(n)$ given by the circles, the signal $z(n)$ resulting from filtering with a three-point MAF ($P = 2$), and the signal $z(n)$ from a nine-point MAF ($P = 8$). The signal gets smoother and more delayed as the number of samples in the MAF increases.

i.e., the FIR filter coefficients are $b_0 = b_1 = \cdots = b_P = 1/(P+1)$. The output $z(n)$ is a smoothed and delayed version of $x(n)$. The more samples $P + 1$ we average over, the smoother and more delayed the output. The delay occurs because the output $z(n)$ is a function of only the current and previous inputs $x(n - j), 0 \leq j \leq P$ (see Figure 22.9).

To find the frequency response of a third-order, four-sample MAF, we test it on some sinusoidal inputs at different frequencies (Figure 22.10). We find that the phase ϕ of the (reconstructed) output sinusoid relative to the input sinusoid, and the ratio G of the amplitude of their amplitudes, depend on the frequency. For the four test frequencies in Figure 22.10, we get the following table:

Frequency	Gain G	Phase ϕ
$0.25f_N$	0.65	−67.5°
$0.5f_N$	0	NA
$0.67f_N$	0.25	0°
f_N	0	NA

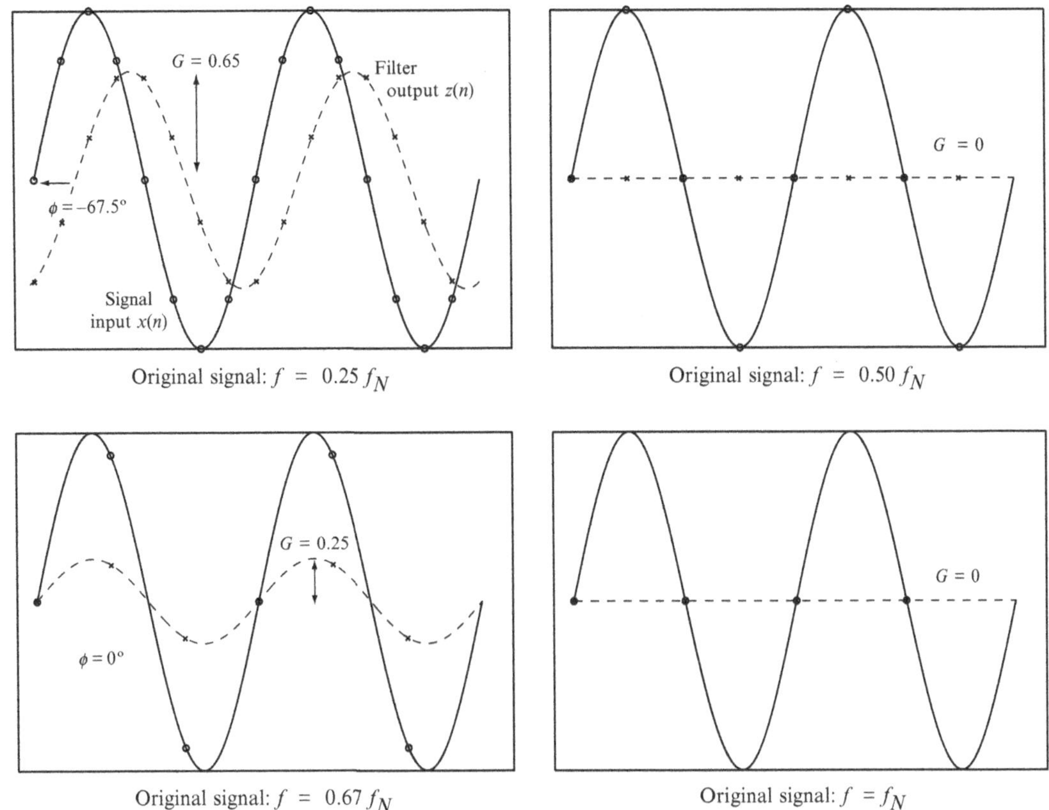

Figure 22.10
Testing the frequency response of a four-sample MAF with different input frequencies.

Testing the response at many different frequencies, we can plot the frequency response in Figure 22.11. Two things to note about the gain plot:

- Gains are usually plotted on a log scale. This allows representation of a much wider range of gains.
- Gains are often expressed in decibels, which are related to gains by the following relationship:

$$M_{\mathrm{dB}} = 20 \log_{10} G.$$

So $G = 1$ corresponds to 0 dB, $G = 0.1$ corresponds to -20 dB, and $G = 0.01$ corresponds to -40 dB.

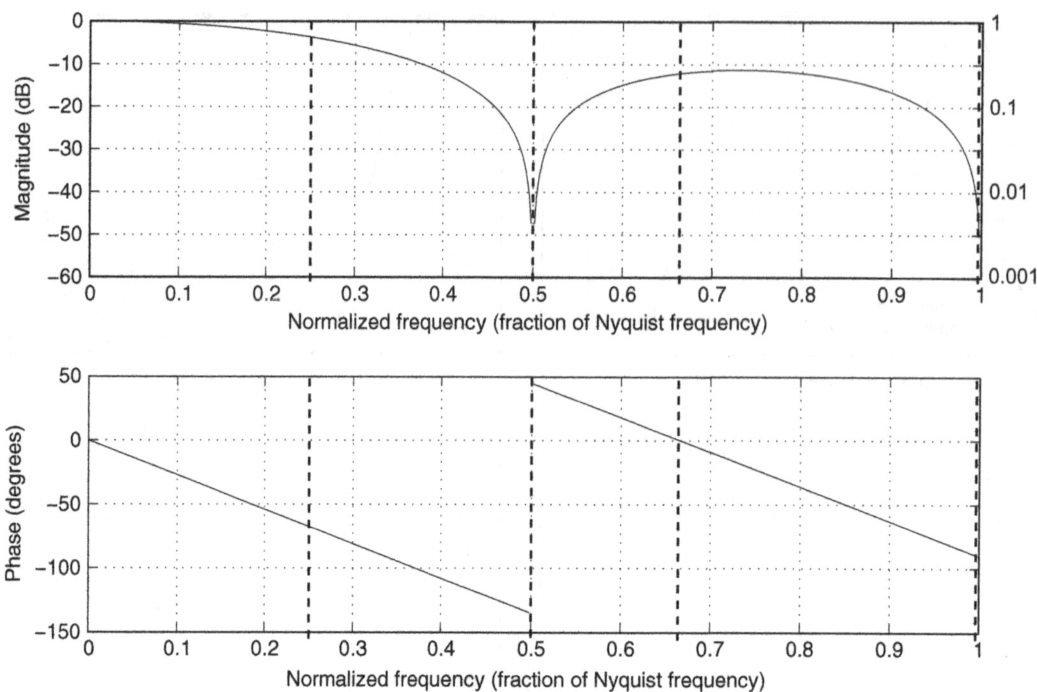

Figure 22.11
The frequency response of a four-sample MAF. Test frequencies are shown as dotted lines.

Examining Figure 22.11 shows that low frequencies are passed with a gain of $G = 1$ and no phase shift. The gain drops monotonically as the frequency increases, until it reaches $G = 0$ ($-\infty$ dB) at input frequencies $f = 0.5f_N$. (The plot truncates the dip to $-\infty$.) The gain then begins to rise again, before falling once more to $G = 0$ at $f = f_N$. The MAF behaves somewhat like a low-pass filter, but not a very good one; high frequency signals can get through with gains of 0.25 or more. Still, it works reasonably well as a signal smoother for input frequencies below $0.5f_N$.

Given a set of filter coefficients `b = [b0 b1 b2 ...]`, we can plot the frequency response in MATLAB using

```
freqz(b);
```

Causal vs. acausal filters

A filter is called *causal* if its output is the result of only current and past inputs, i.e., past inputs "cause" the current output. Causal filters are the only option for real-time implementation. If we are post-processing data, however, we can choose an *acausal* version of the filter, where the outputs at time n are a function of past as well as future inputs. Such acausal filters can eliminate the delay associated with only using past inputs to calculate the current value. For example, a five-sample MAF which calculates the average of the past two inputs, the current input, and the next two inputs is acausal.

Zero padding

When a filter is first initialized, there are no past inputs. In this case we can assume the nonexistent past inputs were all zero. The output transient caused by this assumption will end at the $(P+1)$th input.

22.3.2 FIR Filters Generally

FIR filters can be used for low-pass filtering, high-pass filtering, bandpass filtering, bandstop (notch) filtering, and other designs. MATLAB provides many useful functions for filter design, such as `fir1`, `fdatool`, and `design`. In this section we work with `fir1`. See the MATLAB documentation for more details.

A "good" filter is one that

- passes the frequencies we want to keep with gain close to 1,
- highly attenuates the frequencies we do not want, and
- provides a sharp transition in gain between the passed and attenuated frequencies.

The number of filter coefficients increases with the sharpness of the desired transition and the degree of attenuation needed in the stopped frequencies. This relationship is a general principle: the sharper the transitions in the frequency domain, the smoother and longer the impulse response (i.e., more coefficients are needed in the filter). The converse is also true: the sharper the transition in the impulse response, the smoother the frequency response. We saw this phenomenon with the moving average filter. It has a sharp transition between filter coefficients of 0 and $1/(P+1)$, and the resulting frequency response has only slow transitions.

High-order filters are fine for post-processing data or for non-time-critical applications, but they may be inappropriate for real-time control because of unacceptable delay.

The MATLAB filter design function `fir1` takes the order of the filter, the frequencies you would like to pass or stop (expressed as a fraction of the Nyquist frequency), and other options, and returns a list of filter coefficients. MATLAB considers the cutoff frequency to be where the gain is 0.5 (−6 dB). Here are some examples using `fir1`:

```
b = fir1(10,0.2);                  % 10th-order, 11-sample LPF with cutoff freq
                                     of 0.2 fN
b = fir1(10,0.2,'high');           % HPF cutting off frequencies below 0.2 fN
b = fir1(150,[0.1 0.2]);           % 150th-order bandpass filter with passband 0.1
                                     to 0.2 fN
b = fir1(50,[0.1 0.2],'stop');     % 50th-order bandstop filter with notch at 0.1 to
                                     0.2 fN
```

You can then plot the frequency response of your designed filter using `freqz(b)`.

If the order of your specified filter is not high enough, you will not be able to meet your design criteria. For example, if you want a low-pass filter that cuts off frequencies at 0.1 f_N, and you only allow seven coefficients (sixth order),

```
b = fir1(6,0.1);
```

you will find that the filter coefficients that MATLAB returns do not achieve your aims.

Examples

In the examples in Figures 22.12–22.19, we work with a 1000-sample signal, with components at DC, $0.004f_N$, $0.04f_N$, and $0.8f_N$. The original signal `x` is plotted in Figure 22.12.

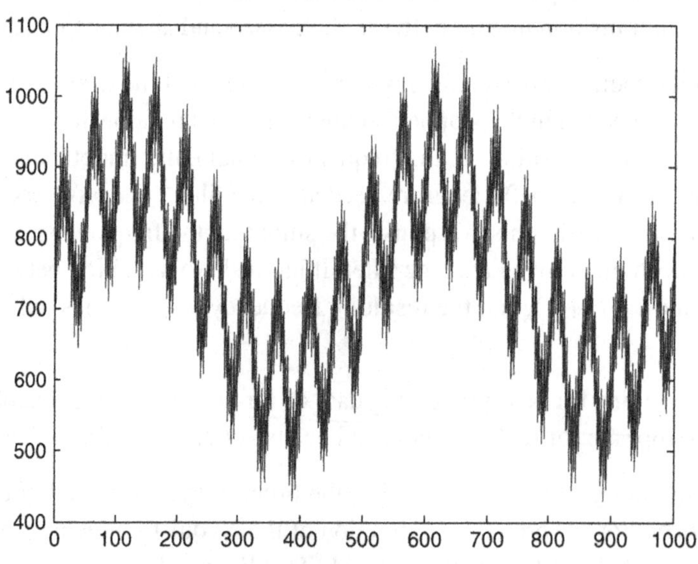

Figure 22.12
The original 1000-sample signal `x`.

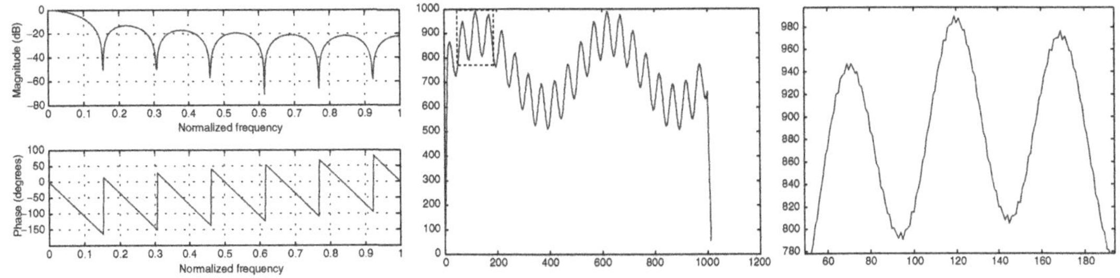

Figure 22.13

Moving average filter: `maf=ones(13,1)/13; freqz(maf); plot(conv(maf,x))`. **Left**: The frequency response of the 12th-order (13-sample) MAF. **Middle**: The result of the MAF applied to (convolved with) the original signal. Since the original signal has 1000 samples, and the MAF has 13 samples, the filtered signal has 1012 samples. (In general, if two signals of length j and k are convolved with each other, the result will have length $j + k - 1$.) This is equivalent to first "padding" the 1000 samples with 12 samples equal to zero on either end (sample numbers -11 to 0, and 1001 to 1012), then applying the 13-sample filter 1012 times, over samples -11 to 1, then -10 to 2, etc., up to samples 1000-1012. This zero-padding explains why the signal drops to close to zero at either end. **Right**: Zoomed in on the smoothed signal.

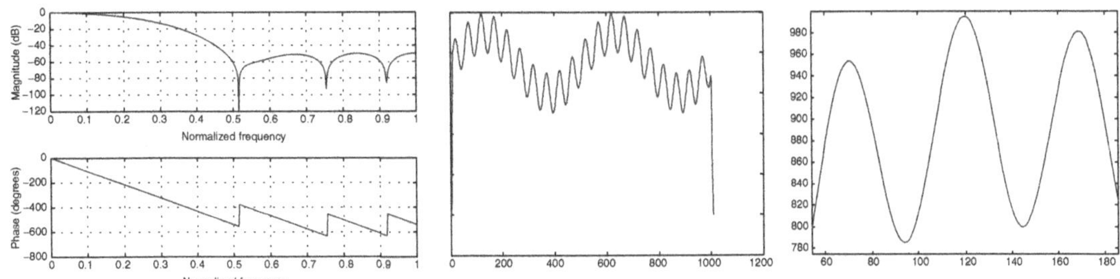

Figure 22.14

`lpf=fir1(12,0.2); freqz(lpf); plot(conv(lpf,x))`. **Left**: The frequency response of a 12th-order LPF with cutoff at $0.2f_N$. **Middle**: The signal smoothed by the LPF. **Right**: A zoomed-in view, showing less high-frequency content than the 12th-order MAF.

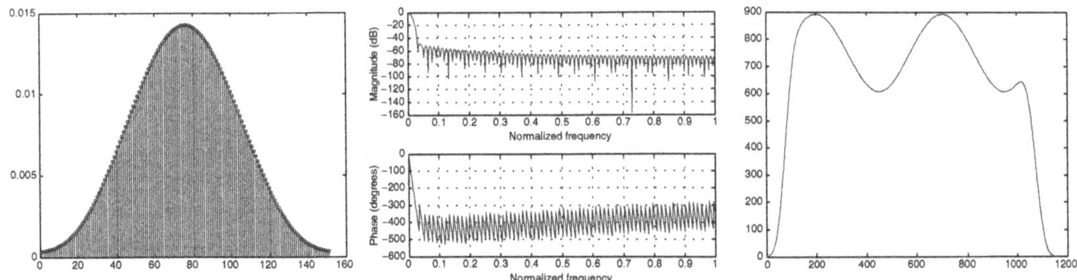

Figure 22.15

`lpf=fir1(150,0.01)`. **Left**: `stem(lpf)`. The coefficients of a 150th-order FIR LPF with a cutoff at $0.01f_N$. **Middle**: `freqz(lpf)`. The frequency response. **Right**: `plot(conv(lpf,x))`. The smoothed signal, where only the $0.004f_N$ and DC components get through.

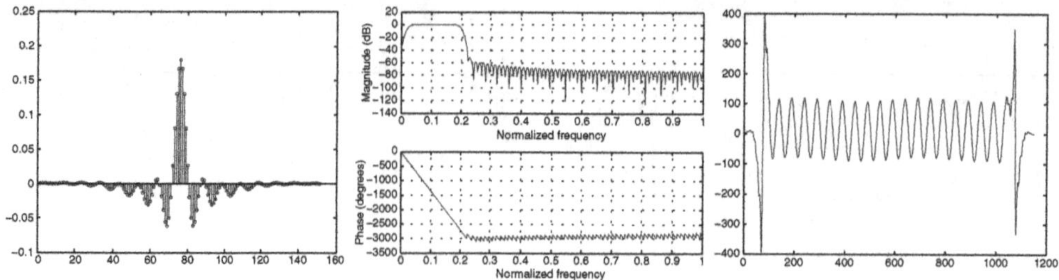

Figure 22.16

`bpf=fir1(150,[0.02,0.2])`. **Left**: `stem(bpf)`. The coefficients of a 150th-order bandpass filter with a passband from $0.02f_N$ to $0.2f_N$. **Middle**: `freqz(bpf)`. The frequency response. **Right**: `plot(conv(bpf,x))`. The signal consisting mostly of the $0.04f_N$ component, with small DC and $0.004f_N$ components.

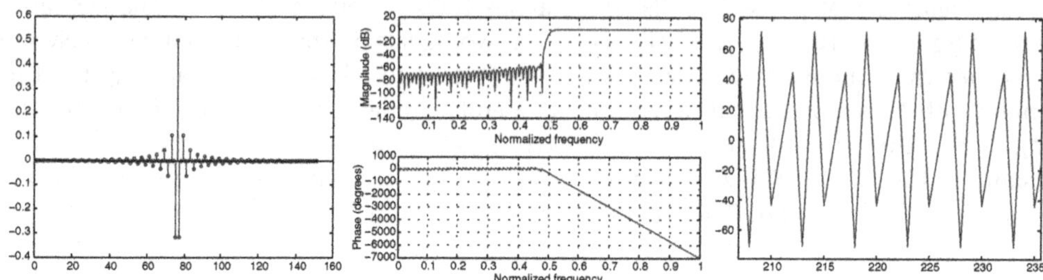

Figure 22.17

`hpf=fir1(150,0.5,'high')`. **Left**: `stem(hpf)`. The coefficients of a 150th-order high-pass filter with frequencies below $0.5f_N$ cut off. **Middle**: `freqz(hpf)`. The frequency response. **Right**: `plot(conv(hpf,x))`. Zoomed in on the high-passed signal.

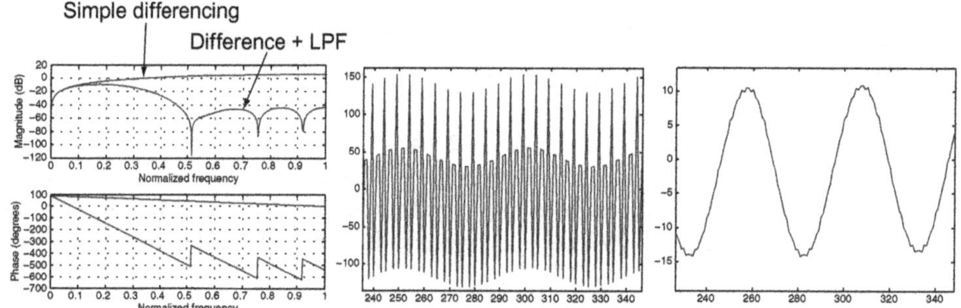

Figure 22.18

A simple differencing (or "velocity") filter has coefficients $b[0] = 1, b[1] = -1$, or written in MATLAB, `b = [1 -1]`. (Note the order: the coefficient that goes with the most recent input is on the left.) A differencing filter responds more strongly to signals with larger slopes (i.e., higher frequency signals) and has zero response to constant (DC) signals. Usually the signal "velocities" we are interested in, though, are those at low frequency; higher-frequency signals tend to come from sensing noise. Thus a better filter is probably a differencing filter convolved with a low-pass filter. **Left**: `b1 = [1 -1]; b2 = conv(b1,fir1(12,0.2)); freqz(b1); hold on; freqz(b2)`. This plot shows the frequency response of the differencing filter, as well as a differencing filter convolved with a 12th-order FIR LPF with cutoff at $0.2f_N$. At low frequencies, where the signals we are interested in live, the two filters have the same response. At high frequencies, the simple differencing filter has a large (unwanted) response, while the other filter attenuates this noise. **Middle**: `plot(conv(b1,x))`. Zoomed in on the signal filtered by the simple difference filter. **Right**: `plot(conv(b2,x))`. Zoomed in on the signal filtered by the difference-plus-LPF.

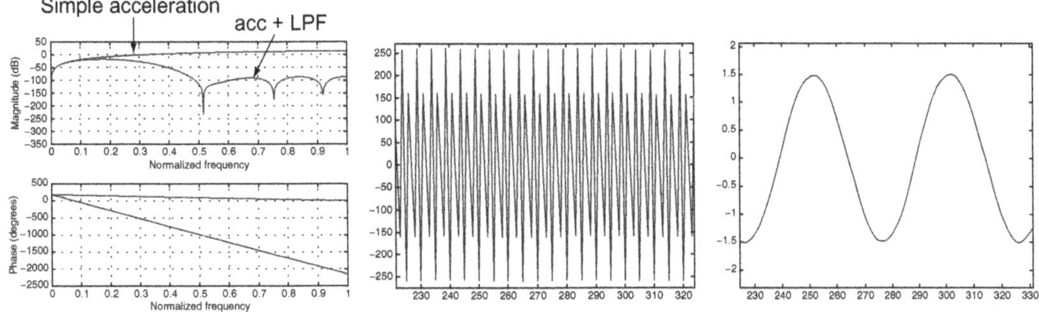

Figure 22.19

We can also make a double-differencing (or "acceleration") filter by taking the difference of two consecutive difference samples, i.e., convolving two differencing filters. This gives a simple filter with coefficients `[1 -2 1]`. This filter amplifies high frequency noise even more than a differencing filter. A better choice would be to use a filter that is the convolution of two difference-plus-low-pass filters from the previous example. **Left**: `bvel=[1 -1]; bacc=conv(bvel,bvel); bvellpf=conv(bvel,fir1(12,0.2)); bacclpf=conv(bvellpf,bvellpf); freqz(bacc); hold on; freqz(bacclpf)`. The low-frequency response of the two filters is identical, while the low-pass version attenuates high frequency noise. **Middle**: `plot(conv(bacc,x))`. Zoomed in on the second derivative of the signal, according to the simple acceleration filter. **Right**: `plot(conv(bacclpf,x))`. Zoomed in on the second derivative of the signal, according to the low-passed version of the acceleration filter.

22.4 Infinite Impulse Response (IIR) Digital Filters

The class of infinite impulse response (IIR) filters generalizes FIR filters to the following form:

$$\sum_{i=0}^{Q} a_i z(n-i) = \sum_{j=0}^{P} b_j x(n-j),$$

or, written in a more useful form for us,

$$z(n) = \frac{1}{a_0}\left(\sum_{j=0}^{P} b_j x(n-j) - \sum_{i=1}^{Q} a_i z(n-i)\right), \tag{22.7}$$

where the output $z(n)$ is a weighted sum of Q past outputs, P past inputs, and the current input. Some differences between FIR and IIR filters are highlighted below:

- IIR filters may be *unstable*, that is, their output may persist indefinitely and grow without bound even if the input is bounded in value and duration. Instability is impossible with FIR filters.
- IIR filters often use fewer coefficients to achieve the same magnitude response transition sharpness. Hence they can be more computationally efficient than FIR filters.

- IIR filters generally have a nonlinear phase response (phase does not change linearly with frequency, as with most FIR filters). This nonlinearity may or may not be acceptable, depending on the application. A linear phase response ensures that the time (not phase) delay associated with signals at all frequencies is the same. A nonlinear phase response, on the other hand, may cause different time delays at different frequencies, which may result in unacceptable distortion (e.g., in an audio application).

Because of roundoff errors in computation, an IIR filter that is theoretically stable may be unstable when implemented directly in the form of (22.7). Because of the possibility of instability, IIR filters with many coefficients are usually implemented as a cascade of filters with $P = 2$ and $Q = 2$. It is relatively easy to ensure that these low-order filters are stable, ensuring the stability of the cascade of filters.

Popular IIR digital filters include Chebyshev and Butterworth filters, which include low-pass, high-pass, bandpass, and bandstop versions. MATLAB offers design tools for these filters; you can refer to the documentation on `cheby1`, `cheby2`, and `butter`. Given a set of coefficients `b` and `a` defining the IIR filter, the MATLAB command `freqz(b,a)` plots the frequency response and `filter(b,a,signal)` returns the filtered version of `signal`.

One of the simplest IIR filters is the integrator

```
z(n) = z(n-1) + x(n)*Ts
```

where `Ts` is the sample time. The coefficients are `a = [1 -1]` and `b = [Ts]`. The behavior of the integrator on the sample signal in Figure 22.12 is shown in Figure 22.20.

22.5 FFT-Based Filters

Another option for filtering signals is to first FFT the signal, then set certain frequency components of the signal to zero, then invert the FFT. Assume we are working with the 256-sample square wave we looked at in Section 22.2.2, and we want to extract only the component at $0.1f_N$. First we build the signal and FFT it:

```
x = 0;
x(1:10) = 2;
x(11:20)= 0;
x = [x x x x x x x x x x x x 2*ones(1,10) zeros(1,6)];
N = length(x);
X = fft(x);
```

The element `X(1)` is the DC component, `X(2)` and `X(256)` correspond to frequency f_s/N, `X(3)` and `X(255)` correspond to frequency $2f_s/N$, `X(4)` and `X(254)` correspond to frequency $3f_s/N$,

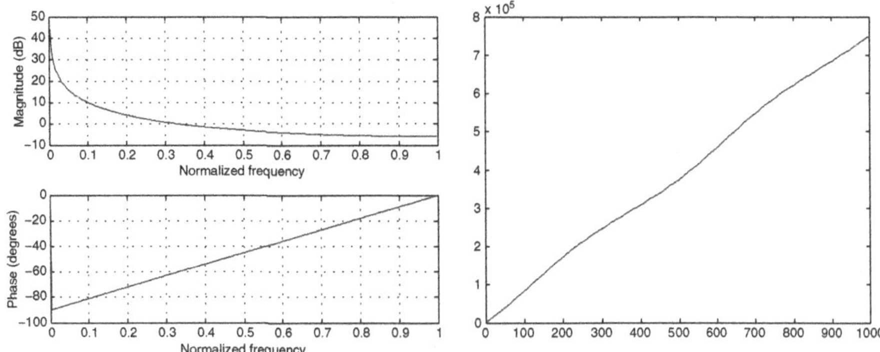

Figure 22.20

Left: `a=[1 -1]; b=[1]; freqz(b,a)`. Note that the frequency response of the integrator is infinite to DC signals (the integral of a nonzero constant signal goes to infinity) and low for high frequency signals. This is opposite of the differencing filter. **Right**: `plot(filter(b,a,x))`. The `filter` command applies the filter with coefficients `b` and `a` to `x`. This generalizes `conv` to IIR filters. (We cannot simply use `conv` for IIR filters, since the impulse response is not finite.) The upward slope of the integral is due to the nonzero DC term. We can also see the wiggle due to the 2 Hz term. It is basically impossible to see the 20 and 400 Hz terms in the signal.

etc., until `X(129)` corresponds to frequency $128f_s/N = f_s/2 = f_N$. So the frequencies we care about are near index 14 and its counterpart $258 - 14 = 244$. To cancel other frequencies, we can do

```
halfwidth = 3;
Xfiltered = zeros(1,256);
Xfiltered(14-halfwidth:14+halfwidth) = X(14-halfwidth:14+halfwidth);
Xfiltered(244-halfwidth:244+halfwidth) = X(244-halfwidth:244+halfwidth);
xrecovered = fft(conj(Xfiltered)/N);
plot(real(xrecovered));
```

The result is the (approximate) sinusoidal component of the square wave at $0.1f_N$, shown in Figure 22.21.

This is simple! (Of course the cost is in computing the FFT and inverse FFT.) FFT-based filter design tools allow you to specify an arbitrary frequency response (e.g., by drawing the magnitude response) and the size of the filter you are willing to accept, then use an inverse FFT to find filter coefficients that best match the desired response. The more coefficients you allow, the closer the approximation.

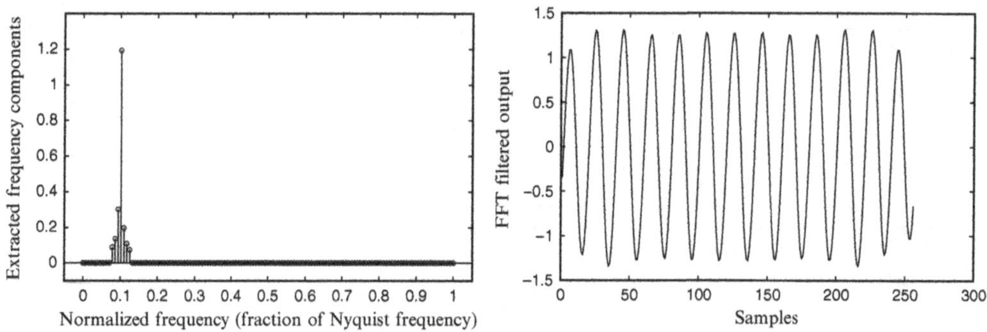

Figure 22.21

(Left) The extracted frequencies from the FFT magnitude plot in Figure 22.6. (Right) The FFT filter output, which is approximately a sinusoid at $0.1f_N$.

22.6 DSP on the PIC32

MIPS provides a DSP library for the MIPS M4K CPU on the PIC32, including FFT, FIR filtering, IIR filtering, and other DSP functions, described in Microchip's "32-Bit Language Tools Libraries" manual. For efficiency, the code is written in assembly language, optimizing the number of instructions. It also uses 16- and 32-bit fixed-point numbers to represent values, unlike MATLAB's double-precision floating point numbers. Fixed-point math is identical to integer math, and, as we have seen, integer math is significantly faster than floating point math. We will revisit fixed-point math soon.

This section presents an example demonstrating FIR filtering and the FFT on the PIC32, comparing the results to results you get in MATLAB. The MATLAB code generates a 1024-sample square wave as well as a 48-coefficient low-pass FIR filter with a cutoff frequency at $0.02f_N$. It sends these to the PIC32, which calculates the FIR-filtered signal, the FFT of the original signal, and the FFT of the filtered signal, and sends the results back to MATLAB for plotting. MATLAB also calculates the filtered signal and the FFTs of the original and filtered signals and compares the results to the PIC32's results. The results are indistinguishable (Figure 22.22). MATLAB also prints out the time it takes to do a 1024-sample FFT on the PIC32, about 13 ms.

The code for this example consists of the following files:

PIC32 code

- `dsp_fft_fir.c`. Communicates with the host and invokes the PIC32 DSP functions in `nudsp.c`.
- `nudsp.c`. Code that calls the MIPS functions.
- `nudsp.h`. Header file with prototypes for functions in `nudsp.c`.

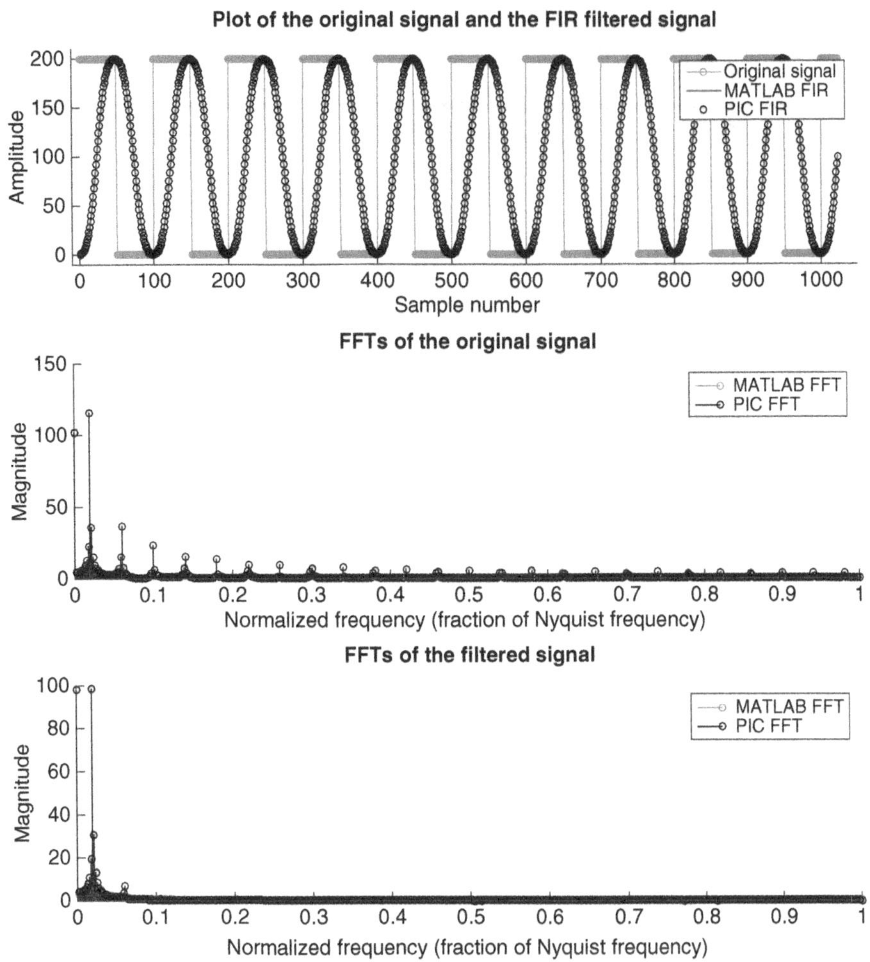

Figure 22.22

(Top) The 1024 samples of the original square wave signal, the MATLAB low-pass-filtered signal, and the PIC32 low-pass-filtered signal. The results are indistinguishable. (Middle) The MATLAB and PIC32 FFTs of the original signal. (Bottom) The MATLAB and PIC32 FFTs of their low-pass-filtered signals. Although it is difficult to see, the FFT results on the two platforms are indistinguishable.

Host computer code. We focus on the MATLAB interface, but we also provide code for Python.

- `sampleFFT.m` if you are using MATLAB, or
- `sampleFFT.py` if you are using Python. Python users need the freely available packages pyserial (for UART communication), matplotlib (for plotting), numpy (for matrix operations), and scipy (for DSP).

The MATLAB client code is given below. The Python code is similar. Load the PIC32 executable, then in MATLAB run `sampleFFT` to do the test.

Code Sample 22.2 `sampleFFT.m`. **MATLAB Client for FIR and FFT.**

```
% Compute the FFT of a signal and FIR filter the signal in both MATLAB and on the PIC32
% and compare the results
% open the serial port
port ='/dev/tty.usbserial-00001014A'; % modify for your own port

if ~isempty(instrfind) % closes the port if it was open
  fclose(instrfind);
  delete(instrfind);
end
fprintf('Opening serial port %s\n',port);
ser = serial(port, 'BaudRate', 230400, 'FlowControl','hardware');
fopen(ser);

% generate the input signal
xp(1:50) = 200;
xp(51:100)= 0;
x = [xp xp xp xp xp xp xp xp xp xp 200*ones(1,24)];

% now, create the FIR filter
Wn = 0.02; % let's see if we can just get the lowest frequency sinusoid

ord = 47; % ord+1 must be a multiple of 4
fir_coeff = fir1(ord,Wn);

N = length(x);
Y = fft(x);    % computer MATLAB's fft

xfil = filter(fir_coeff,1,x); % filter the signal
Yfil = fft(xfil);             % fft the filtered signal

% generate data for FFT plots for the original signal
mag = 2*abs(Y(1:N/2+1))/N;
mag(1) = mag(1)/2;
mag(N/2+1) = mag(N/2+1)/2;

% generate data for FFT plots for the filtered signal
magfil = 2*abs(Yfil(1:N/2+1))/N;
magfil(1) = magfil(1)/2;
magfil(N/2+1) = magfil(N/2+1)/2;

freqs = linspace(0,1,N/2+1);

% send the original signal to the pic32
fprintf(ser,'%d\n',N); % send the length
for i=1:N
  fprintf(ser,'%f\n',x(i)); % send each sample in the signal
end

% send the fir filter coefficients
fprintf(ser,'%d\n',length(fir_coeff));
for i=1:length(fir_coeff)
  fprintf(ser,'%f\n',fir_coeff(i));
end
```

```
% now we can read in the values sent from the PIC.
elapsedns = fscanf(ser,'%d');
disp(['The first 1024-sample FFT took ',num2str(elapsedns/1000.0),' microseconds.']);
Npic = fscanf(ser,'%d');
data = zeros(Npic,4); % the columns in data are
                      % original signal, fir filtered, orig fft, fir fft
for i=1:Npic
  data(i,:) = fscanf(ser,'%f %f %f %f');
end

xpic = data(:,1);            % original signal from the pic
xfirpic = data(:,2);         % fir filtered signal from pic
Xfftpic = data(1:N/2+1,3);   % fft signal from the pic
Xfftfir = data(1:N/2+1,4);   % fft of filtered signal from the pic

                             % used to plot the fft pic signals
Xfftpic = 2*abs(Xfftpic);
Xfftpic(1) = Xfftpic(1)/2;
Xfftpic(N/2+1) = Xfftpic(N/2+1)/2;

Xfftfir = 2*abs(Xfftfir);
Xfftfir(1) = Xfftfir(1)/2;
Xfftfir(N/2+1) = Xfftfir(N/2+1)/2;

% now we are ready to plot
subplot(3,1,1);
hold on;
title('Plot of the original signal and the FIR filtered signal')
xlabel('Sample number')
ylabel('Amplitude')
plot(x,'Marker','o');
plot(xfil,'Color','red','LineWidth',2);
plot(xfirpic,'o','Color','black');
hold off;
legend('Original Signal','MATLAB FIR', 'PIC FIR')
axis([-10,1050,-10,210])
set(gca,'FontSize',18);

subplot(3,1,2);
hold on;
title('FFTs of the original signal')
ylabel('Magnitude')
xlabel('Normalized frequency (fraction of Nyquist Frequency)')
stem(freqs,mag)
stem(freqs,Xfftpic,'Color','black')
legend('MATLAB FFT', 'PIC FFT')
hold off;
set(gca,'FontSize',18);

subplot(3,1,3);
hold on;
title('FFTs of the filtered signal')
ylabel('Magnitude')
xlabel('Normalized frequency (fraction of Nyquist Frequency)')
stem(freqs,magfil)
stem(freqs,Xfftfir,'Color','black')
legend('MATLAB FFT', 'PIC FFT')
hold off;
set(gca,'FontSize',18);

fclose(ser);
```

The PIC32 code `dsp_fft_fir.c` contains the `main` function. It reads the signal and filter information from MATLAB, invokes functions from our `nudsp.{h,c}` library to compute the filtered signal and the FFTs of the original and filtered signals, and sends the data back to the host for plotting.

Code Sample 22.3 `dsp_fft_fir.c`**. Communicates with the Client and Uses** `nudsp` **to Perform Signal Processing Operations.**

```
#include "NU32.h"
#include "nudsp.h"
// Receives a signal and FIR filter coefficients from the computer.
// filters the signal and ffts the signal and filtered signal, returning the results
// We omit error checking for clarity, but always include it in your own code.

#define SIGNAL_LENGTH 1024
#define FFT_SCALE 10.0
#define FIR_COEFF_SCALE 10.0
#define FIR_SIG_SCALE 10.0
#define NS_PER_TICK 25         // nanoseconds per core clock tick

#define MSG_LEN 128

int main(void) {
  char msg[MSG_LEN];                           // communication buffer
  double fft_orig[SIGNAL_LENGTH] = {};         // fft of the original signal
  double fft_fir[SIGNAL_LENGTH] = {};          // fft of the FIR filtered signal
  double xfir[SIGNAL_LENGTH] = {};             // the FIR filtered signal
  double sig[SIGNAL_LENGTH] = {};              // the signal
  double fir[MAX_ORD] = {};                    // the FIR filter coefficients
  int i = 0;
  int slen, clen;                              // signal and coefficient lengths
  int elapsedticks;                            // duration of FFT in core ticks

  NU32_Startup();

  while (1) {
    // read the signal from the UART.
    NU32_ReadUART3(msg, MSG_LEN);
    sscanf(msg,"%d",&slen);
    for(i = 0; i < slen; ++i) {
      NU32_ReadUART3(msg, MSG_LEN);
      sscanf(msg,"%f",&sig[i]);
    }

    // read the filter coefficients from the UART
    NU32_ReadUART3(msg,MSG_LEN);
    sscanf(msg,"%d", &clen);
    for(i = 0; i < clen; ++i) {
      NU32_ReadUART3(msg,MSG_LEN);
      sscanf(msg,"%f",&fir[i]);
    }

    // FIR filter the signal
    nudsp_fir_1024(xfir, sig, fir, clen, FIR_COEFF_SCALE, FIR_SIG_SCALE);

    // FFT the original signal; also time the FFT and send duration in ns
    _CP0_SET_COUNT(0);
```

```
    nudsp_fft_1024(fft_orig, sig, FFT_SCALE);
    elapsedticks = _CP0_GET_COUNT();
    sprintf(msg,"%d\r\n",elapsedticks*NS_PER_TICK);  // the time in ns
    NU32_WriteUART3(msg);
    // FFT the FIR signal
    nudsp_fft_1024(fft_fir, xfir, FFT_SCALE);

    // send the results to the computer
    sprintf(msg,"%d\r\n",SIGNAL_LENGTH);  // send the length
    NU32_WriteUART3(msg);
  for (i = 0; i < SIGNAL_LENGTH; ++i) {
    sprintf(msg,"%12.6f %12.6f %12.6f %12.6f\r\n",sig[i],xfir[i],fft_orig[i],fft_fir[i]);
    NU32_WriteUART3(msg);
   }
  }
  return 0;
}
```

Code Sample 22.4 `nudsp.h`**. Header File for FIR and FFT of Signals Represented as** `double` **Arrays.**

```
#ifndef NU__DSP__H__
#define NU__DSP__H__
// wraps some dsp library functions making it easier to use them with doubles
// all provided operations assume signal lengths of 1024 elements

#define MAX_ORD 128        // maximum order of the FIR filter

// compute a scaling factor for converting doubles into Q15 numbers
double nudsp_qform_scale(double * din, int len, double div);

// FFT a signal that has 1024 samples
void nudsp_fft_1024(double * dout, double * din, double div);

// FIR filter a signal that has 1024 samples
// arguments are dout (output), din (input), c (FIR coeffs), nc (number of coeffs),
//   div_c (coeffs scale factor for Q15), and div_sig (signal scale factor for Q15)
void
  nudsp_fir_1024(double *dout,double *din,double *c,int nc,double div_c,double div_sig);

#endif
```

The code `nudsp.c`, below, uses the MIPS `dsp` library to perform signal processing operations on arrays of type `double`. The `dsp` library, however, represents numbers in a fixed-point fractional format, either Q15 (a 16-bit format) or Q31 (a 32-bit format). The representation is the same as a two's complement integer, with the most significant bit corresponding to the sign, but the values of the bits are interpreted differently. For example, for Q15, bit 14 is the 2^{-1} column, bit 13 is the 2^{-2} column, etc., down to bit 0, the 2^{-15} column (see Table 22.1). This interpretation means that Q15 can represent fractional values from -1 to $1 - 2^{-15}$.

The advantage of a fixed-point format over a floating point format is that all the rules of integer math apply, allowing fast integer operations. The disadvantage is that it covers a smaller range of values than floating point numbers with the same number of bits (though with

Table 22.1: The fixed-point Q15 representation is equivalent to the 16-bit two's complement integer representation, and it uses the same mathematical operations

Binary	`int16` Interpretation	Q15 Interpretation
0000000000000000	0	0
0000000000000001	1	2^{-15}
0000000000000010	2	2^{-14}
0000000000000011	3	$2^{-14}+2^{-15}$
⋮		
0111111111111111	32,767	$1-2^{-15}$
1000000000000000	−32,768	−1
1000000000000001	−32,767	$-1+2^{-15}$
1000000000000010	−32,766	$-1+2^{-14}$
⋮		
1111111111111111	−1	-2^{-15}

The only difference is that consecutive numbers in Q15 are separated by 2^{-15}, covering the range −1 to $1-2^{-15}$, as opposed to the `int16` representation, which has consecutive numbers separated by 1, covering the range −32,768 to 32,767.

uniform resolution over the range, unlike floating point numbers). If a signal is represented as an array of `double`s, before using it in a fixed-point computation the signal should be scaled so that (1) the maximum range of the scaled signal is well less than the fixed-point format's range, to allow headroom for additions and subtractions without causing overflow; and (2) to make sure that there is sufficient resolution in the scaled signal's representation, avoiding quantization effects that significantly alter the shape of the signal.

The `dsp` library defines four data types to hold both real and complex fixed-point Q15 and Q31 numbers: `int16`, `int16c`, `int32`, and `int32c`, where the number indicates the number of bits in the representation and the `c` is added to indicate that the type is a `struct` with both real (`.re`) and imaginary (`.im`) values. Our code only uses Q15 numbers (`int16` and `int16c`), as they provide enough precision for our purposes and some `dsp` functions only accept Q15 arguments. As described above and in Table 22.1, Q15 numbers in the range −1 to $1-2^{-15}$ can be interpreted as `int16` numbers a factor 2^{15} larger. Therefore, in the rest of this section, we refer only to `int16` integers in the range −32,768 to 32,767.

The function `nudsp_qform_scale` computes an appropriate scaling factor for a signal, which is used to convert an array of `double`s into an array of `int16` integers. The scaling normalizes the signal by its largest magnitude value, mapping that value to `1/div`, where `div` is a scaling factor used to provide headroom to prevent overflow. (The sample code `dsp_fft_fir.c` chooses `div = 10.0` for signals used in the FFT and FIR filters, scaling them to the range $[-0.1, 0.1]$.) This scaling factor is then multiplied by `QFORMAT` $= 2^{15}$, which scales the signal up to use the range offered by the `int16` data type. To convert `double`s to `int16`s we multiply by the calculated scaling factor, and to convert back to `double`s we divide by the scaling factor.

The next function, `nudsp_fft_1024`, uses the `dsp` library function `mips_fft16` to perform an FFT on a signal represented as an array of 1024 `double`s. First, the signal must be converted into an array of complex `int16` numbers; we use `nudsp_qform_scale` to compute the scaling factor. Next we copy the *twiddle* factors, parameters used in the FFT algorithm, into RAM. The `dsp` library (through `fftc.h`) provides precomputed factors and places them in an array in flash called `fft16c1024`; we load them into RAM for greater speed. Finally, we call `mips_fft16` with a buffer for the result of the computation, the source signal, the twiddle factors, a scratch array, and the $\log_2$ of the signal length (the signal length must always be a power of 2, and here we assume it is 1024). The scratch array just provides extra memory for the `mips_fft16` function to perform temporary calculations. After computing the FFT, we convert the magnitudes back into `double`s, using the scaling factor computed earlier.

The final function, `nudsp_fir_1024`, applies an FIR filter to a signal represented as an array of 1024 `double`s. The first step prior to using the `dsp` library's FIR function is to scale the signal and the coefficients by `scale_c` and `scale_s`, respectively, converting them to `int16`s. After performing the scaling, we must call `mips_fir16_setup` to initialize a coefficient buffer that is twice as long as the actual number of filter coefficients. Finally, the call to `mips_fir16` performs the filtering operation. In addition to the input and output buffers and prepared coefficients, the filter also requires a buffer long enough to hold the last K samples, where K is the number of filter coefficients. The filter returns its result as `int16` numbers. To convert the result back to `double`s, we must divide by the product of the scaling factors of the coefficients and the signal, because these numbers are multiplied during the filter operation. Since both scaling factors `scale_c` and `scale_s` contain a factor `QFORMAT` to convert to the `int16` range, we eliminate one of these scaling factors by multiplying by `QFORMAT`.

Code Sample 22.5 `nudsp.c`. **Implements FIR and FFT Operations for Arrays of Type** `double`.

```
#include <math.h>       // C standard library math, for sqrt
#include <stdlib.h>     // for max
#include <string.h>     // for memcpy
#include <dsplib_dsp.h> // for int16, int16c data types and FIR and FFT functions
#include <fftc.h>       // for the FFT twiddle factors, stored in flash
#include "nudsp.h"

#define TWIDDLE fft16c1024 // FFT twiddle factors for int16 (Q15), 1024 signal length
#define LOG2N 10           // log base 2 of the length, assumed to be 1024
#define LEN (1 << LOG2N)   // the length of the buffer
#define QFORMAT (1 << 15)  // multiplication factor to map range (-1,1) to int16 range

// compute the scaling factor to convert an array of doubles to int16 (Q15)
// The scaling is performed so that the largest magnitude number in din
// is mapped to 1/div; thus the divisor gives extra headroom to avoid overflow
```

```
double nudsp_qform_scale(double * din, int len, double div) {
  int i;
  double maxm = 0.0;

  for (i = 0; i< len; ++i) {
    maxm = max(maxm, fabs(din[i]));
  }
  return (double)QFORMAT/(maxm * div);
}

// Performs an FFT on din (assuming it is 1024 long), returning its magnitude in dout
// dout - pointer to array where answer will be stored
// din - pointer to double array to be analyzed
// div - input scaling factor.  max magnitude input is mapped to 1/div

void nudsp_fft_1024(double *dout, double *din, double div)
{
  int i = 0;
  int16c twiddle[LEN/2];
  int16c dest[LEN], src[LEN];
  int16c scratch[LEN];
  double scale = nudsp_qform_scale(din,LEN,div);

  for (i=0; i< LEN; i++) {                      // convert to int16 (Q15)
    src[i].re = (int) (din[i] * scale);
    src[i].im = 0;
  }
  memcpy(twiddle, TWIDDLE, sizeof(twiddle));     // copy the twiddle factors to RAM
  mips_fft16(dest, src, twiddle, scratch, LOG2N); // perform FFT
  for (i = 0; i < LEN; i++) {                   // convert the results back to doubles
    double re = dest[i].re / scale;
    double im = dest[i].im / scale;
    dout[i] = sqrt(re*re + im*im);
  }
}

// Perform a finite impulse response filter of a signal that is 1024 samples long
// dout - pointer to result array
// din - pointer to input array
// c - pointer to coefficient array
// nc - the number of coefficients
// div_c - for scaling the coefficients
// The maximum magnitude coefficient is mapped to 1/div_c in int16 (Q15)
// div_sig - for scaling the input signal
// The maximum magnitude input is mapped to 1/div_sig in int16 (Q15)
void
  nudsp_fir_1024(double *dout,double *din,double *c,int nc,double div_c,double div_sig)
{
  int16 fir_coeffs[MAX_ORD], fir_coeffs2x[2*MAX_ORD];
  int16 delay[MAX_ORD] = {};
  int16 din16[LEN], dout16[LEN];
  int i=0;
  double scale_c = nudsp_qform_scale(c, nc, div_c);      // scale coeffs to Q15
  double scale_s = nudsp_qform_scale(din, LEN, div_sig); // scale signal to Q15
  double scale = 0.0;

  for (i = 0; i< nc; ++i) {                              // convert FIR coeffs to Q15
    fir_coeffs[i] = (int) (c[i]*scale_c);
  }
  for (i = 0; i<LEN; i++) {                              // convert input signal to Q15
    din16[i] = (int) (din[i]*scale_s);
```

```
    }
    mips_fir16_setup(fir_coeffs2x, fir_coeffs, nc);        // set up the filter
    mips_fir16(dout16, din16, fir_coeffs2x, delay, LEN, nc, 0);  // run the filter
    scale = (double)QFORMAT/(scale_c*scale_s);             // convert back to doubles
    for (i = 0; i<LEN; i++) {
      dout[i] = dout16[i]*scale;
    }
}
```

22.7 Exercises

1. Use MATLAB to find the coefficients of a 20th-order low-pass FIR filter with a cutoff frequency of $0.5f_N$. Then do the same for a 100th-order low-pass FIR filter. Plot the frequency response of each and discuss the relative merits of each in terms of both the magnitude response and the phase response. For a real-time filter, i.e., one that is performing the filtering on data as it comes in, what are the implications of the different phase responses of the two filters?
2. Experiment with MATLAB's `sound` function, which allows you to play a vector of numbers as a sound waveform through your computer's speakers. Create a one-second signal sampled at 8.192 kHz that is the sum of a 500 and a 2500 Hz sinusoid, each of amplitude 0.5, and play it through your speakers. In MATLAB, design a low-pass FIR filter to extract only the 500 Hz tone and a high-pass FIR filter to extract only the 2500 Hz tone. Plot the frequency response of each filter and verify by audio that the filtered sounds are correct.
3. The PIC32 `dsp` library implements FIR and IIR filters, but it performs a batch filter: all data must be collected, and then you filter it. Often filters must be calculated in real-time for real-time control. For example, noisy sensor data could be low-pass filtered, or position readings could be differenced to get velocity readings.
   Implement your own PIC32 real-time FIR filter library, `FIR.{c,h}`. This library provides variables (e.g., arrays or `structs`) to hold the filter coefficients and the most recently taken samples and calculates the output of the filter. The library should be easy to use and computationally efficient, though you are welcome to use `doubles`, not fixed-point math. How will you let the user "shift in" the next sensor reading while "shifting out" the oldest sensor reading? How will you handle initial conditions when no prior readings have been taken?
   Design a ninth-order low-pass FIR filter with a cutoff frequency at $0.2f_N$ and plot both the input and filtered output for a 1000-input signal consisting of two equal amplitude sinusoids, one at $0.1f_N$ and one at $0.5f_N$.
4. Augment the PIC32 `nudsp.{c,h}` library with an inverse Fast Fourier Transform capability. Test it on a sample signal by taking the FFT and then the inverse FFT and confirming that the original signal is recovered.

5. Use the PIC32 inverse FFT capability to implement an FFT-based filter in `nudsp.{c,h}`. The user specifies the frequency ranges to pass or stop, allowing the FFT-based filter to act as a low-pass, high-pass, bandpass, or bandstop filter. After taking the FFT, the code should zero out the appropriate components and take the inverse FFT, yielding the new filtered signal.

Further Reading

32-Bit language tools libraries. (2012). Microchip Technology Inc.

Oppenheim, A. V., & Schafer, R. W. (2009). *Discrete-time signal processing* (3rd ed.). Upper Saddle River, NJ: Prentice Hall.

Signal processing toolbox help. (2015). The MathWorks Inc.

CHAPTER 23

PID Feedback Control

The cruise control system on a car is an example of a feedback control system. The actual speed v of the car is measured by a sensor (e.g., the speedometer) to yield a sensed value s; the sensed value is compared to a reference speed r set by the driver; and the error $e = r - s$ is fed to a control algorithm that calculates a control for the motor that drives the throttle angle. This control therefore changes the car's speed v and the sensed speed s. The controller tries to drive the error $e = r - s$ to zero.

Figure 23.1 shows a block diagram for a typical feedback control system, like the cruise control system. Typically, a computer (the PIC32 in our case) calculates the error between the reference and the sensor value and implements the control algorithm. The controller produces a control signal that is input into the *plant*, the physical system that we want to control. Often we model the plant's dynamics with differential equations derived from Newton's laws. The plant produces an output that the sensor measures. A system that uses a sensor measurement to determine its control signal is a *closed-loop* control system (the sensor feedback causes the block diagram to form a closed loop).

In this chapter, for simplicity, we assume that the sensor is perfect. Therefore, the sensed output s and the plant's actual output are the same.

A common test of a controller's performance is to start with the reference r equal to zero, then suddenly switch r to 1 and keep it constant for all time. (Equivalently, for the cruise control case, r could start at 50 mph then change suddenly to 51 mph.) Such an input is called a *step input* and the resulting error $e(t)$ is called the *step error response*. Figure 23.2 shows a typical step error response and illustrates three key metrics by which controller performance is measured: overshoot, 2% settling time t_s, and steady-state error e_{ss}. The settling time t_s is the amount of time it takes for the error to settle to within 0.02 of its final value, and the steady-state error e_{ss} is the final error. A controller performs well if the system has a short settling time and zero (or small) overshoot, oscillation, and steady-state error.

Perhaps the most popular feedback control algorithm is the *proportional-integral-derivative* (PID) controller. Entire books are devoted to the analysis and design of PID controllers, but PID control can also be used effectively with a little intuition and experimentation. This chapter provides a brief introduction to help with that intuition.

Embedded Computing and Mechatronics with the PIC32 Microcontroller. http://dx.doi.org/10.1016/B978-0-12-420165-1.00023-8

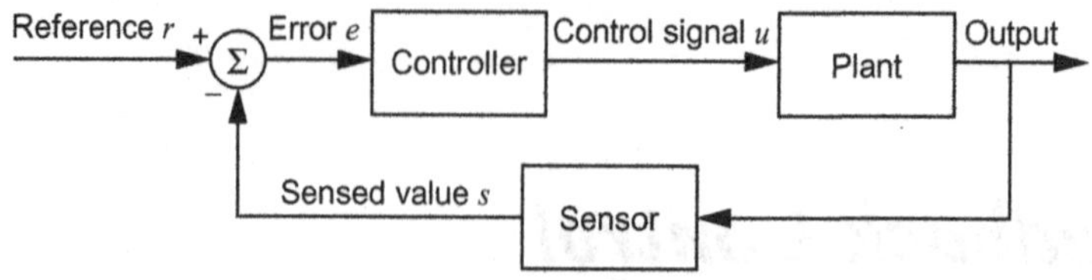

Figure 23.1
A block diagram of a typical control system.

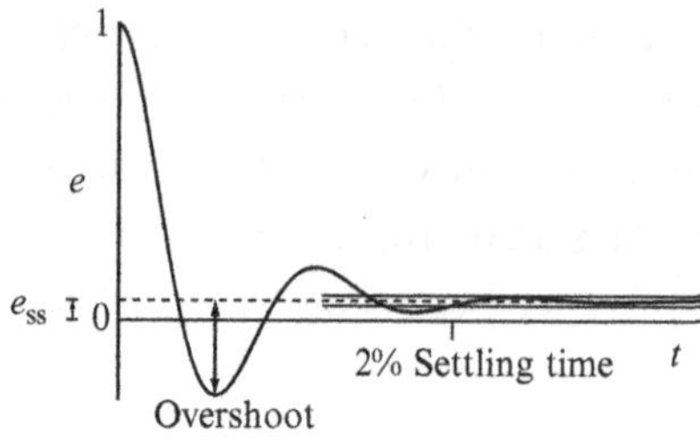

Figure 23.2
A typical error response to a step change in the reference, showing overshoot, 2% settling time t_s, and steady-state error e_{ss}. The error is one immediately after the step change in the reference and converges to e_{ss}.

23.1 The PID Controller

Assume the reference signal is $r(t)$, the sensed output is $s(t)$, the error is $e(t) = r(t) - s(t)$, and the control signal is $u(t)$. Then the PID controller can be written

$$u(t) = K_p e(t) + K_i \int_0^t e(z)\,\mathrm{d}z + K_d \dot{e}(t), \tag{23.1}$$

where K_p, K_i, and K_d are called the proportional, integral, and derivative gains, respectively. To make the discussion of the control law (23.1) concrete, let us assume r is the desired position of a mass moving on a line, s is the sensed position of the mass, and u is the linear force applied to the mass by a motor. Let us look at each of the proportional, derivative, and integral terms individually.

Proportional. The term $K_p e(t)$ creates a force proportional to the distance between the desired and measured position of the mass. This force is exactly what a mechanical spring does: it creates a force that pulls or pushes the mass proportional to a position displacement. Thus the proportional term $K_p e$ acts like a spring with a rest length of zero, with one end attached to the mass at s and the other end attached to the desired position r. The larger K_p, the stiffer the virtual spring.

Because we define the error as $e = r - s$, K_p should be positive. If K_p were negative, then in the case $s > r$ (the mass's position s is "ahead" of the reference r), the force

$K_p(r - s) > 0$ would try to push the mass even further ahead of the reference. Such a controller is called *unstable*, as the actual error tends toward infinity, not zero.

Derivative. The term $K_d\dot{e}(t)$ creates a force proportional to $\dot{e}(t) = \dot{r}(t) - \dot{s}(t)$, the difference between the desired velocity $\dot{r}(t)$ and the measured velocity $\dot{s}(t)$. This force is exactly what a mechanical damper does: it creates a force that tries to zero the relative velocity between its two ends. Thus the derivative term $K_d\dot{e}$ acts like a damper. An example of a spring and a damper working together is an automatic door closing mechanism: the spring pulls the door shut, but the damper acts against large velocities so the door does not slam. Derivative terms are used similarly in PID controllers, to damp overshoot and oscillation typical of mass-spring systems.

As with K_p, K_d should be nonnegative.

Integral. The term $K_i \int_0^t e(z)\,dz$ creates a force proportional to the time integral of the error. This term is less easily explained in terms of a mechanical analog, but we can still use a mechanical example. Assume the mass moves vertically in gravity with a gravitational force $-mg$ acting downward. If the goal is to hold the mass at a constant height r, then a controller using only proportional and derivative terms would bring the mass to rest with a nonzero error e satisfying $K_p e = mg$, the upward force needed to balance the gravitational force. (Note that the derivative term $K_d\dot{e}$ is zero when the mass is at rest.) By increasing the stiffness of K_p, the error e can be made small, but it can never be made zero—nonzero error is always needed for the motor to produce nonzero force.

Using an integral term allows the controller to produce a nonzero force even when the error is zero. Starting from rest with error $e = mg/K_p$, the time-integral of the error accumulates. As a result, the integral term $K_i \int_0^t e(z)\,dz$ grows, pushing the mass upward toward r, and the proportional term $K_p e$ shrinks, due to the shrinking error e. Eventually, the term $K_i \int_0^t e(z)\,dz$ equals mg, and the mass comes to rest at r ($e = 0$). Thus the integral term can drive the steady-state error to zero in systems where proportional and derivative terms alone cannot.

As with K_p and K_d, K_i should be nonnegative.

It is possible to implement a PID controller purely in electronics using op amps. However, nearly all modern PID controllers are implemented digitally on computers.[1] Every `dt` seconds, the computer reads the sensor value and calculates a new control signal. The error derivative $\dot{e}$ becomes an error difference, and the error integral $\int_0^t e(z)\,dz$ becomes an error sum. Pseudocode for a digital implementation of PID control is given below.

```
eprev = 0;                // initial "previous error" is zero
eint = 0;                 // initial error integral is zero
now = 0;                  // "now" tracks the elapsed time
```

[1] A digital PID controller is a type of digital filter, just like those discussed in Chapter 22.

```
every dt seconds do {
s = readSensor();                 // read sensor value
r = referenceValue(now);          // get reference signal for time "now"
e = r - s;                        // calculate the error
edot = e - eprev;                 // error difference
eint = eint + e;                  // error sum
u = Kp*e + Ki*eint + Kd*edot;     // calculate the control signal
sendControl(u);                   // send control signal to the plant
eprev = e;                        // current error is now prev error for next iteration
now = now + dt;                   // update the "now" time
}
```

A few notes about the algorithm:

- **The timestep `dt` and delays.** Generally, the shorter `dt` is, the better. If computing resources are limited, however, it is enough to know that the timestep `dt` should be significantly shorter than time constants associated with the dynamics of the plant. So if the plant is "slow," you can afford a longer `dt`, but if the system can go unstable quickly, a short `dt` is needed. The primary reason is that near the end of a control cycle, the control applied by the controller is in response to old sensor data. Control based on old measurements can cause the system to become unstable.
  For many robot control systems, `dt` is 1 ms.
- **Error difference and sum.** The pseudocode uses an error difference and an error sum. Instead, the error derivative can be approximated as `edot = (e - eprev)/dt` and the error integral could be approximated using `eint = eint + e*dt`. There is no need to do these extra divisions and multiplications, however, which simply scale the results. This scaling can be incorporated in the gains K_d and K_i.
- **Integer math vs. floating point math.** As we have seen, addition, subtraction, multiplication, and division of integer data types is much faster than with floating point types. If we wish to ensure that the control loop runs as quickly as possible, we should use integers where possible. Consider that raw sensor signals (e.g., encoder counts or ADC counts) and control signals (e.g., the period register of an output compare PWM signal) are typically integers anyway. If necessary, control gains can be scaled up or down to maintain good resolution while only using integer values during calculations, while also making sure that integer overflow does not occur. After calculations, the control signal can be scaled back to an appropriate range. The idea of scaling to allow integer math is exactly the same used for fixed-point math in DSP in Chapter 22.6.
  For many applications, since the PID controller involves only a few additions, subtractions, and multiplications, integer math is not necessary (especially on the PIC32, with its relatively fast clock speed).
- **Control saturation.** There are practical limits on the control signal `u`. The function `sendControl(u)` enforces these limits. If the control calculation yields `u=100`, for example, but the maximum control effort available is 50, the value sent by `sendControl(u)` is 50.

If large controller gains `Kp`, `Ki`, and `Kd` are used, the control signal may often be saturated at the limits.

- **Integrator anti-windup.** Imagine that the integrator error `eint` is allowed to build up to a large value. This windup creates a large control signal that tries to create error of the opposite sign, to try to dissipate the integrated error. To limit the oscillation caused by this effect, `eint` can be bounded. This *integrator anti-windup* protection can be implemented by adding the following lines to the code above:

```
eint = eint + e;              // error sum
if (eint > EINTMAX) {         // ADDED: integrator anti-windup
 eint = EINTMAX;
} else if (eint < -EINTMAX) { // ADDED: integrator anti-windup
 eint = -EINTMAX;
}
```

  Choosing `EINTMAX` is a bit of an art, but a good rule of thumb is that `Ki*EINTMAX` should be no more than the maximum control effort available from the actuator.

- **Sensor noise, quantization, and filtering.** The sensor data take discrete or *quantized* values. If this quantization is coarse, or if the time interval `dt` is short, the error `e` is unlikely to change much from one cycle to the next, and therefore `edot = e - eprev` is likely to take only a small set of different values. This effect means that `edot` is likely to be a jumpy, low-resolution signal. The sensor may also be noisy, adding to the jumpiness of the `edot` signal. Digital low-pass filtering, or averaging `edot` over several cycles, yields a smoother signal, at the expense of added delay from considering older `edot` values.

Although the PID control algorithm is quite simple, the challenge is finding control gains that yield good performance. Tuning these gains is the topic of Section 23.3.

23.2 Variants of the PID Controller

Common variants of the PID controller are P, PI, and PD controllers. These controllers are obtained by setting K_i and/or K_d equal to zero. Which variant to use depends on the performance specifications, sensor properties, and the dynamics of the plant, particularly its *order*. The order of a plant is the number of integrations from the control signal to the output. For example, consider the case where the control is a force to drive a mass, and the objective is to control the position of the mass. The force directly creates an acceleration, and the position is obtained from two integrations of the acceleration. Hence this is a second-order system. For such a system, derivative control is often helpful. If the system lacks much natural damping, derivative control can add it, slowing the output as it approaches the desired value. For zeroth-order or first-order systems, derivative control is generally not needed. Integral control should always be considered if it is important to eliminate steady-state error.

Table 23.1: Recommended PID variants based on the order of the plant

Control	Output of Plant	Order	Recommended Controller
Force	Position of mass	2	PD, PID
Force	Velocity of mass	1	P, PI
Current	Voltage across capacitor	1	P, PI
Current	Brightness of LED	0	P, PI

A rough guide to choosing the PID variant is given in Table 23.1. While PID control can be effective for plants of order higher than two, we will not consider such systems, which can have unintuitive behavior. An example is controlling the endpoint location of a flexible robot link by controlling the torque at the joint. The bending modes of the link introduce more state variables, increasing the system's order.

23.3 Empirical Gain Tuning

Although there is no substitute for analytic design techniques using a good model of the system, useful controllers can also be designed using empirical methods.

Empirical gain tuning is the art of experimenting with different control gains on the actual system and choosing gains that give a good step error response. Searching for good gains in the three-dimensional K_p-K_i-K_d space can be tricky, so it is best to be systematic and to obey a few rules of thumb:

- **Steady-state error.** If the steady-state error is too big, consider increasing K_p. If the steady-state error is still unacceptable, consider introducing a nonzero K_i. Be careful with K_i, though, as large K_i can destabilize the system.
- **Overshoot and oscillation.** If there is too much overshoot and oscillation, consider increasing the damping K_d, or the ratio K_d/K_p.
- **Settling time.** If the settling time is too long, consider simultaneously increasing K_p and K_d.

It is a good idea to first get the best possible performance with simple P control ($K_i = K_d = 0$). Then, starting from your best K_p, if you are using PD or PID control, experiment with K_d and K_p simultaneously. Experimenting with K_i should be saved until last, when you have your best P or PD controller, as nonzero K_i can lead to unintuitive behavior and instability.

Assuming you will not break your system by making it unstable, you should experiment with a wide range of control gains. Figure 23.3 shows an example exploration of a PD gain space for a plant with unknown dynamics. The control gains are varied over a few orders or magnitude. For larger control gains, the actuator effort $u(t)$, indicated by dotted lines, is often saturated. As expected, as K_p increases, oscillation and overshoot increases while the steady-state error decreases. As K_d increases, oscillation and overshoot is damped.

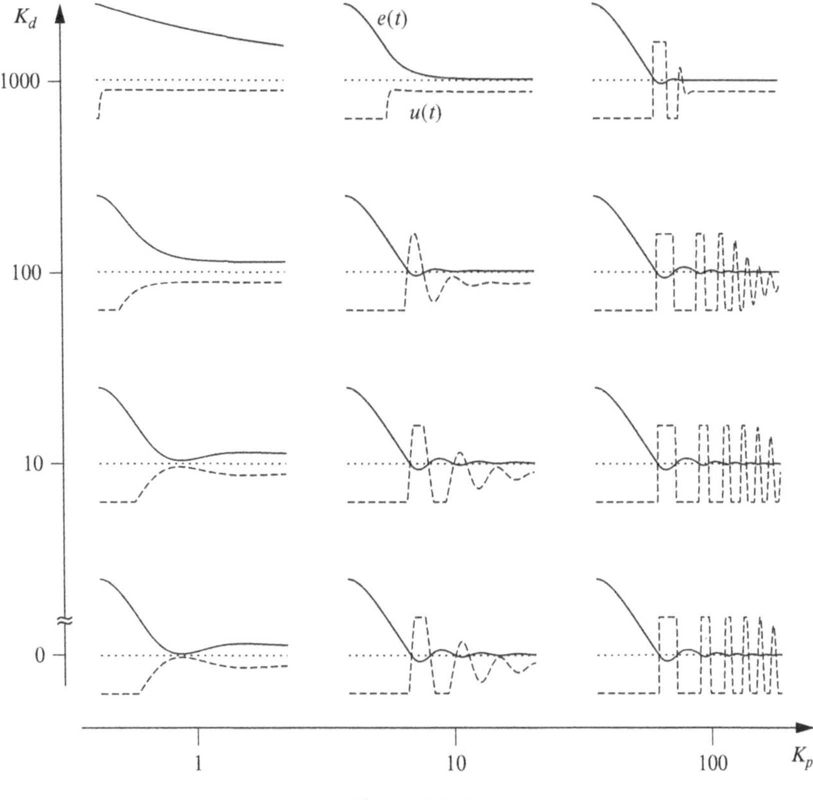

Figure 23.3

The step error response for a mystery system controlled by different PD controllers. The solid lines indicate the error response $e(t)$ and the dotted lines indicate the control effort $u(t)$. Which controller is "best?"

There are practical limits to how large controller gains can be. Large controller gains, combined with noisy sensor measurements or long cycle times `dt`, can lead to instability. They can also result in controls that chatter between the actuator limits.

23.4 Model-Based Control

With a feedback controller, like the PID controller, no control signal is generated unless there is error. If you have a reasonable model of the system's dynamics, why wait for error before applying a control? Using a model to anticipate the control effort needed is called *feedforward control*, because it depends only on the reference trajectory $r(t)$, not sensor feedback.
A model-based feedforward controller can be written as

$$u_{\text{ff}}(t) = f(r(t), \dot{r}(t), \ddot{r}(t), \ldots),$$

where $f(\cdot)$ is a model of the inverse plant dynamics that computes the control effort needed as a function of $r(t)$ and its derivatives (Figure 23.4).

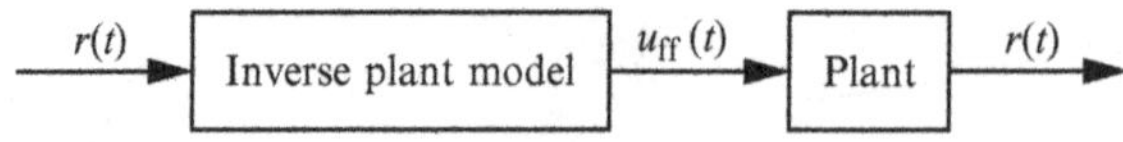

Figure 23.4
An ideal feedforward controller. If the inverse plant model is perfect, the output of the plant exactly tracks the reference $r(t)$.

Since feedforward control is not robust to inevitable model errors, it can be combined with PID feedback control to get the control law

$$u(t) = u_{\mathrm{ff}}(t) + K_p e(t) + K_i \int_0^t e(z)\,\mathrm{d}z + K_d \dot{e}(t). \tag{23.2}$$

In control of a robot arm, this control law is called *computed torque control*, where u is the set of joint torques and the model $f(\cdot)$ computes the joint torques needed given the desired joint angles, velocities, and accelerations.

A related control strategy is to use the reference trajectory $r(t)$ and the error $e(t)$ to calculate a desired change of state. For example, if the plant is a second-order system (the control u directly controls $\ddot{s}$), then the desired acceleration $\ddot{s}_d(t)$ of the plant output can be written as the sum of the planned acceleration $\ddot{r}(t)$ and a PID feedback term to correct for errors:

$$\ddot{s}_d(t) = \ddot{r}(t) + K_p e(t) + K_i \int_0^t e(z)\,\mathrm{d}z + K_d \dot{e}(t).$$

Then the actual control $u(t)$ is calculated using the inverse model,

$$u(t) = f(s(t), \dot{s}(t), \ddot{s}_d(t)). \tag{23.3}$$

If the inverse model is good, an advantage of the control law (23.3) over (23.2) is that the effect of the constant PID gains is the same at different states of the plant. This property can be important for systems like robot arms, where the inertia about a joint can change depending on the angle of outboard joints. For example, the inertia about your shoulder is large when your elbow is fully extended and smaller when your elbow is bent. A shoulder PID controller designed for a bent elbow may not work so well when the elbow is extended if the output of the PID controller is a joint torque. By treating the PID terms as accelerations instead of joint torques, and by passing these accelerations through the inverse model, the shoulder PID controller should have the same performance regardless of the configuration of the elbow.[2]

Feedforward plus feedback control laws like (23.2) and (23.3) provide the advantage of smaller errors with less control effort as compared to feedback control alone. The cost is in

[2] Provided the inverse model is good and the control effort does not saturate.

developing a good model of the plant dynamics and in increased computation time for the controller.

23.5 Chapter Summary

- Performance of a control system is often evaluated by the overshoot, 2% settling time, and steady-state error of the step error response.
- The PID control law is $u(t) = K_p e(t) + K_i \int_0^t e(z)\,dz + K_d \dot{e}(t)$.
- The proportional gain K_p acts like a virtual spring and the derivative gain K_d acts like a virtual damper. The integral gain K_i can be useful for eliminating steady-state error, but large values of K_i may cause the system to become unstable.
- Common variants of PID control are P, PI, and PD control.
- To reduce steady-state error, K_p and K_i can be increased. To reduce overshoot and oscillation, K_d can be increased. To reduce settling time, K_p and K_d can be increased simultaneously. Stability considerations place practical limits on controller gains.
- Feedback control requires error to produce a control signal. Model-based feedforward control can be used in conjunction with feedback control to anticipate the controls needed, thereby reducing errors.

23.6 Exercises

1. Provided with this chapter is a simple MATLAB model of a one-joint revolute robot arm moving in gravity. Perform empirical PID gain tuning by doing tests of the error response

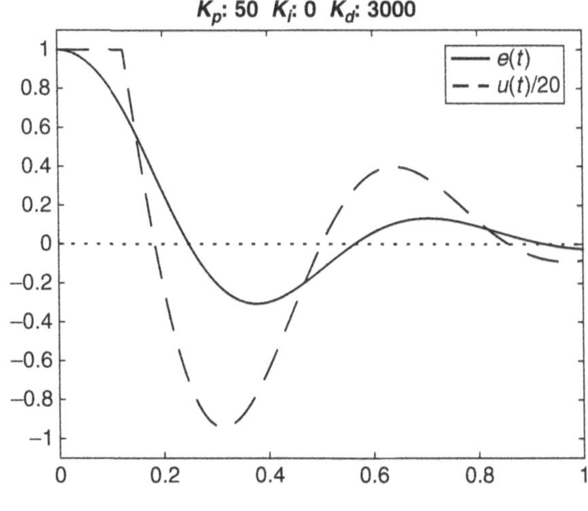

Figure 23.5
The step error and control response of the one-joint robot for $K_p = 50, K_i = 0, K_d = 3000$.

to a step input, where the step input asks the joint to move from $\theta = \dot{\theta} = 0$ (hanging down in gravity) to $\theta = 1$ radian. Tests can be performed using

```
pidtest(Kp, Ki, Kd)
```

Find good gains K_p, K_i, and K_d, and turn in a plot of the resulting step error response. An example output of `pidtest(50, 0, 3000)` is given in Figure 23.5.

Code Sample 23.1 `pidtest.m`. **Empirical Gain Tuning in MATLAB for a Simulated One-Joint Revolute Robot in Gravity.**

```
function pidtest(Kp, Ki, Kd)

INERTIA = 0.5;          % The plant is a link attached to a revolute joint
MASS = 1;               % hanging in GRAVITY, and the output is the angle of the joint.
CMDIST = 0.1;           % The link has INERTIA about the joint, MASS center at CMDIST
DAMPING = 0.1;          % from the joint, and there is frictional DAMPING.
GRAVITY = 9.81;
DT = 0.001;             % timestep of control law
NUMSAMPS = 1001;        % number of control law iterations
UMAX = 20;              % maximum joint torque by the motor

eprev = 0;
eint = 0;
r = 1;                  % reference is constant at one radian
vel = 0;                % velocity of the joint is initially zero
s(1) = 0.0; t(1) = 0; % initial joint angle and time
for i=1:NUMSAMPS
  e = r - s(i);
  edot = e - eprev;
  eint = eint + e;
  u(i) = Kp*e + Ki*eint + Kd*edot;
  if (u(i) > UMAX)
    u(i) = UMAX;
  elseif (u(i) < -UMAX)
    u(i) = -UMAX;
  end
  eprev = e;
  t(i+1) = t(i) + DT;

  % acceleration due to control torque and dynamics
  acc = (u(i) - MASS*GRAVITY*CMDIST*sin(s(i)) - DAMPING*vel)/INERTIA;

  % a simple numerical integration scheme
  s(i+1) = s(i) + vel*DT + 0.5*acc*DT*DT;
  vel = vel + acc*DT;
end

plot(t(1:NUMSAMPS),r-s(1:NUMSAMPS),'Color','black');
hold on;
plot(t(1:NUMSAMPS),u/20,'--','Color','black');
set(gca,'FontSize',18);
legend({'e(t)','u(t)/20'},'FontSize',18);
plot([t(1),t(length(t)-1)],[0,0],':','Color','black');
axis([0 1 -1.1 1.1])
title(['Kp: ',num2str(Kp),'  Ki: ',num2str(Ki),'  Kd: ',num2str(Kd)]);
hold off
```

Further Reading

Franklin, G. F., Powell, D. J., & Emami-Naeini, A. (2014). *Feedback control of dynamic systems* (7th ed.). Upper Saddle River, NJ: Prentice Hall.

Ogata, K. (2002). *Modern control engineering* (4th ed.). Upper Saddle River, NJ: Prentice Hall.

Phillips, C. L., & Troy Nagle, H. (1994). *Digital control system analysis and design* (3rd ed.). Upper Saddle River, NJ: Prentice Hall.

CHAPTER 24

Feedback Control of LED Brightness

In this chapter you will use feedback control to control the brightness of an LED. This project uses counter/timer, output compare, and analog input peripherals, as well as the parallel master port for the LCD screen.

Figure 24.1 shows an example result of this project. The LED's brightness is expressed in terms of the analog voltage of the brightness sensor, measured in ADC counts. The desired brightness alternates between 800 ADC counts (bright) and 200 ADC counts (dim) every half second, shown as a square wave reference in Figure 24.1. A successful feedback controller results in an actual brightness that closely follows the reference brightness.

Figure 24.2 shows the LED and sensor circuits and their connection to the OC1 output and the AN0 analog input. A PWM waveform from OC1 turns the LED on and off at 20 kHz, too fast for the eye to see, yielding an apparent averaged brightness between off and full on. The phototransistor is activated by the LED's light, creating an emitter current proportional to the

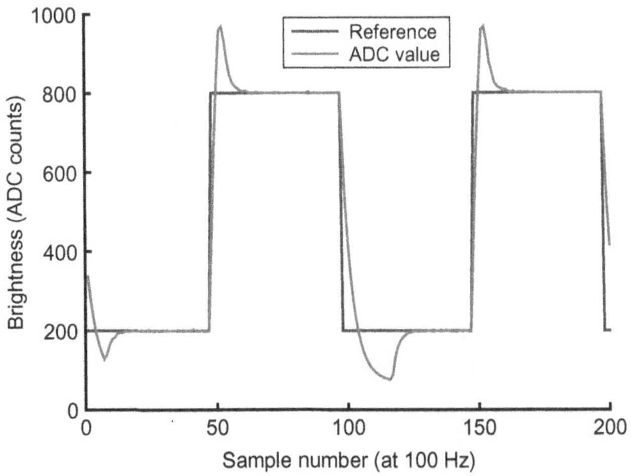

Figure 24.1

A demonstration of feedback control of LED brightness. The square wave is the desired LED brightness, in sensor ADC counts, and the other curve is the actual brightness as measured by the sensor. Samples are taken at 100 Hz, and the duration of the plotted data is 2 s.

Embedded Computing and Mechatronics with the PIC32 Microcontroller. http://dx.doi.org/10.1016/B978-0-12-420165-1.00024-X

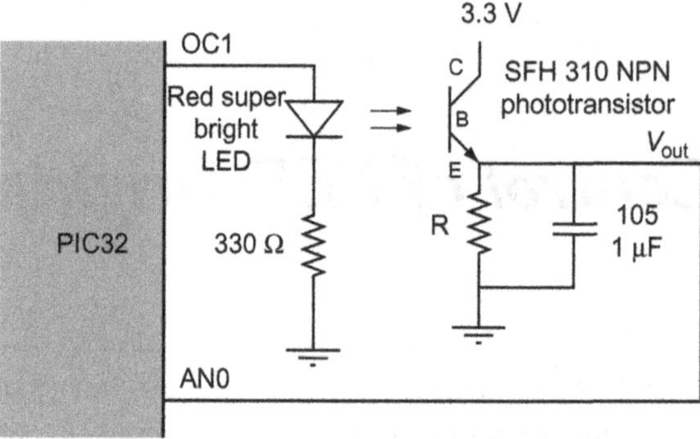

Figure 24.2
The LED control circuit. The long leg of the LED (anode) is connected to OC1 and the short leg (cathode) is connected to the 330 Ω resistor. The short leg of the phototransistor (collector) is attached to 3.3 V and the long leg (emitter) is attached to AN0, the resistor R, and the 1 µF capacitor.

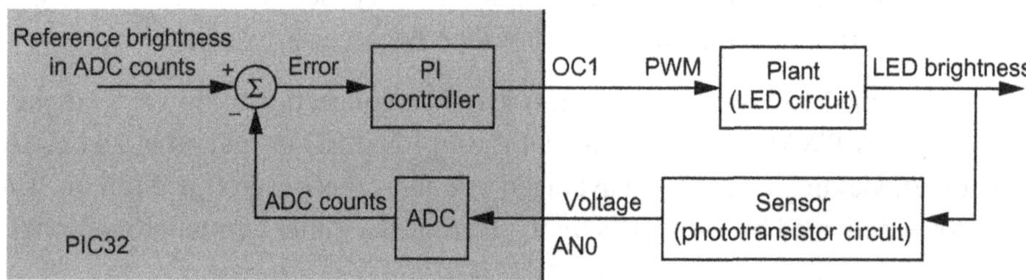

Figure 24.3
A block diagram of the LED brightness control system.

incident light. The resistor R turns this current into a sensed voltage. The 1 µF capacitor in parallel with R creates a low-pass filter with a time constant $\tau = RC$, removing high-frequency components due to the rapidly switching PWM signal and instead giving a time-averaged voltage. This filtering is similar to the low-pass filtering of your visual perception, which does not allow you to see the LED turning on and off rapidly.

A block diagram of the control system is shown in Figure 24.3. The PIC32 reads the analog voltage from the phototransistor circuit, calculates the error as the desired brightness (in ADC counts) minus the measured voltage in ADC counts, and uses a proportional-integral (PI) controller to generate a new PWM duty cycle on OC1. This control signal, in turn, changes the average brightness of the LED, which is sensed by the phototransistor circuit.

Your PIC32 program will generate a reference waveform, the desired light brightness measured in ADC counts, as a function of time. Then a 1 kHz control loop will read the sensor voltage (in ADC counts) and update the duty cycle of the 20 kHz OC1 PWM signal, attempting to make the measured ADC counts track the reference waveform.

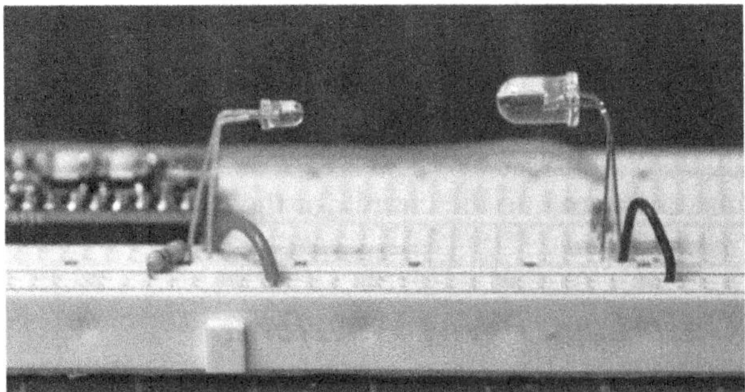

Figure 24.4
The phototransistor (left) and LED (right) pointed toward each other.

This project requires the coordination of many peripherals to work properly. Therefore, it is useful to divide it into smaller pieces and verify that each piece works, rather than attempting the whole project all at once.

24.1 Wiring and Testing the Circuit

1. **LED diode drop.** Connect the LED anode to 3.3 V, the cathode to a 330 Ω resistor, and the other end of the resistor to ground. This is the LED at its maximum brightness. Use your multimeter to record the forward bias voltage drop across the LED. Calculate or measure the current through the LED. Is this current safe for the PIC32 to provide?
2. **Choose** R. Wire the circuit as shown in Figure 24.2, except for the connection from the LED to OC1. The LED and phototransistor should be pointing toward each other, with approximately one inch separation, as shown in Figure 24.4. Now choose R to be as small as possible while ensuring that the voltage V_{out} at the phototransistor emitter is close to 3 V when the LED anode is connected to 3.3 V (maximum LED brightness) and close to 0 V when the LED anode is disconnected (the LED is off). (Something in the 10 kΩ range may work, but use a smaller resistance if you can still get the same voltage swing.) Record your value of R. Now connect the anode of the LED to OC1 for the rest of the project.

24.2 Powering the LED with OC1

1. **PWM calculation.** You will use Timer3 as the timer base for OC1. You want a 20 kHz PWM on OC1. Timer3 takes the PBCLK as input and uses a prescaler of 1. What should PR3 be?
2. **PWM program.** Write a program that uses your previous result to create a 20 kHz PWM output on OC1 (with no fault pin) using Timer3. Set the duty cycle to 75%. Get the following screenshots from your oscilloscope:

a. The OC1 waveform. Verify that this waveform matches your expectations.
b. The sensor voltage V_{out}.
c. Now remove the 1 μF capacitor and get another screenshot of V_{out}. Explain the difference from the previous waveform.

Insert the 1 μF capacitor back into the circuit for the rest of the project.

24.3 Playing an Open-Loop PWM Waveform

Now you will modify your program to generate a waveform stored in an `int` array. This array will eventually be the reference brightness waveform for your feedback controller (the square wave in Figure 24.1), but not yet; here this array will represent a PWM duty cycle as a function of time. Modify your program to define a constant `NUMSAMPS` and the global `volatile int` array `Waveform` by putting the following code near the top of the C file (outside of `main`):

```
#define NUMSAMPS  1000                        // number of points in waveform
static volatile int Waveform[NUMSAMPS];    // waveform
```

Then create a function `makeWaveform()` to store a square wave in `Waveform[]` and call it near the beginning of `main`. The square wave has amplitude `A` centered about the value `center`. Initialize `center` as (PR3+1)/2 and `A` as for the PR3 you calculated in the previous section.

```
void makeWaveform() {
  int i = 0, center = ???, A = ???; // square wave, fill in center value and amplitude
  for (i = 0; i < NUMSAMPS; ++i) {
    if ( i < NUMSAMPS/2) {
      Waveform[i] = center + A;
    } else {
      Waveform[i] = center - A;
    }
  }
}
```

Now configure Timer2 to call an ISR at a frequency of 1 kHz. This ISR will eventually implement the controller that reads the ADC and calculates the new duty cycle of the PWM. For now we will use it to modify the duty cycle according to the waveform in `Waveform[]`. Call the ISR `Controller` and make it interrupt priority level 5. It will use a `static` local `int` that counts the number of ISR entries and resets after 1000 entries.[1] In other words, the ISR should be of the form

[1] Recall that a `static` local variable is only initialized once, not upon every function call, and the value of the variable is retained between function calls. For global variables, the `static` qualifier means that the variable cannot be used in other modules (i.e., other `.c` files).

```
void __ISR(_TIMER_2_VECTOR, IPL5SOFT) Controller(void) { // _TIMER_2_VECTOR = 8
  static int counter = 0;          // initialize counter once

  // insert line(s) to set OC1RS

  counter++;                        // add one to counter every time ISR is entered
  if (counter == NUMSAMPS) {
    counter = 0;                    // roll the counter over when needed
  }
  // insert line to clear interrupt flag
}
```

In addition to clearing the interrupt flag (which we did not show in our example), your `Controller` ISR should set OC1RS to be equal to `Waveform[counter]`. Since your ISR is called every 1 ms, and the period of the square wave in `Waveform[]` is 1000 cycles, your PWM duty cycle will undergo one square wave period every 1 s. You should see your LED become bright and dim once per second.

1. Get a screenshot of your oscilloscope trace of V_{out} showing 2-4 periods of what should be an approximately square-wave sensor reading.
2. Turn in your code.

24.4 Establishing Communication with MATLAB

By establishing communication between your PIC32 and MATLAB, the PIC32 gains access to MATLAB's extensive scientific computing and graphics capabilities, and MATLAB can use the PIC32 as a data acquisition and control device. Refer to Section 11.3.5 for details about how to open a serial port in MATLAB and use it to communicate with `talkingPIC.c`, the basic communication program from Chapter 1.

1. Make sure you can communicate between `talkingPIC` on the PIC32 and `talkingPIC.m` in MATLAB. Do not proceed further until you have verified correct communication.

24.5 Plotting Data in MATLAB

Now that you have MATLAB communication working, you will build on your code from Section 24.3 by sending your controller's reference and sensed ADC data to MATLAB for plotting. This information will help you see how well your controller is working, allowing you to tune the PI gains empirically.

First, add some constants and global variables at the top of your program. The PIC32 program will send to MATLAB `PLOTPTS` data points upon request from MATLAB, where the constant `PLOTPTS` is set to 200. It is unnecessary to record data from every control iteration, so the program will record the data once every `DECIMATION` times, where `DECIMATION` is 10. Since the control loop is running at 1000 Hz, data is collected at 1000 Hz/`DECIMATION` = 100 Hz.

We also define the global `int` arrays `ADCarray` and `REFarray` to hold the values of the sensor signal and the reference signal. The `int StoringData` is a flag that indicates whether data is currently being collected. When it transitions from TRUE (1) to FALSE (0), it indicates that `PLOTPTS` data points have been collected and it is time to send `ADCarray` and `REFarray` to MATLAB. Finally, `Kp` and `Ki` are global `floats` with the PI gains. All of the variables have the specifier `volatile` because they are shared between the ISR and `main` and `static` because they are not needed in other `.c` files (good practice, even though this project only uses one `.c` file).

So you should have the following constants and variables near the beginning of your program:

```
#define NUMSAMPS  1000       // number of points in waveform
#define PLOTPTS   200        // number of data points to plot
#define DECIMATION  10       // plot every 10th point

static volatile int Waveform[NUMSAMPS];     // waveform
static volatile int ADCarray[PLOTPTS];      // measured values to plot
static volatile int REFarray[PLOTPTS];      // reference values to plot
static volatile int StoringData = 0;        // if this flag = 1, currently storing
                                            // plot data
static volatile float Kp = 0, Ki = 0;       // control gains
```

You should also modify your `main` function to define these local variables near the beginning:

```
char message[100];                // message to and from MATLAB
float kptemp = 0, kitemp = 0;     // temporary local gains
int i = 0;                        // plot data loop counter
```

These local variables are used in the infinite loop in `main`, below. This loop is interrupted by the ISR at 1 kHz. The loop waits for a message from MATLAB, which contains the new PI gains requested by the user. When a message is received, the gains from MATLAB are stored into the local variables `kptemp` and `kitemp`. Then interrupts are disabled, these local values are copied into the global gains `Kp` and `Ki`, and interrupts are re-enabled. Interrupts are disabled while `Kp` and `Ki` are assigned to ensure that the ISR does not interrupt in the middle of these assignments, causing it to use the new value of `Kp` but the old value of `Ki`. In addition, since `sscanf` takes longer to execute than simple variable assignments, it is called outside of the period when interrupts are disabled. We want to keep the time that interrupts are disabled as brief as possible, to avoid interfering with the timing of the 1 kHz control loop.

Next, the flag `StoringData` is set to TRUE (1), to tell the ISR to begin storing data. The ISR sets `StoringData` to FALSE (0) when `PLOTPTS` data points have been collected, indicating that it is time to send the stored data to MATLAB for plotting. Your infinite loop in `main` should be the following:

```
while (1) {
  NU32_ReadUART3(message, sizeof(message));        // wait for a message from MATLAB
  sscanf(message, "%f %f" , &kptemp,  &kitemp);
  __builtin_disable_interrupts(); // keep ISR disabled as briefly as possible
  Kp = kptemp;                    // copy local variables to globals used by ISR
  Ki = kitemp;
  __builtin_enable_interrupts();  // only 2 simple C commands while ISRs disabled
  StoringData = 1;                // message to ISR to start storing data
  while (StoringData) {           // wait until ISR says data storing is done
    ; // do nothing
  }
  for (i=0; i<PLOTPTS; i++) {     // send plot data to MATLAB
                             // when first number sent = 1, MATLAB knows we're done
    sprintf(message, "%d %d %d\r\n", PLOTPTS-i, ADCarray[i], REFarray[i]);
    NU32_WriteUART3(message);
  }
}
```

Finally, you will need to write code in your ISR to record data when `StoringData` is TRUE. This code will use the new local `static int` variables `plotind`, `decctr`, and `adcval`. `plotind` is the index, 0 to `PLOTPTS-1`, of the next set of data to collect. `decctr` counts from 1 up to `DECIMATION` to implement the once-every-`DECIMATION` data storing. `adcval` is set to zero for now, until you start reading the ADC.

Your code should look like the following. You only need to insert lines to set OC1RS and to clear the interrupt flag.

```
void __ISR(_TIMER_2_VECTOR, IPL5SOFT) Controller(void) {
  static int counter = 0;          // initialize counter once
  static int plotind = 0;          // index for data arrays; counts up to PLOTPTS
  static int decctr = 0;           // counts to store data one every DECIMATION
  static int adcval = 0;           //

 // insert line(s) to set OC1RS

  if (StoringData) {
    decctr++;
    if (decctr == DECIMATION) {    // after DECIMATION control loops,
      decctr = 0;                  // reset decimation counter
      ADCarray[plotind] = adcval;  // store data in global arrays
      REFarray[plotind] = Waveform[counter];
      plotind++;                   // increment plot data index
    }
    if (plotind == PLOTPTS) {      // if max number of plot points plot is reached,
      plotind = 0;                 // reset the plot index
      StoringData = 0;             // tell main data is ready to be sent to MATLAB
    }
  }
  counter++;                       // add one to counter every time ISR is entered
  if (counter == NUMSAMPS) {
    counter = 0;  // rollover counter over when end of waveform reached
  }

  // insert line to clear interrupt flag
}
```

The MATLAB code to communicate with the PIC32 is below. Load your new PIC32 code and, in MATLAB, use a command like

```
data = pid_plot('COM3', 2.0, 1.0)
```

where you should replace `'COM3'` with the appropriate `COM` port name from your `Makefile`. The `2.0` is your K_p and the `1.0` is your K_i. Since your program does not do anything with the gains yet, it does not matter what gains you type. If all is working properly, MATLAB should plot two cycles of your square wave duty cycle waveform and zero for your measured ADC value (which you have not implemented yet).

Code Sample 24.1 `pid_plot.m`**. MATLAB Code to Plot Data from Your PIC32 LED Control Program.**

```
function data = pid_plot(port,Kp,Ki)
%   pid_plot plot the data from the pwm controller to the current figure
%
%   data = pid_plot(port,Kp,Ki)
%
%   Input Arguments:
%       port - the name of the com port.  This should be the same as what
%               you use in screen or putty in quotes ' '
%       Kp - proportional gain for controller
%       Ki - integral gain for controller
%   Output Arguments:
%       data - The collected data.  Each column is a time slice
%
%   Example:
%       data = pid_plot('/dev/ttyUSB0',1.0,1.0) (Linux)
%       data = pid_plot('/dev/tty.usbserial-00001014A',1.0,1.0) (Mac)
%       data = pid_plot('COM3',1.0,1.0) (PC)
%

%% Opening COM connection
if ~isempty(instrfind)
    fclose(instrfind);
    delete(instrfind);
end
fprintf('Opening port %s....\n',port);
mySerial = serial(port, 'BaudRate', 230400, 'FlowControl', 'hardware');
fopen(mySerial); % opens serial connection
clean = onCleanup(@()fclose(mySerial)); % closes serial port when function exits

%% Sending Data
% Printing to matlab Command window
fprintf('Setting Kp = %f, Ki = %f\n', Kp, Ki);

% Writing to serial port
fprintf(mySerial,'%f %f\n',[Kp,Ki]);

%% Reading data
fprintf('Waiting for samples ...\n');

sampnum = 1; % index for number of samples read
read_samples = 10; % When this value from PIC32 equals 1, it is done sending data
```

```
while read_samples > 1
    data_read = fscanf(mySerial,'%d %d %d'); % reading data from serial port

    % Extracting variables from data_read
    read_samples=data_read(1);
    ADCval(sampnum)=data_read(2);
    ref(sampnum)=data_read(3);

    sampnum=sampnum+1; % incrementing loop number
end
data = [ref;ADCval]; % setting data variable

%% Plotting data
clf;
hold on;
t = 1:sampnum-1;
plot(t,ref);
plot(t,ADCval);
legend('Reference', 'ADC Value')
title(['Kp: ',num2str(Kp),'  Ki: ',num2str(Ki)]);
ylabel('Brightness (ADC counts)');
xlabel('Sample Number (at 100 Hz)');
hold off;
end
```

1. Turn in a MATLAB plot showing `pid_plot.m` is communicating with your PIC32 code.

24.6 Writing to the LCD Screen

1. Write the function `printGainsToLCD()`, and its function prototype `void printGainsToLCD(void);`. This function writes the gains `Kp` and `Ki` on your LCD screen, one per row, like

   ```
   Kp: 12.30
   Ki:  1.00
   ```

   This function should be called by `main` just after `StoringData` is set to 1. Verify that it works before continuing to the next section.

24.7 Reading the ADC

1. Read the ADC value in your ISR, just before the `if (StoringData)` line of code. The value should be called `adcval`, so it will be stored in `ADCarray`. Turn in a MATLAB plot showing the measured `ADCarray` and the `REFArray`. You may wish to use manual sampling and automatic conversion to read the ADC.

24.8 PI Control

Now you will implement the PI controller. Change `makeWaveform` so that `center` is 500 and the amplitude `A` is 300, making a square wave swinging between `200` and `800`. This waveform is now the desired brightness of the LED, in ADC counts. Use the `adcval` read from the ADC and the reference waveform as inputs to the PI controller. Call `u` the output of the PI controller.

The output u may be positive or negative, but the PWM duty cycle can only be between 0 and PR3. If we treat the value u as a percentage, we can make it centered at 50% by adding 50 to the value, then saturate it at 0% and 100%, by the following code:

```
unew = u + 50.0;
if (unew > 100.0) {
  unew = 100.0;
} else if (unew < 0.0) {
  unew = 0.0;
}
```

Finally we must convert the control effort unew into a value in the range 0 to PR3 so that it can be stored in OC1RS:

```
OC1RS = (unsigned int) ((unew/100.0) * PR3);
```

We recommend that you define the integral of the control error, Eint, as a global static volatile int. Then reset Eint to zero in main every time a new Kp and Ki are received from MATLAB. This ensures that this new controller starts fresh, without a potentially large error integral from the previous controller.

1. Using your MATLAB interface, tune your gains K_p and K_i until you get good tracking of the square wave reference. Turn in a plot of the performance.

24.9 Additional Features

Some other features you can add:

1. In addition to plotting the reference waveform and the actual measured signal, plot the OC1RS value, so you can see the control effort.
2. Create a new reference waveform shape. For example, make the LED brightness follow a sinusoidal waveform. You can calculate this reference waveform on the PIC32. You should be able to choose which waveform to use by an input argument in your MATLAB interface. Perhaps even allow the user to specify parameters of the waveform (like center and A).
3. Change the PIC32 and MATLAB code so that MATLAB sends over the 1000 samples of an arbitrary reference trajectory. Then you can use MATLAB code to flexibly create a wide variety of reference trajectories.

24.10 Chapter Summary

- Control of the brightness of an LED can be achieved by a PWM signal at a frequency beyond that perceptible by the eye. The brightness can be sensed by a phototransistor and resistor circuit. A capacitor in parallel with the resistor low-pass filters the sensor signal with a cutoff frequency $f_c = 1/(2\pi RC)$, rejecting the high-frequency components due to the PWM frequency and its harmonics while keeping the low-frequency components (those perceptible by the eye).

- A reference brightness, as a function of time, can be stored in an array with N samples. By cyclically indexing through this array in an ISR invoked at a fixed frequency of f_{ISR}, the reference brightness waveform is periodic with frequency f_{ISR}/N.
- Since LED brightness control by a PWM signal is a zeroth-order system (the PWM voltage directly changes the LED current and therefore brightness, without any integrations), a good choice for a feedback controller is a PI controller.
- When accepting new gains `Kp` and `Ki` from the user, interrupts should be disabled to ensure that the ISR is not called in the middle of updating `Kp` and `Ki`. Interrupts should be disabled as briefly as possible, however, to avoid interfering with expected ISR timing. This can be achieved by keeping the relatively slow `sscanf` outside the period that interrupts are disabled. Interrupts are only disabled during the short period that the values read by `sscanf` are copied to `Kp` and `Ki`.

24.11 Exercises

Complete the LED brightness control project as outlined in the chapter.

CHAPTER 25

Brushed Permanent Magnet DC Motors

Most electric motors operate on the principle that current flowing through a magnetic field creates a force. Because of this relationship between current and force, electric motors can be used to convert electrical power to mechanical power. They can also be used to convert mechanical power to electrical power; as with, for example, generators in hydroelectric dams or regenerative braking in electric and hybrid cars.

In this chapter we study perhaps the simplest, cheapest, most common, and arguably most useful electrical motor: the brushed permanent magnet direct current (DC) motor. For brevity, we refer to these simply as DC motors. A DC motor has two input terminals, and a voltage applied across these terminals causes the motor shaft to spin. For a constant load or resistance at the motor shaft, the motor shaft achieves a speed proportional to the input voltage. Positive voltage causes spinning in one direction, and negative voltage causes spinning in the other.

Depending on the specifications, DC motors cost anywhere from tens of cents up to thousands of dollars. For most small-scale or hobby applications, appropriate DC motors typically cost a few dollars. DC motors are often outfitted with a sensing device, most commonly an encoder, to track the position and speed of the motor, and a gearhead to reduce the output speed and increase the output torque.

25.1 Motor Physics

DC motors exploit the *Lorentz force law*,

$$\mathbf{F} = \ell \mathbf{I} \times \mathbf{B}, \tag{25.1}$$

where $\mathbf{F}, \mathbf{I}$, and $\mathbf{B}$ are three-vectors, $\mathbf{B}$ describes the magnetic field created by permanent magnets, $\mathbf{I}$ is the current vector (including the magnitude and direction of the current flow through the conductor), ℓ is the length of the conductor in the magnetic field, and $\mathbf{F}$ is the force on the conductor. For the case of a current perpendicular to the magnetic field, the force is easily understood using the right-hand rule for cross-products: with your right hand, point your index finger along the current direction and your middle finger along the magnetic field flux lines. Your thumb will then point in the direction of the force (see Figure 25.1).

Embedded Computing and Mechatronics with the PIC32 Microcontroller. http://dx.doi.org/10.1016/B978-0-12-420165-1.00025-1

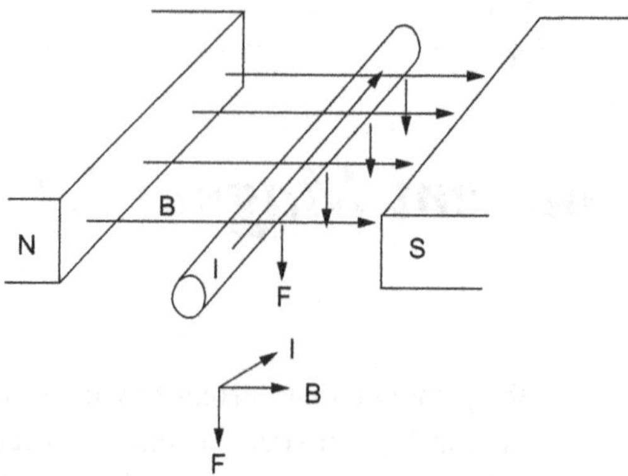

Figure 25.1
Two magnets create a magnetic field **B**, and a current **I** along the conductor causes a force **F** on the conductor.

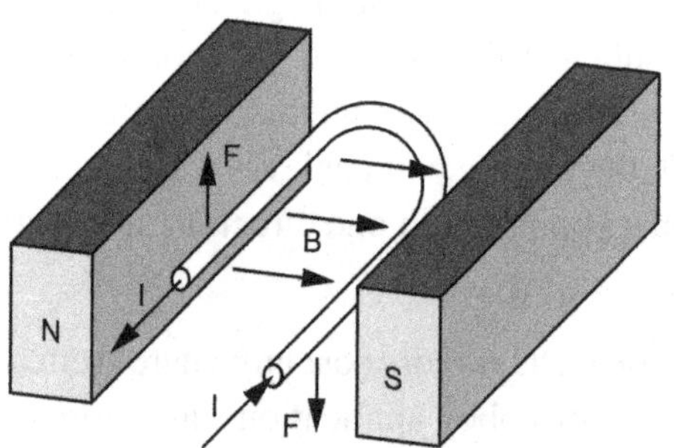

Figure 25.2
A current-carrying loop of wire in a magnetic field.

Now let us replace the conductor by a loop of wire, and constrain that loop to rotate about its center. See Figures 25.2 and 25.3. In one half of the loop, the current flows into the page, and in the other half of the loop the current flows out of the page. This creates forces of opposite directions on the loop. Referring to Figure 25.3, let the magnitude of the force acting on each half of the loop be f, and let d be the distance from the halves of the loop to the center of the loop. Then the total torque acting on the loop about its center can be written

$$\tau = 2df \cos\theta,$$

where θ is the angle of the loop. The torque changes as a function of θ. For $-90° < \theta < 90°$, the torque is positive, and it is maximum at $\theta = 0$. A plot of the torque on the loop as a function of θ is shown in Figure 25.4(a). The torque is zero at $\theta = -90°$ and $90°$, and of these two, $\theta = 90°$ is a stable equilibrium while $\theta = -90°$ is an unstable equilibrium. Therefore, if we send a constant current through the loop, it will likely come to rest at $\theta = 90°$.

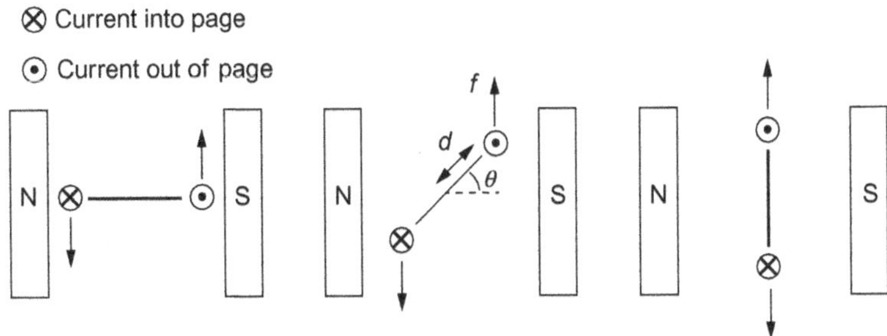

Figure 25.3
A loop of wire in a magnetic field, viewed end-on. Current flows into the page on one side of the loop and out of the page on the other, creating forces of opposite directions on the two halves of the loop. These opposite forces create torque on the loop about its center at most angles θ of the loop.

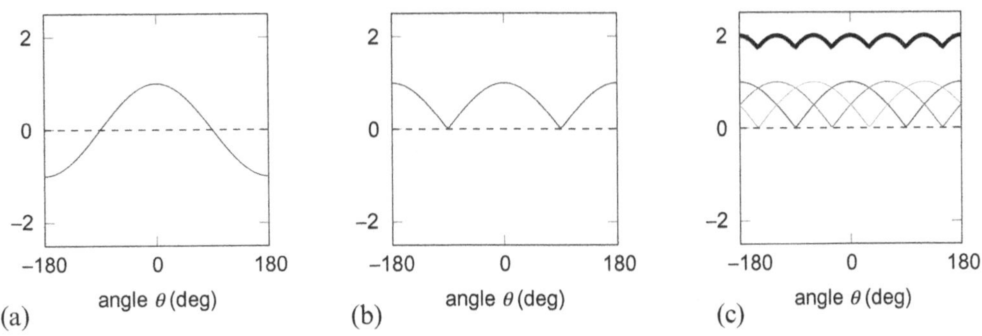

Figure 25.4
(a) The torque on the loop of Figure 25.3 as a function of its angle for a constant current. (b) If we reverse the current direction at the angles $\theta = -90°$ and $\theta = 90°$, we can make the torque nonnegative at all θ. (c) If we use several loops offset from each other, the sum of their torques (the thick curve) becomes more constant as a function of angle. The remaining variation contributes to torque ripple.

To make a more useful motor, we can reverse the direction of current at $\theta = -90°$ and $\theta = 90°$, which makes the torque nonnegative at all angles (Figure 25.4(b)). The torque is still zero at $\theta = -90°$ and $\theta = 90°$, however, and it undergoes a large variation as a function of θ. To make the torque more constant as a function of θ, we can introduce more loops of wire, each offset from the others in angle, and each reversing their current direction at appropriate angles. Figure 25.4(c) shows an example with three loops of wire offset from each other by 120°. Their component torques sum to give a more constant torque as a function of angle. The remaining variation in torque contributes to angle-dependent *torque ripple*.

Finally, to increase the torque generated, each loop of wire is replaced by a coil of wire (also called a winding) that loops back and forth through the magnetic field many times. If the coil consists of 100 loops, it generates 100 times the torque of the single loop for the same current. Wire used to create coils in motors, like magnet wire, is very thin, so there is resistance from one end of a coil to the other, typically from fractions of an ohm up to hundreds of ohms.

As stated previously, the current in the coils must switch direction at the appropriate angle to maintain non-negative torque. Figure 25.5 shows how brushed DC motors accomplish this current reversal. The two input terminals are connected to *brushes*, typically made of a soft conducting material like graphite, which are spring-loaded to press against the *commutator*, which is connected to the motor coils. As the motor rotates, the brushes slide over the commutator and switch between commutator *segments*, each of which is electrically connected to the end of one or more coils. This switching changes the direction of current through the coils. This process of switching the current through the coils as a function of the angle of the motor is called *commutation*. Figure 25.5 shows a schematic of a minimal motor design with three commutator segments and a coil between each pair of segments. Most high quality motors have more commutator segments and coils.

Unlike the simplified example in Figure 25.4, the brush-commutator geometry means that each coil in a real brushed motor is only energized at a subset of angles of the motor. Apart

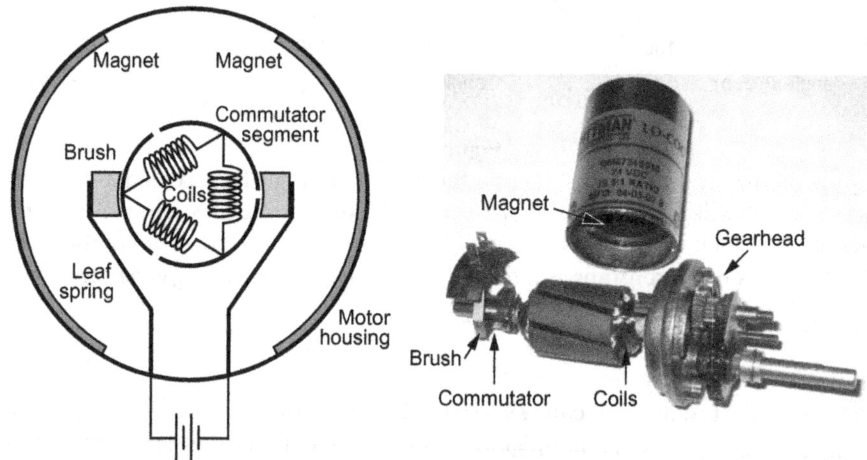

Figure 25.5
(Left) A schematic end-on view of a simple DC motor. The two brushes are held against the commutator by leaf springs which are electrically connected to the external motor terminals. This commutator has three segments and there are coils between each segment pair. The stator magnets are epoxied to the inside of the motor housing. (Right) This disassembled Pittman motor has seven commutator segments. The two brushes are attached to the motor housing, which has otherwise been removed. One of the two permanent magnets is visible inside the housing. The coils are wrapped around a ferromagnetic core to increase magnetic permeability. This motor has a gearhead on the output.

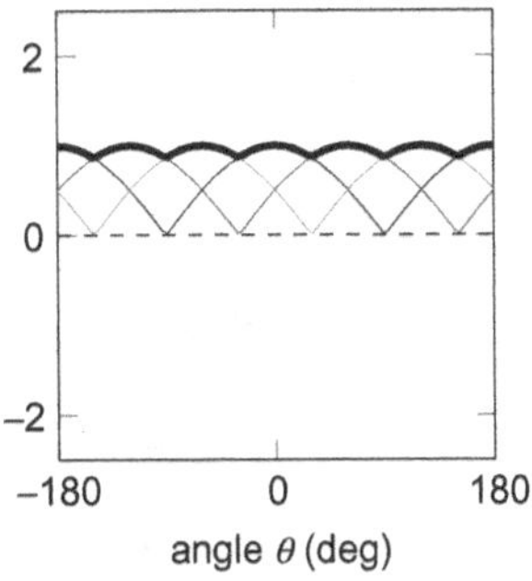

Figure 25.6

Figure 25.4(c) illustrates the sum of the torque of three coils offset by 120° if they are all energized at the same time. The geometry of the brushes and commutator ensure that not all coils are energized simultaneously, however. This figure shows the angle-dependent torque of a three-coil brushed motor that has only one coil energized at a time, which is approximately what happens if the brushes in Figure 25.5 are small. The energized coil is the one at the best angle to create a torque. The result is a motor torque as indicated by the thick curve; the thinner curves are the torques that would be provided by the other coils if they were energized. Comparing this figure to Figure 25.4(c) shows that this more realistic motor produces half the torque, but uses only one-third of the electrical power, since only one of the three coils is energized. Power is not wasted by putting current through coils that would generate little torque.

from being a consequence of the geometry, this has the added benefit of avoiding wasting power when current through a coil would provide little torque. Figure 25.6 is a more realistic version of Figure 25.4(c).

The stationary portion of the motor attached to the housing is called the stator, and the rotating portion of the motor is called the rotor.

Figure 25.7 shows a cutaway of a Maxon brushed motor, exposing the brushes, commutator, magnets, and windings. The figure also shows other elements of a typical motor application: an encoder attached to one end of the motor shaft to provide feedback on the angle and a gearhead attached to the other end of the motor shaft. The output shaft of the gearhead provides lower speed but higher torque than the output shaft of the motor.

Brushless motors are a variant that use electronic commutation as opposed to brushed commutation. For more on brushless DC motors, see Chapter 29.5.

25.2 Governing Equations

To derive an equation to model the motor's behavior, we ignore the details of the commutation and focus instead on electrical and mechanical power. The electrical power into the motor is IV, where I is the current through the motor and V is the voltage across the motor. We know that the motor converts some of this input power to mechanical power $\tau\omega$, where τ

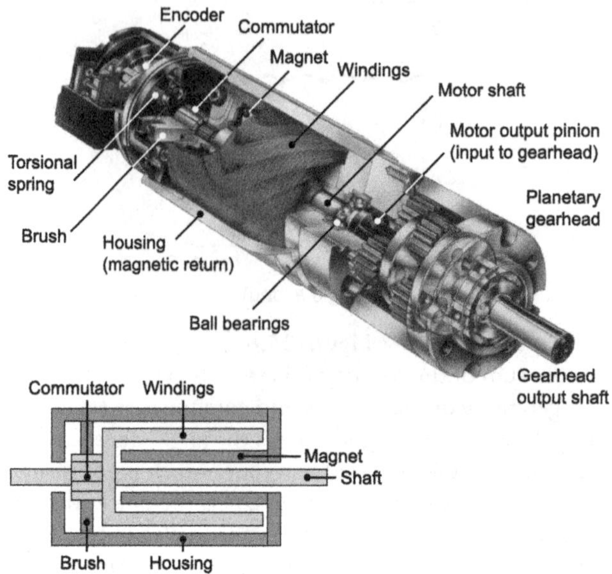

Figure 25.7
A cutaway of a Maxon brushed motor with an encoder and a planetary gearhead. The brushes are spring-loaded against the commutator. The bottom left schematic is a simplified cross-section showing the stationary parts of the motor (the stator) in dark gray and the rotating parts of the motor (the rotor) in light gray. In this "coreless" motor geometry, the windings spin in a gap between the permanent magnets and the housing. (Cutaway image courtesy of Maxon Precision Motors, Inc., maxonmotorusa.com.)

and ω are the torque and velocity of the output shaft, respectively. Electrically, the motor is described by a resistance R between the two terminals as well as an inductance L due to the coils. The resistance of the motor coils dissipates power I^2R as heat. The motor also stores energy $\frac{1}{2}LI^2$ in the inductor's magnetic field, and the time rate of change of this is $LI(\mathrm{d}I/\mathrm{d}t)$, the power into (charging) or out of (discharging) the inductor. Finally, power is dissipated as sound, heat due to friction at the brush-commutator interface and at the bearings between the motor shaft and the housing, etc. In SI units, all these power components are expressed in watts. Combining all of these factors provides a full accounting for the electrical power put into the motor:

$$IV = \tau\omega + I^2R + LI\frac{\mathrm{d}I}{\mathrm{d}t} + \text{ power dissipated due to friction, sound, etc.}$$

Ignoring the last term, we have our simple motor model, written in terms of power:

$$IV = \tau\omega + I^2R + LI\frac{\mathrm{d}I}{\mathrm{d}t}. \tag{25.2}$$

From (25.2) we can derive all other relationships of interest. For example, dividing both sides of (25.2) by I yields

$$V = \frac{\tau}{I}\omega + IR + L\frac{\mathrm{d}I}{\mathrm{d}t}. \tag{25.3}$$

The ratio τ/I is a constant, an expression of the Lorentz force law for the particular motor design. This constant, relating current to torque, is called the *torque constant* k_t. The torque constant is one of the most important properties of the motor:

$$k_t = \frac{\tau}{I} \quad \text{or} \quad \tau = k_t I. \tag{25.4}$$

The SI units of k_t are Nm/A. (In this chapter, we only use SI units, but you should be aware that many different units are used by different manufacturers, as on the speed-torque curve and data sheet in Figure 25.16 in the Exercises.) Equation (25.3) also shows that the SI units for k_t can be written equivalently as Vs/rad, or simply Vs. When using these units, we sometimes call the motor constant the *electrical constant* k_e. The inverse is sometimes called the *speed constant*. You should recognize that these terms all refer to the same property of the motor. For consistency, we usually refer to the torque constant k_t.

We now express the motor model in terms of voltage as

$$V = k_t\omega + IR + L\frac{\mathrm{d}I}{\mathrm{d}t}. \tag{25.5}$$

You should remember, or be able to quickly derive, the power equation (25.2), the torque constant (25.4), and the voltage equation (25.5).

The term $k_t\omega$, with units of voltage, is called the *back-emf*, where emf is short for *electromotive force*. We could also call this "back-voltage." Back-emf is the voltage generated by a spinning motor to "oppose" the input voltage generating the motion. For example, assume that the motor's terminals are not connected to anything (open circuit). Then $I = 0$ and $\frac{\mathrm{d}I}{\mathrm{d}t} = 0$, so (25.5) reduces to

$$V = k_t\omega.$$

This equation indicates that back-driving the motor (e.g., spinning it by hand) will generate a voltage at the terminals. If we were to connect a capacitor across the motor terminals, then spinning the motor by hand would charge the capacitor, storing some of the mechanical energy we put in as electrical energy in the capacitor. In this situation, the motor acts as a generator, converting mechanical energy to electrical energy.

The existence of this back-emf term also means that if we put a constant voltage V across a free-spinning frictionless motor (i.e., the motor shaft is not connected to anything), after some time it will reach a constant speed V/k_t. At this speed, by (25.5), the current I drops to zero, meaning there is no more torque τ to accelerate the motor. This happens because as the motor accelerates, the back-emf increases, countering the applied voltage until no current flows (and hence there is no torque or acceleration).

25.3 The Speed-Torque Curve

Consider a motor spinning a boat's propeller at constant velocity. The torque τ provided by the motor can be written

$$\tau = \tau_{\text{fric}} + \tau_{\text{pushing water}},$$

where τ_{fric} is the torque the motor has to generate to overcome friction and begin to spin, while $\tau_{\text{pushing water}}$ is the torque needed for the propeller to displace water when the motor is spinning at velocity ω. In this section we assume $\tau_{\text{fric}} = 0$, so $\tau = \tau_{\text{pushing water}}$ in this example. In Section 25.4 we consider nonzero friction.

For a motor spinning at constant speed ω and providing constant torque τ (as in the propeller example above), the current I is constant and therefore $\mathrm{d}I/\mathrm{d}t = 0$. Under these assumptions, (25.5) reduces to

$$V = k_t\omega + IR. \tag{25.6}$$

Using the definition of the torque constant, we get the equivalent form

$$\omega = \frac{1}{k_t}V - \frac{R}{k_t^2}\tau. \tag{25.7}$$

Equation (25.7) gives ω as a linear function of τ for a given constant V. This line, of slope $-R/k_t^2$, is called the *speed-torque curve* for the voltage V.

The speed-torque curve plots all the possible constant-current operating conditions with voltage V across the motor. Assuming friction torque is zero, the line intercepts the $\tau = 0$ axis at

$$\omega_0 = V/k_t = \text{no load speed}.$$

The line intercepts the $\omega = 0$ axis at

$$\tau_{\text{stall}} = \frac{k_t V}{R} = \text{Stall torque}.$$

At the no-load condition, $\tau = I = 0$; the motor rotates at maximum speed with no current or torque. At the stall condition, the shaft is blocked from rotating, and the current ($I_{\text{stall}} = \tau_{\text{stall}}/k_t = V/R$) and output torque are maximized due to the lack of back-emf. Which point along the speed-torque curve the motor actually operates at is determined by the load attached to the motor shaft.

An example speed-torque curve is shown in Figure 25.8. This motor has $\omega_0 = 500$ rad/s and $\tau_{stall} = 0.1067$ Nm for a nominal voltage of $V_{nom} = 12$ V. The *operating region* is any point below the speed-torque curve, corresponding to voltages less than or equal to 12 V. If the motor is operated at a different voltage cV_{nom}, the intercepts of the speed-torque curve are linearly scaled to $c\omega_0$ and $c\tau_{stall}$.

Imagine squeezing the shaft of a motor powered by a voltage V and spinning at a constant velocity. Your hand is applying a small torque to the shaft. Since the motor is not accelerating and we are neglecting friction in the motor, the torque created by the motor's coils must be equal and opposite the torque applied by your hand. Thus the motor operates at a specific point on the speed-torque curve. If you slowly squeeze the shaft harder, increasing the torque you apply to the rotor, the motor will slow down and increase the torque it applies, to balance your hand's torque. Assuming the motor's current changes slowly (i.e., $L\mathrm{d}I/\mathrm{d}t$ is negligible), then the operating point of the motor moves down and to the right on the speed-torque curve as you increase your squeeze force. When you squeeze hard enough that the motor can no longer move, the operating point is at the stall condition, the bottom-right point on the speed-torque curve.

The speed-torque curve corresponds to constant V, but not to constant input power $P_{in} = IV$. The current I is linear with τ, so the input electrical power increases linearly with τ. The output mechanical power is $P_{out} = \tau\omega$, and the *efficiency* in converting electrical to mechanical power is $\eta = P_{out}/P_{in} = \tau\omega/IV$. We return to efficiency in Section 25.4.

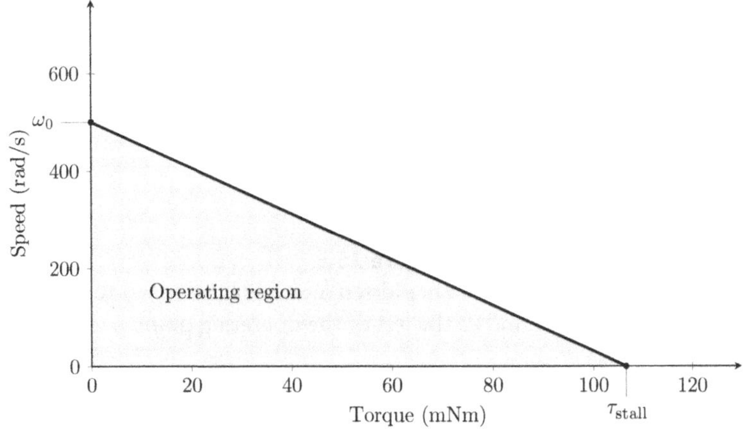

Figure 25.8
A speed-torque curve. Many speed-torque curves use rpm for speed, but we prefer SI units.

To find the point on the speed-torque curve that maximizes the mechanical output power, we can write points on the curve as $(\tau, \omega) = (c\tau_{\text{stall}}, (1-c)\omega_0)$ for $0 \le c \le 1$, so the output power is expressed as

$$P_{\text{out}} = \tau\omega = (c - c^2)\tau_{\text{stall}}\omega_0,$$

and the value of c that maximizes the power output is found by solving

$$\frac{\mathrm{d}}{\mathrm{d}c}\left((c - c^2)\tau_{\text{stall}}\omega_0\right) = (1 - 2c)\tau_{\text{stall}}\omega_0 = 0 \quad \rightarrow \quad c = \frac{1}{2}.$$

Thus the mechanical output power is maximized at $\tau = \tau_{\text{stall}}/2$ and $\omega = \omega_0/2$. This maximum output power is

$$P_{\max} = \left(\frac{1}{2}\tau_{\text{stall}}\right)\left(\frac{1}{2}\omega_0\right) = \frac{1}{4}\tau_{\text{stall}}\omega_0.$$

See Figure 25.9.

Motor current is proportional to motor torque, so operating at high torques means large coil heating power loss I^2R, sometimes called *ohmic heating*. For that reason, motor manufacturers specify a *maximum continuous current* I_{cont}, the largest continuous current such that the coils' steady-state temperature remains below a critical point.[1] The maximum continuous current has a corresponding *maximum continuous torque* τ_{cont}. Points to the left of this torque and under the speed-torque curve are called the *continuous operating region*. The motor can be

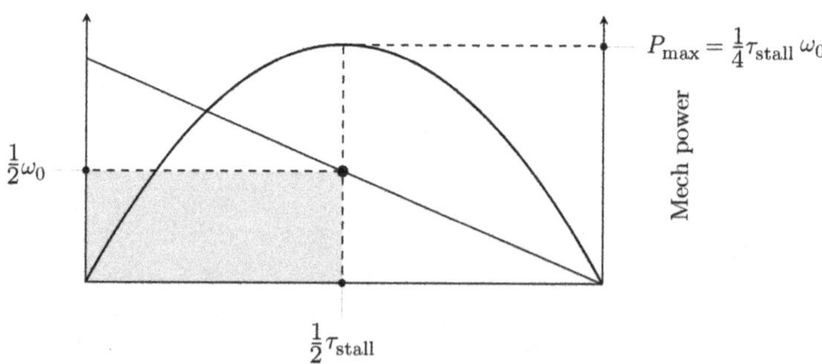

Figure 25.9
The quadratic mechanical power plot $P = \tau\omega$ plotted alongside the speed-torque curve. The area of the speed-torque rectangle below and to the left of the operating point is the mechanical power.

1 The maximum continuous current depends on thermal properties governing how fast coil heat can be transferred to the environment. This depends on the environment temperature, typically considered to be room temperature. The maximum continuous current can be increased by cooling the motor.

operated intermittently outside of the continuous operating region, in the *intermittent operating region*, provided the motor is allowed to cool sufficiently between uses in this region. Motors are commonly rated with a nominal voltage that places the maximum mechanical power operating point (at $\tau_{stall}/2$) outside the continuous operating region.

Given thermal characteristics of the motor of Figure 25.8, the speed-torque curve can be refined to Figure 25.10, showing the continuous and intermittent operating regions of the motor. The point on the speed-torque curve at τ_{cont} is the *rated* or *nominal operating point*, and the mechanical power output at this point is called the motor's *power rating*. For the motor of Figure 25.10, $\tau_{cont} = 26.67$ mNm, which occurs at $\omega = 375$ rad/s, for a power rating of

$$0.02667 \text{ Nm} \times 375 \text{ rad/s} = 10.0 \text{ W}.$$

Figure 25.10 also shows the constant output power hyperbola $\tau\omega = 10$ W passing through the nominal operating point.

The speed-torque curve for a motor is drawn based on a nominal voltage. This is a "safe" voltage that the manufacturer recommends. It is possible to overvolt the motor, however, provided it is not continuously operated beyond the maximum continuous current. A motor also may have a specified *maximum permissible speed* ω_{max}, which creates a horizontal line constraint on the permissible operating range. This speed is determined by allowable brush wear, or possibly properties of the shaft bearings, and it is typically larger than the no-load speed ω_0. The shaft and bearings may also have a maximum torque rating $\tau_{max} > \tau_{stall}$. These

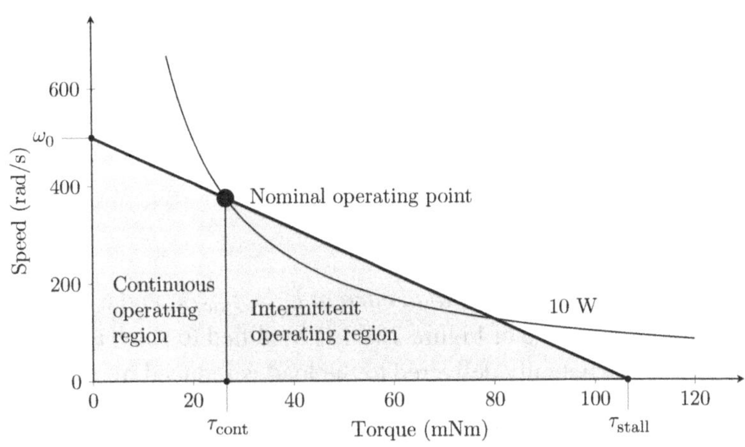

Figure 25.10
The continuous operating region (under the speed-torque curve and left of τ_{cont}) and the intermittent operating region (the rest of the area under the speed-torque curve). The 10 W mechanical power hyperbola is indicated, including the nominal operating point at τ_{cont}.

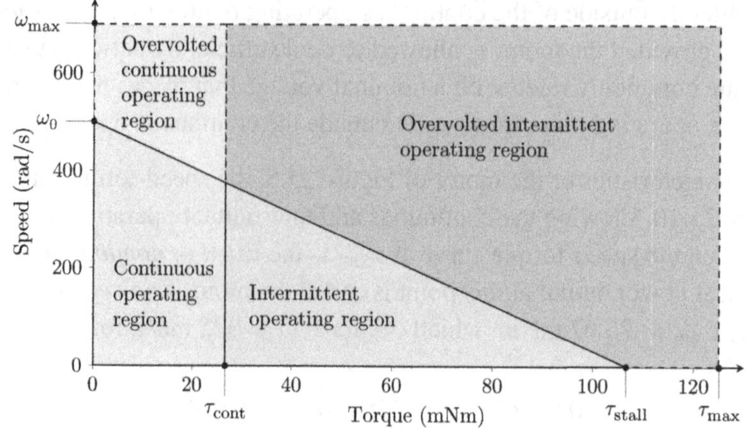

Figure 25.11
It is possible to exceed the nominal operating voltage, provided the constraints $\omega < \omega_{max}$ and $\tau < \tau_{max}$ are respected and τ_{cont} is only intermittently exceeded.

limits allow the definition of overvolted continuous and intermittent operating regions, as shown in Figure 25.11.

25.4 Friction and Motor Efficiency

Until now we have been assuming that the full torque $\tau = k_t I$ generated by the windings is available at the output shaft. In practice, some torque is lost due to friction at the brushes and the shaft bearings. Let us use a simple model of friction: assume a torque $\tau \geq \tau_{fric} > 0$ must be generated to overcome friction and initiate motion, and any torque beyond τ_{fric} is available at the output shaft regardless of the motor speed (e.g., no friction that depends on speed magnitude). When the motor is spinning, the torque available at the output shaft is

$$\tau_{out} = \tau - \tau_{fric}.$$

Nonzero friction results in a nonzero *no-load current* $I_0 = \tau_{fric}/k_t$ and a no-load speed ω_0 less than V/k_t. The speed-torque curve of Figure 25.11 is modified to show a small friction torque in Figure 25.12. The torque actually delivered to the load is reduced by τ_{fric}.

Taking friction into account, the motor's efficiency in converting electrical to mechanical power is

$$\eta = \frac{\tau_{out}\omega}{IV}. \tag{25.8}$$

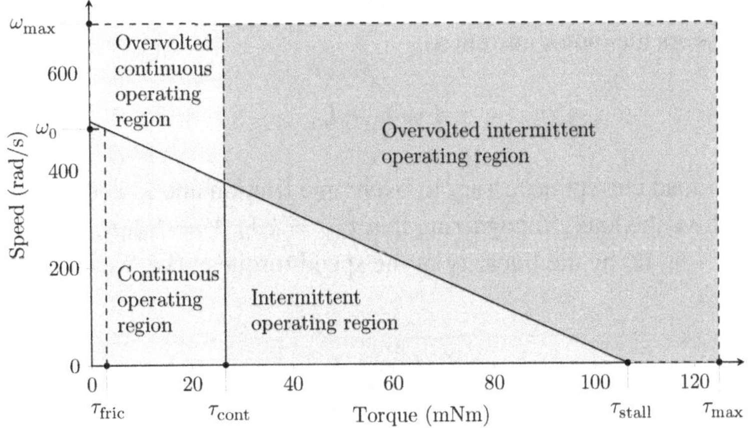

Figure 25.12

The speed-torque curve of Figure 25.11 modified to show a nonzero friction torque τ_{fric} and the resulting reduced no-load speed ω_0.

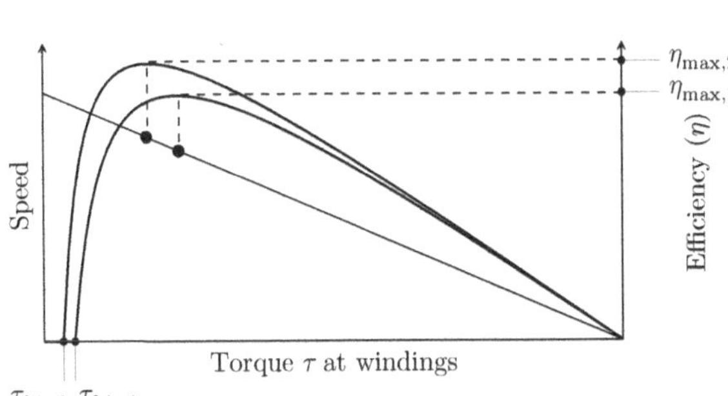

Figure 25.13

The speed-torque curve for a motor and two efficiency plots, one for high friction torque (case 1) and one for low friction torque (case 2). For each case, efficiency is zero for all τ below the level needed to overcome friction. The low friction version of the motor (case 2) achieves a higher maximum efficiency, at a higher speed and lower torque, than the high friction version (case 1).

The efficiency depends on the operating point on the speed-torque curve, and it is zero when either τ_{out} or ω is zero, as there is no mechanical power output. Maximum efficiency generally occurs at high speed and low torque, approaching the limit of 100% efficiency at $\tau = \tau_{out} = 0$ and $\omega = \omega_0$ as τ_{fric} approaches zero. As an example, Figure 25.13 plots efficiency vs. torque for the same motor with two different values of τ_{fric}. Lower friction results in a higher maximum efficiency η_{max}, occurring at a higher speed and lower torque.

To derive the maximally efficient operating point and the maximum efficiency $\eta_{\max}$ for a given motor, we can express the motor current as

$$I = I_0 + I_a,$$

where I_0 is the no-load current necessary to overcome friction and I_a is the added current to create torque to drive the load. Recognizing that $\tau_{\text{out}} = k_t I_a$, $V = I_{\text{stall}} R$, and $\omega = R(I_{\text{stall}} - I_a - I_0)/k_t$ by the linearity of the speed-torque curve, we can rewrite the efficiency (25.8) as

$$\eta = \frac{I_a(I_{\text{stall}} - I_0 - I_a)}{(I_0 + I_a)I_{\text{stall}}}. \tag{25.9}$$

To find the operating point I_a^* maximizing η, we solve $\mathrm{d}\eta/\mathrm{d}I_a = 0$ for I_a^*, and recognizing that I_0 and I_{stall} are nonnegative, the solution is

$$I_a^* = \sqrt{I_{\text{stall}} I_0} - I_0.$$

In other words, as the no-load current I_0 goes to zero, the maximally efficient current (and therefore τ) goes to zero.

Plugging I_a^* into (25.9), we find

$$\eta_{\max} = \left(1 - \sqrt{\frac{I_0}{I_{\text{stall}}}}\right)^2.$$

This answer has the form we would expect: maximum efficiency approaches 100% as the friction torque approaches zero, and maximum efficiency approaches 0% as the friction torque approaches the stall torque.

Choosing an operating point that maximizes motor efficiency can be important when trying to maximize battery life in mobile applications. For the majority of analysis and motor selection problems, however, ignoring friction is a good first approximation.

25.5 Motor Windings and the Motor Constant

It is possible to build two different versions of the same motor by simply changing the windings while keeping everything else the same. For example, imagine a coil of resistance R with N loops of wire of cross-sectional area A. The coil carries a current I and therefore has a voltage drop IR. Now we replace that coil with a new coil with N/c loops of wire with cross-sectional area cA. This preserves the volume occupied by the coil, fitting in the same form factor with similar thermal properties. Without loss of generality, let us assume that the new coil has fewer loops and uses thicker wire ($c > 1$).

The resistance of the new coil is reduced to R/c^2 (a factor of c due to the shorter coil and another factor of c due to the thicker wire). To keep the torque of the motor the same, the new coil would have to carry a larger current cI to make up for the fewer loops, so that the current times the pathlength through the magnetic field is unchanged. The voltage drop across the new coil is $(cI)(R/c^2) = IR/c$.

Replacing the coils allows us to create two versions of the motor: a many-loop, thin wire version that operates at low current and high voltage, and a fewer-loop, thick wire version that operates at high current and low voltage. Since the two motors create the same torque with different currents, they have different torque constants. Each motor has the same *motor constant* k_m, however, where

$$k_m = \frac{\tau}{\sqrt{I^2R}} = \frac{k_t}{\sqrt{R}}$$

with units of $\mathrm{Nm}/\sqrt{\mathrm{W}}$. The motor constant defines the torque generated per square root of the power dissipated by coil resistance. In the example above, the new coil dissipates $(cI)^2(R/c^2) = I^2R$ power as heat, just as the original coil does, while generating the same torque.

Figure 25.16 shows the data sheet for a motor that comes in several different versions, each identical in every way except for the winding. Each version of the motor has a similar stall torque and motor constant but different nominal voltage, resistance, and torque constant.

25.6 Other Motor Characteristics

Electrical time constant

When the motor is subject to a step in the voltage across it, the *electrical time constant* T_e measures the time it takes for the unloaded current to reach 63% of its final value. The motor's voltage equation is

$$V = k_t\omega + IR + L\frac{\mathrm{d}I}{\mathrm{d}t}.$$

Ignoring back-emf (because the motor speed does not change significantly over one electrical time constant), assuming an initial current through the motor of I_0, and an instantaneous drop in the motor voltage to 0, we get the differential equation

$$0 = I_0 R + L\frac{dI}{dt}$$

or

$$\frac{dI}{dt} = -\frac{R}{L} I_0$$

with solution

$$I(t) = I_0 \, e^{-tR/L} = I_0 \, e^{-t/T_e}.$$

The time constant of this first-order decay of current is the motor's electrical time constant, $T_e = L/R$.

Mechanical time constant

When the motor is subject to a step voltage across it, the *mechanical time constant* T_m measures the time it takes for the unloaded motor speed to reach 63% of its final value. Beginning from the voltage equation

$$V = k_t \omega + IR + L\frac{dI}{dt},$$

ignoring the inductive term, and assuming an initial speed ω_0 at the moment the voltage drops to zero, we get the differential equation

$$0 = IR + k_t\omega_0 = \frac{R}{k_t}\tau + k_t\omega_0 = \frac{JR}{k_t}\frac{d\omega}{dt} + k_t\omega_0,$$

where we used $\tau = Jd\omega/dt$, where J is the inertia of the motor. We can rewrite this equation as

$$\frac{d\omega}{dt} = -\frac{k_t^2}{JR}\omega_0$$

with solution

$$\omega(t) = \omega_0 \, e^{-t/T_m},$$

with a time constant of $T_m = JR/k_t^2$. If the motor is attached to a load that increases the inertia J, the mechanical time constant increases.

Short-circuit damping

When the terminals of the motor are shorted together, the voltage equation (ignoring inductance) becomes

$$0 = k_t\omega + IR = k_t\omega + \frac{\tau}{k_t}R$$

or

$$\tau = -B\omega = -\frac{k_t^2}{R}\omega,$$

where $B = k_t^2/R$ is the short-circuit damping. A spinning motor is slowed more quickly by shorting its terminals together, compared to leaving the terminals open circuit, due to this damping.

25.7 Motor Data Sheet

Motor manufacturers summarize motor properties described above in a speed-torque curve and in a data sheet similar to the one in Figure 25.14. When you buy a motor second-hand or surplus, you may need to measure these properties yourself. We will use all SI units, which is not the case on most motor data sheets.

Many of these properties have been introduced already. Below we describe some methods for estimating them.

Experimentally Characterizing a Brushed DC Motor

Given a mystery motor with an encoder, you can use a function generator, oscilloscope, multimeter and perhaps some resistors and capacitors to estimate most of the important properties of the motor. Below are some suggested methods; you may be able to devise others.

Terminal resistance R

You can measure R with a multimeter. The resistance may change as you rotate the shaft by hand, as the brushes move to new positions on the commutator. You should record the minimum resistance you can reliably find. A better choice, however, may be to measure the current when the motor is stalled.

Torque constant k_t

You can measure this by spinning the shaft of the motor, measuring the back-emf at the motor terminals, and measuring the rotation rate ω using the encoder. Or, if friction losses are

Motor Characteristic	Symbol	Value	Units	Comments
Terminal resistance	R		Ω	Resistance of the motor windings. May change as brushes slide over commutator segments. Increases with heat.
Torque constant	k_t		Nm/A	The constant ratio of torque produced to current through the motor.
Electrical constant	k_e		Vs/rad	Same numerical value as the torque constant (in SI units). Also called voltage or back-emf constant.
Speed constant	k_s		rad/(Vs)	Inverse of electrical constant.
Motor constant	k_m		Nm/$\sqrt{\text{W}}$	Torque produced per square root of power dissipated by the coils.
Max continuous current	I_{cont}		A	Max continuous current without overheating.
Max continuous torque	τ_{cont}		Nm	Max continuous torque without overheating.
Short-circuit damping	B		Nms/rad	Not included in most data sheets, but useful for motor braking (and haptics).
Terminal inductance	L		H	Inductance due to the coils.
Electrical time constant	Te		s	The time for the motor current to reach 63% of its final value. Equal to L/R.
Rotor inertia	J		kgm^2	Often given in units gcm^2.
Mechanical time constant	Tm		s	The time for the motor to go from rest to 63% of its final speed under constant voltage and no load. Equal to JR/kt^2.
Friction				Not included in most data sheets. See explanation.
Values at Nominal Voltage				
Nominal voltage	V_{nom}		V	Should be chosen so the no-load speed is safe for brushes, commutator, and bearings.
Power rating	P		W	Output power at the nominal operating point (max continuous torque).
No-load speed	ω_0		rad/s	Speed when no load and powered by $V\text{nom}$. Usually given in rpm (revs/min, sometimes m^{-1}).
No-load current	I_0		A	The current required to spin the motor at the no-load condition. Nonzero because of friction torque.
Stall current	I		A	Same as starting current, $V\text{nom}/R$.
Stall torque	τ_{stall}		Nm	The torque achieved at the nominal voltage when the motor is stalled.
Max mechanical power	P_{max}		W	The max mechanical power output at the nominal voltage (including short-term operation).
Max efficiency	η		%	The maximum efficiency achievable in converting electrical to mechanical power.

Figure 25.14
A sample motor data sheet, with values to be filled in.

negligible, a good approximation is V_{nom}/ω_0. This eliminates the need to spin the motor externally.

Electrical constant k_e

Identical to the torque constant in SI units. The torque constant k_t is often expressed in units of Nm/A or mNm/A or English units like oz-in/A, and often k_e is given in V/rpm, but k_t and k_e have identical numerical values when expressed in Nm/A and Vs/rad, respectively.

Speed constant k_s

Just the inverse of the electrical constant.

Motor constant k_m

The motor constant is calculated as $k_m = k_t/\sqrt{R}$.

Max continuous current I_{cont}

This is determined by thermal considerations, which are not easy to measure. It is typically less than half the stall current.

Max continuous torque τ_{cont}

This is determined by thermal considerations, which are not easy to measure. It is typically less than half the stall torque.

Short-circuit damping B

This is most easily calculated from estimates of R and k_t: $B = k_t^2/R$.

Terminal inductance L

There are several ways to measure inductance. One approach is to add a capacitor in parallel with the motor and measure the oscillation frequency of the resulting RLC circuit. For example, you could build the circuit shown in Figure 25.15, where a good choice for C may be 0.01 or 0.1 μF. The motor acts as a resistor and inductor in series; back-emf will not be an issue, because the motor will be powered by tiny currents at high frequency and therefore will not move.

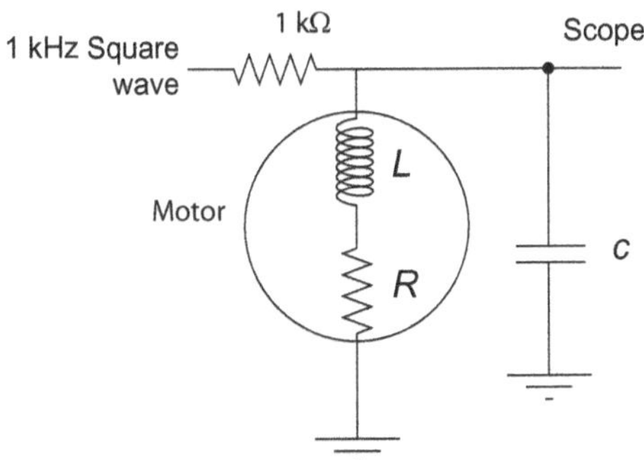

Figure 25.15
Using a capacitor to create an RLC circuit to measure motor inductance.

Use a function generator to put a 1 kHz square wave between 0 and 5 V at the point indicated. The 1 kΩ resistor limits the current from the function generator. Measure the voltage with an oscilloscope where indicated. You should be able to see a decaying oscillatory response to the square wave input when you choose the right scales on your scope. Measure the frequency of the oscillatory response. Knowing C and that the natural frequency of an RLC circuit is $\omega_n = 1/\sqrt{LC}$ in rad/s, estimate L.

Let us think about why we see this response. Say the input to the circuit has been at 0 V for a long time. Then your scope will also read 0 V. Now the input steps up to 5 V. After some time, in steady state, the capacitor will be an open circuit and the inductor will be a closed circuit (wire), so the voltage on the scope will settle to 5 V $\times$ $(R/(1000 + R))$—the two resistors in the circuit set the final voltage. Right after the voltage step, however, all current goes to charge the capacitor (as the zero current through the inductor cannot change discontinuously). If the inductor continued to enforce zero current, the capacitor would charge to 5 V. As the voltage across the capacitor grows, however, so does voltage across the inductor, and therefore so does the rate of change of current that must flow through the inductor (by the relation $V_L + V_R = V_C$ and the constitutive law $V_L = L\,\mathrm{d}I/\mathrm{d}t$). Eventually the integral of this rate of change dictates that all current is redirected to the inductor, and in fact the capacitor will have to provide current to the inductor, discharging itself. As the voltage across the capacitor drops, though, the voltage across the inductor will eventually become negative, and therefore the rate of change of current through the inductor will become negative. And so on, to create the oscillation. If R were large, i.e., if the circuit were heavily damped, the oscillation would die quickly, but you should be able to see it.

Note that you are seeing a damped oscillation, so you are actually measuring a damped natural frequency. But the damping is low if you are seeing at least a couple of cycles of oscillation, so the damped natural frequency is nearly indistinguishable from the undamped natural frequency.

Electrical time constant T_e

The electrical time constant can be calculated from L and R as $T_e = L/R$.

Rotor inertia J

The rotor inertia can be estimated from measurements of the mechanical time constant T_m, the torque constant k_t, and the resistance R. Alternatively, a ballpark estimate can be made based on the mass of the motor, a guess at the portion of the mass that belongs to the spinning rotor, a guess at the radius of the rotor, and a formula for the inertia of a uniform density cylinder. Or, more simply, consult a data sheet for a motor of similar size and mass.

Mechanical time constant T_m

The time constant can be measured by applying a constant voltage to the motor, measuring the velocity, and determining the time it takes to reach 63% of final speed. Alternatively, you could make a reasonable estimate of the rotor inertia J and calculate $T_m = JR/k_t^2$.

Friction

Friction torque arises from the brushes sliding on the commutator and the motor shaft spinning in its bearings, and it may depend on external loads. A typical model of friction includes both Coulomb friction and viscous friction, written

$$\tau_{\text{fric}} = b_0 \operatorname{sgn}(\omega) + b_1\omega,$$

where b_0 is the Coulomb friction torque ($\operatorname{sgn}(\omega)$ just returns the sign of ω) and b_1 is a viscous friction coefficient. At no load, $\tau_{\text{fric}} = k_t I_0$. An estimate of each of b_0 and b_1 can be made by running the motor at two different voltages with no load.

Nominal voltage V_{nom}

This is the specification you are most likely to know for an otherwise unknown motor. It is sometimes printed right on the motor itself. This voltage is just a recommendation; the real issue is to avoid overheating the motor or spinning it at speeds beyond the recommended value for the brushes or bearings. Nominal voltage cannot be measured, but a typical no-load speed for a brushed DC motor is between 3000 and 10,000 rpm, so the nominal voltage will often give a no-load speed in this range.

Power rating P

The power rating is the mechanical power output at the max continuous torque.

No-load speed ω_0

You can determine ω_0 by measuring the unloaded motor speed when powered with the nominal voltage. The amount that this is less than V_{nom}/k_t can be attributed to friction torque.

No-load current I_0

You can determine I_0 by using a multimeter in current measurement mode.

Stall current I_{stall}

Stall current is sometimes called starting current. You can estimate this using your estimate of R. Since R may be difficult to measure with a multimeter, you can instead stall the motor shaft and use your multimeter in current sensing mode, provided the multimeter can handle the current.

Stall torque τ_{stall}

This can be obtained from k_t and I_{stall}.

Max mechanical power P_{max}

The max mechanical power occurs at $\frac{1}{2}\tau_{stall}$ and $\frac{1}{2}\omega_0$. For most motor data sheets, the max mechanical power occurs outside the continuous operation region.

Max efficiency η_{max}

Efficiency is defined as the power out divided by the power in, $\tau_{out}\omega/(IV)$. The wasted power is due to coil heating and friction losses. Maximum efficiency can be estimated using the no-load current I_0 and the stall current I_{stall}, as discussed in Section 25.4.

25.8 Chapter Summary

- The Lorentz force law says that a current-carrying conductor in a constant magnetic field feels a net force according to

$$\mathbf{F} = \ell \mathbf{I} \times \mathbf{B},$$

  where ℓ is the length of the conductor in the field, $\mathbf{I}$ is the current vector, and $\mathbf{B}$ is the (constant) magnetic field vector.
- A brushed DC motor consists of multiple current-carrying coils attached to a rotor, and magnets on the stator to create a magnetic field. Current is transmitted to the coils by two brushes connected to the stator sliding over a commutator ring attached to the rotor. Each coil attaches to two different commutator segments.
- The voltage across a motor's terminals can be expressed as

$$V = k_t\omega + IR + L\frac{dI}{dt},$$

  where k_t is the torque constant and $k_t\omega$ is the back-emf.
- The speed-torque curve is obtained by plotting the steady-state speed as a function of torque for a given motor voltage V,

$$\omega = \frac{1}{k_t}V - \frac{R}{k_t^2}\tau.$$

The maximum speed (at $\tau = 0$) is called the no-load speed ω_0 and the maximum torque (at $\omega = 0$) is called the stall torque τ_{stall}.

- The continuous operating region of a motor is defined by the maximum current I the motor coils can conduct continuously without overheating due to I^2R power dissipation.
- The mechanical power $\tau\omega$ delivered by a motor is maximized at half the stall torque and half the no-load speed, $P_{\max} = \frac{1}{4}\tau_{\text{stall}}\omega_0$.
- A motor's electrical time constant $T_e = L/R$ is the time needed for current to reach 63% of its final value in response to a step input in voltage.
- A motor's mechanical time constant $T_m = JR/k_t^2$ is the time needed for the motor speed to reach 63% of its final value in response to a step change in voltage.

25.9 Exercises

1. Assume a DC motor with a five-segment commutator. Each segment covers 70° of the circumference of the commutator circle. The two brushes are positioned at opposite ends of the commutator circle, and each makes contact with 10° of the commutator circle.
   a. How many separate coils does this motor likely have? Explain.
   b. Choose one of the motor coils. As the rotor rotates 360°, what is the total angle over which that coil is energized? (For example, an answer of 360° means that the coil is energized at all angles; an answer of 180° means that the coil is energized at half of the motor positions.)
2. Figure 25.16 gives the data sheet for the 10 W Maxon RE 25 motor. The columns correspond to different windings.
   a. Draw the speed-torque curve for the 12 V version of the motor, indicating the no-load speed (in rad/s), the stall torque, the nominal operating point, and the rated power of the motor.
   b. Explain why the torque constant is different for the different versions of the motor.
   c. Using other entries in the table, calculate the maximum efficiency $\eta_{\max}$ of the 12 V motor and compare to the value listed.
   d. Calculate the electrical time constant T_e of the 12 V motor. What is the ratio to the mechanical time constant T_m?
   e. Calculate the short-circuit damping B for the 12 V motor.
   f. Calculate the motor constant k_m for the 12 V motor.
   g. How many commutator segments do these motors have?
   h. Which versions of these motors are likely to be in stock?

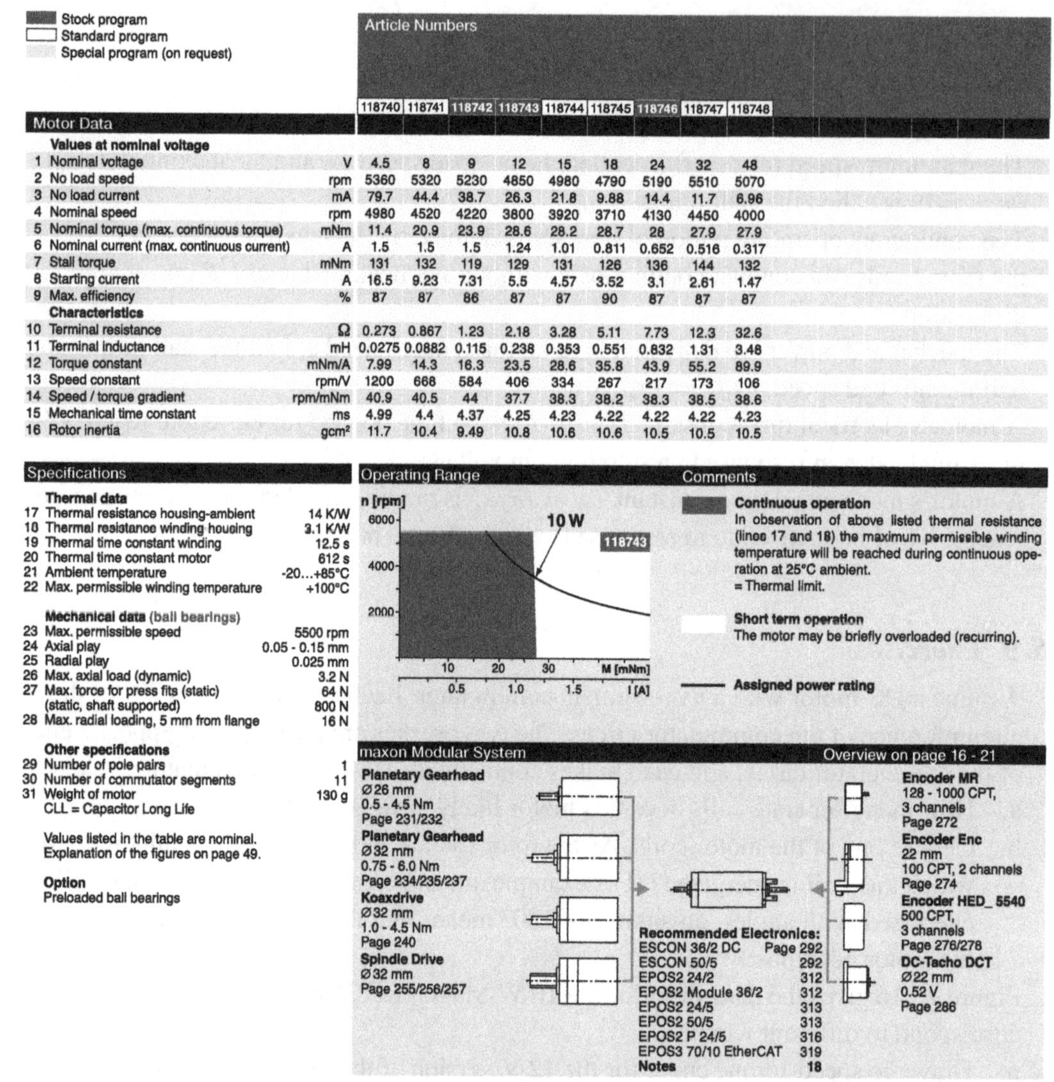

Stock program
Standard program
Special program (on request)

Article Numbers

Motor Data	Unit	118740	118741	118742	118743	118744	118745	118746	118747	118748
Values at nominal voltage										
1 Nominal voltage	V	4.5	8	9	12	15	18	24	32	48
2 No load speed	rpm	5360	5320	5230	4850	4980	4790	5190	5510	5070
3 No load current	mA	79.7	44.4	38.7	26.3	21.8	9.88	14.4	11.7	6.96
4 Nominal speed	rpm	4980	4520	4220	3800	3920	3710	4130	4450	4000
5 Nominal torque (max. continuous torque)	mNm	11.4	20.9	23.9	28.6	28.2	28.7	28	27.9	27.9
6 Nominal current (max. continuous current)	A	1.5	1.5	1.5	1.24	1.01	0.811	0.652	0.516	0.317
7 Stall torque	mNm	131	132	119	129	131	126	136	144	132
8 Starting current	A	16.5	9.23	7.31	5.5	4.57	3.52	3.1	2.61	1.47
9 Max. efficiency	%	87	87	86	87	87	90	87	87	87
Characteristics										
10 Terminal resistance	Ω	0.273	0.867	1.23	2.18	3.28	5.11	7.73	12.3	32.6
11 Terminal inductance	mH	0.0275	0.0882	0.115	0.238	0.353	0.551	0.832	1.31	3.48
12 Torque constant	mNm/A	7.99	14.3	16.3	23.5	28.6	35.8	43.9	55.2	89.9
13 Speed constant	rpm/V	1200	668	584	406	334	267	217	173	106
14 Speed / torque gradient	rpm/mNm	40.9	40.5	44	37.7	38.3	38.2	38.3	38.5	38.6
15 Mechanical time constant	ms	4.99	4.4	4.37	4.25	4.23	4.22	4.22	4.22	4.23
16 Rotor inertia	gcm²	11.7	10.4	9.49	10.8	10.6	10.6	10.5	10.5	10.5

Specifications

Thermal data	
17 Thermal resistance housing-ambient	14 K/W
18 Thermal resistance winding-housing	3.1 K/W
19 Thermal time constant winding	12.5 s
20 Thermal time constant motor	612 s
21 Ambient temperature	-20...+85°C
22 Max. permissible winding temperature	+100°C
Mechanical data (ball bearings)	
23 Max. permissible speed	5500 rpm
24 Axial play	0.05 - 0.15 mm
25 Radial play	0.025 mm
26 Max. axial load (dynamic)	3.2 N
27 Max. force for press fits (static)	64 N
(static, shaft supported)	800 N
28 Max. radial loading, 5 mm from flange	16 N
Other specifications	
29 Number of pole pairs	1
30 Number of commutator segments	11
31 Weight of motor	130 g
CLL = Capacitor Long Life	

Values listed in the table are nominal.
Explanation of the figures on page 49.

Option
Preloaded ball bearings

Comments

Continuous operation
In observation of above listed thermal resistance (lines 17 and 18) the maximum permissible winding temperature will be reached during continuous operation at 25°C ambient.
= Thermal limit.

Short term operation
The motor may be briefly overloaded (recurring).

Assigned power rating

maxon Modular System Overview on page 16 - 21

Planetary Gearhead
Ø26 mm
0.5 - 4.5 Nm
Page 231/232
Planetary Gearhead
Ø32 mm
0.75 - 6.0 Nm
Page 234/235/237
Koaxdrive
Ø32 mm
1.0 - 4.5 Nm
Page 240
Spindle Drive
Ø32 mm
Page 255/256/257

Recommended Electronics:

ESCON 36/2 DC	Page 292
ESCON 50/5	292
EPOS2 24/2	312
EPOS2 Module 36/2	312
EPOS2 24/5	313
EPOS2 50/5	313
EPOS2 P 24/5	316
EPOS3 70/10 EtherCAT	319
Notes	18

Encoder MR
128 - 1000 CPT,
3 channels
Page 272
Encoder Enc
22 mm
100 CPT, 2 channels
Page 274
Encoder HED_ 5540
500 CPT,
3 channels
Page 276/278
DC-Tacho DCT
Ø22 mm
0.52 V
Page 286

Figure 25.16

The data sheet for the Maxon RE 25 motor. The columns correspond to different windings for different nominal voltages. (Image courtesy of Maxon Precision Motors. Motor data is subject to change at any time; consult maxonmotorusa.com for the latest data sheets.)

   i. (Optional) Motor manufacturers may specify slightly different continuous and intermittent operating regions than the ones described in this chapter. For example, the limit of the continuous operating region is not quite vertical in the speed-torque plot of Figure 25.16. Come up with a possible explanation, perhaps using online resources.

3. There are 21 entries on the motor data sheet from Section 25.7. Let us assume zero friction, so we ignore the last entry. To avoid thermal tests, you may also assume a maximum continuous power that the motor coils can dissipate as heat before overheating.

Of the 20 remaining entries, under the assumption of zero friction, how many independent entries are there? That is, what is the minimum number N of entries you need to be able to fill in the rest of the entries? Give a set of N independent entries from which you can derive the other $20 - N$ dependent entries. For each of the $20 - N$ dependent entries, give the equation in terms of the N independent entries. For example, V_{nom} and R will be two of the N independent entries, from which we can calculate the dependent entry $I_{stall} = V_{nom}/R$.

4. This exercise is an experimental characterization of a motor. For this exercise, you need a low-power motor (preferably without a gearhead to avoid high friction) with an encoder. You also need a multimeter, oscilloscope, and a power source for the encoder and motor. Make sure the power source for the motor can provide enough current when the motor is stalled. A low-voltage battery pack is a good choice.
   a. Spin the motor shaft by hand. Get a feel for the rotor inertia and friction. Try to spin the shaft fast enough that it continues spinning briefly after you let go of it.
   b. Now short the motor terminals by electrically connecting them. Spin again by hand, and try to spin the shaft fast enough that it continues spinning briefly after you let go of it. Do you notice the short-circuit damping?
   c. Try measuring your motor's resistance using your multimeter. It may vary with the angle of the shaft, and it may not be easy to get a steady reading. What is the minimum value you can get reliably? To double-check your answer, you can power your motor and use your multimeter to measure the current as you stall the motor's shaft by hand.
   d. Attach one of your motor's terminals to scope ground and the other to a scope input. Spin the motor shaft by hand and observe the motor's back-emf.
   e. Power the motor's encoder, attach the A and B encoder channels to your oscilloscope, and make sure the encoder ground and scope ground are connected together. Do *not* power the motor. (The motor inputs should be disconnected from anything.) Spin the motor shaft by hand and observe the encoder pulses, including their relative phase.
   f. Now power your motor with a low-voltage battery pack. Given the number of lines per revolution of the encoder, and the rate of the encoder pulses you observe on your scope, calculate the motor's no-load speed for the voltage you are using.
   g. Work with a partner. Couple your two motor shafts together by tape or flexible tubing. (This may only work if your motor has no gearhead.) Now plug one terminal of one of the motors (we shall call it the *passive* motor) into one channel of a scope, and plug the other terminal of the passive motor into GND of the same scope. Now power the other motor (the *driving* motor) with a battery pack so that both motors spin. Measure the speed of the passive motor by looking at its encoder count rate on your scope. Also measure its back-emf. With this information, calculate the passive motor's torque constant k_t.
5. Using techniques discussed in this chapter, or techniques you come up with on your own, create a data sheet with all 21 entries for your nominal voltage. Indicate how you

calculated the entry. (Did you do an experiment for it? Did you calculate it from other entries? Or did you estimate by more than one method to cross-check your answer?) For the friction entry, you can assume Coulomb friction only—the friction torque opposes the rotation direction ($b_0 \neq 0$), but is independent of the speed of rotation ($b_1 = 0$). For your measurement of inductance, turn in an image of the scope trace you used to estimate ω_n and L, and indicate the value of C that you used.

If there are any entries you are unable to estimate experimentally, approximate, or calculate from other values, simply say so and leave that entry blank.

6. Based on your data sheet from above, draw the speed-torque curves described below, and answer the associated questions. Do not do any experiments for this exercise; just extrapolate your previous results.
   a. Draw the speed-torque curve for your motor. Indicate the stall torque and no-load speed. Assume a maximum power the motor coils can dissipate continuously before overheating and indicate the continuous operating regime. Given this, what is the power rating P for this motor? What is the max mechanical power P_{max}?
   b. Draw the speed-torque curve for your motor assuming a nominal voltage four times larger than in Exercise 6a. Indicate the stall torque and no-load speed. What is the max mechanical power P_{max}?
7. You are choosing a motor for the last joint of a new direct-drive robot arm design. (A direct-drive robot does not use gearheads on the motors, creating high speeds with low friction.) Since it is the last joint of the robot, and it has to be carried by all the other joints, you want it to be as light as possible. From the line of motors you are considering from your favorite motor manufacturer, you know that the mass increases with the motor's power rating. Therefore you are looking for the lowest power motor that works for your specifications. Your specifications are that the motor should have a stall torque of at least 0.1 Nm, should be able to rotate at least 5 revolutions per second when providing 0.01 Nm, and the motor should be able to operate continuously while providing 0.02 Nm. Which motor do you choose from Table 25.1? Give a justification for your answer.
8. The speed-torque curve of Figure 25.8 is drawn for the positive speed and positive torque quadrant of the speed-torque plane. In this exercise, we will draw the motor's operating region for all four quadrants. The power supply used to drive the motor is 24 V, and assume the H-bridge motor controller (discussed in Chapter 27) can use that power supply to create any average voltage across the motor between −24 and 24 V. The motor's resistance is 1 Ω and the torque constant is 0.1 Nm/A. Assume the motor has zero friction.

   a. Draw the four-quadrant speed-torque operating region for the motor assuming the 24 V power supply (and the H-bridge driver) has no limit on current. Indicate the torque and speed values where the boundaries of the operating region intersect the $\omega = 0$ and $\tau = 0$ axes. Assume there are no other speed or torque constraints on the motor except for the one due to the 24 V limit of the power supply. (Hint: the operating region is unbounded in both speed and torque!)

Table 25.1: Motors to choose from

Assigned power rating	W	3	10	20	90
Nominal voltage	V	15	15	15	15
No load speed	rpm	13,400	4980	9660	7180
No load current	mA	36.8	21.8	60.8	247
Nominal speed	rpm	10,400	3920	8430	6500
Max continuous torque	mNm	2.31	28.2	20.5	73.1
Max continuous current	mA	259	1010	1500	4000
Stall torque	mNm	10.5	131	225	929
Stall current	mA	1030	4570	15,800	47,800
Max efficiency	%	65	87	82	83
Terminal resistance	Ohm	14.6	3.28	0.952	0.314
Terminal inductance	mH	0.486	0.353	0.088	0.085
Torque constant	mNm/A	10.2	28.6	14.3	19.4
Speed constant	rpm/V	932	334	670	491
Mechanical time constant	ms	7.51	4.23	4.87	5.65
Rotor inertia	gcm^2	0.541	10.6	10.4	68.1
Max permissible speed	rpm	16,000	5500	14,000	12,000
Cost	USD	88	228	236	239

Note that sometimes the "Assigned power rating" is different from the mechanical power output at the nominal operating point, for manufacturer-specific reasons. The meanings of the other terms in the table are unambiguous.

b. Update the operating region with the constraint that the power supply can provide a maximum current of 30 A. What is the maximum torque that can be generated using this power supply, and what are the maximum and minimum motor speeds possible at this maximum torque? What is the largest back-emf voltage that can be achieved?
c. Update the operating region with the constraint that the maximum recommended speed for the motor brushes and shaft bearings is 250 rad/s.
d. Update the operating region with the constraint that the maximum recommended torque at the motor shaft is 5 Nm.
e. Update the operating region to show the continuous operating region, assuming the maximum continuous current is 10 A.
f. We typically think of a motor as consuming electrical power ($IV > 0$, or "motoring") and converting it to mechanical power, but it can also convert mechanical power to electrical power ($IV < 0$, or "regenerating"). This occurs in electric car braking systems, for example. Update the operating region to show the portion where the motor is consuming electrical power and the portion where the motor is generating electrical power.

Further Reading

Hughes, A., & Drury, B. (2013). *Electric motors and drives: Fundamentals, types and applications* (4th ed.). Amsterdam: Elsevier.

Maxon DC motor RE 25, ø 25 mm, graphite brushes, 20 Watt. (2015). Maxon.

## CHAPTER 26

Gearing and Motor Sizing

The mechanical power produced by a DC motor is a product of its torque and angular velocity at its output shaft. Even if a DC motor provides enough power for a given application, it may rotate at too high a speed (up to thousands of rpm), and too low a torque, to be useful. In this case, we can add a gearhead to the output shaft to decrease the speed by a factor of $G > 1$ and to increase the torque by a similar factor. In rare cases, we can choose $G < 1$ to actually increase the output speed.

In this chapter we discuss options for gearing the output of a motor, and how to choose a DC motor and gearing combination that works for your application.

26.1 Gearing

Gearing takes many forms, including different kinds of rotating meshing gears, belts and pulleys, chain drives, cable drives, and even methods for converting rotational motion to linear motion, such as racks and pinions, lead screws, and ball screws. All transform torques/forces and angular/linear velocities while ideally preserving mechanical power. For specificity, we refer to a gearhead with an input shaft (attached to the motor shaft) and an output shaft.

Figure 26.1 shows the basic idea. The input shaft is attached to an input gear A with N teeth, and the output shaft is attached to an output gear B with GN teeth, where typically $G > 1$. The meshing of these teeth enforces the relationship

$$\omega_{\text{out}} = \frac{1}{G}\omega_{\text{in}}.$$

Ideally the meshing gears preserve mechanical power, so $P_{\text{in}} = P_{\text{out}}$, which implies

$$\tau_{\text{in}}\omega_{\text{in}} = P_{\text{in}} = P_{\text{out}} = \frac{1}{G}\omega_{\text{in}}\tau_{\text{out}} \quad \rightarrow \quad \tau_{\text{out}} = G\tau_{\text{in}}.$$

It is common to have multiple stages of gearing (Figure 26.2(a)), so the output shaft described above has a second, smaller gear which becomes the input of the next stage. If the gear ratios of the two stages are G_1 and G_2, the total gear ratio is $G = G_1 G_2$.

Embedded Computing and Mechatronics with the PIC32 Microcontroller. http://dx.doi.org/10.1016/B978-0-12-420165-1.00026-3

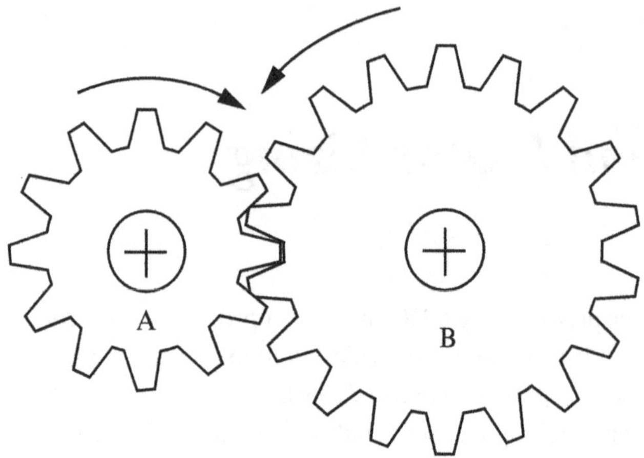

Figure 26.1
The input gear A has 12 teeth and the output gear B has 18, making the gear ratio $G = 1.5$.

Multi-stage gearheads can make huge reductions in speed and increases in torque, up to ratios of hundreds or more.

26.1.1 Practical Issues

Efficiency

In practice, some power is lost due to friction and impacts between the teeth. This power loss is often modeled by an efficiency coefficient $\eta < 1$, such that $P_{\text{out}} = \eta P_{\text{in}}$. Since the teeth enforce the ratio G between input and output velocities, the power loss must appear as a decrease in the available output torque, i.e.,

$$\omega_{\text{out}} = \frac{1}{G}\omega_{\text{in}} \quad \tau_{\text{out}} = \eta G \tau_{\text{in}}.$$

The total efficiency of a multi-stage gearhead is the product of the efficiencies of each stage individually, i.e., $\eta = \eta_1 \eta_2$ for a two-stage gearhead. As a result, high ratio gearheads may have relatively low efficiency.

Backlash

Backlash refers to the angle that the output shaft of a gearhead can rotate without the input shaft moving. Backlash arises due to tolerance in manufacturing; the gear teeth need some play to avoid jamming when they mesh. An inexpensive gearhead may have backlash of a degree or more, while more expensive precision gearheads have nearly zero backlash. Backlash typically increases with the number of gear stages. Some gear types, notably harmonic drive gears (see Section 26.1.2), are specifically designed for near-zero backlash, usually by using flexible elements.

Backlash can be a serious issue in controlling endpoint motions, due to the limited resolution of sensing the gearhead output shaft angle using an encoder attached to the motor shaft (the input of the gearhead).

Backdrivability

Backdrivability refers to the ability to drive the output shaft of a gearhead with an external device (or by hand), i.e., to backdrive the gearing. Typically the motor and gearhead combination is less backdrivable for higher gear ratios, due to the higher friction in the gearhead and the higher apparent inertia of the motor (see Section 26.2.2). Backdrivability also depends on the type of gearing. In some applications we do not want the motor and gearhead to be backdrivable (e.g., if we want the gearhead to act as a kind of brake that prevents motion when the motor is turned off), and in others backdrivability is highly desirable (e.g., in virtual environment haptics applications, where the motor is used to create forces on a user's hand).

Input and output limits

The input and output shafts and gears, and the bearings that support them, are subject to practical limits on how fast they can spin and how much torque they can support. Gearheads will often have maximum input velocity and maximum output torque specifications reflecting these limits. For example, you cannot assume that you get a 10 Nm actuator by adding a $G = 10$ gearhead to a 1 Nm motor; you must make sure that the gearhead is rated for 10 Nm output torque.

26.1.2 Examples

Figure 26.2 shows several different gear types. Not shown are cable, belt, and chain drives, which can also be used to implement a gear ratio while transmitting torques over distances.

Spur gearhead

Figure 26.2(a) shows a multi-stage *spur* gearhead. To keep the spur gearhead package compact, typically each stage has a gear ratio of only 2 or 3; larger gear ratios would require large gears.

Planetary gearhead

A planetary gearhead has an input rotating a *sun* gear and an output attached to a *planet carrier* (Figure 26.2(b)). The sun gear meshes with the planets, which also mesh with an internal gear. An advantage of the planetary gearhead is that more teeth mesh, allowing higher torques.

Bevel gears

Bevel gears (Figure 26.2(c)) can be used to implement a gear ratio as well as to change the axis of rotation.

Worm gears

The screw-like input *worm* interfaces with the output *worm gear* in Figure 26.2(d), making for a large gear ratio in a compact package.

Harmonic drive

The *harmonic drive* gearhead (Figure 26.2(e)) has an elliptical *wave generator* attached to the input shaft and a flexible *flexspline* attached to the output shaft. Ball bearings between the wave generator and the flexibility of the flexspline allow them to move smoothly relative to each other. The flexspline teeth engage with a rigid external *circular spline*. As the wave generator completes a full revolution, the teeth of the flexspline may have moved by as little as one tooth relative to the circular spline. Thus the harmonic drive can implement a high gear ratio (for example, $G = 50$ or 100) in a single stage with essentially zero backlash. Harmonic drives can be quite expensive.

Ball screw and lead screw

A ball screw or lead screw (Figure 26.2(f)) is aligned with the axis of, and coupled to, the motor's shaft. As the screw rotates, a nut on the screw translates along the screw. The nut is prevented from rotating (and therefore must translate) by linear guide rods passing through the nut. The holes in the nuts in Figure 26.2(f) are clearly visible. A lead screw and a ball screw are basically the same, but a ball screw has ball bearings in the nut to reduce friction with the screw.

Ball and lead screws convert rotational motion to linear motion. The ratio of the linear motion to the rotational motion is specified by the *lead* of the screw.

Rack and pinion

The rack and pinion (Figure 26.2(g)) is another way to convert angular to linear motion. The rack is typically mounted to a part on a linear slide.

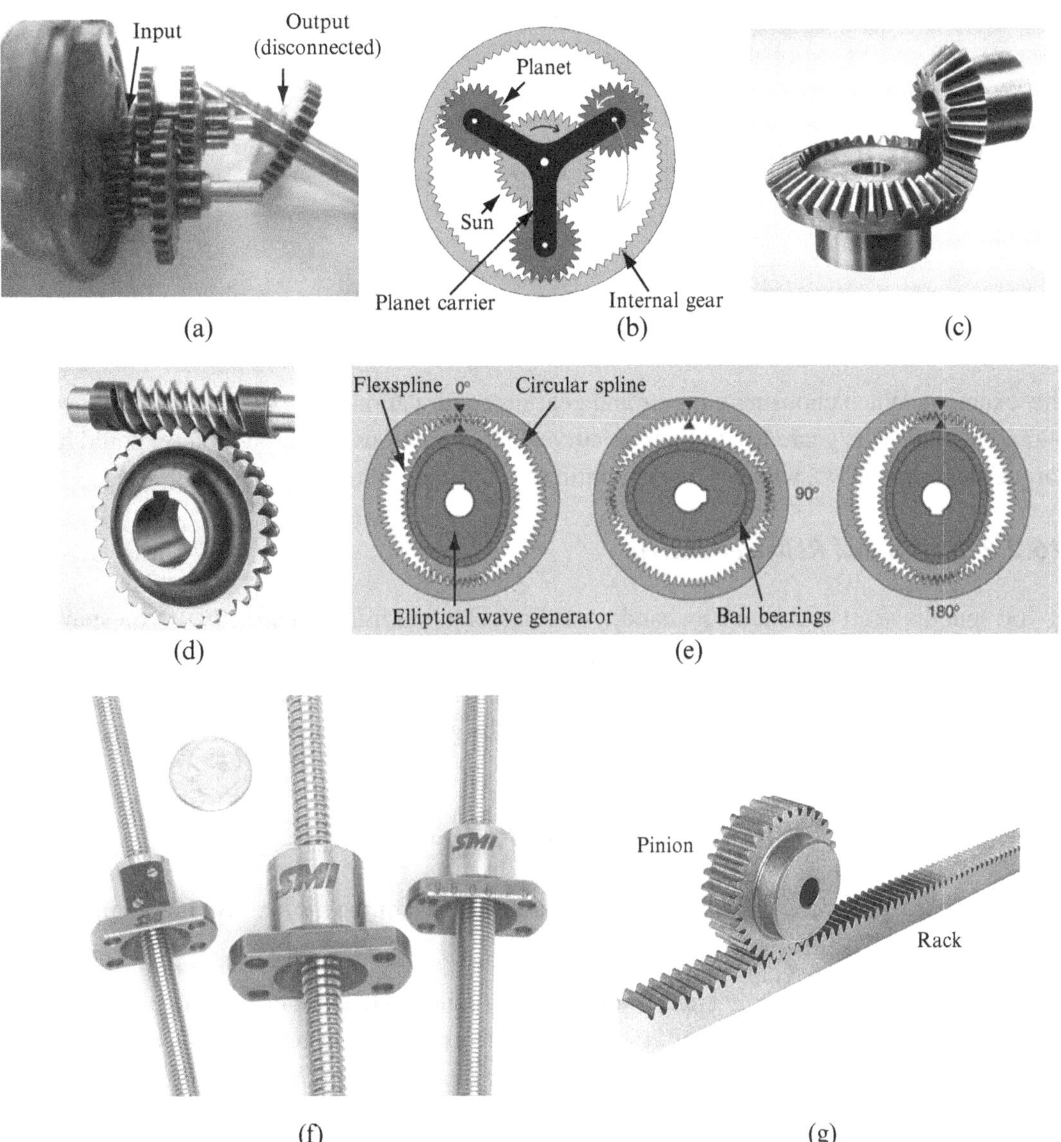

Figure 26.2
(a) Multi-stage spur gearhead. (b) A planetary gearhead. (c) Bevel gears. (d) Worm gears. (e) Harmonic drive gearhead. (f) Ball screws. (g) Rack and pinion.

26.2 Choosing a Motor and Gearhead

26.2.1 Speed-Torque Curve

Figure 26.3 illustrates the effect of a gearhead with $G = 2$ and efficiency $\eta = 0.75$ on the speed-torque curve. The continuous operating torque also increases by a factor ηG, or 1.5 in

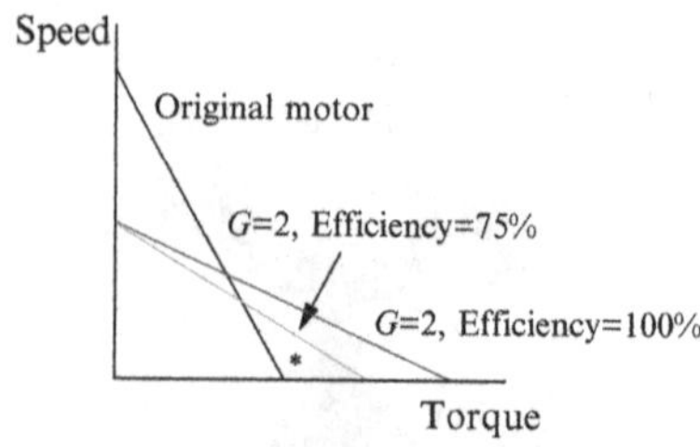

Figure 26.3
The effect of gearing on the speed-torque curve. The operating point * is possible with the gearhead, but not without.

this example. When choosing a motor and gearing combination, the expected operating points should lie under the geared speed-torque curve, and continuous operating points should have torques less than $\eta G\tau_c$, where τ_c is the continuous torque of the motor alone.

26.2.2 Inertia and Reflected Inertia

If you spin the shaft of a motor by hand, you can feel its rotor inertia directly. If you spin the output shaft of a gearhead attached to the motor, however, you feel the *reflected inertia* of the rotor through the gearbox. Say J_m is the inertia of the motor, ω_m is the angular velocity of the motor, and $\omega_{\text{out}} = \omega_m/G$ is the output velocity of the gearhead. Then we can write the kinetic energy of the motor as

$$KE = \frac{1}{2}J_m\omega_m^2 = \frac{1}{2}J_m G^2\omega_{\text{out}}^2 = \frac{1}{2}J_{\text{ref}}\omega_{\text{out}}^2,$$

and $J_{\text{ref}} = G^2 J_m$ is called the reflected (or apparent) inertia of the motor. (We ignore the inertia of the gears.)

Commonly the gearbox output shaft is attached to a rigid-body load. For a rigid body consisting of point masses, the inertia J_{load} about the axis of rotation is calculated as

$$J_{\text{load}} = \sum_{i=1}^{N} m_i r_i^2,$$

where m_i is the mass of point i and r_i is its distance from the axis of rotation. In the case of a continuous body, the discrete sum becomes the integral

$$J_{\text{load}} = \int_V \rho(\mathbf{r})r^2\,\mathrm{d}V(\mathbf{r}),$$

where $\mathbf{r}$ refers to the location of a point on the body, r is the distance of that point to the rotation axis, $\rho(\mathbf{r})$ is the mass density at that point, V is the volume of the body, and $\mathrm{d}V$ is a differential volume element. Solutions to this equation are given in Figure 26.4 for some simple bodies of mass m and uniform density.

If the inertia of a body about its center of mass is J_{cm}, then the inertia J' about a parallel axis a distance r from the center of mass is

$$J' = J_{cm} + mr^2. \tag{26.1}$$

Equation (26.1) is called the *parallel-axis theorem*. With the parallel-axis theorem and the formulas in Figure 26.4, we can approximately calculate the inertia of a load consisting of multiple bodies (Figure 26.5). Typically J_{load} is significantly larger than J_m, but with the gearing, the reflected inertia of the motor $G^2 J_m$ may be as large or larger than J_{load}.

Given a load of mass m and inertia J_{load} (about the gearhead axis) in gravity as shown in Figure 26.6, and a desired acceleration $\alpha > 0$ (counterclockwise), we can calculate the torque needed to achieve the acceleration (using Newton's second law):

$$\tau = (G^2 J_m + J_{load})\alpha + mgr \sin\theta.$$

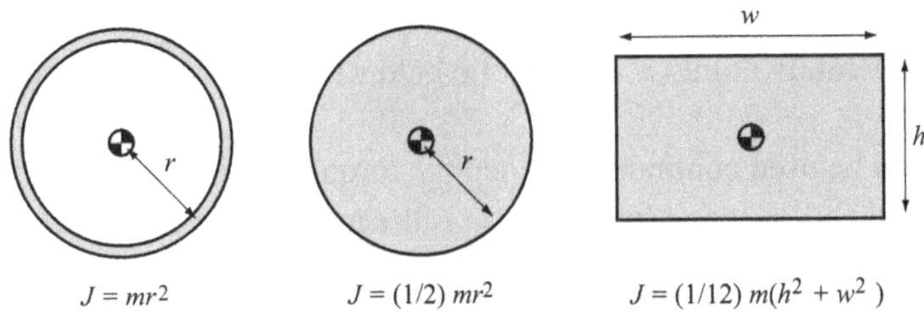

Figure 26.4
Inertia for an annulus, a solid disk, and a rectangle, each of mass m, about an axis out of the page and through the center of mass.

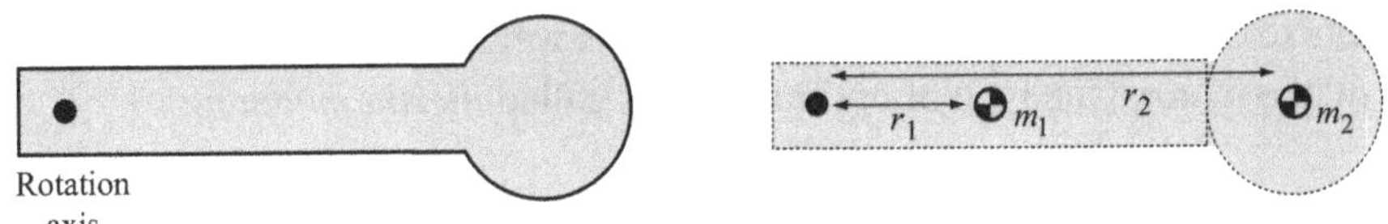

Figure 26.5
The body on the left can be approximated by the rectangle and disk on the right. If the inertias of the two bodies (about their centers of mass) are J_1 and J_2, then the approximate inertia of the compound body about the rotation axis is $J = J_1 + m_1 r_1^2 + J_2 + m_2 r_2^2$ by the parallel-axis theorem.

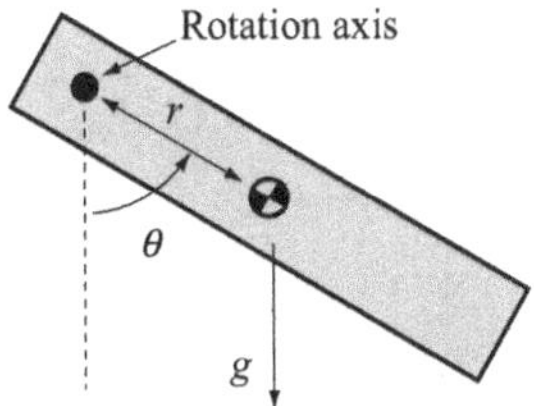

Figure 26.6
A load in gravity.

Given angular velocities at which we would like this acceleration to be possible, we have a set of speed-torque points that must lie under the speed-torque curve (transformed by the gearhead).

26.2.3 Choosing a Motor and Gearhead

To choose a motor and gearing combination, consider the following factors:

- The motor can be chosen based on the peak mechanical power required for the task. If the motor's power rating is sufficient, then theoretically we can follow by choosing a gearhead to give the necessary speed and torque. Our choice of motor might also be constrained by the voltage supply available for the application.
- The maximum velocity needed for the task should be less than ω_0/G, where ω_0 is the no-load speed of the motor.
- The maximum torque needed for the task should be less than $G\tau_{\text{stall}}$, where τ_{stall} is the motor's stall torque.
- Any required operating point (τ, ω) must lie below the gearing-transformed speed-torque curve.
- If the motor will be used continuously, then the torques during this continuous operation should be less than $G\tau_c$, where τ_c is the maximum continuous torque of the motor.

To account for the efficiency η of the gearhead and other uncertain factors, it is a good idea to oversize the motor by a safety factor of 1.5 or 2.

Subject to the hard constraints specified above, we might wish to find an "optimal" design, e.g., to minimize the cost of the motor and gearing, its weight, or the electrical power consumed by the motor. One type of optimization is called *inertia matching*.

Inertia matching

Given the motor inertia J_m and the load inertia J_{load}, the system is inertia matched if the gearing G is chosen so that the load acceleration α is maximized for any given motor torque τ_m. We can express the load acceleration as

$$\alpha = \frac{G\tau_m}{J_{\text{load}} + G^2 J_m}.$$

The derivative with respect to G is

$$\frac{d\alpha}{dG} = \frac{(J_{\text{load}} - G^2 J_m)\tau_m}{(J_{\text{load}} + G^2 J_m)^2}$$

and solving $d\alpha/dG = 0$ yields

$$G = \sqrt{\frac{J_{\text{load}}}{J_m}},$$

or $G^2 J_m = J_{\text{load}}$, hence the term "inertia matched." With this choice of gearing, half of the torque goes to accelerating the motor's inertia and half goes to accelerating the load inertia.

26.3 Chapter Summary

- For gearing with a gear ratio G, the output angular velocity is $\omega_{\text{out}} = \omega_{\text{in}}/G$ and the ideal output torque is $\tau_{\text{out}} = G\tau_{\text{in}}$, where ω_{in} and τ_{in} are the input angular velocity and torque, respectively. If the gear efficiency $\eta < 1$ is taken into account, the output torque is $\tau_{\text{out}} = \eta G \tau_{\text{in}}$.
- For a two-stage gearhead with gear ratios G_1 and G_2 and efficiencies η_1 and η_2 for the individual stages, the total gear ratio is $G_1 G_2$ and total efficiency is $\eta_1 \eta_2$.
- Backlash refers to the amount the output of the gearing can move without motion of the input.
- The reflected inertia of the motor (the apparent inertia of the motor from the output of the gearhead) is $G^2 J_m$.
- A motor and gearing system is inertia matched with its load if

$$G = \sqrt{\frac{J_{\text{load}}}{J_m}}.$$

26.4 Exercises

1. You are designing gearheads using gears with 10, 15, and 20 teeth. When the 10- and 15-teeth gears mesh, you have $\eta = 85\%$. When the 15-and 20-teeth gear mesh, you have $\eta = 90\%$. When the 10- and 20-teeth gear mesh, you have $\eta = 80\%$.
   a. For a one-stage gearhead, what gear ratios $G > 1$ can you achieve, and what are their efficiencies?
   b. For a two-stage gearhead, what gear ratios $G > 1$ can you achieve, and what are their efficiencies? Consider all possible combinations of one-stage gearheads.
2. The inertia of the motor's rotor is J_m, and its load is a uniform solid disk, which will be centered on the gearhead output shaft. The disk has a mass m and a radius R. For what gear ratio G is the system inertia matched?
3. The inertia of the motor's rotor is J_m, and its load is a propeller with three blades. You model the propeller as a simple planar body consisting of a uniform-density solid disk of

radius R and mass M, with each blade a uniform-density solid rectangle extending from the disk. Each blade has mass m, length ℓ, and (small) width w.

a. What is the inertia of the propeller? (Since a propeller must push air to be effective, ideally our model of the propeller inertia would include the *added mass* of the air being pushed, but we leave that out here.)

b. What gear ratio G provides inertia matching?

4. You are working for a startup robotics company designing a small differential-drive mobile robot, and your job is to choose the motors and gearing. A diff-drive robot has two wheels, each driven directly by its own motor, as well as a caster wheel or two for balance. Your design specifications say that the robot should be capable of continuously climbing a 20° slope at 20 cm/s. To simplify the problem, assume that the mass of the whole robot, including motor amplifiers, motors, and gearing, will be 2 kg, regardless of the motors and gearing you choose. Further assume that the robot must overcome a viscous damping force of $(10\ \text{Ns/m})\times v$ when it moves forward at a constant velocity v, regardless of the slope. The radius of the wheels has already been chosen to be 4 cm, and you can assume they never slip. If you need to make other assumptions to complete the problem, clearly state them.

You will choose among the 15 V motors in Table 25.1, as well as gearheads with $G = 1$, 10, 20, 50, or 100. Assume the gearing efficiency η for $G = 1$ is 100%, and for the others, 75%. (Do not combine gearheads! You get to use only one.)

a. Provide a list of all combinations of motor and gearhead that satisfy the specifications, and explain your reasoning. (There are 20 possible combinations: four motors and five gearheads.) "Satisfy the specifications" means that the motor and gearhead can provide at least what is required by the specifications. Remember that each motor only needs to provide half of the total force needed, since there are two wheels.

b. To optimize your design, you decide to use the motor with the lowest power rating, since it is the least expensive. You also decide to use the lowest gear ratio that works with this motor. (Even though we are not modeling it, a lower gear ratio likely means higher efficiency, less backlash, less mass in a smaller package, a higher top-end speed (though lower top-end torque), and lower cost.) Which motor and gearing do you choose?

c. Instead of optimizing the cost, you decide to optimize the power efficiency—the motor and gearing combination that uses the least electrical power when climbing up the 20° slope at a constant 20 cm/s. This is in recognition that battery life is very important to your customers. Which motor and gearhead do you choose?

d. Forget about your previous answers, satisfying the specifications, or the limited set of gear ratios. If the motor you choose has rotor inertia J_m, half of the mass of the robot (including the motors and gearheads) is M, and the mass of the wheels is negligible, what gear ratio would you choose to achieve inertia matching? If you need to make other assumptions to complete the problem, clearly state them.

Further Reading

Budynas, R., & Nisbett, K. (2014). *Shigley's mechanical engineering design.* New York: McGraw-Hill.

CHAPTER 27

DC Motor Control

Driving a brushed DC motor with variable speed and torque requires variable high current. A microcontroller is capable of neither variable analog output nor high current. Both problems are solved through the use of digital PWM and an H-bridge. The H-bridge consists of a set of switches that are rapidly opened and closed by the microcontroller's PWM signal, alternately connecting and disconnecting high voltage to the motor. The effect is similar to the time-average of the voltage. Motion control of the motor is achieved using motor position feedback, typically from an encoder.

27.1 The H-Bridge and Pulse Width Modulation

Let us consider a series of improving ideas for driving a motor. By attempting to fix their problems, we arrive at the H-bridge.

Direct driving from a microcontroller pin (Figure 27.1(a))

The first idea is to simply connect a microcontroller digital output pin to one motor lead and connect the other motor lead to a positive voltage. The pin can alternate between low (ground) and high impedance (disconnected or "tristated"). This connection method is a bad idea, of course, since a typical digital output can only sink a few milliamps, and most motors must draw much more than that to do anything useful.

Current amplification with a switch (Figure 27.1(b))

To increase the current, we can use the digital pin to turn a switch on and off. The switch could be an electromechanical relay or a transistor, and it allows a much larger current to flow through the motor.

Consider, however, what happens when the switch has been closed for a while. A large current is flowing through the motor, which electrically behaves like a resistor and inductor in series. When the switch opens, the current must instantly drop to zero. The voltage across the inductor is governed by

$$V_L = L\frac{\mathrm{d}I}{\mathrm{d}t},$$

Embedded Computing and Mechatronics with the PIC32 Microcontroller. http://dx.doi.org/10.1016/B978-0-12-420165-1.00027-5

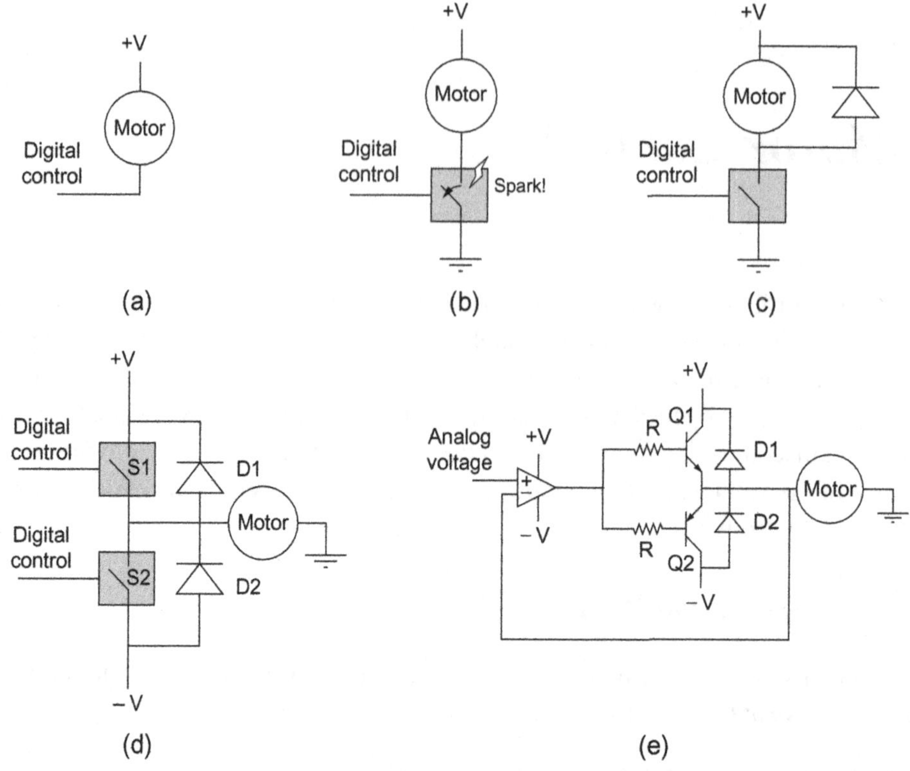

Figure 27.1
A progression of ideas to drive a motor, from worst to better. (a) Attempting to drive directly from a digital output. (b) Using a digital output to control a switch that allows more current to flow. (c) Adding a flyback diode to prevent sparking. (d) Using two switches and a bipolar supply to run the motor bidirectionally. (e) Using an analog control signal, an op-amp, and two transistors to make a linear push-pull amplifier.

so the instantaneous drop in current means that a large (theoretically infinite) voltage develops across the motor leads. The large voltage means that a spark will occur between the switch and the motor lead it was recently attached to. This sparking is certainly bad for the microcontroller.

Adding a flyback diode (Figure 27.1(c))

To prevent the instantaneous change in current and sparking, a *flyback diode* can be added to the circuit. Now when the closed switch is opened, the motor's current has a path to flow through—the diode in parallel with the motor. The voltage across the motor instantaneously changes from +V to the negative of the forward bias voltage of the diode, but that's okay, there is nothing in the resistor-inductor-diode circuit that tries to prevent that. With the switch open, the energy stored in the motor's inductance is dissipated by the current flowing through the motor's resistance, and the initial current I_0 will drop smoothly. Assuming the diode's forward bias voltage is zero, and treating the motor as a resistor and inductor in series, Kirchoff's voltage law tells us that the voltage around the closed loop satisfies

$$L\frac{\mathrm{d}I}{\mathrm{d}t}(t) + RI(t) = 0,$$

and the current through the motor after opening the switch can be solved as

$$I(t) = I_0 \mathrm{e}^{-\frac{R}{L}t},$$

a first-order exponential drop from the initial current I_0 to zero, with time constant equal to the electrical time constant of the motor, $T_e = L/R$.

When the switch is closed, the flyback diode has no effect on the circuit. Flyback diodes should be capable of carrying a lot of current, should be fast switching, and should have low forward bias voltage.

This circuit represents a viable approach to controlling a motor with variable speed: by opening and closing the switch rapidly, it is possible to create a variable average voltage across, and current through, the motor, depending on the duty cycle of the switching. The current always has the same sign, though, so the motor can only be driven in one direction.

Bidirectional operation with two switches and a bipolar supply (Figure 27.1(d))

By using a bipolar power supply (+V, −V, and GND) and two switches, each controlled by a separate digital input, it is possible to achieve bidirectional motion. With the switch S1 closed and S2 open, current flows from +V, through the motor, to GND. With S2 closed and S1 open, current flows through the motor in the opposite direction, from GND, through the motor, to −V. Two flyback diodes are used to provide current paths when switches transition from closed to open. For example, if S2 is open and S1 switches from closed to open, the motor current that was formerly provided by S1 now comes from −V through the diode D2.

To prevent a short circuit, S1 and S2 should never be closed simultaneously.

Bidirectional average voltages between +V and −V are determined by the duty cycle of the rapidly opening and closing switches.

A drawback of this approach is that a bipolar power supply is needed. This issue is solved by the H-bridge. But before discussing the H-bridge, let us consider a commonly used variation of the circuit in Figure 27.1(d).

Linear push-pull amplifier (Figure 27.1(e))

Figure 27.1(e) shows a linear push-pull amplifier. The control signal is a low-current analog voltage to an op-amp configured with negative feedback. Because of the negative feedback and the effectively infinite gain of the op-amp, the op-amp output does whatever it can to make sure that the signals at the inverting and noninverting inputs are equal. Since the inverting input is connected to one of the motor leads, the voltage across the motor is equal to the control voltage at the noninverting input, except with higher current available due to the output transistors. Only one of the two transistors is active at a time: either the NPN bipolar junction transistor Q1 "pushes" current from +V, through the motor, to GND, or the PNP BJT Q2 "pulls" current from GND, through the motor, to −V. Thus the op-amp provides a high impedance input and voltage following of the low-current control voltage, while the transistors provide current amplification.

For example, if +V = 10 V, and the control signal is at 6 V, then the op-amp ensures 6 V across the motor. To double-check that our circuit works as we expect, we calculate the current that would flow through the motor when it is stalled. If the motor's resistance is 6 Ω, then the current $I_e = 6\ \mathrm{V}/6\ \Omega = 1$ A must be provided by the emitter of Q1. If the transistor is capable of providing that much current, we then check if the op-amp is capable of providing the base current $I_b = I_e/(\beta + 1)$ required to activate the transistor, where β is the transistor gain. If so, we are in good shape. The voltage at the base of Q1 is a PN diode drop higher than the voltage across the motor, and the voltage at the op-amp output is that base voltage plus $I_b R$. Note that Q1 is dissipating power approximately equal to the 4 V between the collector and the emitter times the 1 A emitter current, or approximately 4 W. This power goes to heating the transistor, so the transistor must be heat-sinked to allow it to dissipate the heat without burning up.

An example application of a linear push-pull amplifier would be using a rotary knob to control a motor's bidirectional speed. The ends of a potentiometer in the knob would be connected to +V and −V, with the wiper voltage serving as the control signal.

If the op-amp by itself can provide enough current, the op-amp output can be connected directly to the motor and flyback diodes, eliminating the resistors and transistors. Power op-amps are available, but they tend to be expensive relative to using output transistors to boost current.

We could instead eliminate the op-amp by connecting the control signal directly to the base resistors of the transistors. The drawback is that neither transistor would be activated for control signals between approximately −0.7 and 0.7 V, or whatever the base-emitter voltage is when the transistors are activated. We have a "deadband" from the control signal to the motor voltage.

Some issues with the linear push-pull amp, addressed by the H-bridge, include:

- A bipolar power supply is required.
- The control signal is an analog voltage, which is generally not available from a microcontroller.
- The output transistors operate in the linear regime, with significant voltage between the collectors and emitters. A transistor in the linear mode dissipates power approximately equal to the current through the transistor multiplied by the voltage across it. This heats the transistor and wastes power.

Linear push-pull amps are sometimes used when power dissipation and heat are not a concern. They are also common in speaker amplifiers. (A speaker is a current-carrying coil moving in a magnetic field, essentially a linear motor.) There are many improvements to, and variations on, the basic circuit in Figure 27.1(e), and audio applications have raised amplifier circuit design to an art form. You can use a commercial audio amplifier to drive a DC motor, but you would have to remove the high-pass filter on the amplifier input. The high-pass filter is there because we cannot hear sound below 20 Hz, and low-frequency currents simply heat up the speaker coil without producing audible sound.

27.1.1 The H-Bridge

For most motor applications, the preferred amplifier is an H-bridge (Figure 27.2). An H-bridge uses a unipolar power supply (Vm and GND), is controlled by digital pulse width modulation pulse trains that can be created by a microcontroller, and has output transistors (switches) operating in the saturated mode, therefore with little voltage drop across them and relatively little power wasted as heat.

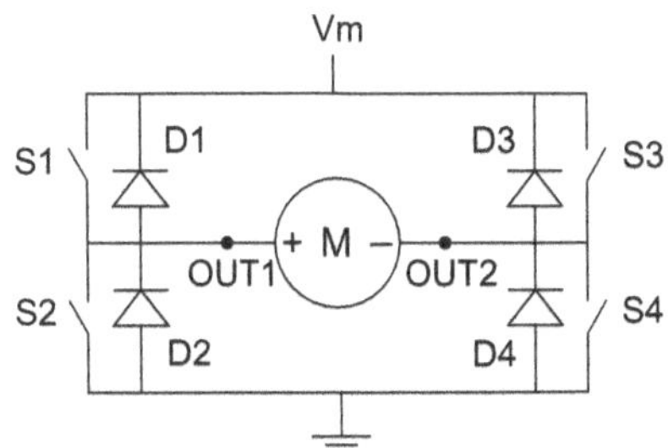

Figure 27.2
An H-bridge constructed of four switches and four flyback diodes. OUT1 and OUT2 are the H-bridge outputs, attached to the two terminals of the DC motor.

An H-bridge consists of four switches, S1–S4, typically implemented by MOSFETs, and four flyback diodes D1–D4.[1] An H-bridge can be used to run a DC motor bidirectionally, depending on which switches are closed:

Closed Switches	Voltage Across Motor
S1, S4	Positive (forward rotation)
S2, S3	Negative (reverse rotation)
S1, S3	Zero (short-circuit braking)
S2, S4	Zero (short-circuit braking)
None or one	Open circuit (coasting)

Switch settings not covered in the table (S1 and S2 closed, or S3 and S4 closed, or any set of three or four switches closed) result in a short circuit and should obviously be avoided!

While you can build your own H-bridge out of discrete components, it is often easier to buy one packaged in an integrated circuit, particularly for low-power applications. Apart from reducing your component count, these ICs also make it impossible for you to accidentally cause a short circuit. An example H-bridge IC is the Texas Instruments DRV8835.

The DRV8835 has two full H-bridges, labeled A and B, each capable of providing up to 1.5 A continuously to power two separate motors. The two H-bridges can be used in parallel to provide up to 3 A to drive a single motor. The DRV8835 works with motor supply voltages (to power the motors) of up to 11 V and logic supply voltages (for interfacing with the microcontroller) between 2 and 7 V. It offers two modes of operation: the IN/IN mode, where the two inputs for each H-bridge control whether the H-bridge is in the forward, reverse, braking, or coasting mode, and the PHASE/ENABLE mode, where one input controls whether the H-bridge is enabled or braking and the other input controls forward vs. reverse if the H-bridge is enabled. We will focus on the PHASE/ENABLE mode.

In the PHASE/ENABLE mode, chosen by setting the MODE pin to logic high, the following truth table determines how the logic inputs (0 and 1) of one H-bridge controls its two outputs:

MODE	PHASE	ENABLE	OUT1	OUT2	Function
1	x	0	L	L	Short-circuit braking (S2, S4 closed)
1	0	1	H	L	Forward (S1, S4 closed)
1	1	1	L	H	Reverse (S2, S3 closed)

[1] MOSFETs themselves allow reverse currents, acting somewhat as flyback diodes, but typically dedicated flyback diodes are incorporated for better performance.

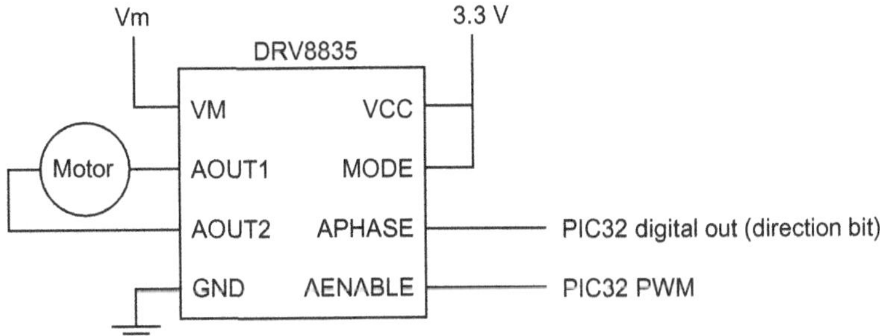

Figure 27.3

Wiring the DRV8835 to use H-bridge A. Not shown is a recommended 10 μF capacitor from VM to GND and a recommended 0.1 μF capacitor from VCC to GND, which may be included already on a DRV8835 breakout board.

When the ENABLE pin is low, OUT1 and OUT2 are held at the same (low) voltage, causing the motor to brake by its own short-circuit damping. When ENABLE is high, then if PHASE is low, switches S1 and S4 are closed, putting a positive voltage across the motor trying to drive it in the forward direction. When ENABLE is high and PHASE is high, switches S2 and S3 are closed, putting a negative voltage across the motor trying to drive it in the reverse direction. PHASE sets the sign of the voltage across the motor, and the duty cycle of the PWM on ENABLE determines the average magnitude of the voltage across the motor, by rapidly switching between approximately +Vm (or −Vm) and zero.

Figure 27.3 shows the wiring of the DRV8835 to use H-bridge A, where Vm is the voltage to power the motor. The logic high voltage VCC is 3.3 V. If the two H-bridges of the DRV8835 are used in parallel for more current, the following pins should be connected to each other: APHASE and BPHASE; AENABLE and BENABLE; AOUT1 and BOUT1; and AOUT2 and BOUT2.

27.1.2 Control with PWM

Rapidly switching ENABLE from high to low can effectively create an average voltage V_{ave} across the motor. Assuming PHASE = 0 (forward), then if DC is the duty cycle of ENABLE, where $0 \le \text{DC} \le 1$, and if we ignore voltage drops due to flyback diodes and resistance at the MOSFETs, then the average voltage across the motor is

$$V_{\text{ave}} \approx \text{DC} * \text{Vm}.$$

Ignoring the details of the motor's inductance charging and discharging, this yields an approximate average current through the motor of

$$I_{\text{ave}} \approx (V_{\text{ave}} - V_{\text{emf}})/R,$$

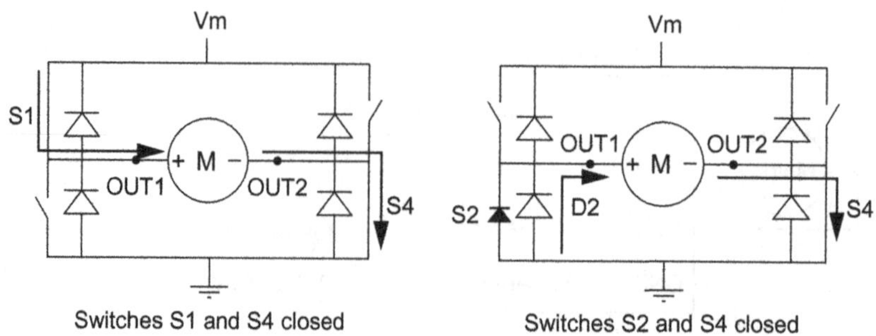

Figure 27.4
Left: The closed switches S1 and S4 provide current to the motor, from Vm to GND. Right: When S1 is opened and S2 is closed, the motor's need for a continuous current is satisfied by the flyback diode D2 and the switch S4. Current could also flow through the MOSFET S2, which acts like a diode to reverse current, but the flyback diode is designed to carry the current.

where $V_{emf} = k_t\omega$ is the back-emf. Since the period of a PWM cycle is typically much shorter than the motor's mechanical time constant T_m, the motor's speed ω (and therefore V_{emf}) is approximately constant during a PWM cycle.

Figure 27.4 shows a motor with positive average current (from left to right). When switches S1 and S4 are closed and S2 and S3 are open (ENABLE is 1, OUT1 is high, and OUT2 is low), S1 and S4 carry current from Vm, through the motor, to ground. When the PWM on ENABLE becomes 0, S2 and S4 are closed and S1 and S3 are open (OUT1 and OUT2 are both low). Because the motor's inductance requires that the current not change instantaneously, the current must flow from ground, through the flyback diode D2, through the motor, then through switch S4 back to ground. (The "closed switch" S2 MOSFET is represented as a diode, since a MOSFET behaves similarly to a diode when current tries to flow in the reverse direction. But the flyback diode is designed to carry this current, so most current flows through D2.)

During the period when OUT1 and OUT2 are low, the voltage across the motor is approximately zero, and the motor current $I(t)$ satisfies

$$0 = L\frac{dI}{dt}(t) + RI(t) + V_{emf},$$

causing an exponential drop in $I(t)$ with time constant $T_e = L/R$, the electrical time constant of the motor. Figure 27.5 shows an example of the voltage across the motor during two cycles of PWM, and the resulting current for two different motors: one with a large T_e and one with a small T_e. A large T_e results in a nearly constant current during a PWM cycle, while a small T_e results in a fast-changing current. The nearly constant motor velocity during a PWM cycle gives a nearly constant V_{emf}.

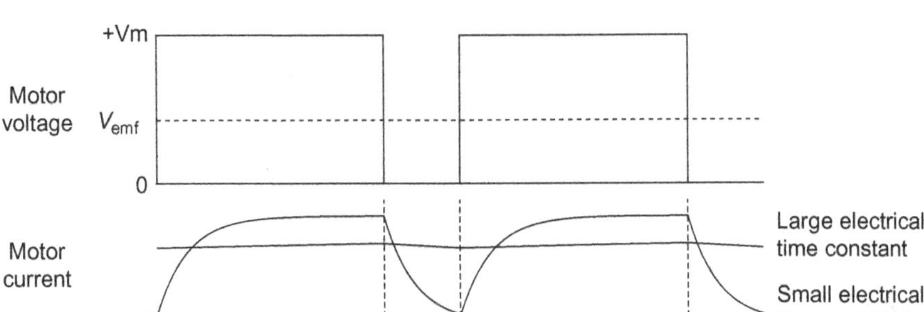

Figure 27.5
A PWM signal puts Vm across the motor for 75% of the PWM cycle and 0 V for 25% of the cycle. This causes a nearly constant positive current for a motor with large L/R, while the current for a motor with small L/R varies significantly.

To understand the behavior of the electromechanical system under PWM control, we need to consider three time scales: the PWM period T, the electrical time constant T_e, and the mechanical time constant T_m. The PWM frequency $1/T$ should be chosen to be much higher than $1/T_m$, to prevent significant speed variation during a PWM cycle. Ideally the PWM frequency would also be much higher than $1/T_e$, to minimize current variation during a cycle. One reason to want little current variation is that more-constant current results in less power wasted heating the motor coils. To see why, consider a motor with resistance $R = 1\ \Omega$ powered by a constant current of 2 A vs. a current alternating with 50% duty cycle between 4 and 0 A, for a time-average of 2 A. Both provide the same average torque, but the average power to heating the coils in the first case is $I^2R = 4$ W while it is $0.5(4\text{ A})^2(1\ \Omega) = 8$ W in the second.

Another consideration is audible noise: to make sure the switching is inaudible, the PWM frequency $1/T$ should be chosen at 20 kHz or larger.

On the other hand, the PWM frequency should not be chosen too high, as it takes time for the H-bridge MOSFETs to switch. During switching, when larger voltages are across the active MOSFETs, more power is wasted to heating. If switching occurs too often, the H-bridge may even overheat. The DRV8835 takes approximately 200 ns to switch, and its maximum recommended PWM frequency is 250 kHz.

Trading off the considerations above, common PWM frequencies are in the range 20-40 kHz.

27.1.3 Regeneration

When the voltage across the motor and the current through the motor have the same sign, the motor is consuming electrical power ($IV > 0$). When the voltage across the motor and the current through the motor have opposite signs ($IV < 0$), then the motor is acting as a generator and is actually producing electrical power. This phenomenon is called *regeneration*.

Regeneration may occur when braking a motor, for example. Regenerative braking is used in hybrid and electric cars to convert some of the car's kinetic energy into battery energy, instead of just wasting it heating the brake pads.

For concreteness, consider the H-bridge of Figure 27.2 powered by 10 V, flyback diodes with a forward bias voltage of 0.7 V, and a motor with a resistance of 1 Ω and a torque constant of 0.01 Nm/A (0.01 Vs/rad). Consider these two examples of regeneration.

1. **Forced motion of the motor output.** Assume all H-bridge switches are open and an external power source spins the motor shaft. The external source could be water falling over the blades of a turbine in a hydroelectric dam, for example. If the motor shaft spins at a constant $\omega = 2000$ rad/s, then the back-emf is $k_t\omega = (0.01\ \text{Vs/rad})(2000\ \text{rad/s}) = 20$ V. The flyback diodes cap the voltage across the motor to the range [−11.4 V, 11.4 V], however, so current must be flowing through the motor. Assuming the flyback diodes D1 and D4 are conducting, we have

$$11.4\ \text{V} = k_t\omega + IR = 20\ \text{V} + I(1\ \Omega).$$

   Solving for I, we get a current of $I = -8.6$ A, flowing from ground through D4, the motor, and D1 to the 10 V supply (Figure 27.6). The power consumed by the motor is (−8.6 A)(11.4 V) = −98.04 W, i.e., the motor is generating 98.04 W. (If we had assumed the flyback diodes D2 and D3 were conducting instead, putting −11.4 V across the motor, and solved for I, we would have seen that the required negative current, from right to left, cannot be provided by D2 and D3. Therefore, D2 and D3 are not conducting.)
2. **Motor braking.** Assume the motor has a positive current of 2 A through it (left to right, carried by switches S1 and S4), then all switches are opened. Immediately after the

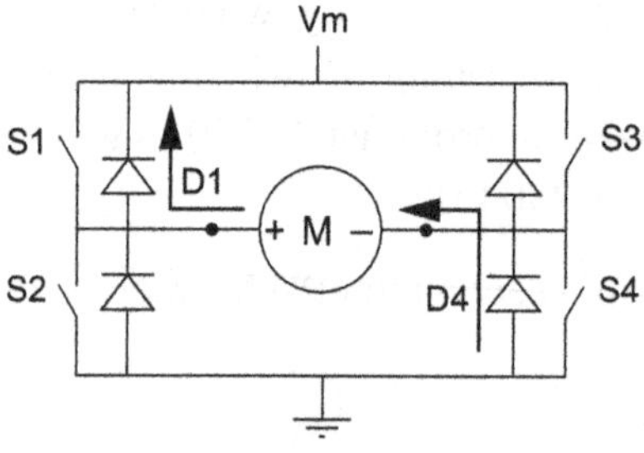

Figure 27.6
One example of regeneration, where the current through and voltage across a motor have opposite signs. The motor is forced to spin forward, by an external source, at a speed ω such that the back-emf $k_t\omega$ is larger than Vm. This forces a negative current to flow through the motor, carried by the flyback diodes D1 and D4. Electrical power is dumped into the power supply.

switches are opened, the only option to continue providing 2 A to the motor is from ground, through D2, the motor, and D3 to the 10 V supply. The voltage across the motor must therefore be −11.4 V: two diode drops and the 10 V supply voltage. The motor consumes (2 A)(−11.4 V) = −22.8 W, i.e., it is acting as a generator, providing 22.8 W of power just after the switches are opened.

As these examples show, regeneration dumps current back into the power supply, charging it up, whether it wants to be charged or not. Some batteries can directly accept the regeneration current. For a wall-powered supply, however, a high-capacitance, high-voltage, typically polarized electrolytic capacitor at the power supply outputs can act as storage for energy dumped back into the power supply. While such a capacitor may be present in a linear power supply, a switched-mode power supply is unlikely to have one, so an external capacitor would have to be added. If the power supply capacitor voltage gets too high, a voltage-activated switch can allow the back-current to be redirected to a "regen" power resistor, which is designed to dissipate electrical energy as heat.

27.1.4 Other Practical Considerations

Motors are electrically noisy devices, creating both electromagnetic interference (EMI), e.g., induced currents on sensitive electronics due to changing magnetic fields induced by large changing motor currents, as well as voltage spikes, due to brush switching and changing PWM current and voltage. These effects can disrupt the functioning of your microcontroller, cause erroneous readings on your sensor inputs, etc. Dealing with EMI is beyond our scope, but it can be minimized by keeping the motor leads short and far from sensitive circuitry, and by using shielded cable or twisted wires for motor and sensor leads.

Optoisolators can be used to separate noisy motor power supplies from clean logic voltage supplies. An optoisolator consists of an LED and a phototransistor. When the LED turns on, the phototransistor is activated, allowing current to flow from its collector to its emitter. Thus a digital on/off signal can be passed between the logic circuit and the power circuit using only an optical connection, eliminating an electrical connection. In our case, the PIC32's H-bridge control signals would be applied to the LEDs and converted by the phototransistors to high and low signals to be passed to the inputs of the H-bridge.

Optoisolators can be bought in packages with multiple optoisolators. Each LED-phototransistor pair uses four pins: two for the internal LED and two for the collector and emitter of the phototransistor. Thus you can get a 16-pin DIP chip with four optoisolators, for example.

It is also common to directly solder a nonpolarized capacitor across the motor terminal leads, effectively turning the motor into a capacitor in parallel with the resistance and inductance of the motor. This capacitor helps to smooth out voltage spikes due to brushed commutation.

Finally, the H-bridge chip should be heat-sinked to prevent overheating. The heat sink dissipates heat due to MOSFET switching and MOSFET output resistance (on the order of hundreds of mΩ for the DRV8835).

27.2 Motion Control of a DC Motor

An example block diagram for control of a DC motor is shown in Figure 27.7.[2] A trajectory generator creates a reference position as a function of time. To drive the motor to follow this reference trajectory, we use two nested control loops: an outer motion control loop and an inner current control loop. These two loops are roughly motivated by the two time scales of the system: the mechanical time constant of the motor and load and the electrical time constant of the motor.

- **Outer motion control loop.** This outer loop runs at a lower frequency, typically a few hundred Hz to a few kHz. The motion controller takes as input the desired position and/or velocity, as well as the motor's current position, as measured by an encoder or potentiometer, and possibly the motor's current velocity, as measured by a tachometer. The output of the controller is a commanded current I_c. The current is directly proportional to the torque. Thus the motion control loop treats the mechanical system as if it has direct control of motor torque.
- **Inner current control loop.** This inner loop typically runs at a higher frequency, from a few kHz to tens of kHz, but no higher than the PWM frequency. The purpose of the current controller is to deliver the current requested by the motion controller. To do this, it

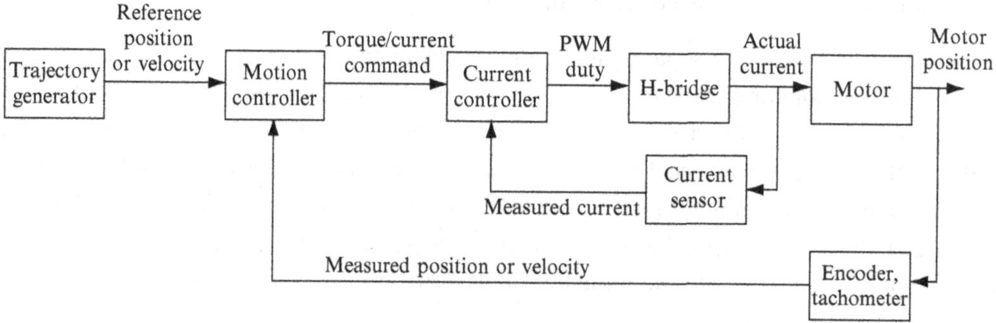

Figure 27.7
A block diagram for motion control.

[2] A simpler block diagram would have the motion controller block directly output a PWM duty cycle to an H-bridge, with no inner-loop control of the actual motor current, which would be sufficient for many applications. However, the block diagram in Figure 27.7 is more typical of industrial implementations.

monitors the actual current flowing through the motor and outputs a commanded average voltage V_c (expressed as a PWM duty cycle).

Traditionally a mechanical engineer might design the motion control loop, and an electrical engineer might design the current control loop. But you are a mechatronics engineer, so you will do both.

27.2.1 Motion Control

Feedback control

Let θ and $\dot{\theta}$ be the actual position and velocity of the motor, and r and $\dot{r}$ be the desired position and velocity. Define the error $e = r - \theta$, error rate of change $\dot{e} = \dot{r} - \dot{\theta}$, and error sum (approximating an integral) $e_{\text{int}} = \sum_k e(k)$. Then a reasonable choice of controller would be a PID controller (Chapter 23),

$$I_{\text{c,fb}} = k_p e + k_i e_{\text{int}} + k_d \dot{e}, \tag{27.1}$$

where $I_{\text{c,fb}}$ is the commanded current by the feedback controller. The $k_p e$ term acts as a virtual spring that creates a force proportional to the error, pulling the motor to the desired angle. The $k_d \dot{e}$ term acts as a virtual damper that creates a force proportional to the "velocity" of the error, driving the error rate of change toward zero. The $k_i e_{\text{int}}$ term creates a force proportional to the time integral of error. See Chapter 23 for more on PID control.

In the absence of a model of the motor's dynamics, a reasonable commanded current I_c is simply $I_c = I_{\text{c,fb}}$.

An alternative form of the feedback controller (27.1) is

$$\ddot{\theta}_d = k_p e + k_i e_{\text{int}} + k_d \dot{e}, \tag{27.2}$$

where the feedback gains set a desired corrective acceleration of the motor $\ddot{\theta}_d$ instead of a current. This alternative form of the PID controller is used in conjunction with a system model in the next section.

Feedforward plus feedback control

If you have a decent model of the motor and its load, a model-based controller can be combined with a feedback controller to yield better performance. For example, for an unbalanced load as in Figure 27.8, you could choose a feedforward current command to be

$$I_{\text{c,ff}} = \frac{1}{k_t}(J\ddot{r} + mgd\sin\theta + b_0\,\text{sgn}(\dot{\theta}) + b_1\dot{\theta}),$$

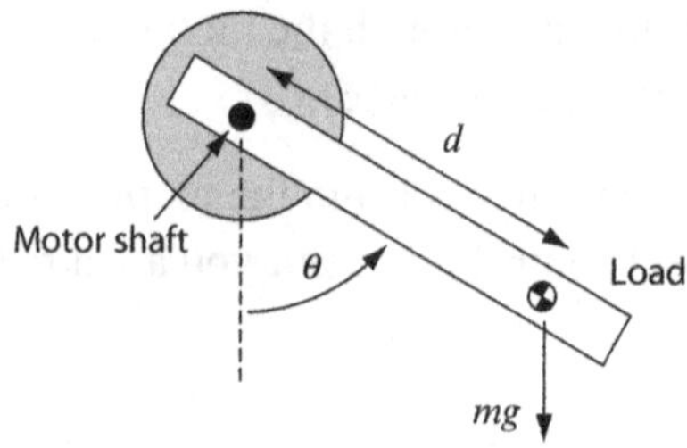

Figure 27.8
An unbalanced load in gravity.

where k_t is the torque constant, J is the motor and load inertia about the motor's axis, the planned motor acceleration $\ddot{r}$ can be obtained by finite differencing the desired trajectory, mg is the weight of the load, d is the distance of the load center of mass from the motor axis, θ is the angle of the load from vertical, b_0 is Coulomb friction torque, and b_1 is a viscous friction coefficient. To compensate for errors, the feedback current command $I_{c,fb}$ can be combined with the feedforward command to yield

$$I_c = I_{c,ff} + I_{c,fb}. \tag{27.3}$$

Alternatively, the motor acceleration feedback law (27.2) could be combined with the system model to yield the controller

$$I_c = \frac{1}{k_t}(J(\ddot{r} + \ddot{\theta}_d) + mgd\sin\theta + b_0\,\mathrm{sgn}(\dot{\theta}) + b_1\dot{\theta}), \tag{27.4}$$

an implementation of a model-based controller from Chapter 23.4.

27.2.2 Current Control

To implement a current controller, a current sensor is required. In this chapter we assume a current-sense resistor with a current-sense amplifier, as described in Chapter 21.10.1 and Figure 21.22.

The output of the current controller is V_c, the commanded average voltage (to be converted to a PWM duty cycle). The simplest current controller would be

$$V_c = k_V I_c$$

for some gain k_V. This controller would be a good choice if your load were only a resistance R, in which case you would choose $k_V = R$. Even if not, if you do not have a good mechanical model of your system, achieving a particular current/torque may not matter anyway. You can just tune your motion control PID gains, use $k_V = 1$, and not worry about what the actual current is, eliminating the inner control loop.

On the other hand, if your battery pack voltage changes (due to discharging, or changing batteries, or changing from a 6 to a 12 V battery pack), the change in behavior of your overall controller will be significant if you do not measure the actual current in your current controller. More sophisticated current controller choices might be a mixed model-based and integral feedback controller

$$V_c = I_c R + k_t\dot{\theta} + k_{I,i}\, e_{I,\mathrm{int}}$$

or, recognizing that the electrical system is a first-order system (using voltage to control a current through a resistor and inductor), a PI feedback controller

$$V_c = k_{I,p}\, e_I + k_{I,i}\, e_{I,\mathrm{int}},$$

where e_I is the error between the commanded current I_c and the measured current, $e_{I,\mathrm{int}}$ is the integral of current error, R is the motor resistance, k_t is the torque constant, $k_{I,p}$ is a proportional current control gain, and $k_{I,i}$ is an integral current control gain. A good current controller would closely track the commanded current.

27.2.3 An Industrial Example: The Copley Controls Accelus Amplifier

Copley Controls is a well known manufacturer of amplifiers for brushed and brushless motors for industrial applications and robotics. One of their models is the Accelus, pictured in Figure 27.9. The Accelus supports many different operating modes. Examples include control of motor current or velocity to be proportional to either an analog voltage input or the duty cycle of a PWM input. A microcontroller on the Accelus interprets the analog input or PWM duty cycle and implements a controller similar to that in Figure 27.7.

Figure 27.9
The Copley Controls Accelus amplifier. (Image courtesy of Copley Controls, copleycontrols.com.)

The mode most relevant to us is the Programmed Position mode. In this mode, the user specifies a few parameters to describe a desired rest-to-rest motion of the motor. The controller's job is to drive the motor to track this trajectory.

When the amplifier is first paired with a motor, some initialization steps must be performed. A GUI interface on the host computer, provided by Copley, communicates with the microcontroller on the Accelus using RS-232.

1. **Enter motor parameters.** From the motor's data sheet, enter the inertia, peak torque, continuous torque, maximum speed, torque constant, resistance, and inductance. These values are used for initial guesses at control gains for motion and current control. Also enter the number of lines per revolution of the encoder.
2. **Tune the inner current control loop.** Set a limit on the recent current to avoid overheating the motor. This limit is based on the integral $\int_{T_1}^{T_2} I^2(t)\,dt$, which is a measure of how much energy the motor coils have dissipated recently. (When this limit is exceeded, the motor current is limited to the continuous operating current until the history of currents indicates that the motor has cooled.) Also, tune the values of P and I control gains for a PI current controller. This tuning is assisted by plots of reference and actual currents as a function of time. See Figure 27.10, which shows a square wave reference current of amplitude 1 A and frequency 100 Hz. The zero average current and high frequency of the reference waveform ensure that the motor does not move during current tuning, which focuses on the electrical properties of the motor.
   The current control loop executes at 20 kHz, which is also the PWM frequency (i.e., the PWM duty cycle is updated every cycle).

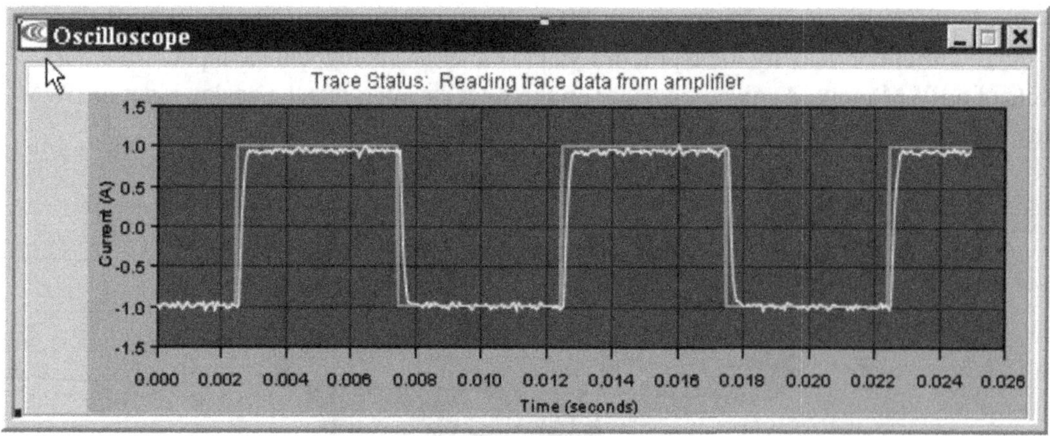

Figure 27.10
A plot of the reference square wave current and the actual measured current during PI current controller tuning.

3. **Tune the outer motion control loop with the load attached.** Attach the load to the motor and tune PID feedback control gains, a feedforward acceleration term, and a feedforward velocity term to achieve good tracking of sample reference trajectories. This process is assisted by plots of reference and actual positions and velocities as a function of time. The motion control loop executes at 4 kHz.

Once the initial setup procedures have been completed, the Accelus microcontroller saves all the motor parameters and control gains to nonvolatile flash memory. These tuned parameters then survive power cycling and are available the next time you power up the amplifier.

Now the amplifier is ready for use. The user specifies a desired trajectory using any of a number of interfaces (RS-232, CAN, etc.), and the amplifier uses the saved parameters to drive the motor to track the trajectory.

27.3 Chapter Summary

- An H-bridge amplifier allows bidirectional control of a DC motor based on a PWM control signal.
- Flyback diodes are used with an H-bridge to provide a current path for the inductive load (the motor) at all times as the H-bridge transistors switch on and off.
- A motor acts as an electrical generator, generating electrical power instead of consuming it, when the voltage across the motor and the current through it have opposite signs. An example is regenerative braking in electric and hybrid vehicles.
- A typical motor control system has a nested structure: an outer motion-control loop, which commands a torque (or current) from the inner current-control loop, which attempts to deliver the current requested by the outer-loop controller. Typically the inner-loop controller executes at a higher frequency than the outer-loop controller.

27.4 Exercises

1. The switch in Figure 27.1(b), with no flyback diode, has been closed for a long time, and then it is opened. The voltage supply is 10 V, the motor's resistance is $R = 2\ \Omega$, the motor's inductance is $L = 1$ mH, and the motor's torque constant is $k_t = 0.01$ Nm/A. Assume the motor is stalled.
   a. What is the current through the motor just before the switch is opened?
   b. What is the current through the motor just after the switch is opened?
   c. What is the torque being generated by the motor just before the switch is opened?
   d. What is the torque being generated by the motor just after the switch is opened?
   e. What is the voltage across the motor just before the switch is opened?
   f. What is the voltage across the motor just after the switch is opened?

2. The switch in Figure 27.1(c), with the flyback diode, has been closed for a long time, and then it is opened. The voltage supply is 10 V, the motor's resistance is $R = 2\ \Omega$, the motor's inductance is $L = 1$ mH, and the motor's torque constant is $k_t = 0.01$ Nm/A. The flyback diode has a forward bias voltage drop of 0.7 V. Assume the motor is stalled.
   a. What is the current through the motor just before the switch is opened?
   b. What is the current through the motor just after the switch is opened?
   c. What is the torque being generated by the motor just before the switch is opened?
   d. What is the torque being generated by the motor just after the switch is opened?
   e. What is the voltage across the motor just before the switch is opened?
   f. What is the voltage across the motor just after the switch is opened?
   g. What is the rate of change of current through the motor dI/dt just after the switch is opened? (Make sure to use a correct sign, relative to your current answers above.)
3. In Figure 27.1(d), the voltage supplies are ± 10 V, the motor's resistance is $R = 5\ \Omega$, the motor's inductance is $L = 1$ mH, and the motor's torque constant is $k_t = 0.01$ Nm/A. The flyback diodes have a forward bias voltage drop of 0.7 V. Switch S1 has been closed for a long time, with no voltage drop across it, and the motor is stalled. Switch S2 is open. Then switch S1 is opened while switch S2 remains open. Immediately after S1 opens, which flyback diode conducts current? What is the voltage across the motor? What is the current through the motor? What is the rate of change of the current through the motor?
4. Give some advantages of driving a DC motor using an H-bridge with PWM over a linear push-pull amplifier with an analog control input. Give at least one advantage of using a linear push-pull amplifier over an H-bridge. (Hint: consider the case of low PWM frequency or low motor inductance.)
5. Explain why an initially spinning motor comes to rest faster if the two motor leads are shorted to each other rather than left disconnected. Derive the result from the motor voltage equation.
6. Provide a circuit diagram showing the DRV8835 configured to drive a single motor with more than 2 A continuous.
7. Consider a motor connected to an H-bridge with all switches opened (motor is unpowered). The motor rotor is rotated by external forces (e.g., rushing water in a hydroelectric dam spinning the blades of a turbine). If the H-bridge is connected to a battery supply of voltage Vm, and the forward bias voltage of the flyback diodes is Vd, give a mathematical expression for the rotor speed at which the battery begins to charge. When this speed is exceeded, and assuming that the battery voltage Vm is constant (e.g., it acts somewhat like a very high capacitance capacitor, accepting current without changing voltage), give an expression for the current through the motor as a function of the rotor speed. Also give an expression for the power lost due to the heating the windings.

   How does the presence of the hydrogenerator affect the total energy of a bucket of water at the top of the dam compared to the total energy just before that water splashes into the river at the bottom of the dam? (The total energy is the potential energy plus the kinetic energy.)

8. To create an average current I through a motor, you could send I constantly, or you could alternate quickly between kI for $100\%/k$ of a cycle and zero current for the rest of the cycle. Provide an expression for the average power dissipated by the motor coils for each of these cases.
9. Imagine a motor with a 500 line incremental encoder on one end of the motor shaft and a gearhead with $G = 50$ on the other end. If the encoder is decoded in 4x mode, how many encoder counts are counted per revolution of the gearhead output shaft?
10. A simple outer-loop motion controller is to command a torque (current) calculated by a PID controller. A more sophisticated controller would attempt to use a model of the motor-load dynamics to get better trajectory tracking. One possibility is the control law (27.3). Another possibility is the control law (27.4). Describe any advantages of (27.4) over (27.3).
11. Choose R2/R1 for the current-sense amplifier in Figure 21.22 in Chapter 21 so the output voltage OUT swings full range (0-3.3 V) for a current range of ± 1 A.
12. Choose an example R and C for the current-sense amplifier in Figure 21.22 in Chapter 21 to create a cutoff frequency of 200 Hz.
13. For the current-sense amplifier in Figure 21.22 in Chapter 21, the reference voltage at REFIN should be constant. We know that the currents into and out of the high-impedance op-amp inputs REFIN and FB are negligible, as is the current into the PIC32 analog input. But the currents through the external resistors R1 and R2 may not be small. As a result, REFIN actually varies as a function of the sensed input voltage and the output voltage OUT. We should choose the resistances R1, R2, and R3 so the voltage variation at REFIN is small.

    Pay attention only to the op-amp portion of the circuit in the bottom of Figure 21.22 in Chapter 21. Now assume that OUT is 3.3 V. What is the voltage at REFIN as a function of R1, R2, and R3? (Note that it is not exactly 1.65 V.) Use your equation to comment on how to choose the relative values of R1, R2, and R3 to make sure that REFIN is close to 1.65 V. Explain other practical constraints on the absolute values (minimum and maximum values) of R1, R2, and R3. (You may focus on slowly changing signals at the + input of the op amp. For high-frequency inputs, using large resistances R1 and R2 may combine with small parasitic capacitances in the circuit to create RC time delays that adversely affect the response.)
14. You decide on a current amplifier gain of $G = 101$. Using your result from the previous exercise, choose R1, R2, and R3 to achieve the gain G while ensuring that REFIN does not vary by more than 1 mV for any voltage at OUT in the range [0, 3.3 V]. For your choice of R3, indicate how much power is used in the R3-R3 voltage divider.
15. Due to natural variations in resistor values within their tolerance ranges, the gain G, and the voltage at REFIN, that you design for your current-sense circuit in Figure 21.22 in

Chapter 21 will not be exactly realized. Due to this and other variations, you need to calibrate your current sensor by running some experiments. Clearly explain what experiments you would do, and how you would use the results to interpret analog voltages read at the PIC32 as motor currents. Be specific; the interpretation should be easy to implement in software.

16. Clearly outline how you would implement the motor controller in Figure 27.7 in software on a PIC32. Indicate what peripherals, ISRs, NU32 functions, and global variables you might use. You may use concepts from the LED brightness control project, if it is helpful. In particular, indicate how you would implement:
    - **The trajectory generator.** Assume that desired trajectories are sent from MATLAB on the host computer, and trajectory tracking results are plotted in MATLAB.
    - **The outer-loop motion controller.** Assume that an external decoder/counter chip maintains the encoder count, and that communication with the chip occurs using SPI. How would you collect trajectory tracking data to be plotted in MATLAB?
    - **The inner-loop current controller**, using the current sensor described in this chapter. How would you collect data on tracking the desired current to be plotted in MATLAB?

Further Reading

Accelus panel: Digital servoamplifier for brushless/brush motors. (2011). Copley Controls.

DRV8835: Dual low voltage H-bridge IC. (2014). Texas Instruments.

DRV8835 dual motor driver carrier. (2015). `https://www.pololu.com/product/2135` (Accessed: May 6, 2015).

MAX9918/MAX9919/MAX9920 −20V to +75V input range, precision uni-/bidirectional, current-sense amplifiers. (2015). Maxim Integrated.

CHAPTER 28

A Motor Control Project

Imagine combining the power and convenience of a personal computer with the peripherals of a microcontroller. Motors moving in response to keyboard commands. Plots showing the motor's positional history. Interactive controller tuning. By combining your knowledge of microcontrollers, motors, and controls you can accomplish these goals.

It starts with a menu. This menu will provide a simple user interface: a list of commands that you can choose by pressing a key. We will begin with an empty menu. By the end of this project, it will have nearly 20 options: everything from reading encoders to setting control gains to plotting trajectories. Much work lies ahead, but by breaking the project into subgoals and using discipline in your coding, you will complete it successfully.

28.1 Hardware

The major hardware components for the project are listed below. Data sheets for the chips can be found on the book's website.

- brushed DC motor with encoder, including a mounting bracket and a bar for the motor load (Figure 28.1)
- the NU32 microcontroller board
- a printed circuit board with an encoder counting chip counting in 4x mode (quadrature inputs from the encoder and an SPI interface to the NU32)
- a printed circuit board with the MAX9918 current-sense amplifier and a built-in 15 mΩ current-sense resistor, to provide a voltage proportional to the motor current (this voltage output is attached to an ADC input on the NU32)
- a printed circuit board breaking out the DRV8835 H-bridge (attached to a digital output and a PWM/output compare channel on the NU32)
- a battery pack to provide power to the H-bridge
- resistors and capacitors

We discuss the hardware in greater detail after an overview of the software you will develop.

Embedded Computing and Mechatronics with the PIC32 Microcontroller. http://dx.doi.org/10.1016/B978-0-12-420165-1.00028-7

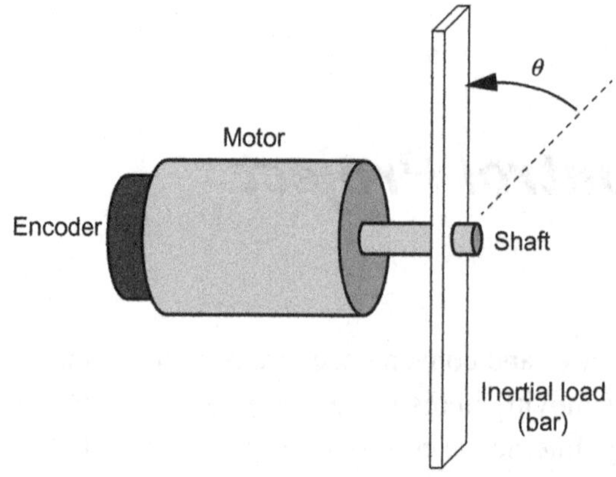

Figure 28.1
A bar attached as the motor's load. Positive rotation θ is counterclockwise when viewing along the axis of the motor shaft toward the motor.

28.2 Software Overview

The motivation for this project is to build an intelligent motor driver, with a subset of the features found in industrial products like the Copley Accelus drive described in Chapter 27. The drive incorporates the amplifier to drive the motor as well as feedback control to track a reference position, velocity, or torque. Many robots and computer-controlled machine tools have drives like this, one for each joint.

Your motor drive system should accept a desired motor trajectory, execute that trajectory, and send the results back to your computer for plotting (Figure 28.2). We break this project into several supporting features, accessible from a menu on your computer. This approach enables you to build and test incrementally.

This project requires you to develop two different pieces of software that communicate with each other: the PIC32 code for the motor driver and the *client* user interface that runs on the host computer. In this chapter we assume that the client is developed for MATLAB, taking advantage of its plotting capabilities. You could easily use another programming language (e.g., python with its plotting capability), provided you can establish communication with the NU32 via the UART (Chapter 11).

For convenience, we will refer to the code on the host computer as the "client" and the code on the PIC32 as the "PIC32."

The PIC32 will implement the control strategy of Chapter 27, specifically Figure 27.7, consisting of a low-frequency position control loop and a nested high-frequency current control loop. This is a common industrial control scheme. In this project, the outer position

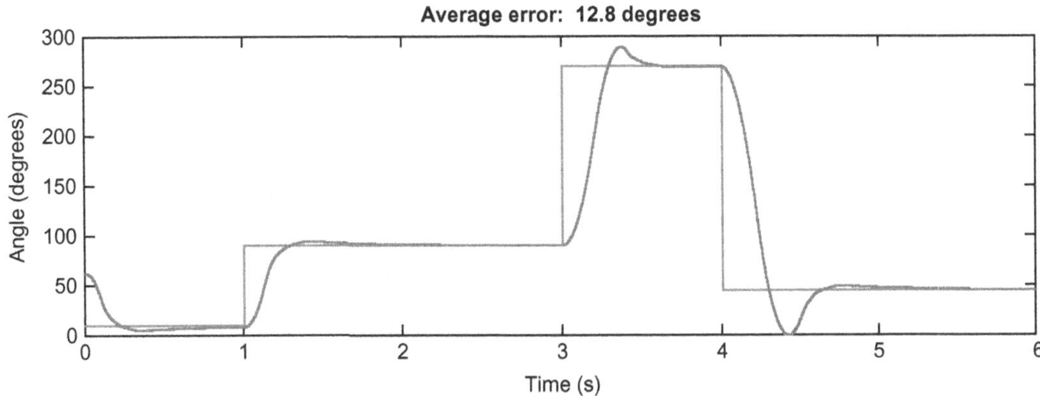

Figure 28.2
An example result of this project. The user specifies a position reference trajectory for the motor and the system attempts to follow it. An average error is calculated as an indication of how well the motor controller achieved its objective.

control loop runs at 200 Hz and the inner current control loop runs at 5 kHz. PWM is at 20 kHz.

Prior to implementing these control loops, we will implement some more basic features. The current control loop must output a PWM signal to the motor and read values from the current sensor, and the position control loop needs encoder feedback. Therefore we begin with menu options that allow the user to directly view sensor readings and specify the PWM duty cycle. Another option allows the user to tune the current control feedback loop independently of the position controller. The ability to interactively specify control gains and see the resulting control performance simplifies this process. All of the features will be accessible from the client menu that you create, depicted in Figure 28.3.

Below we describe the purpose of the individual commands, in order, but let us start with the last one, `r: Get mode`. The PIC32 can operate in one of five *modes*: IDLE, PWM, ITEST, HOLD, and TRACK. When it first powers on, the PIC32 should be in IDLE mode. In IDLE mode, the PIC32 puts the H-bridge in brake mode, with zero voltage across the motor. In PWM mode, the PIC32 implements a fixed PWM duty cycle, as requested by the user. In ITEST mode, the PIC32 tests the current control loop, without caring about the motion of the motor. In HOLD mode, the PIC32 attempts to hold the motor at the last position commanded by the user. In TRACK mode, the PIC32 attempts to track a reference motor trajectory specified by the user (Figure 28.2). Some of the menu commands cause the PIC32 to change mode, as noted below.

`a:` `Read current sensor (ADC counts)`. Print the motor current as measured using the current sensor, in ADC counts (0-1023). For example, the result could look like

```
PIC32 MOTOR DRIVER INTERFACE

    a: Read current sensor (ADC counts)     b: Read current sensor (mA)
    c: Read encoder (counts)                d: Read encoder (deg)
    e: Reset encoder                        f: Set PWM (-100 to 100)
    g: Set current gains                    h: Get current gains
    i: Set position gains                   j: Get position gains
    k: Test current control                 l: Go to angle (deg)
    m: Load step trajectory                 n: Load cubic trajectory
    o: Execute trajectory                   p: Unpower the motor
    q: Quit client                          r: Get mode

ENTER COMMAND:
```

Figure 28.3

The final menu. The user enters a single character, which may result in the user being prompted for more information. Additional options are possible, but these are a minimum.

```
ENTER COMMAND: a
The motor current is 903 ADC counts.
```

After printing the current sensor reading, the client should display the full menu again, to help the user enter the next command. Alternatively, to save screen space, you could just keep a printout of the commands handy and not print the menu to the screen each time. Although the "a" command is somewhat redundant with the next command, which returns the current in mA, it is sometimes useful for debugging to see the raw ADC data, rather than the scaled version in more familiar units.

`b:` `Read current sensor (mA)`. Print the motor current as measured using the current sensor, in mA. The output could look like

```
The motor current is 763 mA.
```

By a convention we will adopt, the current is positive when trying to drive the motor counterclockwise, in the direction of increasing motor angle (Figure 28.1). If you are getting the opposite sign, you can swap the wires to the current sensor or handle it in software.

`c:` `Read encoder (counts)`. Print the encoder angle, in encoder counts. By convention, the encoder count increases as the motor rotates counterclockwise. An example output:

```
The motor angle is 314 counts.
```

`d:` `Read encoder (deg)`. Print the encoder angle, in degrees. By convention, the encoder angle increases as the motor rotates counterclockwise. If the encoder gives 1000 counts per revolution in 4x decoding mode, then 314 counts would give an output

```
The motor angle is 113.0 degrees.
```

since $360°(314/1000) \approx 113°$. In this project, 0 degrees, 360 degrees, 720 degrees, etc., are all treated differently, allowing us to control multiple turns of the motor.

e: `Reset encoder`. The present angle of the motor is defined as the zero position. No output is necessary. Optionally, command `d` could be called automatically after zeroing the encoder, to confirm the result to the user.

f: `Set PWM (-100 to 100)`. Prompt the user for a desired PWM duty cycle, specified in the range [−100%, 100%]. The PIC32 switches to PWM mode and implements the constant PWM until overridden. An input −100 means that the motor is full on in the clockwise direction, 100 means full on in the counterclockwise direction, and 0 means that the motor is unpowered. An example prompt, user reply, and result is

```
What PWM value would you like [-100 to 100]?  73
PWM has been set to 73% in the counterclockwise direction.
```

with the motor speeding up to its 73% steady-state speed. The PIC32 should saturate values outside the range [−100, 100] and convert values in the range [−100, 100] to an appropriate PWM duty cycle and direction bit to the DRV8835.

g: `Set current gains`. Prompt for the current loop control gains and send them to the PIC32, to be implemented immediately. (This command does not change the mode of the PIC32, however; the gains only affect the motor's behavior when the PIC32 is in the ITEST, HOLD, or TRACK mode.) We suggest two gains, for a PI controller. An example interface:

```
Enter your desired Kp current gain [recommended: 4.76]:  3.02
Enter your desired Ki current gain [recommended: 0.32]:  0
Sending Kp = 3.02 and Ki = 0 to the current controller.
```

It is not necessary to recommend values in the prompting text, but if you find good values, you may wish to put them into your client, so you remember for the future. (The values given here are just for demonstration, not actual recommendations!) You might also wish to use only integers for your gains, depending on the implicit units, to allow for more time-efficient math.

h: `Get current gains`. Print the current controller gains. Example output:

```
The current controller is using Kp = 3.02 and Ki = 0.
```

i: `Set position gains`. Prompt for the position loop control gains and send them to the PIC32, to be implemented immediately. (This command does not change the mode of the PIC32, however; the gains only affect the motor's behavior if the PIC32 is in the HOLD or TRACK mode.) If you use PID control, the interface could be:

```
Enter your desired Kp position gain [recommended: 4.76] :  7.34
Enter your desired Ki position gain [recommended: 0.32] :  0
Enter your desired Kd position gain [recommended: 10.63]:  5.5
Sending Kp = 3.02, Ki = 0, and Kd = 5.5 to the position controller.
```

Other motion controllers (Chapter 27) require specification of other gains and parameters.

j: `Get position gains`. Print the position controller gains. Example output:

```
The position controller is using Kp = 7.34, Ki = 0, and Kd = 5.5.
```

k: `Test current control`. Test the current controller, which uses the gains set using the "g" command. This command puts the PIC32 in ITEST mode. In this mode, the 5 kHz current control loop uses a 100 Hz, 50% duty cycle, ±200 mA square wave reference current. Each time through the loop, the reference and actual current are stored in data arrays for later plotting. After 2-4 cycles of the 100 Hz current reference, the PIC32 switches to IDLE mode, and the data collected during the ITEST mode is sent back to the client, where it is plotted. An example of reasonably good tracking is shown in Figure 28.4. The plots help the user to choose good current loop gains.
During the current control test, the motor may move a little bit, but it is important to remember that we are not interested in the motor's motion, only the performance of the current controller. The motor should not move far, since the actual current should be changing quickly with zero average.
The client should calculate an average of the absolute value of the current error over the samples. This score allows you to compare the performance of different current controllers, in addition to the eyeball test based on the plots.

l: `Go to angle (deg)`. Prompt for an angle, in degrees, that the motor should move to. The PIC32 switches to HOLD mode, and the motor should move to the specified position immediately and hold the position. Example interface:

```
Enter the desired motor angle in degrees:  90
Motor moving to 90 degrees.
```

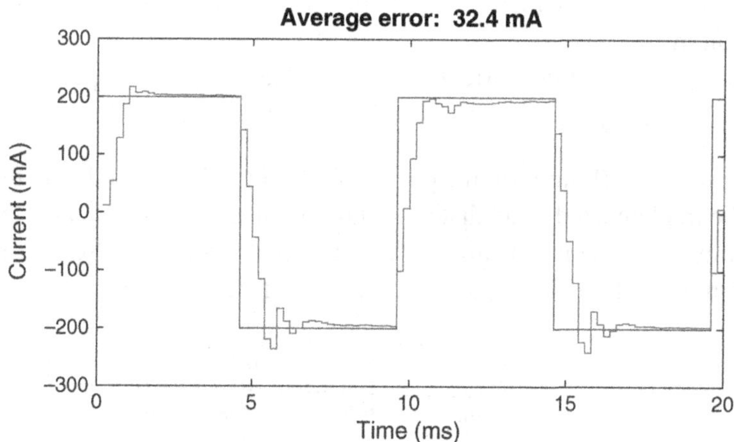

Figure 28.4
A result of a current controller test. The square wave is the reference current and the other plot is the measured current.

m: `Load step trajectory.` Prompt for the time and angle parameters describing a series of one or more steps in the motor position. Plot the reference trajectory on the client so the user can see if the trajectory was entered correctly. Convert the reference trajectory to a series of setpoint positions at 200 Hz (each point is separated by 5 ms), store it in an array, and send it to the PIC32 for future execution. In addition, set the number of samples the PIC32 should record in the position control loop according to the duration of the trajectory.

The PIC32 should discard any trajectory data that exceeds the maximum trajectory length. This command does not change the mode of the PIC32.

An example interface is shown below. First the user requests a trajectory that starts at angle 0 degrees at time 0 s; then steps to 90 degrees at 1 s; and holds at 90 degrees until 500 s have passed. This duration is too long: the PIC32 RAM cannot store 500 s × 200 samples/s = 100,000 samples of the reference trajectory. In the second try, the user requests the step trajectory shown in Figure 28.5. Note that the client only plots the reference trajectory that is sent to the PIC32; no data is received from the PIC32, because the trajectory has not been executed yet.

```
Enter step trajectory, in sec and degrees [time1, ang1; time2, ang2; ...]:
  [0, 0; 1, 90; 500, 90]
Error:  Maximum trajectory time is 10 seconds.

Enter step trajectory, in sec and degrees [time1, ang1; time2, ang2; ...]:
  [0, 0; 1, 180; 3, -90; 4, 0; 5, 0]
```

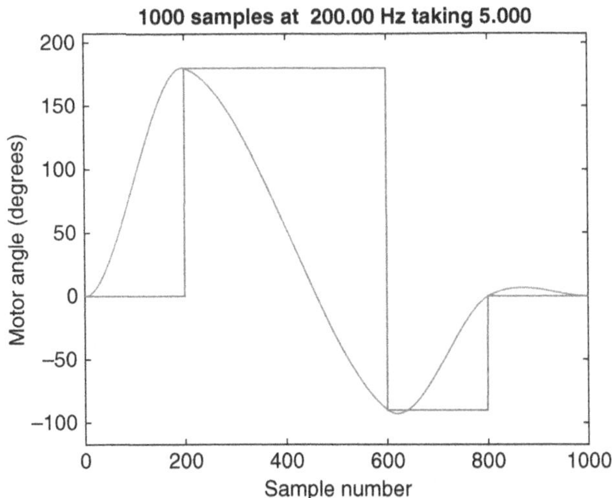

Figure 28.5

Sample reference trajectories for the position controller. The step trajectory is generated in MATLAB using the command `ref = genRef([0,0; 1,180; 3,-90; 4,0; 5,0], 'step')` and the smooth cubic curve is generated by replacing `'step'` with `'cubic'`.

```
Plotting the desired trajectory and sending to the PIC32 ... completed.
```

Now the PIC32 has a reference trajectory waiting to be executed. A MATLAB function `genRef.m` is provided with this chapter to generate the sample points of step trajectories, and is invoked using

```
ref = genRef([0,0; 1,180; 3,-90; 4,0; 5,0], 'step');
```

The resulting trajectory is shown in Figure 28.5, as a plot of the motor angle in degrees vs. the sample number.

n: `Load cubic trajectory`. Prompt for the time and angle parameters describing a set of via points for a cubic interpolation trajectory for the motor. This command is similar to the step trajectory command, except a smooth trajectory is generated through the specified points, with zero velocity at the beginning and end. The same MATLAB function `genRef` calculates the cubic trajectory, using the option `'cubic'` instead of `'step'` (Figure 28.5).

```
Enter cubic trajectory, in sec and degrees [time1, ang1; time2, ang2; ...]:
  [0, 0; 1, 180; 3, -90; 4, 0; 5, 0]
Plotting the desired trajectory and sending to the PIC32 ... completed.
```

o: `Execute trajectory`. Execute the trajectory stored on the PIC32. This command changes the PIC32 to TRACK mode, and the PIC32 attempts to track the reference trajectory previously stored on the PIC32. After the trajectory has finished, the PIC32 switches to HOLD mode, holding the last position of the trajectory, then sends the reference and actual position data back to the client for plotting, as shown in Figure 28.2 for a step trajectory. The client should calculate an average position error, similar to the current control test. Because step trajectories request unrealistic step changes of the motor angle, the average error for a step trajectory could be large, while the average error for a smooth cubic trajectory can be quite small.

p: `Unpower the motor`. The PIC32 switches to IDLE mode.

q: `Quit client`. The client should release the communication port so other applications can communicate with the NU32. The PIC32 should be set to IDLE mode.

r: `Get mode`. Described above. Example output:

```
The PIC32 controller mode is currently HOLD.
```

Once the circuits have been built, the functions above have been fully implemented and tested (both the client and the PIC32), and control gains have been found that yield good performance of the motor, the project is complete.

To help you finish this project, Section 28.3 gives some recommendations on how to develop maintainable, modular PIC32 code. Section 28.4 breaks the project down into a series of steps that you should complete, in order. Finally, Section 28.5 describes a number of extensions to go further with the project.

28.3 Software Development Tips

Since this is a big project, it is a good idea to be disciplined in the development of your PIC32 code. This discipline will make the code easier to maintain and build upon. Here are some tips. We recommend you read this section to help you better understand the project, but if you are confident in your software skills and your understanding of the project, you need not adhere to the advice.

Debugging

Run-time errors in your PIC32 code are inevitable. The ability to locate bugs quickly is crucial to keeping development time reasonable. Typically debugging consists of having your program provide feedback at breakpoints, to pinpoint where the unexpected behavior occurs.[1] This type of debugging can be challenging for embedded code, since there is no `printf` to a monitor. You should be comfortable using the following tools, however:

- **Terminal emulator.** Using a terminal emulator to connect to the PIC32, instead of your client, allows you to send simple commands and see exactly what information is being sent back by the PIC32, without the complication of potential errors in the client code. Just be aware that there is only one communication port with the NU32, so either the client or the terminal emulator can be used, not both simultaneously.
- **LCD screen.** You can use the LCD screen to display information about the state of the PIC32 without using the UART communication with the host.
- **LEDs and digital outputs.** You can use LEDs or digital outputs connected to an oscilloscope as other ways of getting feedback from the PIC32. To verify that your current control and position control ISRs are operating at the correct frequencies, for example, you could toggle a digital output in each ISR and look at the resulting square wave "heartbeats" on an oscilloscope.

Modularity

Instead of writing one large `.c` file to do everything, consider breaking the project into multiple *modules*. A module consists of one `.c` and one `.h` file. The `.h` file contains the module's *interface* and the `.c` file contains the module's *implementation*. The `.c` file should always `#include` the corresponding `.h` file.

A module's interface (the `.h` file) consists of function prototypes and data types that other modules can use. You can use functions from module `A` in module `B` by having `B.c` include `A.h`.

[1] The PICkit 3 and other Microchip programming tools can be used to set breakpoints and step through code as it executes. These features can be accessed through MPLAB X but cannot be used in conjunction with the NU32's bootloader.

A module's implementation (the `.c` file) contains the code that makes the interface functions work. It also contains any functions and variables that the module uses but wants to hide from other modules. These hidden functions and variables should be declared as `static`, limiting their scope to the module's `.c` file. By hiding implementations in the `.c` file and only exposing certain functions in the `.h` file, you decrease the dependencies between modules, making maintenance easier and helping prevent bugs. For those readers familiar with an object-oriented programming language such as C++ or Java, these concepts roughly mirror the ideas of public and private class members.

In an embedded system, peripherals can be accessed by code in any module; therefore, dividing the code into modules is only the first step in a good design. To maintain proper module separation, you should document which modules *own* which peripherals. If a module owns a peripheral, only it should access that peripheral directly. If a module needs to access a peripheral it does not own, it must call a function in the owning module. These rules, enforced by your vigilance rather than the compiler, will help you more easily isolate bugs and fix them as they occur.

All `.c` and `.h` files should be in the same directory, allowing the project to be built by compiling and linking all `.c` files in the directory.

One module your project will certainly use is `NU32`, for communication with the client. Other modules we suggest are the following. Some example interface functions are suggested for each module, but feel free to use your own functions.

- **main.** This module is the only one with no `.h` file, since no other module needs to call functions in the `main` module. The `main.c` file is the only one with a `main` function. `main.c` calls appropriate initialization functions for the other modules and then enters an infinite loop. The infinite loop waits for commands from the client, then interprets the user's input and responds appropriately.
- **encoder.** This module owns an SPI peripheral, to communicate with the encoder counting chip. The interface `encoder.h` should provide functions to (1) do a one-time setup or initialization of the encoder module, (2) read the encoder in encoder counts, (3) read the encoder in degrees, and (4) reset the encoder position so that the present angle of the encoder is treated as the zero angle.
- **isense.** This module owns the ADC peripheral, used to sense the motor current. The interface `isense.h` should provide functions for a one-time setup of the ADC, to provide the ADC value in ADC counts (0-1023), and to provide the ADC value in terms of milliamps of current through the motor.
- **currentcontrol.** This module implements the 5 kHz current control loop. It owns a timer to implement the fixed-frequency control loop, an output compare and another timer to implement a PWM signal to the H-bridge, and one digital output controlling the motor's

direction (see the description of the DRV8835 in Chapter 27). Depending on the PIC32 operating mode, the current controller either brakes the motor (IDLE mode), implements a constant PWM (PWM mode), uses the current control gains to try to provide a current specified by the position controller (HOLD or TRACK mode), or uses the current control gains to try to track a 100 Hz ±200 mA square wave reference (ITEST mode). The interface `currentcontrol.h` should provide functions to initialize the module, receive a fixed PWM command in the range [−100, 100], receive a desired current (from the `positioncontrol` module), receive current control gains, and provide the current control gains.
- **positioncontrol.** This module owns a timer to implement the 200 Hz position control loop. The interface `positioncontrol.h` should provide functions to initialize the module, load a trajectory from the client, load position control gains from the client, and send position control gains back to the client.
- **utilities.** This module is used for various bookkeeping tasks. It maintains a variable holding the operating mode (IDLE, PWM, ITEST, HOLD, or TRACK) and arrays (buffers) to hold data collected during trajectory tracking (TRACK) or current tracking (ITEST). The interface `utilities.h` provides functions to set the operating mode, return the current operating mode, receive the number N of samples to save into the data buffers during the next TRACK or ITEST, write data to the buffers if N is not yet reached, and to send the buffer data to the client when N samples have been collected (TRACK or ITEST has completed).

Variables

Variables that are shared by ISRs and mainline code should be declared `volatile`. You may consider disabling interrupts before using a shared variable in the mainline code, then re-enabling interrupts afterward, to make sure a read or write of the variable in the mainline code is not interrupted. If you do disable interrupts, make sure not to leave interrupts disabled for more than a few simple lines of code; otherwise you are defeating the purpose of having interrupts in the first place.

Function local variables that need to keep their value from one invocation of the function to the next should be declared `static`. A variable `foo` that is only meant to be used in one `.c` file, but should be available to several functions in that file (i.e., "global" within the file), can be defined at the beginning of the file but outside any function. To prevent this `foo` from colliding with a variable of the same name defined in another `.c` file, `foo` should be declared `static`, limiting its scope to the file it is defined in.

Consider writing your code so that no variable is used outside the `.c` file it is defined in. To share the value of a variable in module `A` with other modules, provide the prototype of a public accessor function in `A.h`, like `int get_value_from_A()`. Do not define variables in header files.

New data types

The five operating modes (IDLE, etc.) can be represented by a variable of type `int` (or `short`), taking values 0 to 4. Instead you could consider using `enum`, allowing you to create a data type `mode_t` with only five possible values, named IDLE, PWM, ITEST, HOLD, and TRACK.

For your data buffers, consider creating a new data type `control_data_t` using a `struct`, where the `struct` has several members (e.g., the current reference, the actual current, the position reference, and the actual position). This way you can make a single data array of type `control_data_t` instead of having several arrays.

Integer math

The PIC32 control laws begin by sensing an integral number of encoder counts and ADC counts, and end by storing an integer to the period register of an output compare module (PWM). Thus it is possible, though not necessary, to calculate all control laws using integer math with integer-valued gains. Using integer math ensures that the controllers run as quickly as possible. If you do use only integers, however, you are responsible for making sure that there are no unacceptable roundoff or overflow errors.

Degrees and milliamps are the most natural units to display to the user on the client interface, however.

28.4 Step by Step

This section provides a step-by-step guide to building the project, testing and debugging as you go. Make sure that each step works properly before moving on to the next step, and be sure to perform all the numbered action items associated with each subsection. Turn in your answers for the items in **bold**.

28.4.1 Decisions, Decisions

Before writing any code, you must decide which peripherals you will use in the context of the controller block diagram in Figure 27.7. The PIC32 will connect to three external circuit boards: the H-bridge, the motor encoder counter, and the motor current sensor. Record your answers to the following questions:

1. The NU32 communicates with the encoder counter by an SPI channel. Which SPI channel will you use? Which NU32 pins does it use?
2. The NU32 reads the MAX9918 current sensor using an ADC input. Which ADC input will you use? Which NU32 pin is it?

3. The NU32 controls the DRV8835 H-bridge using a direction bit (a digital output) and PWM (an output compare and a timer). Which peripherals will you use, and which NU32 pins?
4. Which timers will you use to implement the 200 Hz position control ISR and the 5 kHz current control ISR? What priorities will you use?
5. Based on your answers to these questions, and your understanding of the project, annotate the block diagram of Figure 27.7. Each block should clearly indicate which devices or peripherals perform the operation in the block, and each signal line should clearly indicate how the signal is carried from one block to the other. (After this step, there should be no question about the hardware involved in the project. The details of wiring the H-bridge, current sensor, and encoder are left to later.)
6. Based on which circuit boards need to be connected to which pins of the NU32, and the connections of the circuit boards to the motor and encoder, sketch a proposed layout of the circuit boards relative to the NU32 so that wire crossing is approximately minimized. (Do not make a full circuit diagram at this time.)
7. **Turn in your answers for items 1-6.**

28.4.2 Establishing Communication with a Terminal Emulator

The role of the `main` function is to call any functions initializing peripherals or modules and to enter an infinite loop, dispatching commands that the PIC32 receives from the client.

Code Sample 28.1 `main.c`. **Basic Code for Setup and Communication.**

```
#include "NU32.h"          // config bits, constants, funcs for startup and UART
// include other header files here

#define BUF_SIZE 200

int main()
{
  char buffer[BUF_SIZE];
  NU32_Startup(); // cache on, min flash wait, interrupts on, LED/button init, UART init
  NU32_LED1 = 1;  // turn off the LEDs
  NU32_LED2 = 1;
  __builtin_disable_interrupts();
  // in future, initialize modules or peripherals here
  __builtin_enable_interrupts();

  while(1)
  {
    NU32_ReadUART3(buffer,BUF_SIZE); // we expect the next character to be a menu command
    NU32_LED2 = 1;                   // clear the error LED
    switch (buffer[0]) {
      case 'd':                      // dummy command for demonstration purposes
```

```
      {
        int n = 0;
        NU32_ReadUART3(buffer,BUF_SIZE);
        sscanf(buffer, "%d", &n);
        sprintf(buffer,"%d\r\n", n + 1); // return the number + 1
        NU32_WriteUART3(buffer);
        break;
      }
      case 'q':
      {
        // handle q for quit. Later you may want to return to IDLE mode here.
        break;
      }
      default:
      {
        NU32_LED2 = 0;  // turn on LED2 to indicate an error
        break;
      }
    }
  }
  return 0;
}
```

The infinite `while` loop reads from the UART, expecting a single character. That character is processed by a `switch` statement to determine what action to perform. If the character matches a known menu entry, we may have to retrieve additional parameters from the client. The specific format for these parameters depends on the particular command, but in this case, after receiving a "d," we expect an integer and store it in `n`. The command then increments the integer and sends the result to the client.

Note that each `case` statement ends with a `break`. The `break` prevents the code from falling through to the next case. Make sure that every `case` has a corresponding `break`.

Unrecognized commands trigger the default case. The default case illuminates LED2 to indicate an error. We could have designed the communication protocol to allow the PIC32 to return error conditions to the client. Although crucial in the real world, proper error handling complicates the communication protocol and obscures the goals of this project. Therefore we omit it.

Before proceeding further, test the PIC32 menu code:

1. Compile and load `main.c`, then connect to the PIC32 with a terminal emulator.
2. Enter the "d" command, then enter a number. Ensure that the command works as expected.
3. Issue an unknown command and verify that LED2 illuminates.

28.4.3 Establishing Communication with the Client

We have also provided you with some basic MATLAB client code. This code expects a single character of keyboard input and sends it to the PIC32. Then, depending on the command sent, the user can be prompted to enter more information and that information can be sent to the PIC32.

Code Sample 28.2 `client.m`. **MATLAB Client Code.**

```
function client(port)
%   provides a menu for accessing PIC32 motor control functions
%
%   client(port)
%
%   Input Arguments:
%       port - the name of the com port.  This should be the same as what
%               you use in screen or putty in quotes ' '
%
%   Example:
%       client('/dev/ttyUSB0') (Linux/Mac)
%       client('COM3') (PC)
%
%   For convenience, you may want to change this so that the port is hardcoded.

% Opening COM connection
if ~isempty(instrfind)
    fclose(instrfind);
    delete(instrfind);
end

fprintf('Opening port %s....\n',port);

% settings for opening the serial port. baud rate 230400, hardware flow control
% wait up to 120 seconds for data before timing out
mySerial = serial(port, 'BaudRate', 230400, 'FlowControl', 'hardware','Timeout',120);
% opens serial connection
fopen(mySerial);
% closes serial port when function exits
clean = onCleanup(@()fclose(mySerial));

has_quit = false;
% menu loop
while ~has_quit
    fprintf('PIC32 MOTOR DRIVER INTERFACE\n\n');
    % display the menu options; this list will grow
    fprintf('     d: Dummy Command     q: Quit\n');
    % read the user's choice
    selection = input('\nENTER COMMAND: ', 's');

    % send the command to the PIC32
    fprintf(mySerial,'%c\n',selection);

    % take the appropriate action
    switch selection
        case 'd'                         % example operation
            n = input('Enter number: '); % get the number to send
            fprintf(mySerial, '%d\n',n); % send the number
```

```
            n = fscanf(mySerial,'%d');    % get the incremented number back
            fprintf('Read: %d\n',n);      % print it to the screen
        case 'q'
            has_quit = true;              % exit client
        otherwise
            fprintf('Invalid Selection %c\n', selection);
    end
end

end
```

Test the client as follows. Do not move on until you have completed each step.

1. Exit the terminal emulator (if it is still open).
2. Run `client.m` in MATLAB and confirm that you are connected to the PIC32. Issue the "d" command, and verify that it works as you expect. (The client prompts you for a number and sends it to the PIC32. The PIC32 increments it and sends the value back. The client prints out the return value.)
3. Quit the client by using "q."
4. Implement a command "x" on the PIC32 that accepts two integers, adds them, and returns the sum to the client.
5. Test the new command using the terminal emulator. Once you have verified that the PIC32 code works, quit the emulator.
6. Add an entry "x" to the client menu and verify that the client's new menu command works as expected.

After completing the tasks above, you should be familiar with the menu system. Adding menu entries and determining a communication protocol for these commands will become routine as you proceed through this project. If you encounter problems, you can open the terminal emulator and enter the commands manually to narrow down whether the issue is on the client or on the PIC32. Remember, do not attempt to simultaneously open the serial port in the terminal emulator and the client. Also, a mismatch between the data that the PIC32 expects and what the client sends, or vice versa, may cause one or the other to freeze while waiting for data. You can force-quit the client in MATLAB by typing CTRL-C.

When you are comfortable with how the menu system works, you can remove the "d" and "x" commands, as they are no longer needed.

28.4.4 Testing the Encoder

1. Power the encoder with 3.3 V and GND. Be absolutely certain of your wiring before applying power! Some encoders are easy to destroy with the wrong power and ground connections.

2. Attach two oscilloscope probes to the A and B encoder outputs.
3. Twist the motor in both directions and make sure you see out-of-phase square waves from the two encoder channels.

28.4.5 Adding Encoder Menu Options

In this section you will implement the menu items "c" (`Read encoder (counts)`), "d" (`Read encoder (deg)`), and "e" (`Reset encoder`).

Figure 28.6 shows the wiring of the decoder PCB to both the motor encoder and the NU32, based on the assumption that the PIC32 uses SPI channel 4 to communicate with the decoder. The decoder uses 4x decoding of the quadrature encoder inputs and keeps a 16-bit count, 0 to 65,535. The decoder is actually a dedicated PIC microcontroller, programmed only to count encoder pulses and to send the count to the PIC32 when requested. To verify that the decoder chip is programmed, you can look for the 1 kHz "heartbeat" square wave on pin B3.

Once you have connected all the components, it is time to add an encoder reading option to the PIC32 menu code:

```
case 'c':
{
  sprintf(buffer,"%d", encoder_counts());
  NU32_WriteUART3(buffer);  // send encoder count to client
  break;
}
```

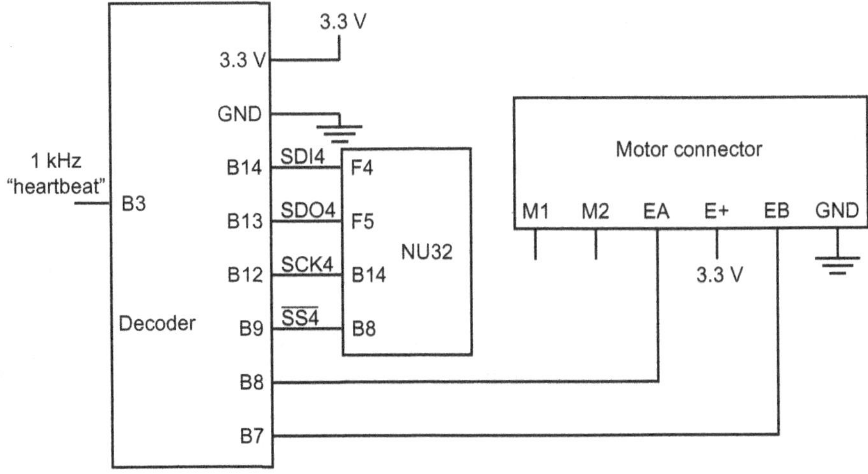

Figure 28.6
Encoder counter circuit.

This code invokes a function called `encoder_counts()`, which apparently returns an integer. Below we give you an implementation of `encoder_counts()`. Here we assume this function lives in a module consisting of `encoder.c` and `encoder.h` (Section 28.3), but it could also be included directly in the `main.c` file with no header `encoder.h`. The function `encoder_counts()` relies on the helper function `encoder_command()`.

Code Sample 28.3 `encoder.c`. **Implementation of Some Encoder Functions.**

```
#include "encoder.h"
#include <xc.h>

static int encoder_command(int read) { // send a command to the encoder chip
                                       // 0 = reset count to 32,768, 1 = return the count
  SPI4BUF = read;                      // send the command
  while (!SPI4STATbits.SPIRBF) { ; }   // wait for the response
  SPI4BUF;                             // garbage was transferred, ignore it
  SPI4BUF = 5;                         // write garbage, but the read will have the data
  while (!SPI4STATbits.SPIRBF) { ; }
  return SPI4BUF;
}

int encoder_counts(void) {
  return encoder_command(1);
}

void encoder_init(void) {
  // SPI initialization for reading from the decoder chip
  SPI4CON = 0;              // stop and reset SPI4
  SPI4BUF;                  // read to clear the rx receive buffer
  SPI4BRG = 0x4;            // bit rate to 8 MHz, SPI4BRG = 80000000/(2*desired)-1
  SPI4STATbits.SPIROV = 0;  // clear the overflow
  SPI4CONbits.MSTEN = 1;    // master mode
  SPI4CONbits.MSSEN = 1;    // slave select enable
  SPI4CONbits.MODE16 = 1;   // 16 bit mode
  SPI4CONbits.MODE32 = 0;
  SPI4CONbits.SMP = 1;      // sample at the end of the clock
  SPI4CONbits.ON = 1;       // turn SPI on
}
```

The function `encoder_init()` initializes SPI4. This function should be called at the beginning of `main`, while interrupts are disabled. The SPI peripheral uses a baud of 8 MHz, 16-bit operation, sampling on the falling edge, and automatic slave detect.

The function `encoder_counts()` uses `encoder_command()` to send a command to the decoder chip and return a response. Valid commands to `encoder_command()` are 0x01, which reads the decoder count, and 0x00, which resets the count to 32,768, halfway through the count range 0 to 65,535. Note that every time you write to SPI you must also read, even if you do not need the data.

If you use a separate `encoder` module (Section 28.3), `encoder_init()` and `encoder_counts()` should have prototypes in `encoder.h` so that they are available to other modules that `#include` "encoder.h". The function encoder_command() is private to encoder.c, and therefore should be declared static and should have no prototype in encoder.h.

In addition to encoder_counts(), you will implement a function to reset the encoder count (e.g., encoder_reset()) using encoder_command().

1. Implement the PIC32 code to read the encoder counts (the "c" command).
2. Use a terminal emulator to issue commands and display the results from the PIC32. Twist the motor shaft, issue the "c" command, and ensure that the count increases as the shaft rotates counterclockwise and decreases as it rotates clockwise. If you get opposite results, swap the encoder inputs to the decoder PCB.
3. Implement PIC32 code for the "e" command to reset the encoder count to 32,768. Using the terminal emulator and the "c" command, verify that resetting the encoder works as expected.
4. Knowing the 4x resolution of the encoder, implement PIC32 code for the "d" command to read the encoder in degrees. Using the terminal emulator, verify that the angle reads as zero degrees after the reset command "e." Also verify that rotating the shaft 180 or −180 degrees results in appropriate readings with the "d" command.
5. Close the terminal emulator and update the client to handle the "c" (Read encoder (counts)), "d" (Read encoder (deg)), and "e" (Reset encoder) commands. For example,

```
case 'c'
  counts = fscanf(mySerial,'%d');
  fprintf('The motor angle is %d counts.', counts)
```

   Verify that the three commands work as expected on the client.

From now on, when you are asked to implement menu items, you should implement them on both the PIC32 and client. We will not ask you to test with the terminal emulator first. You can always fall back to the terminal emulator for debugging purposes.

28.4.6 PIC32 Operating Mode

In this section you will implement the menu item "r" (Get mode).

The PIC32 can be in one of five operating modes: IDLE, PWM, ITEST, HOLD, and TRACK. You will create functions to both set and query the mode. If you are following the modular design suggested in Section 28.3, these functions, and the variable holding the mode, will likely be in the utilities module.

1. Implement PIC32 functions to set the mode and to query the mode.
2. Use the mode-setting function at the beginning of main to put the PIC32 in IDLE mode.
3. Implement the "r" menu item (Get mode) on the PIC32 and the client. Verify that it works.
4. Update the "q" (quit) menu entry to set the PIC32 to the IDLE state prior to exiting the menu.

Note that there is no client menu command that only sets the mode.

28.4.7 Current Sensor Wiring and Calibration

The current sensor detects the amount of current flowing through the motor. We use a PCB breaking out the MAX9918 current-sense amplifier and an onboard 15 mΩ current-sense resistor, as described in Chapter 21.10.1.

In this section, you will first use the information in Chapter 21.10.1 to set up and calibrate your current sensor, independent of the NU32 and the motor. The questions refer to the circuit in Figure 21.22.

1. Choose the voltage divider resistors R3 to be a few hundred ohms (e.g., 330 Ω).
2. Find the maximum current you expect to sense. If the H-bridge's battery voltage is V and the motor resistance is R_{motor}, then the maximum current you can expect to see is approximately $I_{\max} = 2V/R_{\text{motor}}$. This occurs when the motor is spinning at no-load speed in reverse, with essentially zero current and $-V$ across the motor terminals,

   $$-V = k_t \omega_{\text{rev}} \quad \rightarrow \quad \omega_{\text{rev}} = -V/k_t$$

   and then the control voltage switches suddenly to V, yielding (ignoring inductance)

   $$V = k_t \omega_{\text{rev}} + I_{\max} R_{\text{motor}} \quad \rightarrow \quad I_{\max} = 2V/R_{\text{motor}}.$$

   Record your calculated $I_{\max}$ for your battery and motor.
3. Calculate the voltage across the 15 mΩ sense resistor if $I_{\max}$ flows through it. Call this $V_{\max}$.
4. Choose resistors R1 and R2 so the current-sense amplifier gain $G = 1 + (\text{R2}/\text{R1})$ approximately satisfies

   $1.65\ \text{V} = G \times V_{\max}$.

   This ensures that the maximum positive motor current yields a 3.3 V output from the current sensor and the maximum negative motor current yields a 0 V output from the current sensor, utilizing the full range of the ADC input. Choose R1 and R2 to be in the range of 10^4-10^6 Ω.
5. Choose a resistor R and a capacitor C to make an RC filter on the MAX9918 output with a cutoff frequency $f_c = 1/(2\pi RC)$ in the neighborhood of 200 Hz, to suppress high-frequency components due to the 20 kHz PWM.
6. Build the circuit as shown in Figure 21.22, but do not connect to the motor or the PIC32. You will calibrate the circuit using resistors, an ammeter, and an oscilloscope or voltmeter. Figure 28.7 shows how to use a resistor R0 to provide controlled positive and negative test currents to the current sensor. You will choose different values of R0 to create test currents over the range of likely currents. For example, if you have two 20 Ω resistors, you can use

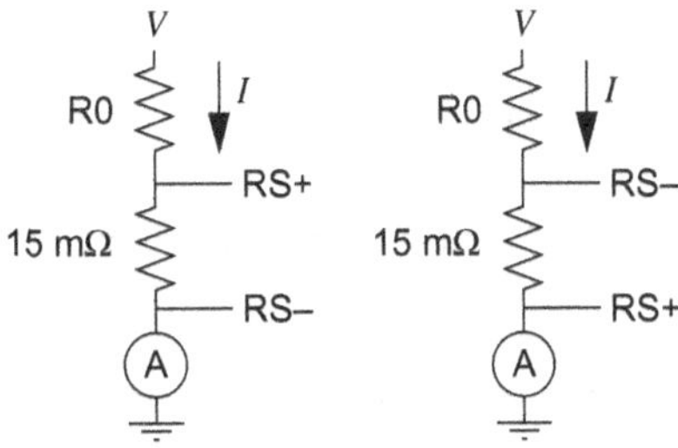

Figure 28.7
Test circuits for calibrating the current sensor. The circuit on the left provides positive currents through the current-sense resistor; switching the RS+ and RS– connections in the right circuit results in negative currents.

them to create an R0 of 20 Ω, 40 Ω (two resistors in series), or 10 Ω (two resistors in parallel). If the battery voltage V is 6 V, this results in expected currents of ±300, ±150, and ±600 mA, respectively.

Important: The calibration resistors must be rated to handle high currents without burning up. For example, a 20 Ω resistor with 300 mA through it dissipates $(300\ \text{mA})^2(20\ \Omega) = 1.8$ W, more than a typical 1/4 W resistor can dissipate.

With different resistances R0, use an ammeter to measure the actual current and a voltmeter or oscilloscope to measure the output of the current sensor. Fill out a table similar to the table below, for your particular resistances and battery. If you built your current sensor circuit correctly, zero current should give approximately 1.65 V at the sensor output, and the data points (sensor voltage as a function of the measured current) should agree with the amplifier gain G you designed. If not, time to fix your circuit.

R0 (Ω)	Expected *I* (mA)	Measured *I* (mA)	Sensor (V)	ADC (counts)
10 (to RS+)	600	587	2.82	
20 (to RS+)	300	295	2.24	
40 (to RS+)	150	140	1.93	
Open circuit	0	0	1.63	
40 (to RS–)	−150	−147	1.34	
20 (to RS–)	−300	−322	1.01	
10 (to RS–)	−600	−605	0.45	

Leave the column "ADC (counts)" blank; you will fill in that column in the next section. As a sanity check, you can replace R0 with your motor, stalled, and make sure that the sensor voltage makes sense.

7. **Turn in your answers for items 2-6.**

28.4.8 ADC for the Current Sensor

In this section you will implement the menu items “a” (`Read current sensor (ADC counts)`) and “b” (`Read current sensor (mA)`).

The ADC reads the voltage from the motor current sensor. See Chapter 10 for information on setting up and using the ADC.

1. Create a PIC32 function to initialize the ADC, and call it at the beginning of `main`.
2. Create a PIC32 function that reads the ADC and returns the value, 0-1023. Consider reading the ADC a few times and averaging for a more stable reading.
3. Add the menu item “a” (`Read current sensor (ADC counts)`) to the PIC32 and client and verify that it works. Use simple voltage dividers at the analog input to make sure that the readings make sense.
4. Hook up the current sensor output from the previous section to the analog input. Provide a circuit diagram showing all connections to the current sensor PCB.
5. Using the power-resistor voltage dividers from the previous section, and the “a” menu command, fill in the last column of the current sensor calibration table. (It is a good idea to update your ammeter-measured currents, too, just in case they have changed due to a draining battery.)
6. Plot your data points, measured current in mA as a function of ADC counts. Find the equation of a line that best fits your data (e.g., using least-squares fitting in MATLAB).
7. Use the line equation in a PIC32 function that reads the ADC counts and returns the calibrated measured current, in mA. Add the menu item “b” (`Read current sensor (mA)`) and verify that it works as expected.

28.4.9 PWM and the H-Bridge

In this section you will implement the menu items “f” (`Set PWM (-100 to 100)`) and “p” (`Unpower the motor`).

By now you should have control of both the current sensor and the encoder. The next step is to provide low-level motor control. First you will implement part of the software associated with the current control loop. Next you will connect the H-bridge and the motor. When you finish this section you will be able to control the motor PWM signal from the client.

1. The current controller uses a timer for the 5 kHz ISR, another timer and an output compare to generate a 20 kHz PWM signal, and a digital output to control the motor direction. Write a PIC32 function that initializes these peripherals and call it from `main`.
2. Write the 5 kHz ISR. It should set the PWM duty cycle to 25% and invert the motor direction digital output. Look at the digital output and the PWM output on an oscilloscope

and confirm that you see a 2.5 kHz "heartbeat" square wave for the ISR and a 25% duty cycle 20 kHz PWM signal. Remember to clear the interrupt flag.

3. Now modify the ISR to choose the PWM duty cycle and direction bit depending on the operating mode. You should use a `switch-case` construct, similar to the `switch–case` in `main`, except the value in question here is the operating mode, as returned by the mode-querying function developed in Section 28.4.6. There will eventually be five modes to handle—IDLE, PWM, ITEST, HOLD, and TRACK—but in this section we focus on IDLE and PWM. If the operating mode is IDLE, the PWM duty cycle and direction bit should put the H-bridge in brake mode. If the operating mode is PWM, the duty cycle and direction bit are set according to the value −100 to 100 specified by the user through the client. This leads to the next action item...
4. Implement the menu item "f" (`Set PWM (-100 to 100)`). The PIC32 switches to PWM mode, and in this mode the 5 kHz ISR creates a 20 kHz PWM pulse train of the specified duty cycle and a digital output with the correct direction bit.
5. Implement the menu item "p" (`Unpower the motor`). The PIC32 switches to IDLE mode.
6. Test whether the mode is being changed properly in response to the new "f" and "p" commands by using the menu item "r" (`Get mode`).
7. Set the PWM to 80%. Verify the duty cycle with an oscilloscope and record the value of the direction pin. Then set the PWM to −40%. Verify the new duty cycle and that the direction pin has changed.
8. Now that the PWM output appears to be working, it is time to wire up the DRV8835 H-bridge circuit, as discussed in Chapter 27.1.1, to the motor and the PIC32 outputs (Figure 28.8). Notice that the 15 mΩ resistor on the current-sense PCB is in series with the motor. **Turn in a circuit diagram showing all connections of the H-bridge to the NU32, motor, and current sensor PCB.**

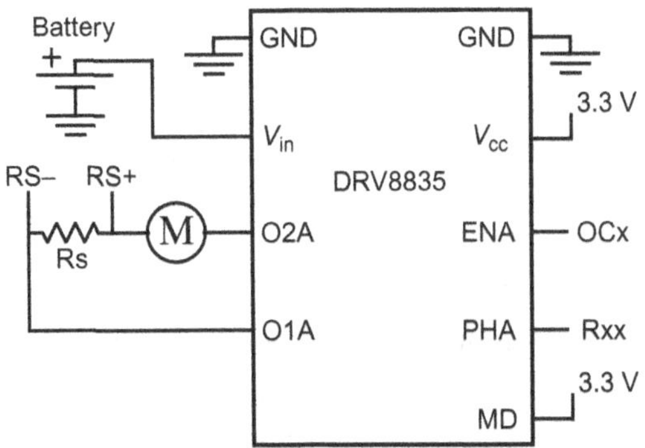

Figure 28.8
H-bridge circuit.

9. Verify the following:
   a. Set the PWM to 100%. Make sure that the motor rotates counterclockwise, that the angle returned by the encoder is increasing, and that the measured current is positive. You may have to swap the motor terminals or the encoder channels if not.
   b. Stall the motor at 100% PWM and see that the current is greater than during free running, and check that the measured current is consistent with your estimate of the resistance of the motor. (Note that the voltage at the H-bridge outputs will be somewhat lower than the voltage of the battery, due to voltage drops at the output MOSFETs.)
   c. Set the PWM to 50% and make sure that the motor spins slower than at 100%.
   d. Repeat the steps above for negative values of PWM.
   e. Make sure the motor stops when you issue the "p" (`Unpower the motor`) command.
   f. Attach the bar to the motor to increase the inertia, if it was not attached already. Get the motor spinning at its max negative speed with PWM set at −100%. Then change the PWM to 100% and quickly query the motor current ("a") several times as the motor slows down and then reverses direction on its way to its max positive speed. You should see the motor current is initially very large due to the negative back-emf, and drops continuously as the back-emf increases toward its maximum positive value (when the motor is at full speed in the forward direction).

You now have full control of the low-level features of the hardware!

28.4.10 PI Current Control and ITEST Mode

In this section you will implement menu items "g" (`Set current gains`), "h" (`Get current gains`), and "k" (`Test current control`).

The PI current controller tries to make the current sensor reading match a reference current by adjusting the PWM signal. For details about PI controllers, see Chapters 23 and 27.2.2.

In this section we focus particularly on the ITEST mode in the 5 kHz current control ISR.

1. Implement the menu items "g" (`Set current gains`) and "h" (`Get current gains`) to set and read the current loop's proportional and integral gains. You can use either floating point or integer gains. Verify the menu items by setting and reading some gains.
2. In the 5 kHz current control ISR, add a case to the `switch-case` to handle the ITEST mode. When in ITEST mode, the following should happen in the ISR:
   - In this mode, the current controller attempts to track a ±200 mA 100 Hz square wave reference current (Figure 28.4). Since a half-cycle of a 100 Hz signal is 5 ms, and 5 ms at 5000 samples/s is 25 samples, this means that the reference current toggles between +200 and −200 mA every 25 times through the ISR. To implement two full cycles of the current reference, you could use a `static int` variable in the ISR that counts

from 0 to 99. At 25, 50, and 75, the reference current changes sign. When the counter reaches 99, the PIC32 mode should be changed to IDLE—the current loop test is over.

- A PI controller reads the current sensor, compares it to the square wave reference, and calculates a new PWM duty cycle and motor direction bit.
- The reference and actual current data are saved in arrays for later plotting (e.g., Figure 28.4).

3. Add the menu item "k" (`Test current gains`). This puts the PIC32 in ITEST mode, triggering the ITEST case in the `switch-case` in the current control ISR. When the PIC32 returns to IDLE mode, indicating that the ITEST has completed, the data saved during the ITEST should be sent back to the client for plotting (e.g., Figure 28.4). This data transfer should not occur in an ISR, as it will be slow. See below for sample MATLAB code for plotting the ITEST data.
4. Experiment by setting different current gains ("g") and testing them with the square wave reference current ("k"). Verify that the measured current is approximately zero if both PI gains are zero. See how good a response you can get with a proportional gain only. Finally, get the best response possible using both the P and I gains.
5. **Turn in your best ITEST plot, and indicate the control gains you used, as well as their units.**

We provide you with a sample MATLAB function to read and plot the ITEST data (Figure 28.4). This code assumes that the PIC32 first sends the number of samples N it will be sending, then sends N pairs of integers: the reference and the actual current, in mA.

Code Sample 28.4 `read_plot_matrix.m`. **Reads a Matrix of Current Data from the PIC32 and Plots the Results. It also Computes an Average Tracking Error, to Help You Evaluate the Current Controller.**

```
function data = read_plot_matrix(mySerial)
  nsamples = fscanf(mySerial,'%d');       % first get the number of samples being sent
  data = zeros(nsamples,2);               % two values per sample:  ref and actual
  for i=1:nsamples
    data(i,:) = fscanf(mySerial,'%d %d'); % read in data from PIC32; assume ints, in mA
    times(i) = (i-1)*0.2;                 % 0.2 ms between samples
  end
  if nsamples > 1
    stairs(times,data(:,1:2));            % plot the reference and actual
  else
    fprintf('Only 1 sample received\n')
    disp(data);
  end
  % compute the average error
  score = mean(abs(data(:,1)-data(:,2)));
  fprintf('\nAverage error: %5.1f mA\n',score);
  title(sprintf('Average error: %5.1f mA',score));
  ylabel('Current (mA)');
  xlabel('Time (ms)');
end
```

28.4.11 Position Control

In this section you will implement the menu items "i" (`Set position gains`), "j" (`Get position gains`), and "l" (`Go to angle (deg)`).

Of the five PIC32 operating modes, only HOLD and TRACK are relevant to the position controller. If in either of these modes, the 200 Hz position control ISR specifies the reference current for the current controller to track.

Complete the following steps in order.

1. Implement the menu items 'i" (`Set position gains`) and "j" (`Get position gains`). The type of controller is up to you, but a reasonable choice is a PID controller, requiring three numbers from the user. Verify that the "i" and "j" menu items work by setting and reading gains from the client.
2. The position controller uses a timer to implement a 200 Hz ISR. Write a position controller initialization function that sets up the timer and ISR and call it at the beginning of `main`.
3. Write the 200 Hz position control ISR. In addition to clearing the interrupt flag, have it toggle a digital output, and verify the 200 Hz frequency of the ISR with an oscilloscope.
4. Implement the menu item "l" (`Go to angle (deg)`). The user enters the desired angle of the motor, in degrees, and the PIC32 switches to HOLD mode. In the 200 Hz position control ISR, check if the PIC32 is in the HOLD mode. When in the HOLD mode, the ISR should read the encoder, compare the actual angle to the desired angle set by the user, and calculate a reference current using the PID control gains. It is up to you whether to compare the angles in terms of encoder counts or degrees or some other unit (e.g., an integer number of tenths or hundredths of degrees).
   You also need to add the HOLD case to the 5 kHz current control ISR. When in the HOLD mode, the current controller uses the current commanded by the position controller as the reference for the PI current controller.
5. With the bar on the motor, verify that the "l" (`Go to angle (deg)`) command works as expected. Find control gains that hold the bar stably at the desired angle, and move the bar to the new desired angle on the next "l" command. Try to choose gains that give a quick motion with little overshoot. It may be easiest to use zero integral gain.

28.4.12 Trajectory Tracking

All that remains is to implement the menu items "m" (`Load step trajectory`), "n" (`Load cubic trajectory`), and "o" (`Execute trajectory`). These commands allow us to design a reference trajectory for the motor, execute it, and see the position controller tracking results.

We have provided a MATLAB function `genRef.m`, below, to generate and visualize step and cubic trajectories similar to those seen in Figure 28.5. For trajectories that are T seconds long, `genRef.m` generates an array of $200T$ reference angles, one per 200 Hz control loop iteration, and plots it in MATLAB. Before continuing, try generating some sample trajectories using `genRef.m`. See the description of menu items "m" and "n" in Section 28.2 for more information on `genRef.m`.

Code Sample 28.5 `genRef.m`**. MATLAB Code to Generate a Step or Cubic Reference Trajectory.**

```
function ref = genRef(reflist, method)

% This function takes a list of "via point" times and positions and generates a
% trajectory (positions as a function of time, in sample periods) using either
% a step trajectory or cubic interpolation.
%
%   ref = genRef(reflist, method)
%
%   Input Arguments:
%       reflist: points on the trajectory
%       method: either 'step' or 'cubic'
%
%   Output:
%       An array ref, each element representing the reference position, in degrees,
%       spaced at time 1/f, where f is the frequency of the trajectory controller.
%       Also plots ref.
%
%   Example usage:  ref = genRef([0, 0; 1.0, 90; 1.5, -45; 2.5, 0], 'cubic');
%   Example usage:  ref = genRef([0, 0; 1.0, 90; 1.5, -45; 2.5, 0], 'step');
%
%   The via points are 0 degrees at time 0 s; 90 degrees at time 1 s;
%   -45 degrees at 1.5 s; and 0 degrees at 2.5 s.
%
%   Note:  the first time must be 0, and the first and last velocities should be 0.

MOTOR_SERVO_RATE = 200;     % 200 Hz motion control loop
dt = 1/MOTOR_SERVO_RATE;    % time per control cycle

[numpos,numvars] = size(reflist);

if (numpos < 2) || (numvars ~= 2)
  error('Input must be of form [t1,p1; ... tn,pn] for n >= 2.');
end
reflist(1,1) = 0;           % first time must be zero
for i=1:numpos
  if (i>2)
    if (reflist(i,1) <= reflist(i-1,1))
      error('Times must be increasing in subsequent samples.');
    end
  end
end

if strcmp(method,'cubic')  % calculate a cubic interpolation trajectory

  timelist = reflist(:,1);
  poslist = reflist(:,2);
```

```
    vellist(1) = 0; vellist(numpos) = 0;
    if numpos >= 3
      for i=2:numpos-1
        vellist(i) = (poslist(i+1)-poslist(i-1))/(timelist(i+1)-timelist(i-1));
      end
    end

    refCtr = 1;
    for i=1:numpos-1              % go through each segment of trajectory
      timestart = timelist(i); timeend = timelist(i+1);
      deltaT = timeend - timestart;
      posstart = poslist(i); posend = poslist(i+1);
      velstart = vellist(i); velend = vellist(i+1);
      a0 = posstart;              % calculate coeffs of traj pos = a0+a1*t+a2*t^2+a3*t^3
      a1 = velstart;
      a2 = (3*posend - 3*posstart - 2*velstart*deltaT - velend*deltaT)/(deltaT^2);
      a3 = (2*posstart + (velstart+velend)*deltaT - 2*posend)/(deltaT^3);
      while (refCtr-1)*dt < timelist(i+1)
        tseg = (refCtr-1)*dt - timelist(i);
        ref(refCtr) = a0 + a1*tseg + a2*tseg^2 + a3*tseg^3;  % add an element to ref array
        refCtr = refCtr + 1;
      end
    end

  else  % default is step trajectory

  % convert the list of times to a list of sample numbers
    sample_list = reflist(:,1) * MOTOR_SERVO_RATE;
    angle_list = reflist(:,2);
    ref = zeros(1,max(sample_list));
    last_sample = 0;
    samp = 0;
    for i=2:numpos
      if (sample_list(i,1) <= sample_list(i-1,1))
        error('Times must be in ascending order.');
      end
      for samp = last_sample:(sample_list(i)-1)
        ref(samp+1) = angle_list(i-1);
      end
      last_sample = sample_list(i)-1;
    end
    ref(samp+1) = angle_list(end);

  end

  str = sprintf('%d samples at %7.2f Hz taking %5.3f sec', ...
        length(ref),MOTOR_SERVO_RATE,reflist(end,1));
  plot(ref);
  title(str);
  border = 0.1*(max(ref)-min(ref));
  axis([0, length(ref), min(ref)-border, max(ref)+border]);
  ylabel('Motor angle (degrees)');
  xlabel('Sample number');
  set(gca,'FontSize',18);
```

In the MATLAB client, you can use code of the form

```
A = input('Enter step trajectory: ');
```

to read the user's matrix of times and positions into the variable `A`, to send to `genRef.m`. The user would simply type a string similar to `[0,0; 1,180; 3,-90; 4,0; 5,0]` in response.

1. Implement the menu item "m" (`Load step trajectory`). The client should prompt the user for the trajectory parameters for `genRef.m`. Provided the duration of the trajectory is not too long to store in the PIC32's data array, the client first sends the number of samples N to the PIC32, then sends N reference positions. It is up to you whether the reference positions are sent as floating point numbers (e.g., degrees) or integers (e.g., tenths of degrees, hundredths of degrees, or encoder counts). But the user should only ever have to deal with degrees, in specifying angles and looking at plots.
2. Implement the menu item "n" (`Load cubic trajectory`). This entry is very similar to the previous item, except `genRef.m` is invoked with the `'cubic'` option.
3. Implement the menu item "o" (`Execute trajectory`). The PIC32 is placed in TRACK mode. In the 200 Hz position control ISR, check if the mode is TRACK. If so, then the ISR increments an index into the reference trajectory array, and the indexed trajectory position is used as the reference to the PID controller, which calculates a commanded current for the current control ISR.
   The position control ISR should also collect motor angle data for later plotting.
   When the array index reaches N, the operating mode is switched to HOLD, and the last angle of the reference trajectory array is used as the holding angle. The collected data is sent back to the client for plotting, similar to the ITEST case. The MATLAB client plotting code should be similar to Code Sample 28.4, but using the appropriate data types and scaling of the sample times, so the user sees the results in terms of time, not samples.
   You also need to add the TRACK case to the 5 kHz current control ISR. To the current controller, the TRACK case is identical to the HOLD case: in both cases, the current controller attempts to match the current commanded by the position controller.
4. Experiment with tracking different trajectories (such as the ones in Figure 28.5) with different position control gains until you get good performance. For example, the performance in Figure 28.2 is reasonable, though a larger derivative gain would help to eliminate the overshoot. In experimentally tuning your gains, it is easiest to start with proportional (P) control alone, then add derivative (D) control. Finally, tune your PD gains simultaneously. You may not need an integral (I) term for good performance.
5. **Turn in your best plots of following the step and cubic trajectories in Figure 28.5 with the load attached. Indicate the control gains you used, as well as their units.**

Congratulations! You now have a full motor control system.

28.5 Extensions

Saving gains in flash memory

Currently you must re-enter gains every time you reset the PIC32; this quickly becomes annoying. You can, however, store the gains in flash memory, which allows them to persist, even when the PIC32 is powered off. To see how to access flash memory, refer to Chapter 18. Add a menu item that saves gains to flash memory, and modify the startup code in `main` to first read any stored gains.

LCD display

Connect an LCD display to the PIC32. Use it to show information about the current system state, the gains, or whatever else you want.

Model-based control

In Chapter 25, we learned how to characterize a motor. Notice that we have mostly ignored the motor parameters when implementing the control law. Our ability to do this demonstrates the power of feedback to compensate for a lack of a system model. However, incorporating a model of the motor and load into the control loop could improve the motor's performance. See Chapter 27.2.1, for example.

To test your model-based control, try adding a small weight to one end of the bar and track the trajectories in Figure 28.5. Either hardcode the mass, center of mass, and inertia of the total load in your PIC32 code, or allow the user to enter information about the load. See if your model-based control can provide better tracking than your original controller.

As an even more advanced project, you could identify the properties of the motor and the load by spinning the motor, measuring the current (torque) as a function of the position, velocity, and acceleration of the motor, and using the data to fit parameters of a model.

Anytime data collection

Instead of only collecting data during ITEST or TRACK modes, the user could have the option to collect and plot a specified duration of data at any time. This would allow collecting data during a HOLD while the user perturbs the motor, for example.

Real-time data

Currently we employ batch processing to retrieve motor data. This severely limits how long you can collect data before running out of memory. What if you want to see the motor data in real-time? A data structure called a circular (a.k.a. ring) buffer can help. The circular buffer has two position indexes; one for reading and one for writing. Data is added to the array at the write position, which is subsequently advanced. If the end of the array is reached, the write position wraps around to the beginning. Data is read from the read position and the read index advanced, also wrapping around. In one style of circular buffer, if the read position and write position are the same, the buffer is empty. If the write position is one behind the read position, the buffer is full. When using a circular buffer in this project, either the current loop or

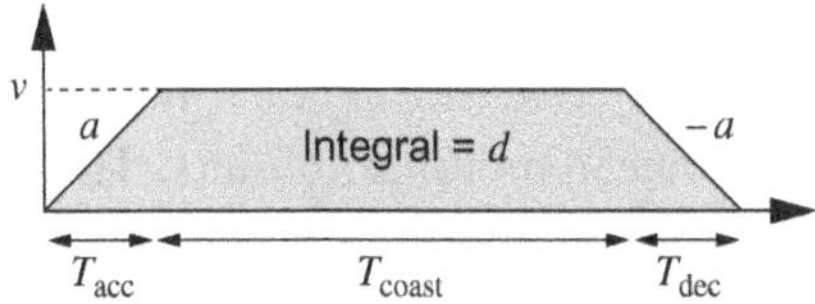

Figure 28.9
A trapezoidal move of length d.

position loop will add data to the buffer. Rather than waiting for the buffer to be full, data can be sent back any time the buffer is not empty. It is up to the client how to handle this continuous stream of data, perhaps by using an oscilloscope-style display. See Chapter 11 for more details about circular buffers.

If the communication cannot keep up with the data, then the stored data can be decimated, sending back only every nth data sample, $n > 1$.

Trapezoidal moves

The reference trajectories in this chapter are rest-to-rest step or cubic motions in position. One type of rest-to-rest trajectory that is common in machine tools is the trapezoidal move. The name comes from the fact that the trajectory, represented in the velocity space, is a trapezoid (Figure 28.9): the motor accelerates with a constant acceleration a for a time T_{acc} until it reaches a maximum velocity v; it coasts with a constant velocity v for a time T_{coast}; and then it comes to a stop by decelerating at $-a$ for a time $T_{dec} = T_{acc}$. The user should specify the total move distance d and either (a) the total time $T = T_{acc} + T_{coast} + T_{dec}$ and the maximum velocity v, (b) the acceleration a and the maximum velocity v, or (c) T and a. The other parameters are calculated so that the integral of the velocity trapezoid is equal to d.

Add a menu item that allows the user to generate a trapezoidal reference trajectory based on either (a), (b), or (c).

28.6 Chapter Summary

A typical commercial motor amplifier consists of at least a microcontroller, a high-current H-bridge output, a current sensor, and some type of motion feedback input (e.g., from an encoder or a tachometer). The amplifier is connected to a high-power power supply to power the H-bridge. The amplifier accepts a velocity or current/torque input from an external controller (e.g., a PC), either as an analog signal or the duty cycle of a PWM signal. When the amplifier is in current mode, current sensor feedback is used to alter the PWM input to the H-bridge to achieve close tracking of the commanded current. In velocity mode, the amplifier uses motion feedback to alter the PWM input to achieve close tracking of the desired velocity.

More advanced motor amplifiers, like the Copley Accelus, offer other features, like client graphical user interfaces and trajectory tracking. The project in this chapter emulates some of the capabilities that come with advanced motor amplifiers. This project brings together your knowledge of the PIC32 microcontroller, C programming, brushed DC motors, feedback control, and interfacing a microcontroller with sensors and actuators to achieve capabilities found in every robot and computer-controlled machine tool.

28.7 Exercises

Complete the motor control project. Complete all of the numbered action items in each subsection. Turn in your answers for the **action items in bold** as well as your well-commented code.

CHAPTER 29

Other Actuators

In addition to brushed permanent magnet DC motors, many other types of electric actuators exist. In this chapter we discuss some of the most commonly used electric actuators in mechatronics.

29.1 Solenoids

A solenoid is an on-off linear actuator. It consists of a stationary wire coil around a tube and a plunger that moves in the tube, typically until it rests against a stop at one end or moves completely out of the tube at the other end. When current flows through the coil it becomes an electromagnet, pulling the plunger into the tube and against the stop. When current is off, the plunger moves freely.

Figure 29.1 shows an example solenoid, the Pontiac Coil F0411A. Note that the plunger is not physically connected to the coil housing; the plunger returns to a home position when the current is off either by a return spring or some other external means (e.g., the influence of gravity).

A solenoid draws more current, at a higher voltage, than the PIC32 output can provide. For example, the F0411A is rated for 12 V and the coil has a resistance of 36 Ω; therefore, it draws 1/3 A and dissipates 4 W when energized. To provide this current and voltage, an H-bridge can be used (Chapter 27.1.1). Figure 29.1 shows an alternative circuit which uses a transistor to interface the PIC32 with a solenoid. When the PIC32's digital output is high, the TIP120 transistor turns on, pulling current from the 12 V supply through the solenoid coil. When the PIC32 digital output drops low, the TIP120 turns off and current that is already flowing through the inductive solenoid flows through the 1N4001 flyback diode until the initial energy $\frac{1}{2}LI^2$ in the solenoid is dissipated.

To see that the circuit in Figure 29.1 is adequate, we consult the data sheet for the TIP120 Darlington NPN bipolar junction transistor. The TIP120 can provide up to 5 A of current, more than the 1/3 A needed by the solenoid. The DC current gain β (see Appendix B.3.2) for the TIP120 is at least 1000. When the transistor is in the linear mode, it enforces the equation $I_C = \beta I_B$, where I_C is the collector current (flowing through the solenoid) and I_B is the base

Embedded Computing and Mechatronics with the PIC32 Microcontroller. http://dx.doi.org/10.1016/B978-0-12-420165-1.00029-9

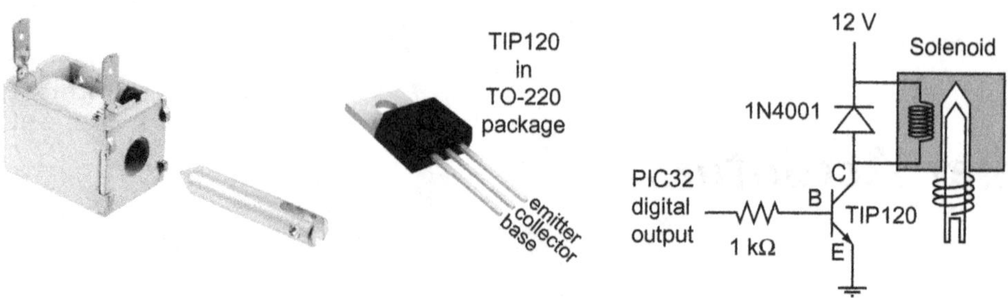

Figure 29.1
(Left) The Pontiac Coil F0411A 12 V solenoid, with the plunger separated from the coil. (Image courtesy of Digi-Key Electronics, digikey.com.) (Middle) The TIP120 Darlington NPN bipolar junction transistor in a TO-220 package. (Right) A circuit that allows a PIC32 digital output to activate the solenoid (plunger equipped with a return spring).

current. Since we want $I_C = 1/3$ A, rounded up to $I_C = 0.5$ A to be safe, then the base current should be at least $0.5\ \text{A}/\beta = 0.5$ mA. Since PIC32 digital outputs can produce a few mA, this current is safe to draw from a digital output. Again consulting the TIP120 data sheet, we see that the maximum base-emitter voltage when the transistor is on, $V_{BE,on}$, is 2.5 V. The high digital output is 3.3 V, so to get a base current of 0.5 mA, we have

$$I_B = (3.3\ \text{V} - 2.5\ \text{V})/R = 0.5\ \text{mA}.$$

Solving, we get a base resistance $R = 1600\ \Omega$. We round this down to 1 kΩ as a safety factor to ensure enough base current. Finally, we need to ensure that the 1N4001 diode can handle the current from the solenoid. Consulting its data sheet, we see that it can handle 1 A continuously, more than enough.

Electrical characteristics of solenoids are their rated voltage and coil resistance, as well as a specification of whether they can be activated continuously without overheating the coil. For example, the F0411A can be activated continuously. Its sister 12 V solenoid, the F0412A, has a resistance of only 16.9 Ω, so it heats up faster. The F0412A can only be activated intermittently, e.g., one minute on and three minutes off.

Mechanical characteristics of a solenoid include the maximum force when the coil is energized and the stroke. The maximum force is obtained when the plunger is fully inserted into the coil tube. As the plunger is displaced from its retracted position, the electromagnetic force drops. The stroke of the solenoid is the distance from the fully retracted position to a position where the force has dropped below a minimum threshold.

Although all solenoids operate by pulling the plunger into the coil, some solenoids are called *push* type solenoids. These solenoids have a rod attached to the plunger that protrudes from the back of the solenoid. When the plunger is pulled into the tube, the end of the rod is pushed away from the tube.

29.2 Speakers and Voice Coil Actuators

A speaker is an electromagnetic actuator used to generate sound. It consists of magnets to create a magnetic field and a coil (commonly referred to as a voice coil) attached to a speaker cone. Current through the coil creates a force on the coil/cone assembly due to the Lorentz force law, and alternating current causes the coil/cone to alternate motion directions (up/down). The resulting vibration of the cone creates pressure waves that carry sound. Since human hearing is sensitive only to frequencies in the range 20 Hz to 20 kHz, the frequency content of the alternating current through the speaker coils is generally limited to that range. In particular, DC current would simply heat the coils without producing any vibration that would generate sound. Speaker sizes and shapes are designed to produce audible sound in particular frequency ranges, from the lowest audible frequencies for subwoofers, to low frequencies for woofers, to mid-range and high frequencies for tweeters.

Speakers can be used as inexpensive linear actuators, driven by AC or DC currents. The linear force generated by a speaker for a given current is maximized when the coil is centered in the magnet's magnetic field, and drops quickly as the coil moves away from the center. A large subwoofer, with its corresponding large magnets, may allow significant forces to be generated at displacements of a centimeter or more, while a small tweeter has a much smaller operating range.

Voice coil actuators operate on the same principle as speakers, except they are designed to provide a nearly constant magnetic field over a larger displacement of the coil (stroke). Just as a rotational brushed permanent magnet DC motor has a motor constant (in $\text{Nm}/\sqrt{\text{W}}$) and a torque constant (in Nm/A), a voice coil actuator has a motor constant (in $\text{N}/\sqrt{\text{W}}$) and a force constant (in N/A). For example, the H2W Technologies NCC05-11-011-1PBS (Figure 29.2) has a stroke of 1.27 cm, a motor constant of $2.98\ \text{N}/\sqrt{\text{W}}$, and a force constant of 5.2 N/A (and therefore an electrical constant, or back-emf constant, of 5.2 Vs/m). Other relevant characteristics of the NCC05-11-011-1PBS are a peak force of 14.7 N, a maximum continuous force of 4.9 N without overheating, a resistance of 3 Ω, and an inductance of 1 mH. For a given voltage, a speed-force curve could be drawn, analogous to the speed-torque curve for a brushed DC motor, but linear voice coil actuators are rarely run at constant speed for significant time due to their limited stroke.

The NCC05-11-011-1PBS is a moving coil actuator. There are also moving magnet voice coil actuators, where the coil is attached to the stator. Voice coil actuators can either come with linear bearings installed or require the user to supply the bearings.

Figure 29.2
The NCC05-11-011-1PBS moving coil voice coil actuator by H2W Technologies, Inc. The magnetic stator is on the left and the moving coil is on the right. (Image courtesy of H2W Technologies, h2wtech.com.)

To drive a voice coil actuator with a PIC32, an H-bridge can be used, just as with brushed DC motors. Position feedback can be obtained using a linear potentiometer, a linear encoder, or other position sensor, as described in Chapter 21.

Another option to drive a voice coil actuator is to use an analog speaker amplifier, since voice coil actuators are essentially speakers.[1] This method, however, requires an analog voltage from the controller, and you would have to remove the amplifier's high-pass filter that prevents DC currents.

29.3 RC Servos

An RC servo consists of a DC motor, gearhead, angle sensor such as a potentiometer, and a feedback control circuit (typically implemented by a microcontroller), all in a single package (Figure 29.3). The integration of these components makes RC servos an excellent choice for high-torque approximate positioning without requiring your own feedback controller. The servo's output shaft typically has a total rotation angle of less than 360 degrees, with 180 degrees being common.

An RC servo has three input connections: power (typically 5 or 6 V), GND, and a digital control signal. The power line must provide enough current to drive the motor. By convention, the control signal consists of a high pulse every 20 ms, and the duration of this pulse indicates the desired output shaft angle. A typical design has a 0.5 ms pulse mapping to one end of the rotation range and a 2.5 or 3 ms pulse mapping to the other end of the range, with a linear relationship between pulse length and rotation angle in between (Figure 29.4).

[1] Speaker amplifiers can also be used to drive a DC motor.

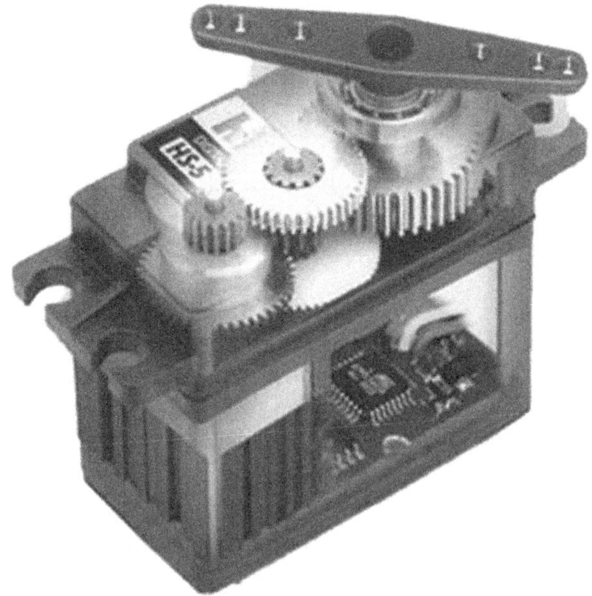

Figure 29.3

The interior of a Hitec RC servo showing the gears and the microcontroller that receives the control signal, reads the potentiometer, and implements the feedback controller. (Image courtesy of Hitec RCD USA, Inc., hitecrcd.com.)

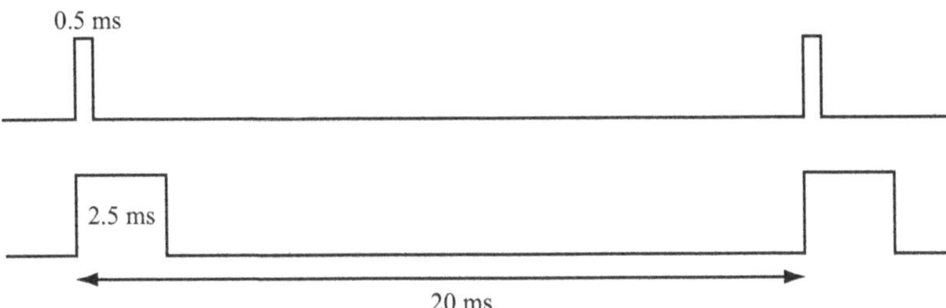

Figure 29.4

Typical RC servo control waveforms. These pulse widths drive the servo output shaft near the two ends of its rotation range.

29.4 Stepper Motors

A stepper motor is designed to progress through a revolution in a series of small increments or steps. There are several types of stepper motors with different stator and rotor configurations. The easiest type to understand is the permanent magnet stepper motor shown in Figure 29.5. Permanent magnet stepper motors have permanent magnets on the rotor. Current flows through coils 1 and 2 of the stator. Current through coil 1 acts as an electromagnet of one sign

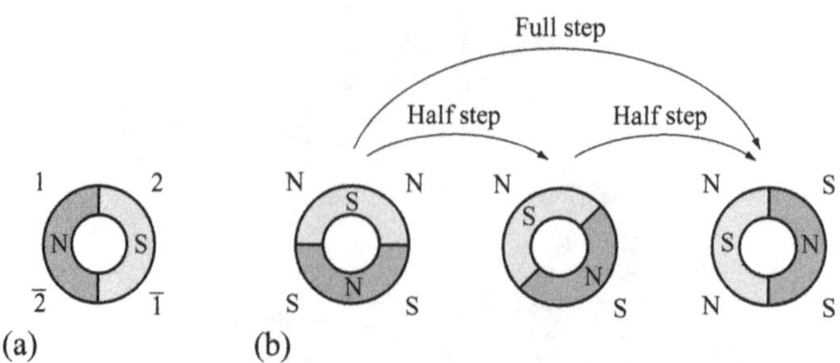

Figure 29.5
(a) A stepper motor with two rotor poles and four stator poles. The four stator poles correspond to two independent coils, 1 and 2. (b) The stepping sequence. As the coils change from current flowing one way, to off, to the other way, the rotor rotates to new equilibrium positions. Note that the holding torque at the intermediate half-step may be lower than at the full-step positions.

(e.g., N) on coil 1 and the opposite sign (e.g., S) on its partner coil $\bar{1}$; coil 2 operates similarly. These magnetic fields attract permanent magnets on the rotor, causing the rotor to rotate (step) to a new equilibrium position. By switching the direction of the current through a coil, the sign of the electromagnet changes.

As the currents through the electromagnets proceed through a fixed sequence, the motor rotates, one step at a time. A complete full-stepping sequence for the coils is shown below:

coil 1	+	−	−	+	+ (back to the beginning)
coil 2	+	+	−	−	+ (back to the beginning)

Another type of stepper motor is the variable reluctance stepper motor. These motors have no permanent magnets; rather their rotor is made of a ferrous material. Both the stator poles and the rotor have teeth, and when the stator coils are energized, the rotor rotates so the rotor teeth are positioned to minimize magnetic reluctance.[2] Hybrid stepper motors incorporate principles of both permanent magnet and variable reluctance motors. Regardless, all motors progress through their revolution by energizing the coils in the correct sequence.

Stepper motors can also be controlled using *half-stepping*, to increase their position resolution. With half-stepping, the current to one of the coils is turned off before transitioning to the next full-step state. Half-stepping doubles the number of steps available, but results in reduced holding torque at the half-steps. A complete half-stepping sequence is given below:

[2] The relationship of reluctance to magnetic flux is analogous to the relationship of resistance to current.

coil 1	+	0	−	−	−	0	+	+	+ (back to the beginning)
coil 2	+	+	+	0	−	−	−	0	+ (back to the beginning)

Figure 29.5(b) shows two half-steps making a full step.

Microstepping refers to partially energizing the coils, instead of using only full current in one direction, full current in the other direction, or off. Microstepping allows for increased resolution in the control of the motor's angle, but the holding torque at each microstep is reduced even further beyond that for half-steps.

The simple stepper shown in Figure 29.5(b) has only two *rotor poles*, and a full rotation of the motor consists of only four full steps (or eight half steps). Real stepper motors have many more rotor poles and *stator poles*, but usually only two independently controlled sets of coils that energize these electromagnetic stator poles. The rotor poles are typically toothed and are attracted to stator pole teeth, and a large number of rotor teeth create much finer step sizes than the 90° of our simple example. Figure 29.6 shows a hybrid stepper motor that has been opened, exposing the stator poles and rotor poles.

A stepper motor is characterized by the number of steps per revolution (e.g., 100 steps, or 3.6° per step); the resistance of the coils; the current each coil can carry continuously without overheating (usually implicit in the motor's voltage rating and the coil resistances, since stepper motors are intended to be powered continuously); the *holding torque*, the maximum restoring force as the rotor is forced away from an equilibrium when the coils are energized; and the *detent torque*, the amount of torque needed to move the rotor out of an equilibrium position when the coils are off. The detent torque comes from the attraction of the rotor's permanent magnets to the stator teeth.

When a stepper is stepped slowly, it settles into its equilibrium position between steps. When stepped quickly, however, it may never settle before the coils change, leading to continuous

Figure 29.6
(Left) A stepper motor stator, showing the eight stator coils and the stator teeth. (Right) The stepper motor rotor, removed from the motor, showing many magnetized rotor pole teeth.

motion. If a stepper motor is stepped too quickly, or if the motor's load or the acceleration of the stepping sequence is too high, the stepper will be unable to keep up and will simply vibrate. Similarly, if a stepper is moving at a high speed and the stepping sequence stops suddenly, inertia may cause the stepper to continue to rotate. Once either of these events occurs, it is impossible to know the angle of the rotor without an external sensing device, like an encoder. The speed at which the motor can be stepped reliably decreases as the torque it must provide increases. To ensure that the stepping sequence is followed, particularly in the presence of a significant load, it is a good idea to ramp up the stepping frequency at the beginning of a motion and ramp down the frequency at the end, which limits the torque required to perform the motion.

To save power when no motion is required, the coils can be turned off, provided the detent torque is sufficient to prevent any unwanted motion.

Steppers are generally used for precise position control of known loads, such as printer heads, without the need for feedback control. If the load is well known, and if velocity and acceleration limits are not exceeded, you can be certain of the stepper's position, within a tight tolerance, based solely on your commanded stepping sequence.

29.4.1 Stepper Motor Coil Configurations

Stepper motors come in two types of coil configurations, bipolar and unipolar (Figure 29.7). With bipolar stepper motors, current can flow in either direction (hence bipolar) through each coil. With unipolar stepper motors, each coil is broken into two subcoils, and current typically only flows one direction (hence unipolar) through each subcoil.

Bipolar stepper motor

A bipolar stepper motor has four external wires to connect: one at each end of coil 1 and one at each end of coil 2. Let us call the two ends of coil 1 1A and 1B. To switch coil 1's current,

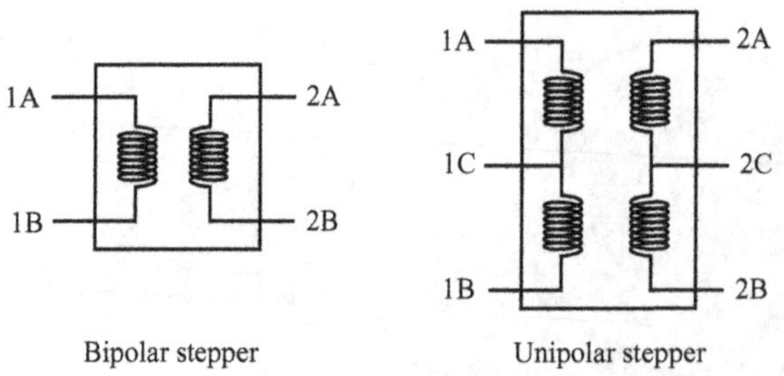

Figure 29.7
(Left) A bipolar stepper motor has four connections to the two coils. (Right) A unipolar stepper motor has six connections to the two coils, or five if 1C and 2C are combined.

we apply V and GND to 1A and 1B, respectively, then switch to GND and V at 1A and 1B, respectively. A bipolar stepper motor can be driven using two H-bridges, treating each coil separately as a DC motor that is driven either full forward or full backward for full stepping. Many H-bridge chips, like the DRV8835 (Chapter 27.1.1), come with two H-bridges.

There are also microstepping stepper motor drivers, such as the Texas Instruments DRV8825. The DRV8825 has two H-bridges, and each H-bridge has two wires (AOUT and BOUT) connected to the two ends of one of the stepper's coils. The DRV8825 has a DIR digital input, which specifies whether the motor steps clockwise or counterclockwise, and a STEP digital input. On a rising edge of STEP, the motor moves to the next position as indicated by DIR. There are also three digital inputs, MODE0-MODE2, that determine whether the next position is at a full step, 1/2 step, 1/4 step, 1/8 step, 1/16 step, or 1/32 step. To achieve microstepping, the coils are partially energized, using PWM, to a fixed percentage of the full current. To ensure that holding torque is approximately the same at every position of the motor, when both coils are energized equally, each coil is only energized to 71% full current; when one coil is at 0%, the other is energized to 100% full current.

If you are using an unknown bipolar stepper motor, you can determine which wires are connected to which coil using a multimeter in resistance-testing mode.

Unipolar stepper motor

A unipolar stepper motor typically has six external wires to connect, three for each coil. For coil 1, for example, the wires 1A and 1B are at either end of coil 1, as before, and the connection 1C is a "center tap." This center tap is commonly connected to the voltage V, and we switch current through the coil by alternating between grounding 1A while leaving 1B floating, and grounding 1B while leaving 1A floating.

Five-wire unipolar stepper motors have a single center tap that is common to both coils.

One method for driving a unipolar stepper motor is to ignore the center taps, treating it as a bipolar stepper, and use one H-bridge for each coil.

If you are using an unknown unipolar stepper motor, you can determine which wires are connected to which coil using a multimeter set to resistance-testing mode. The center taps are at half the resistance of the full coil resistance.

29.5 Brushless DC Motors

A brushless DC motor (BLDC) is similar to a permanent magnet stepper motor, but with some properties of a brushed DC motor. A BLDC has coils on the stator and permanent magnets on the rotor, like a permanent magnet stepper motor. Like a brushed DC motor, the coils are driven using PWM to H-bridges, and the coils are commutated as the rotor rotates. Unlike a

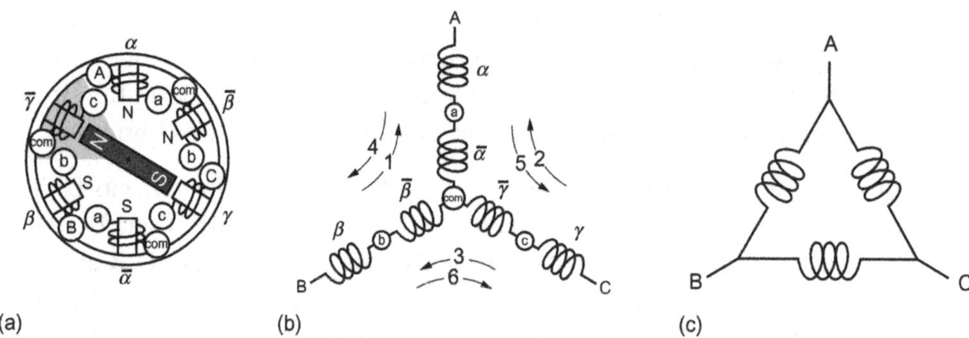

Figure 29.8

(a) A simple three-phase BLDC motor with two rotor poles and six stator poles. Current flowing from B to A (current path 1) creates N magnetic poles on the coils α and $\overline{\beta}$ and S magnetic poles on the coils $\overline{\alpha}$ and β, driving the rotor counterclockwise. (b) The electrical wiring of the BLDC in (a), and the six possible current paths. This wiring is known as a "wye" configuration, since it looks like a Y. (c) Some BLDCs have a delta wiring configuration, which we do not consider further in this chapter.

brushed DC motor, however, the commutation of the stator coils is performed electronically; there are no brushes.

BLDCs are often used as replacements for brushed DC motors, and they are popular in applications like computer fans and quadcopters. Advantages of BLDCs over brushed motors include:

- longer lifetime and higher speeds since there is no brush friction and wear;
- no particles due to wearing brushes;
- less electrical noise, since there are no abrupt brush-commutation switching events; and
- higher continuous current, and therefore continuous torque, because the coils are attached to the stator which serves as a heat sink to cool the coils.

A disadvantage of BLDCs is the more complex drive circuitry required.

Figure 29.8(a) illustrates the basic operation of a three-phase BLDC, where a *phase* is one of the three inputs, A, B, or C. The permanent magnet rotor has two poles. The stator has six poles, wired electrically as shown in Figure 29.8(b). Current is flowing from B to A, indicated as current path 1 in Figure 29.8(b). The current from B to b creates an S pole at coil β, as seen by the permanent magnet, while the current from b to com creates an N pole at $\overline{\beta}$ on the opposite side of the stator. The current continues from com to a, creating another S pole at $\overline{\alpha}$ and from a to A, creating an N pole at α on the opposite side of the stator. The net result is that the permanent magnet experiences a torque to rotate it counterclockwise. In fact, as long as the N pole of the permanent magnet is in the angular region shaded gray, the current should either flow from B to A (current path 1) to rotate the rotor counterclockwise, or from A to B (current path 4) to rotate the rotor clockwise. No current should flow through the C coils, as current through these coils simply wastes power while creating little or zero torque on the rotor.

If the rotor exits the gray shaded region in the clockwise direction, then A should float, as the coils α and $\overline{\alpha}$ cannot create much torque on the rotor, and current should flow between B and C (current path 3 or 6) to generate torque. If the rotor exits the gray shaded region in the counterclockwise direction, then B should be left floating while current should flow between A and C (current path 2 or 5).

Thus the coils are commutated much like they are for a brushed DC motor. Every sixty degrees of rotation of the rotor, the energized coils change. During a full counterclockwise revolution of the rotor, starting from the configuration shown in Figure 29.8(a), the current flows from B to A (current path 1), C to A (path 2), C to B (path 3), A to B (path 4), A to C (path 5), and B to C (path 6) before repeating. In other words, the sequence is 1-2-3-4-5-6. During a clockwise revolution, the sequence goes 4-3-2-1-6-5 before repeating. A full commutation cycle, from current path 1 to current path 6, is called an *electrical cycle* or *electrical revolution*. For the two-pole-rotor BLDC, one mechanical revolution corresponds to one electrical revolution.

A BLDC rotor can have more than two magnetic poles; BLDCs with four and eight poles are common, for example. For every pair of rotor poles, there is one electrical revolution per mechanical revolution. Thus there is one electrical revolution per mechanical revolution for a two-pole (one pair) rotor, two electrical revolutions per mechanical revolution for a four-pole (two pair) rotor (twelve different commutation periods, or 30 mechanical degrees between coil changes) and four electrical revolutions (15 mechanical degrees between coil changes) for an eight-pole (four pair) rotor.

To know when to switch which coils are energized, position feedback is required. BLDCs are typically equipped with three digital Hall effect sensors that sense the position in an electrical revolution, and therefore which of A, B, and C should be left floating while the other two are powered.

Let us say that the Hall effect sensors indicate that phase C should be left floating. Driving A and B is exactly the same as driving the two terminals of a brushed DC motor. An H-bridge and PWM is used to create variable, bidirectional voltage across the terminals. When the Hall sensor state switches, you simply switch the terminal left floating and drive the other two terminals in the same way. Flyback diodes provide a path for current through the newly inactivated coil to begin to dissipate. We will return shortly to details of interfacing a BLDC with a PIC32.

An example BLDC is the Pittman 24 V ELCOM SL 4443S013. This motor is a three-phase, four-rotor-pole wye-style BLDC with six stator poles (Figure 29.9). Because it has four rotor poles, there are two electrical revolutions per mechanical revolution, i.e., 30 mechanical

Figure 29.9
Left to right: the Pittman ELCOM SL 4443S013, with its three motor connections A, B, and C and its five Hall sensor wires (three Hall sensor outputs and two wires for power (5 V) and ground); the three-phase, six-pole stator; and the rotor with four poles created by bar magnets along the axis of the shaft.

degrees between commutation changes. This is illustrated in the motor's data sheet, the relevant section of which is reproduced in Figure 29.10. The three Hall sensor outputs are concatenated to make a three-bit number $S_1S_2S_3$, and the six electrical steps are represented by the Hall sensor readings 100, 110, 010, 011, 001, and 101. These correspond to the current paths 1-6 from Figure 29.8(b), respectively, when the motor rotates counterclockwise. To reverse the direction, at each sensor state you swap which phase is high and which is grounded.

The coil resistance of the 24 V ELCOM SL 4443S013 is 0.89 Ω, indicating a stall current of 24 V/0.89 Ω = 27 A. The torque constant is 0.039 Nm/A, indicating a stall torque of 0.039 Nm/A × 27 A = 1.05 Nm. The maximum continuous torque is 0.135 Nm/A and the terminal inductance is 0.19 mH.

29.5.1 Interfacing the PIC32 to a BLDC

While it is possible to drive a BLDC with standard H-bridge chips, it is easier to use an IC designed specifically to drive BLDCs, such as the STMicroelectronics L6234 three-phase motor driver. The L6234 has three half H-bridges which are connected to the three-phase motor as shown in Figure 29.11. One of the three half H-bridges always has both switches open, leaving its motor terminal floating. Typically one of the other half H-bridges holds its output low, while the last half H-bridge is driven by PWM that alternates its output between V_s and GND. Each half H-bridge has two flyback diodes to provide a current path for an energized coil that is suddenly left floating.

The L6234 allows a supply voltage V_s between 7 and 52 V, can handle up to 5 A peak current, and allows PWM frequencies up to 150 kHz. Each of the three half H-bridge outputs, OUTx, has two associated digital inputs: ENx and INx. If ENx is low, OUTx is floating (both switches in the half H-bridge are off/open). If ENx is high and INx is low, the lower switch is

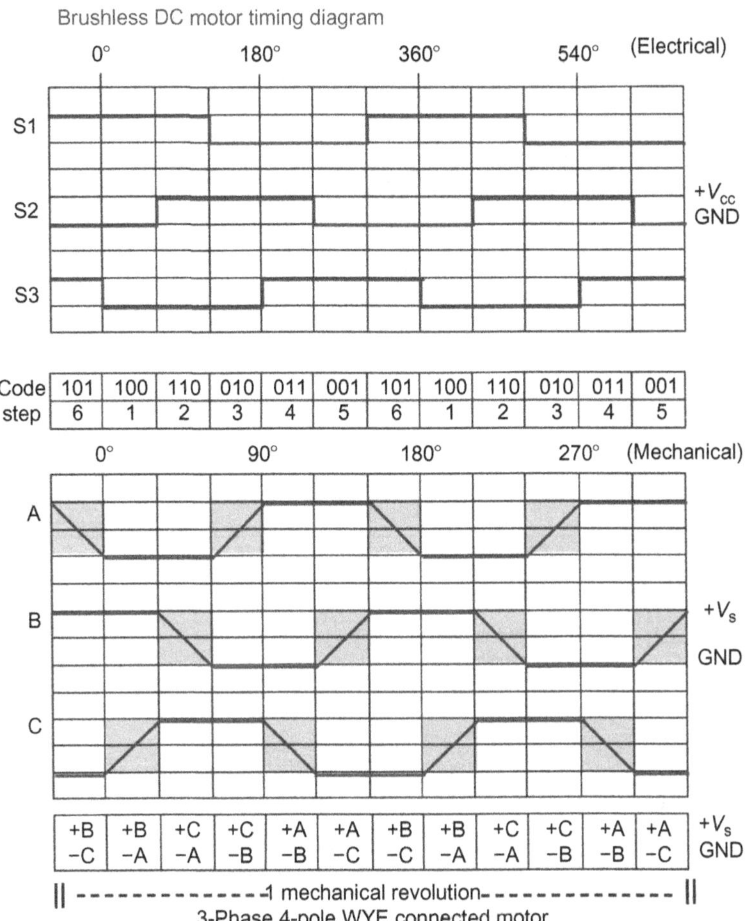

Figure 29.10

Bottom: The motor inputs as a function of the mechanical angle of the rotor. In regions shaded gray, the corresponding input is left floating at those rotor angles while the other two inputs are powered. For example, between 0 and 30 degrees, input B is powered by $+V_S$, A is grounded, and C is left floating. The voltages shown here are to drive the motor counterclockwise at maximum speed; to drive the motor clockwise, the voltages should be reversed (e.g., between 0 and 30 degrees, A should be $+V_S$ and B should be grounded). Instead of using the maximum voltage V_S at all times, PWM can be used to achieve variable speeds. Top: The digital Hall sensor readings as a function of electrical angle. Note there are two electrical revolutions per mechanical revolution. The steps 1-6 correspond to the counterclockwise-driving current paths in Figure 29.8.

closed, pulling OUTx close to GND. If ENx is high and INx is high, the upper switch is closed, pulling OUTx close to V_S. For other details on the operation of the L6234, consult the data sheet and associated application notes.

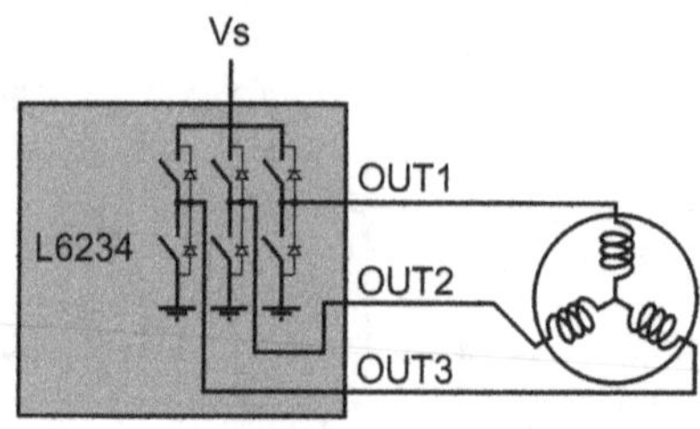

Figure 29.11
Connecting the L6234 three-phase motor driver to a BLDC.

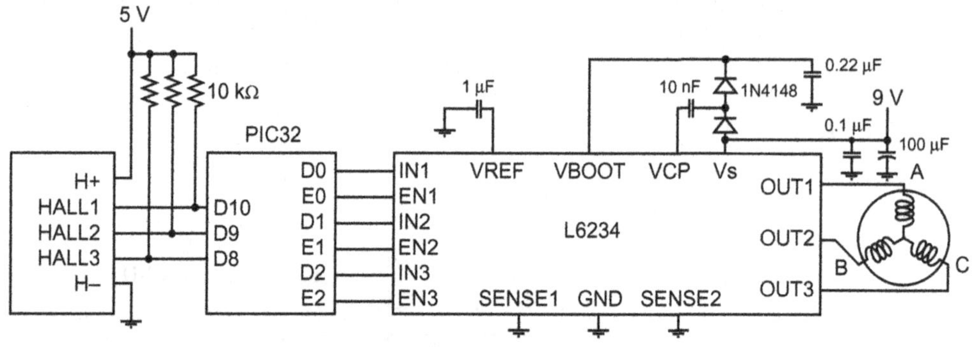

Figure 29.12
Interfacing a BLDC with Hall sensors to a PIC32 using an L6234.

Figure 29.12 illustrates a circuit using the L6234 to interface the PIC32 with the Pittman ELCOM SL 4443S013. We choose $V_s = 9$ V. While this can create a stall current of 9 V/0.89 A = 10 A, beyond the rated peak current of the L6234, this 5 A rating applies to supply voltages up to 52 V. In practice we have not encountered an issue with stall currents at 9 V, although we do not suggest stalling the motor for long periods of time.

29.5.2 A BLDC Library

The `bldc.{c,h}` library implements functions to handle commutation based on the sensor state. It is written for the L6234 circuit shown in Figure 29.12. The `bldc` library uses three output compare (PWM) modules, OC1 (D0), OC2 (D1), and OC3 (D2), and three digital outputs, E0, E1, and E2. The PWMs connect to the INx inputs for the three phases, and the digital outputs connect to the ENx inputs for the three phases.

The function `bldc_get_pwm` prompts the user for a signed PWM duty cycle percentage. The function `bldc_commutate` accepts a three-bit Hall sensor reading, as in Figure 29.10, as well as the signed PWM duty cycle percentage, and determines which of the three motor phases should be floating, which should be grounded, and which should be PWMed between V_S and GND. This function offloads the commutation details from user programs that use the `bldc` library.

Code Sample 29.1 `bldc.h`. **BLDC Library Header.**

```
#ifndef COMMUTATION_H__
#define COMMUTATION_H__

// Set up three PWMs and three digital outputs to control a BLDC via
// the STM L6234.  The three PWM channels (OC1 = D0, OC2 = D1, OC3 = D2) control the
// INx pins for phases A, B, and C, respectively.  The digital outputs
// E0, E1, and E2 control the ENx pins for phases A, B, and C, respectively.
// The PWM uses timer 2.
void bldc_setup(void);

// Perform commutation, given the PWM percentage and the current sensor state.
void bldc_commutate(int pwm, unsigned int state);

// Prompt the user for a signed PWM percentage.
int bldc_get_pwm(void);

#endif
```

Code Sample 29.2 `bldc.c`. **BLDC Library C Implementation.**

```
#include "NU32.h"

void bldc_setup() {
  TRISECLR = 0x7;            // E0, E1, and E2 are outputs

  // Set up timer 2 to use as the PWM clock source.
  T2CONbits.TCKPS = 1;       // Timer2 prescaler N=2 (1:2)
  PR2 = 1999;                // period = (PR2+1) * N * 12.5 ns = 50 us, 20 kHz

  // Set up OC1, OC2, and OC3 for PWM mode; use defaults otherwise.
  OC1CONbits.OCM = 0b110;  // PWM mode without fault pin; other OC1CON bits are defaults
  OC2CONbits.OCM = 0b110;
  OC3CONbits.OCM = 0b110;

  T2CONbits.ON = 1;          // turn on Timer2
  OC1CONbits.ON = 1;         // turn on OC1
  OC2CONbits.ON = 1;         // turn on OC2
  OC3CONbits.ON = 1;         // turn on OC3
}

// A convenient new type to use the mnemonic names PHASE_A, etc.
// PHASE_NONE is not needed, but there for potential error handling.

typedef enum {PHASE_A = 0, PHASE_B = 1, PHASE_C = 2, PHASE_NONE = 3} phase;

// Performs the actual commutation: one phase is PWMed, one is grounded, the third floats.
// The sign of the PWM determines which phase is PWMed and which is low, so the motor can
// spin in either direction.
```

```
static void phases_set(int pwm, phase p1, phase p2) {

  // an array of the addresses of the PWM duty cycle-controlling SFRs
  static volatile unsigned int * ocr[] = {&OC1RS, &OC2RS, &OC3RS};

  // If p1 and p2 are the energized phases, then floating[p1][p2] gives
  // the floating phase.  Note p1 should not equal p2 (that would be an error), so
  // the diagonal entries in the 2d matrix are the bogus PHASE_NONE.
  static phase floating[3][3] = {{PHASE_NONE, PHASE_C,    PHASE_B},
                                 {PHASE_C,    PHASE_NONE, PHASE_A},
                                 {PHASE_B,    PHASE_A,    PHASE_NONE}};

  // elow_bits[pfloat] takes the floating phase pfloat (e.g., pfloat could be PHASE_A)
  // and returns a 3-bit value with a zero in the column corresponding to the
  // floating phase (0th column = A, 1st column = B, 2nd column = C).
  static int elow_bits[3] = {0b110, 0b101, 0b011};

  phase pfloat = floating[p1][p2]; // the floating phase
  phase phigh, plow;               // phigh is the PWMed phase, plow is the grounded phase
  int apwm;                        // magnitude of the pwm count

  // choose the appropriate direction
  if(pwm > 0) {
    phigh = p1;
    plow =  p2;
    apwm = pwm;
  } else {
    phigh = p2;
    plow = p1;
    apwm = -pwm;
  }
  // Pin E0 controls enable for phase A; E1 for B; E2 for C.
  // The pfloat phase should have its pin be 0; other pins should be 1.
  LATE = (LATE & ~0x7) | elow_bits[pfloat];

  // set the PWM's appropriately by setting the OCxRS SFRs
  *ocr[pfloat] = 0;   // floating pin has 0 duty cycle
  *ocr[plow] = 0;     // low will always be low, 0 duty cycle
  *ocr[phigh] = apwm; // the high phase gets the actual duty cycle
}

// Given the Hall sensor state, use the mapping between sensor readings and
// energized phases given in the Pittman ELCOM motor data sheet to determine the
// correct phases to PWM and GND and the phase to leave floating.

void bldc_commutate(int pwm, unsigned int state) {
  pwm = ((int)PR2 * pwm)/100; // convert pwm to ticks
  switch(state) {
    case 0b100:
      phases_set(pwm,PHASE_B, PHASE_A); // if pwm > 0, phase A = GND and B is PWMed
      break;                            // if pwm < 0, phase B = GND and A is PWMed
    case 0b110:
      phases_set(pwm,PHASE_C, PHASE_A);
      break;
    case 0b010:
      phases_set(pwm, PHASE_C, PHASE_B);
      break;
    case 0b011:
      phases_set(pwm, PHASE_A, PHASE_B);
      break;
    case 0b001:
```

```
      phases_set(pwm,PHASE_A,PHASE_C);
      break;
    case 0b101:
      phases_set(pwm,PHASE_B, PHASE_C);
      break;
    default:
      NU32_WriteUART3("ERROR: Read the state incorrectly!\r\n"); // ERROR!
  }
}

// Get the signed PWM duty cycle from the user.

int bldc_get_pwm() {
  char msg[100];
  int newpwm;
  NU32_WriteUART3("Enter signed PWM duty cycle (-100 to 100): ");
  NU32_ReadUART3(msg, sizeof(msg));
  NU32_WriteUART3(msg);
  NU32_WriteUART3("\r\n");
  sscanf(msg,"%d", &newpwm);
  return newpwm;
}
```

29.5.3 Commutation Using Hall Sensor Feedback

In Code Sample 29.3 `closed_loop.c`, which uses the `bldc` library, the user is prompted for a signed PWM duty cycle percentage (an integer in the range −100 to 100). Three input capture modules (IC1 to IC3, corresponding to digital input pins D8 to D10) are used to detect changes in the Hall sensor state, as shown in Figure 29.12. After receiving a PWM value, the `main` function initiates a commutation, causing the proper coils to be energized and the proper coil to float, based on the Hall effect sensor input. When one of the Hall sensors changes value, the associated input capture ISR is invoked. The ISR reads the three-bit Hall sensor state and calls `bldc_commutate` from the `bldc` library to choose which motor phase gets the PWM input, which motor phase is grounded, and which motor phase is left floating. Notice that PORTD (input capture) bits 10, 9, and 8, written $D_{10}D_9D_8$, directly correspond to the Hall sensor code $S_1S_2S_3$ from the motor's data sheet, depicted in Figure 29.10.

Using input capture modules allows the possibility for velocity estimation (see Exercise 5). Since `closed_loop.c`, below, does not implement velocity estimation, it could have used change notification instead of input capture.

Code Sample 29.3 `closed_loop.c`**. Using the Hall Effect Sensors to Implement Brushless Commutation. The User Specifies the PWM Level.**

```
#include "NU32.h"  // constants, funcs for startup and UART
#include "bldc.h"
// Using Hall sensor feedback to commutate a BLDC.
// Pins E0, E1, E2 correspond to the connected state of
// phase A, B, C.  When low, the respective phase is floating
// and when high, the voltage on the phase is determined by
```

```
// the value of pins D0 (OC1), D1 (OC2), and D2 (OC3), respectively.
//
// Pins IC1 (RD8), IC2 (RD9), and IC3 (RD10) read the Hall effect sensor output
// and provide velocity feedback using the input capture peripheral.
// When one of the Hall sensor values changes, an interrupt is generated
// by the input capture, and a call to the bldc library updates the
// commutation.

// read the hall effect sensor state

inline unsigned int state() {
  return (PORTD & 0x700) >> 8;
}

static volatile int pwm = 0;    // current PWM percentage

void __ISR(_INPUT_CAPTURE_1_VECTOR,IPL6SRS) commutation_interrupt1(void) {
  IC1BUF;                        // clear the input capture buffer
  bldc_commutate(pwm,state());   // update the commutation
  IFS0bits.IC1IF = 0;            // clear the interrupt flag
}

void __ISR(_INPUT_CAPTURE_2_VECTOR,IPL6SRS) commutation_interrupt2(void) {
  IC2BUF;                        // clear the input capture buffer
  bldc_commutate(pwm,state());   // update the commutation
  IFS0bits.IC2IF = 0;            // clear the interrupt flag
}

void __ISR(_INPUT_CAPTURE_3_VECTOR,IPL6SRS) commutation_interrupt3(void) {
  IC3BUF;                        // clear the input capture buffer
  bldc_commutate(pwm,state());   // update the commutation
  IFS0bits.IC3IF = 0;            // clear the interrupt flag
}

int main(void) {
  NU32_Startup(); // cache on, interrupts on, LED/button init, UART init
  TRISECLR = 0x7; // E0, E1, and E2 are outputs

  bldc_setup();   // initialize the bldc

  // set up input capture to generate interrupts on Hall sensor changes
  IC1CONbits.ICTMR = 1; // use timer 2
  IC1CONbits.ICM = 1;   // interrupt on every rising or falling edge
  IFS0bits.IC1IF = 0;   // enable interrupt for IC1
  IPC1bits.IC1IP = 6;
  IEC0bits.IC1IE = 1;

  IC2CONbits.ICTMR = 1; // use timer 2
  IC2CONbits.ICM = 1;   // interrupt on every rising or falling edge
  IC2CONbits.ON = 1;    // enable interrupt for IC2
  IFS0bits.IC2IF = 0;
  IPC2bits.IC2IP = 6;
  IEC0bits.IC2IE = 1;

  IC3CONbits.ICTMR = 1; // use timer 2
  IC3CONbits.ICM = 1;   // interrupt on every rising or falling edge
  IFS0bits.IC3IF = 0;   // enable interrupt for IC3
  IPC3bits.IC3IP = 6;
  IEC0bits.IC3IE = 1;
```

```
    IC1CONbits.ON = 1;      // start the input capture modules
    IC2CONbits.ON = 1;
    IC3CONbits.ON = 1;

    while(1) {
      int newpwm = bldc_get_pwm();    // prompt user for PWM duty cycle
      __builtin_disable_interrupts();
      pwm = newpwm;
      __builtin_enable_interrupts();
      bldc_commutate(pwm, state());   // ground, PWM, and float the right phases
    }
}
```

For proper commutation, it is important to handle Hall sensor changes in a timely manner. If you would like to avoid using computation time for your own Hall-sensor-based commutation, you can buy an external chip, like the Texas Instruments UC3625N, which uses a PWM and direction input from your the PIC32 and the three Hall sensor readings to drive three external half H-bridges connected to the motor's phases.

29.5.4 Synchronous Driving Using Open-Loop Control

Many inexpensive brushless motors do not provide Hall sensor feedback.[3] These motors have only three wires, for the phases A, B, and C. Such motors are often used in quadcopters, for example. It is possible to drive these motors to achieve a desired angular velocity by commutating open-loop (no Hall sensor feedback).

To understand how a BLDC motor can run without sensing, consider that if we kept a constant voltage at A, ground at B, and left C floating, the rotor might first experience a positive torque, accelerating in the positive direction until it reaches the point where the magnetic poles at $\alpha, \overline{\alpha}, \beta$, and $\overline{\beta}$ provide zero torque. Then, after overshooting the zero torque position, it would experience a negative torque, decelerating it. With damping, the rotor would eventually rest at a position where it experiences zero torque, much like the operation of a stepper motor.

The eventual negative torque that comes from not commutating can be viewed as a type of built-in stabilizing feedback. In the extreme stepper-like case above, the feedback causes the rotor to stop. Instead, consider what happens when the commutation cycles through the steps 1-6 at a fixed frequency. If the frequency is slightly too low for the voltage (or PWM level) used, the rotor may experience deceleration before the next commutation step occurs and it begins to accelerate it again. Although somewhat inefficient, this deceleration acts as negative feedback, causing the rotor to slow its travel around its mechanical revolution to better match the speed of the electrical revolution imposed by the commutation.

[3] Hobby BLDCs are often classified as *inrunners* or *outrunners*. An inrunner has its rotor inside the stator coils, as discussed so far in this chapter. An outrunner has the stator coils on the outer circumference of the stator cylinder and the rotor permanent magnet poles are around an even larger circumference, outside the stator.

In summary, if the commutation frequency is neither too fast nor too slow for the given PWM level, the rotor will synchronize with the commutation due to the built-in negative feedback. If the commutation frequency is too fast, then, much like a stepper, the rotor will fall behind, become desynchronized, and likely end up vibrating. Similarly, if the acceleration or deceleration of the commutation is too high, the inertia of the BLDC and its load will prevent the rotor from following the commutation. To achieve open-loop control of a BLDC from rest, to a high velocity, and back to rest, the acceleration, deceleration, and maximum velocity must be limited.

Open-loop driving of a BLDC only works when the loads on the BLDC are relatively well known. If a large torque is suddenly applied to the BLDC rotor, the rotor is likely to desynchronize with the commutation.

Code Sample 29.4 `open_loop.c` demonstrates open-loop control of the unloaded Pittman ELCOM SL 4443S013 using the 9 V motor supply in Figure 29.12. The user specifies the desired PWM duty cycle as a signed percentage (−100 to 100). Using information from the motor's data sheet, the relationship between the average voltage (and therefore PWM duty cycle) and the unloaded motor's speed is derived. Knowing the motor speed corresponding to the user's requested PWM, the duration of each commutation segment is calculated. For example, since there are 12 commutation phases per mechanical revolution, at 500 RPM, there are $500 \times 12 = 6000$ commutation phases per minute, or 100 commutation phases per second, so each commutation phase lasts 1 s/100 = 10 ms.

The duration of each commutation phase is counted in increments of 50 μs by using the ISR of Timer2, which is generating the 20 kHz (= 1/50 μs) PWM. When the duration of the commutation phase is exceeded, `bldc_commutate` (in the `bldc` library, above) is called with the virtual Hall sensor state of the next commutation segment.

Finally, when the user asks to change the PWM duty cycle, the code ramps up or ramps down the PWM in increments of 1%, which yields an acceleration rate found empirically to prevent desynchronization.

Code Sample 29.4 `open_loop.c`. **Driving a BLDC Without Hall Effect Sensors.**

```
#include "NU32.h" // constants, funcs for startup and UART
#include "bldc.h"

// Open-loop control of the Pittman ELCOM SL 4443S013 BLDC operating at no load and 9 V.
// Pins E0, E1, E2 correspond to the connected state of
// phase A, B, C.  When low, the respective phase is floating,
// and when high, the voltage on the phase is determined by
// the value of pins D0 (OC1), D1 (OC2), and D2 (OC3), respectively.

// Some values below have been tuned experimentally for the Pittman motor.

#define MAX_RPM 1800 // the max speed of the motor (no load speed), in RPM, for 9 V
#define MIN_PWM 3    // the min PWM required to overcome friction, as a percentage
```

```
#define SLOPE (MAX_RPM/(100 - MIN_PWM)) // slope of the RPM vs PWM curve
#define OFFSET (-SLOPE*MIN_PWM) // RPM where the RPM vs PWM curve intersects the RPM axis
                               // NOTE: RPM = SLOPE * PWM + OFFSET
#define TICKS_MIN 1200000  // number of 20 kHz (50 us) timer ticks in a minute
#define EMREV 2     // number of electrical revs per mechanical revs (erev/mrev)
#define PHASE_REV 6 // the number of phases per electrical revolution (phase/erev)

// convert minutes/mechanical revolution into ticks/phase (divide TP_PER_MPR by the rpm)
#define TP_PER_MPR (TICKS_MIN/(PHASE_REV *EMREV))

#define ACCEL_PERIOD 200000 // Time in 40 MHz ticks to wait before accelerating to next
                            // PWM level (i.e., the higher this value, the slower
                            // the acceleration).  When doing open-loop control, you
                            // must accel/decel slowly enough, otherwise you lose sync.
                            // The PWM is adjusted by 1 percent in each accel period,
                            // but the deadband where the motor does not move is skipped.

static volatile int pwm = 0;    // current PWM as a percentage
static volatile int period = 0; // commutation period, in 50 us (1/20 kHz) ticks

void __ISR(_TIMER_2_VECTOR, IPL6SRS) timer2_ISR(void) { // entered every 50 us (1/20 kHz)
  // the states, in the order of rotation through the phases (what we'd expect to read)
  static unsigned int state_table[] = {0b101,0b001,0b011,0b010,0b110,0b100};

  static int phase = 0;
  static int count = 0; // used to commutate when necessary
  if(count >= period) {
    count = 0;
    if(pwm > MIN_PWM) {
      ++phase;
    } else if (pwm < -MIN_PWM){
      --phase;
    }
    if(phase == 6) {
      phase = 0;
    } else if (phase == -1) {
      phase = 5;
    }
    bldc_commutate(pwm, state_table[phase]);
  } else {
    ++count;
  }
  IFS0bits.T2IF = 0;
}

// return true if the PWM percentage is in the deadband region
int in_deadband(int pwm) {
  return -MIN_PWM <= pwm && pwm <= MIN_PWM;
}

int main(void) {
  char msg[100];

  NU32_Startup(); // cache on, interrupts on, LED/button init, UART init
  bldc_setup();

  // Set up Timer2 interrupts (the bldc already uses Timer2 for the PWM).
  // We just reuse it for our timer here.
  IPC2bits.T2IP = 6;
  IFS0bits.T2IF = 0;
  IEC0bits.T2IE = 1;
```

```
    while(1) {
      int newpwm = bldc_get_pwm();  // get new PWM from user
      if(in_deadband(newpwm)) {     // if PWM is in deadband where motor doesn't move, pwm=0
        __builtin_disable_interrupts();
        period = 0;
        pwm = 0;
        __builtin_enable_interrupts();
      } else {
        // newpwm is not in the deadband
        int curr_pwm = pwm;
        _CP0_SET_COUNT(0);

        while(curr_pwm != newpwm) {  // ramp the PWM up or down, respecting accel limits
          int comm_period;

          if(curr_pwm > newpwm) {
            --curr_pwm;
            // skip the deadband
            if(in_deadband(curr_pwm)) {
              curr_pwm = - MIN_PWM - 1;
            }
          } else if(curr_pwm < newpwm) {
            ++curr_pwm;
            if(in_deadband(curr_pwm)) {
              curr_pwm = MIN_PWM + 1;
            }
          }
          // divide T_PER_MPR by the RPM to get the commutation period
          // We compute the RPM based on the RPM vs pwm curve RPM = SLOPE pwm + OFFSET
          comm_period =  (TP_PER_MPR/(SLOPE*abs(curr_pwm)+ OFFSET));
          while(_CP0_GET_COUNT() < ACCEL_PERIOD) { ; } // delay until accel period over
          __builtin_disable_interrupts();
          period = comm_period;
          pwm = curr_pwm;
          __builtin_enable_interrupts();
          _CP0_SET_COUNT(0); // we just moved to a new pwm, reset the acceleration period
        }
        sprintf(msg,"PWM Percent: %d, PERIOD: %d\r\n",pwm,period);
        NU32_WriteUART3(msg);
      }
    }
  }
```

Another method for estimating the angle of a BLDC without Hall sensors involves monitoring the back-emf of the windings, but we do not pursue this method in this chapter.

29.6 Linear Brushless Motors

Linear brushless motors are "unrolled" BLDCs. Often the track consists of a number of permanent magnetic poles while the moving body consists of the energized coils. Hall effect sensors are often used to determine the current electrical step.

29.7 Chapter Summary

- In addition to brushed DC motors, other electric actuators include solenoids, speakers and voice coil actuators, RC servos, stepper motors, and brushless DC motors.
- An RC servo incorporates a DC motor, gearhead, angle sensor, and feedback controller in a single package. A PWM control signal is used to specify the desired angle of the RC servo output.
- A stepper motor rotates in fixed increments in response to digital pulses. A stepper is typically used in open-loop applications.
- For a brushless DC motor, commutation is performed electronically based on the angle of the motor indicated by three digital Hall sensors.

29.8 Exercises

1. Obtain a solenoid and its data sheet. Confirm the coil's resistance using a multimeter. Then design a circuit to power the solenoid and write a simple PIC32 program that periodically activates the solenoid.
2. Get a speaker, preferably a small woofer, and measure the coil resistance. Wire an H-bridge circuit to drive the speaker, ensuring that the maximum current through the speaker does not exceed the H-bridge's capability. Write a program for the PIC32 that creates a 100 Hz tone through the speaker by creating a 100 Hz sinusoidal average H-bridge output voltage, centered at 0 V. To do this, you can use 40 kHz or higher PWM, above the audible range, to drive the speaker. Your program should use an ISR at 20-40 kHz to vary the duty cycle of the PWM. If you use a 20 kHz ISR, this means that the duty cycle goes through a full sinusoidal waveform after 20 kHz/100 Hz = 200 times through the ISR.
3. Write a program for the PIC32 that uses the NU32 library to allow the user to control an RC servo from a computer interface. The user types an angle for the output shaft of the RC servo, and the RC servo goes there.
4. Wire a stepper motor to the PIC32 using a dual H-bridge chip. (a) Write a program that uses the NU32 library to allow you to use your computer to specify a desired motion of the motor, in degrees. The program then executes the motion. (b) Let the user also specify the time of the motion. (c) What is the maximum speed, in degrees per second, that the motor can follow, according to your experiments? How would you alter your program to support acceleration and deceleration, to allow speeding up to higher speeds than you can obtain instantaneously? If your program does not use interrupts, how might you alter it to use interrupts to drive the motor, so the CPU is available for other tasks between ISRs?
5. Modify Code Sample 29.3 (`closed_loop.c`) to calculate the velocity of the motor in the ISRs. This is possible because the code uses input capture modules to detect Hall sensor state changes, and input capture can be used to measure the time between Hall state changes. Periodically write the velocity to the user's screen.

6. Modify an inexpensive three-phase BLDC without Hall sensor feedback to use an external encoder to provide the feedback needed for closed-loop commutation. Demonstrate closed-loop commutation, e.g., using `closed_loop.c` in this chapter.
7. Implement a motion feedback controller for a BLDC on top of the closed-loop commutation.
8. Modify Code Sample 29.4 (`open_loop.c`) to prompt the user for a desired velocity, not a desired PWM.

Further Reading

Arrigo, D. (2001). *AN1088: L6234 three phase motor driver.* STMicroelectronics.

AVR446: Linear speed control of stepper motor. ATMEL.

Brown, W. (2011). *AN857 brushless DC motor control made easy.* Microchip Technology Inc.

Hughes, A., & Drury, B. (2013). *Electric motors and drives: Fundamentals, types and applications* (4th ed.). Amsterdam: Elsevier.

L6234 three phase motor driver. (2011). STMicroelectronics.

Pittman Express LO-COG brushed and ELCOM brushless motor data sheet. (2007). Ametek Precision Motion Control.

APPENDIX A

A Crash Course in C

This appendix provides an introduction to C for readers with no C experience but some experience in another programming language. It is not intended as a complete reference; plenty of great C resources already exist and provide a more complete picture. This appendix applies to C in general, not just C on the Microchip PIC32. We recommend that you start by programming your computer so you can experiment with C without needing extra equipment or complication.

A.1 Quick Start in C

To start with C, you need a computer, a text editor, and a C compiler. You use the text editor to write your C program, a plain text file with a name ending with the extension `.c` (e.g., `myprog.c`). The C compiler converts this program into machine code that your computer can execute. There are many free C compilers available; we recommend the `gcc` C compiler, which is part of the GNU Compiler Collection (GCC, found at `http://gcc.gnu.org`). GCC is available for Windows, Mac OS, and Linux. For Windows, you can download the GCC collection in MinGW.[1] If the installation asks you about what tools to install, make sure to include the `make` tools. For Mac OS, you can download the full Xcode environment from the Apple Developers website. This installation is multiple gigabytes; however, you can opt to install only the "Command Line Tools for Xcode," which is smaller and more than sufficient for getting started with C (and for this appendix).

Many C installations come with an Integrated Development Environment (IDE) complete with text editor, menus, and graphical tools to assist you with your programming projects. Every IDE is different, and what we cover in this appendix does not require a sophisticated IDE. We therefore use only *command line tools*, meaning that we initiate compilation and run the program by typing at the command line. In Mac OS, the command line can be accessed from the Terminal program. In Windows, you can access the command line by searching for `cmd` or `command prompt`. Linux users should run a shell such as bash.

---

[1] You are also welcome to use Visual C from Microsoft. The command line compile command will look a bit different than what you see in this appendix.

To use the command line, you must learn some command line instructions. The Mac operating system is built on top of Unix, which is similar to Linux, so Mac/Unix/Linux use the same syntax. Windows uses slightly different commands. See the table of a few useful commands below. You can find more information on how to use these commands as well as others by searching for command line commands in Unix, Linux, or Windows.

Function	Mac/Unix/Linux	Windows
Show current directory	`pwd`	`cd`
List directory contents	`ls`	`dir`
Make subdirectory `newdir`	`mkdir newdir`	`mkdir newdir`
Change to subdirectory `newdir`	`cd newdir`	`cd newdir`
Move "up" to parent directory	`cd ..`	`cd ..`
Copy `file` to `filenew`	`cp file filenew`	`copy file filenew`
Delete file `file`	`rm file`	`del file`
Delete directory `dir`	`rmdir dir`	`rmdir dir`
Help on using command `cmd`	`man cmd`	`cmd /?`

Following the long-established programming tradition, your first C program will simply print "Hello world!" to the screen. Use a text editor to create the file `HelloWorld.c`:

```
#include <stdio.h>
int main(void) {
  printf("Hello world!\n");
  return 0;
}
```

Possible text editors include Notepad++ for Windows, TextWrangler for Mac OS, and Gedit for Linux. You can also try vim or emacs, though they are not easy to get started with! Whichever editor you use, you should save your file as plain text, not rich text or any other formatted text.

To compile your program, navigate from the command line to the directory where the program sits. Then, assuming your command prompt appears as >, type the following at the prompt:

```
> gcc HelloWorld.c -o HelloWorld
```

This command should create the executable file `HelloWorld` in the same directory. (The argument after the `-o` output flag is the name of the executable file to be created from `HelloWorld.c`.) Now, to execute the program, type

Windows: `> HelloWorld`

Linux/MacOS: `> ./HelloWorld`

For Linux/MacOS users, the "." is shorthand for "current directory," and the `./` tells your computer to look in the current directory for `HelloWorld`. Windows implicitly searches the current directory for executables, so you need not explicitly specify it.

If you have succeeded in getting this far, your C installation works and you can proceed. If not, you may need to get help from friends or the internet.

A.2 Overview

If you are familiar with a high-level language like MATLAB or Python, you may know about loops, functions, and other programming constructs. You will see that although C's syntax is different, the same concepts translate to C. Rather than starting with basic loops, if statements, and functions, we begin by focusing on important concepts that you must master in C but which you probably have not dealt with in a language such as MATLAB or Python.

- **Memory, addresses, and pointers.** Variables are stored at particular *addresses* in memory as bits (0's and 1's). In C, unlike in MATLAB or Python, it is often useful to access a variable's memory address. Special variables called *pointers* contain the address of another variable and can be used to access the contents of that address. Although powerful, pointers can also be dangerous; misusing them can cause all sorts of bugs, which is why many higher-level languages forgo them completely.
- **Data types.** In MATLAB, for example, you can simply type `a = 1; b = [1.2 3.1416]; c = [1 2; 3 4]; s = 'a string'`. MATLAB determines that `a` is a scalar, `b` is a vector with two elements, `c` is a 2×2 matrix, and `s` is a string; automatically tracks the variable's type (e.g., a list of numbers for a vector or a list of characters for a string); and sets aside, or *allocates*, memory to store them. In C, on the other hand, you must first *define* the variable before you ever use it. To use a vector, for example, you must specify the number and *data type* of its elements—integers or decimal numbers (floating point). The variable definition tells the C compiler how much memory it needs to store the vector, the address of each element, and how to interpret the bits of each element (as integers or floating point numbers, for example).
- **Compiling.** MATLAB programs are typically *interpreted*: the commands are converted to machine code and executed while the program is running. C programs, on the other hand, are *compiled*, i.e., converted to machine-code in advance. This process consists of several steps whose purpose is to turn your C program into machine-code before it ever runs. The compiler can identify some errors and warn you about other questionable code. Compiled code typically runs faster than interpreted code, since the translation to machine code is done in advance.[2]

Each of these concepts is described in Section A.3 without going into detail on C syntax. In Section A.4 we look at sample programs to introduce syntax, then offer more detailed explanations.

---

[2] The distinction between compiled and interpreted programs is narrowing: many interpreted languages are actually just-in-time (JIT) compiled, that is program chunks are compiled in advance right before they are needed.

A.3 Important Concepts in C

We begin our discussion of C with this caveat:

> C consists of an evolving set of standards for a programming language, and any specific C installation is an "implementation" of C. While C standards require certain behavior from all implementations, some details are implementation-dependent. For example, the number of bytes used for some data types is non-standard. We sometimes ignore these details in favor of clarity and succinctness. Platform- and compiler-specific results are from gcc 4.9.2 running on an x86_64 compatible processor.

A.3.1 Data Types

Binary and hexadecimal

On a computer, programs and data are represented by sequences of 0's and 1's. A 0 or 1 may be represented by two different voltages (low and high) held and controlled by a logic circuit, for example. Each of these units of memory is called a **bit**.

A sequence of bits may be interpreted as a base-2 or **binary** number, just as a sequence of digits in the range 0 to 9 is commonly treated as a base-10 or **decimal** number.[3] In the decimal numbering system, a multi-digit number like 793 is interpreted as $7 * 10^2 + 9 * 10^1 + 3 * 10^0$; the rightmost column is the 10^0 (or 1's) column, the next column to the left is the 10^1 (or 10's) column, the next column to the left is the 10^2 (or 100's) column, and so on. Similarly, the rightmost column of a binary number is the 2^0 column, the next column to the left is the 2^1 column, etc. Converting the binary number `00111011` to its decimal representation, we get

$$0*2^7+0*2^6+1*2^5+1*2^4+1*2^3+0*2^2+1*2^1+1*2^0 = 32+16+8+2+1 = 59.$$

The leftmost digit in a multi-digit number is called the **most significant digit**, and the rightmost digit, corresponding to the 1's column, is called the **least significant digit**. For binary representations, these are often called the **most significant bit (msb)** and **least significant bit (lsb)**, respectively.

We specify that a sequence of numbers is base-2 by writing it as $\texttt{00111011}_2$ or `0b00111011`, where the `b` stands for "binary."

To convert a base-10 number x to binary:

1. Initialize the binary result to all zeros and k to a large integer, such that 2^k is known to be larger than x.

[3] Bit is a portmanteau of binary and digit.

2. If $2^k \leq x$, place a 1 in the 2^k column of the binary number and set x to $x - 2^k$.
3. If $x = 0$ or $k = 0$, we are finished. Else set k to $k - 1$ and go to line 2.

An alternative base-10 to binary conversion algorithm builds the binary number from the rightmost to leftmost bit.

1. Divide x by 2.
2. The next digit (from right to left) is the remainder (so 1 if x is odd, 0 if x is even).
3. x = the quotient. (So if x were 5, the new x is 2, and if x were 190 the new x is 95).
4. Repeat process until $x = 0$.

Compared to base-10, base-2 has a closer connection to actual hardware. Binary can be inconvenient for human reading and writing, however, due to the large number of digits. Therefore we often group four binary digits together (taking values `0b0000` to `0b1111`, or 0 to 15 in base-10) and represent them with a single character using the numbers 0 to 9 or the letters A to F. This base-16 representation is called **hexadecimal** or hex for short:Thus we can

base-2 (binary)	base-16 (hex)	base-10 (decimal)	base-2 (binary)	base-16 (hex)	base-10 (decimal)
0000	0	0	1000	8	8
0001	1	1	1001	9	9
0010	2	2	1010	A	10
0011	3	3	1011	B	11
0100	4	4	1100	C	12
0101	5	5	1101	D	13
0110	6	6	1110	E	14
0111	7	7	1111	F	15

write the eight-digit binary number `0b00111011`, or `0011 1011`, more succinctly in hex as `3B`, or $\texttt{3B}_{16}$ or `0x3B` to clarify that it is a hex number. The corresponding decimal number is $3 * 16^1 + 11 * 16^0 = 59$.

Bits, bytes, and data types

Bits of memory are grouped together in groups of eight called **bytes**. A byte can be written in binary or hexadecimal (e.g., `0b00111011` or `0x3B`), and can represent values between 0 and $2^8 - 1 = 255$. Sometimes the four bits represented by a single hex digit are referred to as a **nibble**. (Get it?)

A **word** is a grouping of multiple bytes. The number of bytes in a word depends on the processor, but four-byte words are common, as with the PIC32. A word `01001101 11111010 10000011 11000111` in binary can be written in hexadecimal as `0x4DFA83C7`. The **most significant byte (MSB)** is the leftmost byte, `0x4D` in this case, and the **least significant byte**

(LSB) is the rightmost byte `0xC7`. The msb is the leftmost bit of the MSB, a `0` in this case, and the lsb is the rightmost bit of the LSB, a `1` in this case.

A byte is the smallest unit of memory that has its own **address**. The address of the byte is a number that represents the byte's location in memory. Suppose your computer has 4 gigabytes (GB), or $4 \times 2^{30} = 2^{32}$ bytes, of RAM.[4] Then to find the value stored in a particular byte, you need at least 32 binary digits (8 hex digits or 4 bytes) to specify the address.

An example showing the first eight addresses in memory is given below. Here we show the lowest address on the right, but we could have made the opposite choice.

...	7	6	5	4	3	2	1	0	Address
	11001101	00100111	01110001	01010111	01010011	00011110	10111011	01100010	Value

Assume that the byte at address 4 is part of the representation of a variable. Do these 0's and 1's represent an integer, or part of an integer? A number with a fractional component? Something else?

The answer lies in the **type** of the variable at that address. In C, before you use a variable, you must *define* it and its type, telling the compiler how many bytes to allocate for the variable (its size) and how to interpret the bits.[5]

The most common data types come in two flavors: integers and floating point numbers (numbers with a decimal point). Of the integers, the two most common types are `char`, often used to represent keyboard characters, and `int`.[6] Of the floating point numbers, the two most common types are `float` and `double`. As we will see shortly, a `char` uses 1 byte and an `int` usually uses 4, so two possible interpretations of the data held in the eight memory addresses could be

...	7	6	5	4	3	2	1	0	Address
	11001101	00100111	01110001	01010111	01010011	00011110	10111011	01100010	Value
		int						char	

4 In common usage, a kilobyte (KB) is $2^{10} = 1024$ bytes, a megabyte (MB) is $2^{20} = 1,048,576$ bytes, a gigabyte is $2^{30} = 1,073,741,824$ bytes, and a terabyte (TB) is $2^{40} = 1,099,511,627,776$ bytes. To remove confusion with the common SI prefixes that use powers of 10 instead of powers of 2, these are sometimes referred to instead as kibibyte, mebibyte, gibibyte, and tebibyte, where the "bi" refers to "binary."

5 In C you can *declare* or *define* a variable. They use similar syntax, but a declaration simply gives the name and the type of the variable, while a definition also allocates the memory to hold it. We avoid using the distinction for now and just call everything a definition.

6 `char` is derived from the word "character." People pronounce `char` variously as "car" (as in "driving the car"), "care" (a shortening of "character"), and "char" (as in charcoal), and some just punt and say "character."

where byte 0 is used to represent a `char` and bytes 4-7 are used to represent an `int`, or

...	7	6	5	4	3	2	1	0	Address
	11001101	00100111	01110001	01010111	01010011	00011110	10111011	01100010	Value
				char		int			

where bytes 0-3 are used to represent an `int` and byte 4 represents a `char`. Fortunately we do not usually have to worry about how variables are packed into memory.

Below we describe the common data types. Although the number of bytes used for each type is not the same for every processor, the numbers given are common on modern computers. (Differences for the PIC32 are noted in Table A.1.) Example syntax for defining variables is also given. Note that most C statements end with a semicolon.

`char`
Example definition:
```
char ch;
```

This syntax defines a variable named `ch` to be of type `char`. `char`s are the smallest data type, using only one byte. They are often interpreted according to the "ASCII table" (pronounced "ask-key"), the American Standard Code for Information Interchange, which maps the values 0 to 127 to letters, numbers, and other characters (Figure A.1). (The values 128 to 255 map to an "extended" ASCII table.) For example, the values 48 to 57 map to the characters '0' to '9', 65 to 90 map to the uppercase letters 'A' to 'Z', and 97 to 122 map to the lowercase letters 'a' to 'z'. The assignments

```
ch = 'a';
```

and

```
ch = 97;
```

are equivalent, as C equates characters inside single quotation marks with their ASCII table numerical value.

Depending on the C implementation, `char` may be treated by default as `unsigned`, i.e., taking values from 0 to 255, or `signed`, taking values from −128 to 127. If you use the `char` to represent a standard ASCII character, the distinction does not matter. If, however, you use the `char` data type for integer math on small integers, you should use the specifier `signed` or `unsigned`, as appropriate. For example, we could use the following definitions, where everything after `//` is a comment:

ASCII Table

0 NULL	16 DLE	32 space	48 0	64 @	80 P	96 `	112 p
1 SOH	17 DC1	33 !	49 1	65 A	81 Q	97 a	113 q
2 STX	18 DC2	34 "	50 2	66 B	82 R	98 b	114 r
3 ETX	19 DC3	35 #	51 3	67 C	83 S	99 c	115 s
4 EOT	20 DC4	36 $	52 4	68 D	84 T	100 d	116 t
5 ENQ	21 NAK	37 %	53 5	69 E	85 U	101 e	117 u
6 ACK	22 SYN	38 &	54 6	70 F	86 V	102 f	118 v
7 BELL	23 ETB	39 '	55 7	71 G	87 W	103 g	119 w
8 BACKSPACE	24 CAN	40 (	56 8	72 H	88 X	104 h	120 x
9 TAB	25 EM	41)	57 9	73 I	89 Y	105 i	121 y
10 NEWLINE	26 SUB	42 *	58 :	74 J	90 Z	106 j	122 z
11 VT	27 ESC	43 +	59 ;	75 K	91 [	107 k	123 {
12 FORMFEED	28 FS	44 ,	60 <	76 L	92 \	108 l	124 |
13 RETURN	29 GS	45 -	61 =	77 M	93]	109 m	125 }
14 SO	30 RS	46 .	62 >	78 N	94 ^	110 n	126 ~
15 SI	31 US	47 /	63 ?	79 O	95 _	111 o	127 DEL

Figure A.1

The 128 standard ASCII characters. The first 32 characters are non-printing characters and the names of most of them are obscure. Values 128 to 255 (or −128 to −1) correspond to the extended ASCII table.

```
unsigned char ch1; // ch1 can take values 0 to 255
signed char ch2;   // ch2 can take values -128 to 127
```

`int` (also known as `signed int` or `signed`)

Example definition:

```
int i,j;
signed int k;
signed n;
```

`int`s typically use four bytes (32 bits) and take values from $-(2^{31})$ to $2^{31} - 1$ (approximately ±2 billion). In the example syntax, each of `i`, `j`, `k`, and `n` are defined to be the same data type.

We can use specifiers to get the following integer data types: `unsigned int` or simply `unsigned`, a four-byte integer taking nonnegative values from 0 to $2^{32} - 1$; `short int`, `short`, `signed short`, or `signed short int`, all meaning the same thing: a two-byte integer taking values from $-(2^{15})$ to $2^{15} - 1$ (i.e., $-32,768$ to $32,767$); `unsigned short int` or `unsigned short`, a two-byte integer taking nonnegative values from 0 to $2^{16} - 1$ (i.e., 0 to $65,535$); `long int`, `long`, `signed long`, or `signed long int`, often consisting of eight bytes and representing values from $-(2^{63})$ to $2^{63} - 1$; and `unsigned long int` or `unsigned long`, an eight-byte integer taking nonnegative values from 0 to $2^{64} - 1$. A `long long int` data type may also be available.

```
float
```
Example definition:
```
float x;
```

This syntax defines the variable x to be a four-byte "single-precision" floating point number.

```
double
```
Example definition:
```
double x;
```

This syntax defines the variable x to be an eight-byte "double-precision" floating point number. The data type `long double` (quadruple precision) may also be available, using 16 bytes (128 bits). These types allow the representation of larger numbers, to more decimal places, than single-precision `float`s.

The sizes of the data types, both on a typical x86_64 computer with gcc and on the PIC32, are summarized in Table A.1. Note the differences; C lets the compiler determine these details. The C99 standard introduces data types such as `int32_t` (32-bit signed integer) and `unit8_t` (8-bit unsigned integer) which are guaranteed to be the specified size across platforms and compilers.

Table A.1: Data type sizes on two different machines

Type	# bytes on x86_64	# bytes on PIC32
char	1	1
short int	2	2
int	4	4
long int	8	4
long long int	8	8
float	4	4
double	8	4
long double	16	8

Using the data types

If your program requires floating point calculations, you can choose between `float`, `double`, and `long double` data types. The advantages of smaller types are that they use less memory and computations with them (e.g., multiplies, square roots, etc.) may be faster. The advantage of the larger types is the greater precision in the representation (e.g., smaller roundoff error).

If your program needs integer calculations, you should use integer rather than floating point data types due to the higher speed of integer math and the ability to represent a larger range of integers for the same number of bytes.[7] You can decide whether to use `signed` or `unsigned char`s, or {`signed/unsigned`} {`short/long`} `int`s. The considerations are memory usage, possibly the time of the computations, and whether the type can represent a sufficient range of integer values.[8] For example, if you decide to use `unsigned char`s for integer math to save memory, and you add two of them with values 100 and 240 and assign to a third `unsigned char`, you will get a result of 84 due to *integer overflow*. This example is illustrated in the program `overflow.c` in Section A.4.

Representations of data types

A simple representation for integers is the *sign and magnitude* representation. In this representation, the msb represents the sign of the number (0 = positive, 1 = negative), and the remaining bits represent the magnitude of the number. The sign and magnitude method represents zero twice (positive and negative zero) and is not often used.

A more common representation for integers is called *two's complement*. This method also uses the msb as a sign bit, but it only has a single representation of zero. The two's complement representation of an 8-bit `char` is given below:

binary	`signed char`, **base-10**	`unsigned char`, **base-10**
00000000	0	0
00000001	1	1
00000010	2	2
00000011	3	3
⋮		
01111111	127	127
10000000	−128	128
10000001	−127	129
⋮		
11111111	−1	255

As the binary representation is incremented, the two's complement (signed) interpretation of the binary number also increments, until it "wraps around" to the most negative value when the msb becomes 1 and all other bits are 0. The signed value then resumes incrementing until

[7] Just as a four-byte `float` can represent fractional values that a four-byte `int` cannot, a four-byte `int` can represent more integers than a four-byte `float` can. See the type conversion example program `typecast.c` in Section A.4 for an example.

[8] Computations with smaller data types are not always faster than with larger data types. It depends on the architecture.

it reaches −1 when all bits are 1. To negate a number using two's complement arithmetic, invert all of the bits and add one. For example, 1 (0b00000001) becomes −1 (0b11111111). What happens when you perform the negation procedure on zero?

Another representation choice is *endianness*. The *little-endian* representation of an `int` stores the least significant byte at the lowest address (`ADDRESS`) and the most significant byte at highest (`ADDRESS+3`) (remember, little-lowest-least), while the *big-endian* convention is the opposite, storing the most significant byte at the lowest address (`ADDRESS`) and the least significant byte at the highest address (`ADDRESS+3`).[9] The convention used depends on the processor. For definiteness in this appendix, we always assume little-endian representation, which is used by x86_64 (most likely your computer's architecture) and the PIC32.

`float`s, `double`s, and `long double`s are commonly represented in the IEEE 754 floating point format

$$\text{value} = (-1)^s \times b \times 2^c, \tag{A.1}$$

where one bit is used to represent the sign ($s = 0$ for positive, $s = 1$ for negative); $m = 23/52/112$ bits are used to represent the significand b (also known as the mantissa) in the range 1 to $2 - 2^{-m}$; and $n = 8/11/15$ bits are used to represent the exponent c in the range $-(2^{n-1}) + 2$ to $2^{n-1} - 1$, where n and m depend on whether the type uses 4/8/16 bytes. Certain exponent and significand combinations are reserved for representing special cases like positive and negative infinity and "not a number" (`NaN`).

Specifically for a four-byte `float`, the 32 bits of the IEEE 754 representation are

$$\underbrace{s}_{\text{sign bit}}\ \underbrace{e_7\,e_6\,e_5\,e_4\,e_3\,e_2\,e_1\,e_0}_{\text{8 bits of exponent } c}\ \underbrace{f_{23}f_{22}\ldots f_2 f_1}_{\text{23 bits of significand } b}\ .$$

The exponent c in (A.1) is equal to $e - 127$, where e is the unsigned integer value of the eight bits of e, ranging from 0 to $2^8 - 1 = 255$. (The values $e = 0$ and $e = 255$ are reserved to represent special cases, like ±infinity and "not a number.") The significand b in (A.1) is given by

$$b = 1 + \sum_{i=1}^{23} f_i \quad 2^{-i},$$

so b ranges from 1 to $2 - 2^{-23}$.

[9] These phrases come from *Gulliver's Travels* by Jonathan Swift, where Lilliputians fanatically divide themselves according to which end of a soft-boiled egg they crack open.

See Exercise 14 to experiment with two's complement and IEEE 754 floating point representations.

Rarely do you need to worry about the specific bit-level representation of the different data types: endianness, two's complement, IEEE 754, etc. You tell the compiler to store values and retrieve values, and it handles implementing the representations.

A.3.2 Memory, Addresses, and Pointers

Consider the following C syntax:

```
int i;
int *ip;
```

These definitions may appear to define the variables `i` and `*ip` of type `int`; however, the character `*` cannot be part of a variable name. The variable name is actually `ip`, and the special character `*` means that `ip` is a **pointer** to something of type `int`. Pointers store the address of another variable; in other words, they "point" to another variable. We often use the words "address" and "pointer" interchangeably.

When the compiler sees the definition `int i`, it allocates four bytes of memory to hold the integer `i`. When the compiler sees the definition `int *ip`, it creates the variable `ip` and allocates to it whatever amount of memory is needed to hold an address, a platform-dependent quantity.[10] The compiler also remembers the data type that `ip` points to, `int` in this case, so when you use `ip` in a context that requires a pointer to a different data type, the compiler may generate a warning or an error. Technically, the type of `ip` is "pointer to type `int`."

> **Important!** Defining a pointer only allocates memory to hold the pointer. It does **not** allocate memory for a pointee variable to be pointed at. Also, simply defining a pointer does not initialize it to point to anything valid. Do not use pointers without explicitly initializing them!

When we want the address of a variable, we apply the **address** (or **reference**) **operator** to the variable, which returns a pointer to the variable (its address). In C, the address operator is written `&`. Thus the following command assigns the address of `i` to the pointer `ip`:

```
ip = &i;  // ip now holds the address of i
```

[10] When computers switched from the 32-bit x86 to the 64-bit x86_64 architecture, code that relied on pointers being 32 bits long were in trouble; x86_64 uses 64-bit long pointers!

The address operator always returns the lowest address of a multi-byte type. For example, if the four-byte `int i` occupies addresses 0x0004 to 0x0007 in memory, `&i` will return 0x0004.[11]

If we have a pointer (an address) and we want the contents at that address, we apply the **dereference operator** to the pointer. In C, the dereference operator is written `*`. Thus the following command stores the value at the address pointed to by `ip` in `i`:

```
i = *ip;  // i now holds the contents at the address ip
```

However, you should never dereference a pointer until it has been initialized to point to something using a statement such as `ip = &i`.

As an analogy, consider the pages of a book. A page number can be considered a pointer, while the text on the page can be considered the contents of a variable. So the notation `&text` would return the page number (pointer or address) of the text, while `*page_number` would return the text on that page (but only after `page_number` is initialized to point at a page of text).

Even though we are focusing on the concept of pointers, and not C syntax, let us look at some sample C code, remembering that everything after `//` on the same line is a comment:

```
int i,j;        // define i, j as type int
int *ip;        // define ip as type "pointer to type int"
ip = &i;        // set ip to the address of i (& "references" i)
i = 100;        // put the value 100 in the location allocated by the compiler for i
j = *ip;        // set j to the contents of the address ip (* dereferences ip),
                // i.e., 100
j = j+2;        // add 2 to j, making j equal to 102
i = *(&j);      // & references j to get the address, then * gets contents; i is set
                // to 102
*(&j) = 200;  // content of the address of j (j itself) is set to 200; i is unchanged
```

The use of pointers can be powerful, but also dangerous. For example, you may accidentally try to access an illegal memory location. The compiler is unlikely to recognize this during compilation, and you may end up with a "segmentation fault" when you execute the code.[12] This kind of bug can be difficult to find, and dealing with it is a C rite of passage. More on pointers in Section A.4.8.

[11] The address my actually be a "virtual" address rather than a physical location in memory. The computer automatically translates the value of `&i` to an actual physical address, when needed.

[12] A good name for a program like this is `coredumper.c`.

A.3.3 Compiling

The process loosely referred to as "compilation" actually consists of four steps:

1. **Preprocessing.** The preprocessor takes the `program.c` source code and produces an equivalent `.c` source code, performing operations such as removing comments. Section A.4.3 discusses the preprocessor in more detail.
2. **Compiling.** The compiler turns the preprocessed code into *assembly* code for the specific processor. The C code becomes a set of instructions that directly correspond to actions that the processor can perform. The compiler can be configured with several options that impact the assembly code generated. For example, the compiler can generate assembly code that increases execution time to reduce the amount of memory needed to store the code. Assembly code generated by a compiler can be inspected with a standard text editor. Coding directly in assembly is still a popular, if painful (or fun), way to program microcontrollers.
3. **Assembling.** The assembler converts the assembly instructions into processor-dependent machine-level binary *object* code. This code cannot be examined using a text editor.[13] Object code is called *relocatable*, in that the exact memory addresses for the data and program statements are not specified.
4. **Linking.** The linker takes one or more object code files and produces a single executable file. For example, if your code includes pre-compiled libraries, such as the C standard library that allows you to print to the screen (described in Sections A.4.3 and A.4.14), this code is included in the final executable. The data and program statements in the various object code files are assigned to specific memory locations.

In our `HelloWorld.c` program, this entire process is initiated by the single command line statement

```
> gcc HelloWorld.c -o HelloWorld
```

If our `HelloWorld.c` program used any mathematical functions in Section A.4.7, the compilation would be initiated by

```
> gcc HelloWorld.c -o HelloWorld -lm
```

where the `-lm` flag tells the linker to link the math library, which may not be linked by default like other libraries are.

If you want to see the intermediate results of the preprocessing, compiling, and assembling steps, Exercise 42 provides an example.

For more complex projects requiring compilation of several files into a single executable or specifying various options to the compiler, it is common to create a `Makefile` that specifies

---

[13] Well, you can view it in a text editor, but it will be incomprehensible.

how the compilation should be performed, and to then use the command `make` to actually execute the commands to create the executable. Details on the use of `Makefiles` is beyond the scope of this appendix; however, we use one extensively when programming the PIC32. Section A.4.15 gives a simple example of compiling multiple C files into single executable using a `Makefile`.

A.4 C Syntax

So far we have seen only glimpses of C syntax. Let us begin our study of C syntax with a few simple programs. We then jump to a more complex program, `invest.c`, that demonstrates many of the major elements of C structure and syntax. If you can understand `invest.c` and can create programs using similar elements, you are well on your way to mastering C. We defer the more detailed descriptions of the syntax until after introducing `invest.c`.

Printing to screen

Because it is the simplest way to see the results of a program, as well as a useful tool for debugging, let us start with the function `printf` for printing to the screen.[14] We have already seen it in `HelloWorld.c`. Here's a slightly more interesting example. Let us call this program file `printout.c`.

```
#include <stdio.h>

int main(void) {

  int i;
  float f;
  double d;
  char c;

  i = 32;
  f = 4.278;
  d = 4.278;
  c = 'k'; // or, by ASCII table, c = 107;

  printf("Formatted output:\n");
  printf(" i = %4d  c = '%c'\n",i,c);
  printf(" f = %19.17f\n",f);
  printf(" d = %19.17f\n",d);
  return 0;
}
```

[14] Programs called debuggers (such as gdb) also help you debug, allowing you to step through your program line by line as it runs.

The first line of the program

```
#include <stdio.h>
```

tells the preprocessor that the program will use functions from the "standard input and output" library, one of many code libraries provided in standard C installations that extend the power of the language. The `stdio.h` function used in `printout.c` is `printf`, covered in more detail in Section A.4.14.

The next line

```
int main(void) {
```

starts the block of code that defines the `main` function. The `main` code block is closed by the final closing brace `}`. Each C program has exactly one `main` function, and program execution begins there. The type of `main` is `int`, meaning that the function should end by returning a value of type `int`. In our case, it returns a `0`, which indicates to the operating system that the program has terminated successfully.

The next four lines define and allocate memory for four variables with four different types. The following lines assign values to those variables. The `printf` lines will be discussed after we look at the output.

Now that you have created `printout.c`, you can create the executable file `printout` and run it from the command line. Make sure you are in the directory containing `printout.c`, then type the following:

```
> gcc printout.c -o printout
> printout
```

(Again, you may have to use `./printout` to tell your computer to look in the current directory.) Here is the output:

```
Formatted output:
  i =   32  c = 'k'
  f = 4.27799987792968750
  d = 4.27799999999999958
```

The main purpose of this program is to demonstrate formatted output from the code

```
printf("Formatted output:\n");
printf(" i = %4d  c = '%c'\n",i,c);
printf(" f = %19.17f\n",f);
printf(" d = %19.17f\n",d);
```

Inside the function call to `printf`, everything inside the double quotation marks is printed to the screen, but some character sequences have special meaning. The `\n` sequence creates a newline. The `%` is a special character, indicating that some data will be printed, and for each `%`

in the double quotes, there must be a variable or other expression in the comma-separated list at the end of the `printf` call. The `%4d` means that an `int` type variable is expected, and it will be displayed right-justified using four spaces. (If the number has more than four digits, it will take as much space as is needed.) The `%c` means that a `char` is expected. The `%19.17f` means that a `float` or `double` will be printed right-justified over 19 spaces with 17 spaces after the decimal point. If you did not care how many decimal places were displayed, you could have simply used `%f` and let the C implementation default make the choice for you.[15] More details on `printf` can be found in Section A.4.14.

The output of the program also shows that neither the `float f` nor the `double d` can represent 4.278 exactly, though the double-precision representation comes closer.

Data sizes

Since we have focused on data types, our next program measures how much memory is used by different data types. Create a file called `datasizes.c` that looks like the following:

```
#include <stdio.h>

int main(void) {
  char a;
  char *bp;
  short c;
  int d;
  long e;
  float f;
  double g;
  long double h;
  long double *ip;

  printf("Size of char:                 %2ld bytes\n",sizeof(a)); // "% 2 ell d"
  printf("Size of char pointer:         %2ld bytes\n",sizeof(bp));
  printf("Size of short int:            %2ld bytes\n",sizeof(c));
  printf("Size of int:                  %2ld bytes\n",sizeof(d));
  printf("Size of long int:             %2ld bytes\n",sizeof(e));
  printf("Size of float:                %2ld bytes\n",sizeof(f));
  printf("Size of double:               %2ld bytes\n",sizeof(g));
  printf("Size of long double:          %2ld bytes\n",sizeof(h));
  printf("Size of long double pointer:  %2ld bytes\n",sizeof(ip));
  return 0;
}
```

The first lines in the `main` function define nine variables, telling the compiler to allocate space for these variables. Two of these variables are pointers. The `sizeof()` operator returns the number of bytes allocated in memory for its argument. You can use `sizeof()` on either a variable or a type (i.e., `sizeof(int)`); here we use it exclusively on variables.

[15] `printf` does not distinguish between doubles and floats, so use `%f` for both.

Here is the output of the program:

```
Size of char:                     1 bytes
Size of char pointer:             8 bytes
Size of short int:                2 bytes
Size of int:                      4 bytes
Size of long int:                 8 bytes
Size of float:                    4 bytes
Size of double:                   8 bytes
Size of long double:             16 bytes
Size of long double pointer:      8 bytes
```

We see that, on an x86_64 computer with `gcc`, `int`s and `float`s use four bytes, `short int`s two bytes, `long int`s and `double`s eight bytes, and `long double`s 16 bytes. Regardless of whether it points to a `char` or a `long double`, a pointer (address) uses eight bytes, meaning we can address a maximum of $(2^8)^8 = 256^8$ bytes of memory. Considering that corresponds to almost 18 quintillion bytes, or 18 billion gigabytes, we should have enough available addresses (at least for the time-being)!

Overflow

Now let us try the program `overflow.c`, which demonstrates the issue of integer overflow mentioned in Section A.3.1.

```
#include <stdio.h>

int main(void) {
  char i = 100, j = 240, sum;
  unsigned char iu = 100, ju = 240, sumu;
  signed char is = 100, js = 240, sums;

  sum = i+j;
  sumu = iu+ju;
  sums = is+js;

  printf("char:           %d + %d = %3d or ASCII %c\n",i,j,sum,sum);
  printf("unsigned char:  %d + %d = %3d or ASCII %c\n",iu,ju,sumu,sumu);
  printf("signed char:    %d + %d = %3d or ASCII %c\n",is,js,sums,sums);
  return 0;
}
```

In this program we initialize the values of some of the variables when they are defined. You might also notice that we are assigning a `signed char` a value of 240, even though the range for that data type is −128 to 127. So something fishy is happening. The program outputs:

```
char:           100 + -16 =  84 or ASCII T
unsigned char:  100 + 240 =  84 or ASCII T
signed char:    100 + -16 =  84 or ASCII T
```

Notice that, with our C compiler, chars are the same as signed chars. Even though we assigned the value of 240 to js and j, they contain the value −16 because the binary representation of 240 has a 1 in the 2^7 column. For the two's complement representation of a signed char, this column indicates whether the value is positive or negative. Finally, we notice that the unsigned char ju is successfully assigned the value 240 since its range is 0 to 255, but the addition of iu and ju leads to an overflow. The correct sum, 340, has a 1 in the 2^8 (or 256) column, but this column is not included in the 8 bits of the unsigned char. Therefore we see only the remainder of the number, 84. The number 84 is assigned the character T in the standard ASCII table.

Type conversion

Continuing our focus on data types, we try another simple program that illustrates what happens when you mix data types in mathematical expressions. This program uses a helper function in addition to the main function. We name this program typecast.c.

```
#include <stdio.h>

void printRatio(int numer, int denom) {  // printRatio is a helper function
  double ratio;

  ratio = numer/denom;
  printf("Ratio, %1d/%1d:                           %5.2f\n",numer,denom,ratio);
  ratio = numer/((double) denom);
  printf("Ratio, %1d/((double) %1d):                %5.2f\n",numer,denom,ratio);
  ratio = ((double) numer)/((double) denom);
  printf("Ratio, ((double) %1d)/((double) %1d):  %5.2f\n",numer,denom,ratio);
}

int main(void) {
  int num = 5, den = 2;

  printRatio(num,den);                     // call the helper function
  return(0);
}
```

The helper function printRatio "returns" type void since it does not return a value. It takes two ints as arguments and calculates their ratio in three different ways. In the first, the two ints are divided and the result is assigned to a double. In the second, the integer denom is **typecast** or **cast** to double before the division occurs, so an int is divided by a double and the result is assigned to a double.[16] In the third, both the numerator and denominator are cast as doubles before the division, so two doubles are divided and the result is assigned to a double.

[16] The typecasting does not change the variable denom itself; it simply creates a temporary double version of denom which is lost as soon as the division is complete.

The `main` function defines two `int`s, `num` and `den`, and passes their values to `printRatio`, where those values are copied to `numer` and `denom`, respectively. The variables `num` and `den` are only available to `main`, and the variables `numer` and `denom` are only available to `printRatio`, since they are defined inside those functions.

Execution of any C program always begins with the `main` function, regardless of where it appears in the file.

After compiling and running, we get the output

```
Ratio, 5/2:                         2.00
Ratio, 5/((double) 2):              2.50
Ratio, ((double) 5)/((double) 2):   2.50
```

The first answer is "wrong," while the other two answers are correct. Why?

The first division, `numer/denom`, is an *integer* division. When the compiler sees that there are `int`s on either side of the divide sign, it assumes you want integer math and produces a result that is an `int` by simply truncating any remainder (rounding toward zero). This value, 2, is then converted to the floating point number 2.0 so it can be assigned to the double-precision floating point variable `ratio`. On the other hand, the expression `numer/((double) denom)`, by virtue of the parentheses, first produces a `double` version of `denom` before performing the division. The compiler recognizes that you are dividing two different data types, so it temporarily **coerces** the `int` to a `double` so it can perform a floating point division. This is equivalent to the third and final division, except that the typecast of the numerator to `double` is explicit in the code for the third division.

Thus we have two kinds of type conversions:

- **Implicit** type conversion, or **coercion**. This occurs, for example, when a type has to be converted to carry out a variable assignment or to allow a mathematical operation. For example, dividing an `int` by a `double` will cause the compiler to treat the `int` as a `double` before carrying out the division.
- **Explicit** type conversion. An explicit type conversion is coded using a casting operator, e.g., (double) <expression> or (char) <expression>, where <expression> may be a variable or mathematical expression.

Certain type conversions may result in a change of value. For example, assigning the value of a `float` to an `int` results in truncation of the fractional portion; assigning a `double` to a `float` may result in roundoff error; and assigning an `int` to a `char` may result in overflow. Here's a less obvious example:

```
float f;
int i = 16777217;
f = i;             // f now has the value 16,777,216.0, not 16,777,217!
```

It turns out that $16,777,217 = 2^{24} + 1$ is the smallest positive integer that cannot be represented by a 32-bit `float`. On the other hand, a 32-bit `int` can represent all integers in the range -2^{31} to $2^{31} - 1$.

Some type conversions, called **promotions**, never result in a change of value because the new type can represent all possible values of the original type. Examples include converting a `char` to an `int` or a `float` to a `long double`.

As with pointers, typecasts are dangerous and should be used sparingly. Knowing where type coercion occurs, however, can be crucial. In the example below, the first line performs integer division and then converts the result to a double, whereas the second line performs floating point division.

```
double f = 3/2;       // yields 1.0!
double g = 3.0/2.0;   // yields 1.5
```

We will see more on use of parentheses (Section A.4.1), the scope of variables (Section A.4.5), and defining and calling helper functions (Section A.4.6).

> **Advanced:** Pointers can be used in conjunction with typecasts to view the same data in different ways. For example, the declaration `unsigned short s = 0xAB12` stores `0xAB12` in memory as two consecutive bytes: `0x12 0xAB` (remember, the LSB is in the lowest address on a little-endian processor). Performing `(*&s)` dereferences the address `&s` and treats the memory location as an `unsigned short` because `&s` has type `unsigned short *`; thus, the expression yields `0xAB12`. Performing `*(unsigned char *)&s` yields `0x12` because the typecast converts the pointer `&s` into a pointer to an `unsigned char *`. Dereferencing such a pointer yields an `unsigned char`, which is only one byte long. Pointers always refer to the lowest address of the variable, so the result is `0x12` not `0xAB`. On a big-endian system, however, the result would be `0xAB`.

A more complete example: `invest.c`

Until now we have been dipping our toes in the C pool. Now let us dive in headfirst.

Our next program is called `invest.c`, which takes an initial investment amount, an expected annual return rate, and a number of years, and returns the growth of the investment over the years. After performing one set of calculations, it prompts the user for another scenario, and continues this way until the data entered is invalid. The data is invalid if, for example, the initial investment is negative or the number of years to track is outside the allowed range.

The real purpose of `invest.c`, however, is to demonstrate the syntax and several useful features of C.

Here's an example of compiling and running the program. The only data entered by the user are the three numbers corresponding to the initial investment, the growth rate, and the number of years.

```
> gcc invest.c -o invest
> invest
Enter investment, growth rate, number of yrs (up to 100): 100.00 1.05 5
Valid input?  1

RESULTS:

Year   0:       100.00
Year   1:       105.00
Year   2:       110.25
Year   3:       115.76
Year   4:       121.55
Year   5:       127.63

Enter investment, growth rate, number of yrs (up to 100): 100.00 1.05 200
Valid input?  0
Invalid input; exiting.
>
```

Before we look at the full `invest.c` program, let us review two principles that should be adhered to when writing a longer program: modularity and readability.

- *Modularity.* You should break your program into a set of functions that perform specific, well-defined tasks, with a small number of inputs and outputs. As a rule of thumb, no function should be longer than about 20 lines. (Experienced programmers often break this rule of thumb, but if you are a novice and are regularly breaking this rule, you are likely not thinking modularly.) Almost all variables you define should be "local" to (i.e., only recognizable by) their particular function. Global variables, which can be accessed by all functions, should be minimized or avoided altogether, since they break modularity, allowing one function to affect the operation of another without the information passing through the well-defined "pipes" (input arguments to a function or its returned results). If you find yourself typing the same (or similar) code more than once, that's a good sign you should determine how to write a single function and just call that function from multiple places. Modularity makes it much easier to develop large programs and track down the inevitable bugs.
- *Readability.* You should use comments to help other programmers, and even yourself, understand the purpose of the code you have written. Variable and function names should be chosen to indicate their purpose. Be consistent in how you name variables and functions. Any "magic number" (constant) used in your code should be given a name and defined at the beginning of the program, so if you ever want to change this number, you can just change it at one place in the program instead of every place it is used. Global variables and constants should be written in a way that easily distinguishes them from more common local variables; for example, you could WRITE CONSTANTS IN UPPERCASE and Capitalize Globals. You should use whitespace (blank lines, spaces, tabbing, etc.) consistently to make it easy to read the program. Use a fixed-width font

(e.g., `Courier`) so that the spacing/tabbing is consistent. Modularity (above) also improves readability.

The program `invest.c` demonstrates readable modular code using the structure and syntax of a typical C program. In the program's comments, you will see references of the form `==SecA.4.3==` that indicate where you can find more information in the review of syntax that follows the program.

```
/*****************************************************************************
 * PROGRAM COMMENTS (PURPOSE, HISTORY)
 *****************************************************************************/

/*
 * invest.c
 *
 * This program takes an initial investment amount, an expected annual
 * return rate, and the number of years, and calculates the growth of
 * the investment.  The main point of this program, though, is to
 * demonstrate some C syntax.
 *
 * References to further reading are indicated by ==SecA.B.C==
 *
 */

/*****************************************************************************
 * PREPROCESSOR COMMANDS   ==SecA.4.3==
 *****************************************************************************/

#include <stdio.h>       // input/output library
#define MAX_YEARS 100    // constant indicating max number of years to track

/*****************************************************************************
 * DATA TYPE DEFINITIONS (HERE, A STRUCT)  ==SecA.4.4==
 *****************************************************************************/

typedef struct {
  double inv0;                      // initial investment
  double growth;                    // growth rate, where 1.0 = zero growth
  int years;                        // number of years to track
  double invarray[MAX_YEARS+1];     // investment array   ==SecA.4.9==
} Investment;                       // the new data type is called Investment

/*****************************************************************************
 * GLOBAL VARIABLES   ==SecA.4.2, A.4.5==
 *****************************************************************************/

// no global variables in this program

/*****************************************************************************
 * HELPER FUNCTION PROTOTYPES  ==SecA.4.2==
 *****************************************************************************/

int getUserInput(Investment *invp);     // invp is a pointer to type ...
```

```
void calculateGrowth(Investment *invp); // ... Investment ==SecA.4.6, A.4.8==
void sendOutput(double *arr, int years);

/******************************************************************************
 * MAIN FUNCTION   ==SecA.4.2==
 ******************************************************************************/

int main(void) {

  Investment inv;                  // variable definition, ==SecA.4.5==

  while(getUserInput(&inv)) {      // while loop ==SecA.4.13==
    inv.invarray[0] = inv.inv0;    // struct access ==SecA.4.4==
    calculateGrowth(&inv);         // & referencing (pointers) ==SecA.4.6, A.4.8==
    sendOutput(inv.invarray,       // passing a pointer to an array ==SecA.4.9==
               inv.years);         // passing a value, not a pointer ==SecA.4.6==
  }
  return 0;                        // return value of main ==SecA.4.6==
} // ***** END main *****

/******************************************************************************
 * HELPER FUNCTIONS   ==SecA.4.2==
 ******************************************************************************/

/* calculateGrowth
 *
 * This optimistically-named function fills the array with the investment
 * value over the years, given the parameters in *invp.
 */
void calculateGrowth(Investment *invp) {

  int i;

  // for loop ==SecA.4.13==
  for (i = 1; i <= invp->years; i= i + 1) {   // relational operators ==SecA.4.10==
                                              // struct access ==SecA.4.4==
    invp->invarray[i] = invp->growth * invp->invarray[i-1];
  }
} // ***** END calculateGrowth *****


/* getUserInput
 *
 * This reads the user's input into the struct pointed at by invp,
 * and returns TRUE (1) if the input is valid, FALSE (0) if not.
 */
int getUserInput(Investment *invp) {

  int valid;       // int used as a boolean ==SecA.4.10==

  // I/O functions in stdio.h ==SecA.4.14==
  printf("Enter investment, growth rate, number of yrs (up to %d): ",MAX_YEARS);
  scanf("%lf %lf %d", &(invp->inv0), &(invp->growth), &(invp->years));

  // logical operators ==SecA.4.11==
```

```
  valid = (invp->inv0 > 0) && (invp->growth > 0) &&
    (invp->years > 0) && (invp->years <= MAX_YEARS);
  printf("Valid input?  %d\n",valid);

  // if-else ==SecA.4.12==
  if (!valid) { // ! is logical NOT ==SecA.4.11==
    printf("Invalid input; exiting.\n");
  }
  return(valid);
} // ***** END getUserInput *****


/* sendOutput
 *
 * This function takes the array of investment values (a pointer to the first
 * element, which is a double) and the number of years (an int).  We could
 * have just passed a pointer to the entire investment record, but we decided
 * to demonstrate some different syntax.
 */
void sendOutput(double *arr, int yrs) {

  int i;
  char outstring[100];      // defining a string ==SecA.4.9==

  printf("\nRESULTS:\n\n");
  for (i=0; i<=yrs; i++) {  // ++, +=, math in ==SecA.4.7==
    sprintf(outstring,"Year %3d:  %10.2f\n",i,arr[i]);
    printf("%s",outstring);
  }
  printf("\n");
} // ***** END sendOutput *****
```

A.4.1 Basic Syntax

Comments

Everything after a `/*` and before the next `*/` is a comment. Comments are removed during the preprocessing step of compilation. They help make the purpose of the program, function, loop, or statement clear to yourself or other programmers.[17] Keep the comments neat and concise for program readability. Some programmers use extra asterisks or other characters to make the comments stand out (see the examples in `invest.c`), but all that matters is that `/*` starts the comment and the next `*/` ends it.

If your comment is short, you can use `//` instead. Everything after `//` and before the next carriage return will be ignored. The `//` style comments originate from C++ but most modern C compilers support them.

[17] Reading your own code after several months away from it is often like reading someone else's code!

Semicolons

A code statement must be completed by a semicolon. Some exceptions to this rule include preprocessor commands (see `PREPROCESSOR COMMANDS` in the program and Section A.4.3) and statements that end with blocks of code enclosed by braces `{ }`. A single code statement may extend over multiple lines of the program listing until it is terminated by a semicolon (see, for example, the assignment to `valid` in the function `getUserInput`).

Braces and blocks of code

Blocks of code are enclosed in braces `{ }`. Examples include entire functions (see the definition of the `main` function and the helper functions), blocks of code executed inside of a `while` loop (in the `main` function) or `for` loop (in the `calculateGrowth` and `sendOutput` functions), as well as other examples. In `invest.c`, braces are placed as shown here

```
while (<expression>) {
  /* block of code */
}
```

but this style is equivalent

```
while (<expression>)
  {
    /* block of code */
  }
```

as is this

```
while (<expression>) { /* block of code */ }
```

Which brings us to...

Whitespace

Whitespace, such as spaces, tabs, and carriage returns, is only required where it is needed to recognize keywords and other syntax. The whole program `invest.c` could be written on a single line, for example. Indentations and line breaks should be used consistently, however, to make the program readable. Insert line breaks after each semicolon. Statements within the same code block should be left-justified with each other and statements in a code block nested within another code block should be indented with respect to the parent code block. Text editors should use a fixed-width font so that alignment is clear. Most editors provide fixed-width fonts and automatic indentation to enhance readability.

Parentheses

C has rules defining the order in which operations in an expression are evaluated, much like standard math rules that say $3 + 5 * 2$ evaluates to $3 + (10) = 13$, not $(8) * 2 = 16$. If uncertain about the default order of operations, use parentheses `( )` to enclose sub-expressions. For example, $3 + (40/(4 * (3 + 2)))$ evaluates to $3 + (40/(4 * 5)) = 3 + (40/20) = 3 + 2 = 5$, whereas $3 + 40/4 * 3 + 2$ evaluates to $3 + 30 + 2 = 35$.

A.4.2 Program Structure

`invest.c` demonstrates a typical structure for a program written in one `.c` file. When you write larger programs, you may wish to divide your program into multiple files, to increase modularity. Section A.4.15 discusses C programs that consist of multiple source code files.

Let us consider the seven major sections of the program in order of appearance. `PROGRAM COMMENTS` describe the purpose of the program. `PREPROCESSOR COMMANDS` define constants and "header" files that should be included, giving the program access to library functions that extend the power of the C language. This section is described in more detail in Section A.4.3. In some programs, it may be helpful to define a new data type, as shown in `DATA TYPE DEFINITIONS`. In `invest.c`, several variables are packaged together in a single record or `struct` data type, as described in Section A.4.4. Any `GLOBAL VARIABLES` are then defined. These are variables that are available for use by all functions in the program. Because of this special status, the names of global variables could be Capitalized or otherwise written in a way to remind the programmer that they are not local variables (Section A.4.5). Generally, global variables should be avoided because they violate modularity.

The next section of the program contains the `HELPER FUNCTION PROTOTYPES` of the various helper functions. A prototype of a function declares the name, argument types, and return types of a function that will be defined later. Prototypes are used to allow code to call functions that have not yet been fully defined or are defined elsewhere (perhaps in another source file). For example, the function `printRatio` has a return type of `void`, meaning that it does not return a value. It takes two arguments, each of type `int`. The function `getUserInput` returns an `int` and takes a single argument: a pointer to a variable of type `Investment`, a data type defined a few lines above the `getUserInput` prototype.

The next section of the program, `MAIN FUNCTION`, is where the `main` function is defined. Every program has exactly one `main` function, where the program starts execution. The `main` function returns an `int`. By convention, it returns 0 if it executes successfully, and otherwise returns a nonzero value. In `invest.c`, `main` takes no arguments, hence the `void` in the argument list. Some programs accept arguments on the command line; these can be passed as arguments to `main`. For example, we could have written `invest.c` to run with a command such as this:

```
> invest 100.0 1.05 5
```

To allow this, `main` would have been defined with the following syntax:

```
int main(int argc, char *argv[]) {
```

Then when the program is invoked as above, the integer `argc` would be set to four, the number of whitespace-separated strings on the command line, and `argv` would point to a array of four strings, where the string `argv[0]` is 'invest', `argv[1]` is '100.0', etc. You can learn more about arrays and strings in Section A.4.9.

Finally, the last section of the program is the definition of the `HELPER FUNCTIONS` whose prototypes were given earlier. It is not strictly necessary that the helper functions have prototypes, but if not, every function should be defined before it is used by any other function. For example, none of the helper functions uses another helper function, so they could have all been defined before the `main` function, in any order, and their function prototypes eliminated. The names of the variables in a function prototype and in the actual definition of the function need not be the same; for example, the prototype of `sendOutput` uses variables named `arr` and `years`, whereas the actual function definition uses `arr` and `yrs`. What matters is that the prototype and actual function definition have the same number of arguments, of the same types, and in the same order. In fact, in the arguments of the function prototypes, you can leave out variable names altogether, and just keep the comma separated list of argument data types; however, including the names serves as additional documentation and is generally a good practice.

A.4.3 Preprocessor Commands

In the preprocessing stage of compilation, all comments are removed from the program. Additionally, the preprocessor performs actions when encountering the following preprocessor commands, recognizable by the # character:

```
#include <stdio.h>       // input/output header
#define MAX_YEARS 100    // constant indicating max number of years to track
```

Include files

The first preprocessor command in `invest.c` indicates that the program will use standard C input/output functions. The file `stdio.h` is called a **header** file for the library. This file is readable by a text editor and contains constants, function prototypes, and other included headers that are made available to the program. The preprocessor replaces the

#include <stdio.h> command with the contents of the header file stdio.h.[18] Examples of function prototypes that are included are

```
int printf(const char *Format, ...);
int sprintf(char *Buffer, const char *Format, ...);
int scanf(const char *Format, ...);
```

Each of these three functions is used in invest.c. If the program were compiled without including stdio.h, the compiler would generate a warning or an error due to the lack of function prototypes. See Section A.4.14 for more information on using the stdio input and output functions.

During the linking stage, the object code of invest.c is linked with the object code for printf, sprintf, and scanf in your C installation. Libraries like the C standard library provide access to functions beyond the basic C syntax. Other useful libraries (and header files for the C standard library) are briefly described in Section A.4.14.

Constants

The second line defines the constant MAX_YEARS to be equal to 100. The preprocessor searches for each instance of MAX_YEARS in the program and replaces it with 100. If we later decide that the maximum number of years to track investments should be 200, we can change the definition of this constant in one place, instead of in several places. Since MAX_YEARS is constant, not a variable, it can never be assigned another value somewhere else in the program. To indicate that it is not a variable, a common convention is to write constants in UPPERCASE. This is not required by C, however. We should emphasize that the preprocessor performs a text substitution, as if you used the "find and replace" feature of your text editor. So, in this example, the preprocessor literally replaces every occurrence of MAX_YEARS with the number 100.

Macros

One more use of the preprocessor is to define simple function-like *macros* that you may use in more than one place in your program. Constants, as described above, are technically a simple macro. Here's an example that converts radians to degrees:

```
#define RAD_TO_DEG(x) ((x) * 57.29578)
```

[18] The preprocesser searches for header files in directories specified by the "include path." If the header file header.h sits in the same directory as invest.c, we would write #include "header.h" instead of #include <header.h>.

The preprocessor searches for any instance of `RAD_TO_DEG(x)` in the program, where `x` can be any text, and replaces it with `((x) * 57.29578)`. For example, the initial code

```
angle_deg = RAD_TO_DEG(angle_rad);
```

is replaced by

```
angle_deg = ((angle_rad) * 57.29578);
```

Note the importance of the outer parentheses in the macro definition. If we had instead used the preprocessor command

```
#define RAD_TO_DEG(x) (x) * 57.29578   // don't do this!
```

then the code

```
answer = 1.0 / RAD_TO_DEG(3.14);
```

would be replaced by

```
answer = 1.0 / (3.14) * 57.29578;
```

which is very different from

```
answer = 1.0 / ((3.14) * 57.29578);
```

Moral: if the expression you are defining is anything other than a single constant, enclose it in parentheses, to tell the compiler to evaluate the expression first.

As a second example, the macro

```
#define MAX(A,B)       ((A) > (B) ? (A):(B))
```

returns the maximum of two arguments. The `?` is the *ternary operator* in C, which has the form

```
<test> ? return_value_if_test_is_true : return_value_if_test_is_false
```

The preprocessor replaces

```
maxval = MAX(13+7,val2);
```

with

```
maxval = ((13+7) > (val2) ? (13+7):(val2));
```

Why define a macro instead of just writing a function? One reason is that the macro may execute slightly faster, since no passing of control to another function and no passing of variables is needed. Most of the time, you should use functions.

A.4.4 Typedefs, Structs, and Enums

In simple programs, you will do just fine with the data types `int`, `char`, `float`, `double`, and variations. Sometimes you may find it useful to create an alias for a data type, using the following syntax:

```
typedef <type> newtype;
```

where `<type>` is an existing data type and `newtype` is its alias. Then, you can define a new variable `x` of type `newtype` by

```
newtype x;
```

For example, you could write

```
typedef int days_of_the_month;
days_of_the_month day;
```

You might find it satisfying that your variable `day` (taking values 1 to 31) is of type `days_of_the_month`, but the compiler will still treat it as an `int`. However, if you use this type a lot and later want to change it, using the `typedef` provides one location to make the change rather than needing to go through your whole program: you can think of a `typedef` as a constant but for data types.

In addition to aliasing existing data types, you can also create new types that combine several variables into a single record or `struct`. We gather investment information into a single `struct` in `invest.c`:

```
typedef struct {
  double inv0;                    // initial investment
  double growth;                  // growth rate, where 1.0 = zero growth
  int years;                      // number of years to track
  double invarray[MAX_YEARS+1];   // investment values
} Investment;                     // the new data type is called Investment
```

Notice how the `struct {...}` replaces the data type `int` in our previous `typedef` example. This syntax creates a new data type `Investment` with a record structure, with *fields* named `inv0` and `growth` of type `double`, `years` of type `int`, and `invarray`, an array of `doubles`.[19] (Arrays are discussed in Section A.4.9.) With this new type definition, we can define a variable named `inv` of type `Investment`:

```
Investment inv;
```

This definition allocates sufficient memory to hold the two `doubles`, the `int`, and the array of `doubles`. We can access the contents of the `struct` using the "`.`" operator:

```
int yrs;
yrs = inv.years;
inv.growth = 1.1;
```

An example of this kind of usage is seen in `main`.

Referring to the discussion of pointers in Sections A.3.2 and A.4.8, if we are working with a pointer `invp` that points to `inv`, we can use the "`->`" operator to access the contents of the record `inv`:

```
Investment inv;      // allocate memory for inv, an investment record
Investment *invp;    // invp will point to something of type Investment
int yrs;
invp = &inv;         // invp points to inv
inv.years = 5;       // setting one of the fields of inv
yrs = invp->years;   // inv.years, (*invp).years, and invp->years are all identical
invp->growth = 1.1;
```

Examples of this usage are seen in `calculateGrowth()` and `getUserInput()`. Using the operator `a->b` is equivalent to doing `(*a).b`, dereferencing the `struct` pointer and accessing a specific field.

Another data type you can create is called an `enum`, short for "enumeration." Although `invest.c` does not use an enumeration, they can be useful for describing a type that can take one of a limited set of values. For example, if you wanted a function to use a cardinal direction you could define an enumeration as follows:

```
typedef enum {NORTH, SOUTH, EAST, WEST} Direction;
```

[19] The `typedef` is actually aliasing an anonymous struct with the name `Investment`. You can omit the `typedef`, but then you create a type that must be referred to as `struct Investment` rather than `Investment`. The `typedef` provides a more convenient syntax when you use the type.

Each item in the `enum` gets assigned a constant numerical value. You can explicitly state this value, or use the default compiler-provided values, which start at zero and increment by one for each element. For example, the declaration above is equivalent to

```
typedef enum {NORTH = 0, SOUTH = 1, EAST = 2, WEST = 3} Direction;
```

You can use an `enum` as you would any other data type.

A.4.5 Defining Variables

Variable names

Variable names can consist of uppercase and lowercase letters, numbers, and underscore characters '_'. You should generally use a letter as the first character; `var`, `Var2`, and `Global_Var` are all valid names, but `2var` is not. C is case sensitive, so the variable names `var` and `VAR` are different. A variable name cannot conflict with a reserved keyword in C, like `int` or `for`. Names should be succinct but descriptive. The variable names `i`, `j`, and `k` are often used for integer counters in `for` loops, and pointers often begin with `ptr_`, such as `ptr_var`, or end with `p`, such as `varp`, to remind you that they are pointers. Regardless of how you choose to name your variables, adopting a consistent naming convention throughout a program aids readability.

Scope

The **scope** of a variable refers to where it can be used in the program. A variable may be *global*, i.e., usable by any function, or *local* to a specific function or piece of a function. A global variable is one that is defined in the `GLOBAL VARIABLES` section, outside of and before any function that uses it. Such variables can be referred to or altered in any function.[20] Because of this special status, global variables are often Capitalized. Global variable usage should be minimized for program modularity and readability.

A local variable is one that is defined in a function. Such a variable is only usable inside that function, after the definition.[21] If you choose a local variable name `var` that is also the name of a global variable, inside that function `var` will refer to the local variable, and the global variable will not be available. It is not good practice to choose local variable names to be the

[20] You could also define a variable outside of any function definition but *after* some of the function definitions. This quasi-global variable would be available to all functions defined after the variable is defined, but not those before. This practice is discouraged, as it makes the code harder to read.

[21] Since we recommend that each function be brief, you can define all local variables in that function at the beginning of the function, so we can see in one place what local variables the function uses. Some programmers prefer instead to define variables just before their first use, to minimize their scope. Older C specifications required that all local variables be defined at the beginning of a code block enclosed by braces `{ }`.

same as global variable names, as it makes the program confusing to understand and is often the source of bugs.

The parameters to a function are local to that function's definition, as in `sendOutput` at the end of `invest.c`:

```
void sendOutput(double *arr, int yrs) {    // ...
```

The variables `arr` and `yrs` are local to `sendOutput`.

Otherwise, local variables are defined at the beginning of the function code block by syntax similar to that shown in the function `main`.

```
int main(void) {
  Investment inv;     // Investment is a variable type we defined
  // ... rest of the main function ...
```

Since this definition appears within the function, `inv` is local to `main`. Had this definition appeared before any function definition, `inv` would be a global variable.

A global variable can be declared as `static`:

```
static int i;
```

The `static` specifier means that the global variable can only be used from within the given `.c` file; other files cannot use the variable. If you must use a global variable, you should declare it `static` if possible. Preventing other `.c` files in a multi-file C program from accessing the variable helps increase modularity and reduce bugs.

Variables can also have *qualifiers* attached to their types. There are two main qualifiers in C, `const` and `volatile`. The `const` qualifier prevents the "variable" from being modified. The definition

```
const int i = 3;
```

sets `i` to 3, and `i` can never be changed after that. The `volatile` qualifier is used extensively in embedded programming, and indicates that the variable may change outside of the normal flow of the program (i.e., due to interrupts, Chapter 6) that the compiler cannot know about in advance. Therefore, the compiler should not assume anything about the current value of a `volatile` variable. Here is an example usage:

```
volatile int i;
```

We discuss `volatile` more fully in Chapter 6.

Qualifiers can also be applied to pointers. The syntax, however, can be a bit tricky. The first two definitions below make `cp` a `const` pointer, meaning that the content pointed to by `cp` (i.e., `*cp`) is constant: the value of `cp`, however, may change. The third definition makes the pointer `cp` itself constant, but the contents pointed to by `cp` may change. The fourth line makes both the pointer and the data that the pointer references constant.

```
const int * cp; // pointer to const data
int const * cp; // pointer to const data
int * const cp; // pointer to data, value of cp is const
const int * const cp; // pointer and data are const
```

Definition and initialization

When a variable is defined, memory for the variable is allocated. In general, you cannot assume anything about the contents of the variable until you have initialized it. For example, if you want to define a `float x` with value 0.0, the command

```
float x;
```

is insufficient. The memory allocated may have random 0's and 1's already in it, and the allocation of memory does not generally change the current contents of the memory. Instead, you can use

```
float x = 0.0;
```

to initialize the value of `x` when you define it. Equivalently, you could use

```
float x;
x = 0.0;
```

but, when possible, it is better to initialize a variable when you define it, so you do not accidentally use an uninitialized value.

Static local variables

Each time a function is called, its local variables are allocated space in memory. When the function completes, its local variables are discarded, freeing memory. If you want to keep the results of some calculation by the function after the function completes, you could return the results from the function or store them in a global variable. Sometimes, a better alternative is to use the `static` modifier in the local variable definition, as in the following program:

```
#include <stdio.h>
```

```
void myFunc(void) {
  static char ch='d';     // this local variable is static, allocated and initialized
                          // only once during the entire program
  printf("ch value is %d, ASCII character %c\n",ch,ch);
  ch = ch+1;
}

int main(void) {
  myFunc();
  myFunc();
  myFunc();
  return 0;
}
```

The `static` modifier in the definition of `ch` in `myFunc` means that `ch` is only allocated, and initialized to `'d'`, the first time `myFunc` is called during the execution of the program. This allocation persists after the function returns, and the value of `ch` is remembered. The output of this program is

```
ch value is 100, ASCII character d
ch value is 101, ASCII character e
ch value is 102, ASCII character f
```

Numerical values

Just as you can assign an integer a base-10 value using commands like `ch=100`, you can assign a number written in hexadecimal notation by putting "`0x`" at the beginning of the digit sequence, e.g.,

```
unsigned char ch = 0x64;  // ch now has the base-10 value 100
```

This form may be convenient when you want to directly control bit values. This is often useful in microcontroller applications. Some C compilers, including the PIC32 C compiler, allow specifying bits directly using the following syntax:

```
unsigned char ch = 0b1100100;  // ch now has the base-10 value 100
```

A.4.6 Defining and Calling Functions

A function definition consists of the function's return type, function name, argument list, and body (a block of code). Allowable function names follow the same rules as variable names. The function name should make the purpose of the function clear; for example, `getUserInput` (which gets input from the user) in `invest.c`.

If the function does not return a value, it has return type `void`, as with `calculateGrowth`. If it does return a value, such as `getUserInput` which returns an `int`, the function should end with the command

```
return val;
```

where `val` is a variable with the same type as the function's return type. The `main` function returns an `int` and should return 0 upon successful completion.

The function definition

```
void sendOutput(double *arr, int yrs) {  // ...
```

indicates that `sendOutput` returns nothing and takes two arguments, a pointer to type `double` and an `int`. When the function is called with the statement

```
sendOutput(inv.invarray, inv.years);
```

the `invarray` and `years` fields of the `inv` structure in `main` are copied to `sendOutput`, which now has its own local copies of these variables, stored in `arr` and `yrs`. The difference between the two is that `yrs` is just data, while `arr` is a pointer, holding the address of the first element of `invarray`, i.e., `&(inv.invarray[0])`. (Arrays will be discussed in more detail in Section A.4.9.) Since `sendOutput` now has the memory address of the beginning of this array, *it can directly access, and potentially change, the original array seen by* `main`. On the other hand, `sendOutput` cannot, by itself change the value of `inv.years` in `main`, since it only has a copy of that value, not the actual memory address of `main`'s `inv.years`. `sendOutput` takes advantage of its direct access to the `inv.invarray` to print out all the values stored there, eliminating the need to copy all the values of the array from `main` to `sendOutput`. To prevent the function from changing the contents of `arr` we could have (and probably should have) declared the parameter `const`, as in `double const * arr`.

The function `calculateGrowth`, which is called with a pointer to `main`'s `inv` data structure, takes advantage of its direct access to the `invarray` field to change the values stored there.

When a function is passed a pointer argument, it is called a *pass by reference*; the argument is a reference (address, or pointer) to data. When a function is passed non-pointer data, it is called a *pass by value*; data is copied but not an address.

If a function takes no arguments and returns no value, we can define it as `void myFunc(void)`. The function is called using

```
myFunc();
```

A.4.7 Math

Standard *binary* math operators (operators on two operands) include `+`, `-`, `*`, and `/`. These operators take two operands and return a result, as in

```
ratio = a/b;
```

If the operands are the same type, then the CPU carries out a division (or add, subtract, multiply) specific for that type and produces a result of the same type. In particular, if the operands are integers, the result will be an integer, even for division (fractions are rounded toward zero). If one operand is an integer type and the other is a floating point type, the integer type will generally be coerced to a floating point to allow the operation (see the `typecast.c` program description of Section A.4).

The modulo operator `%` takes two integers and returns the remainder of their division, i.e.,

```
int i;
i = 16 % 7;  // i is now equal to 2
```

C also provides `+=, -=, *=, /=, %=` to simplify some expressions, as shown below:

```
x = x * 2;  // these two lines
x *= 2;     // are equivalent

y = y + 7;  // these two lines
y += 7;     // are equivalent
```

Since adding one to an integer or subtracting one from an integer are common operations in loops, these have a further simplification. For an integer `i`, we can write

```
++i;  // adds 1 to i, equivalent to i = i+1;
--i;  // equivalent to i = i-1;
```

In fact we also have the syntax `i++` and `i-`. If the `++` or `-` come in front of the variable, the variable is modified before it is used in the rest of the expression. If they come after, the variable is modified after the expression has been evaluated. So

```
int i = 5, j;
j = (++i)*2;    // after this line, i is 6 and j is 12
```

but

```
int i = 5,j;
j = (i++)*2;   // after this line, i is 6 and j is 10
```

But your code would be much more readable if you just wrote `i++` before or after the `j = i*2` line.

If your program includes the C math library with the preprocessor command `#include <math.h>`, you have access to a much larger set of mathematical operations, some of which are listed here:

```
int     abs     (int x);               // integer absolute value
double  fabs    (double x);            // floating point absolute value
double  cos     (double x);            // all trig functions work in radians,
                                          not degrees
double  sin     (double x);
double  tan     (double x);
double  acos    (double x);            // inverse cosine
double  asin    (double x);
double  atan    (double x);
double  atan2   (double y, double x);  // two-argument arctangent
double  exp     (double x);            // base e exponential
double  log     (double x);            // natural logarithm
double  log2    (double x);            // base 2 logarithm
double  log10   (double x);            // base 10 logarithm
double  pow     (double x, double y);  // raise x to the power of y
double  sqrt    (double x);            // square root of x
```

These functions also have versions for `float`s. The names of those functions are identical, except with an `'f'` appended to the end, e.g., `cosf`.

When compiling programs using `math.h`, you may need to include the linker flag `-lm`, e.g.,

```
gcc myprog.c -o myprog -lm
```

to tell the linker to link with the math library.

A.4.8 Pointers

It's a good idea to review the introduction to pointers in Section A.3.2 and the discussion of call by reference in Section A.4.6. In summary, the operator `&` references a variable, returning a pointer to (the address of) that variable, and the operator `*` dereferences a pointer, returning the contents of the address.

These statements define a variable `x` of type `float` and a pointer `ptr` to a variable of type `float`:

```
float x;
float *ptr;
```

At this point, the assignment

```
*ptr = 10.3;
```

would result in undefined behavior, because the pointer ptr does not currently point to anything. The following code would be valid:

```
ptr = &x;          // assign ptr to the address of x; x is the "pointee" of ptr
*ptr = 10.3;       // set the contents at address ptr to 10.3; now x is equal to 10.3
*(&x) = 4 + *ptr; // the * and & on the left cancel each other; x is set to 14.3
```

Since ptr is an address, it is an integer (technically the type is "pointer to type float"), and we can add and subtract integers from it. For example, say that ptr contains the value n, and then we execute the statement

```
ptr = ptr + 1;    // equivalent to ptr++;
```

If we now examined ptr, we would find that it has the value $n + 4$. Why? Because the compiler knows that ptr points to the type float, so when we add 1 to ptr, the assumption is that we want to increment by one float in memory, not one byte. Since a float occupies four bytes, the address ptr must increase by 4 to point to the next float. The ability to increment a pointer in this way can be useful when dealing with arrays, next.

A.4.9 Arrays and Strings

One-dimensional arrays

An array of five floats can be defined by

```
float arr[5];
```

We could also initialize the array at the time we define it:

```
float arr[5] = {0.0, 10.0, 20.0, 30.0, 40.0};
```

Each of these definitions allocates five floats in memory, accessed by arr[0] (initialized to 0.0 above) through arr[4] (initialized to 40.0). The elements are stored consecutively, as per Figure A.2. The assignment

```
arr[5] = 3.2;
```

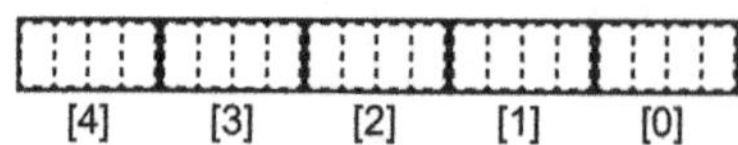

Figure A.2
A float array with five elements, as stored in memory. The dashed lines separate bytes and the solid lines separate floats. Each float has four bytes.

is a mistake, since only `arr[0..4]` have been allocated. This statement will compile successfully because compilers do not check for indexing arrays out of bounds. The best result at this point would be for your program to crash, to alert you to the fact that you are overwriting memory that may be allocated for another purpose. More insidiously, the program could seem to run just fine, but with difficult-to-debug erratic behavior.[22] Bottom line: never access arrays out of bounds!

In the expression `arr[i]`, `i` is an integer called the *index*, and `arr[i]` is of type `float`. The variable `arr` by itself points to the first element of the array, and is equivalent to `&(arr[0])`. The address `&(arr[i])` is at `arr` plus `i*4` bytes, since the elements of the array are stored consecutively, and a `float` uses four bytes. Both `arr[i]` and `*(arr+i)` are correct syntax to access the `i`th element of the array. Since the compiler knows that `arr` is a pointer to the four-byte type `float`, the address represented by `(arr+i)` is `i*4` bytes higher than the address `arr`.

Consider the following code:

```
float arr[5] = {0.0, 10.0, 20.0, 30.0, 40.0};
float *ptr;
ptr = arr + 3;
// arr[0] contains 0.0 and ptr[0] = arr[3] = 30.0
// arr[0] is equivalent to *arr; ptr[0] is equivalent to *ptr and *(arr+3);
// ptr is equivalent to &(arr[3])
```

If we would like to pass the array `arr` to a function that initializes each element of the array, we could call

```
arrayInit(arr,5);
```

or

```
arrayInit(&(arr[0]),5);
```

The function definition for `arrayInit` might look like

```
void arrayInit(float *vals, int length) {

  int i;

  for (i=0; i<length; i++) vals[i] = i*10.0;
  // equivalently, we could substitute the line below for the line above
  // for (i=0; i<length; i++) {*vals = i*10.0; vals++;}
}
```

[22] The mistake could potentially be exploited as a security flaw.

The pointer `vals` in `arrayInit` is set to point to the same location as `arr` in the calling function. Therefore `vals[i]` refers to the same memory contents that `arr[i]` does.

Note that `arr` does not carry any information on the length of the array, so we must send the length separately to `arrayInit`.

Strings

A string is an array of `char`s. The definition

```
char s[100];
```

allocates memory for 100 `char`s, `s[0]` to `s[99]`. We could initialize the array with

```
char s[100] = "cat"; // note the double quotes
```

This places a '`c`' (integer value 99) in `s[0]`, an '`a`' (integer value 97) in `s[1]`, a '`t`' (integer value 116) in `s[2]`, and a value of 0 in `s[3]`, corresponding to the NULL ('`\0`') character and indicating the end of the string. (You could also do this, less elegantly, by initializing just those four elements using braces as we did with the `float` array above.)

You notice that we allocated more memory than was needed to hold "cat." Perhaps we will append something to the string in future, so we might want to allocate that extra space just in case. But if not, we could have initialized the string using

```
char s[] = "cat";
```

and the compiler would only assign the minimum memory needed (four bytes in this case, three for each character and one for the NULL character).

The function `sendOutput` in `invest.c` shows an example of constructing a string using `sprintf`, a function provided by `stdio.h`. Other functions for manipulating strings are provided in `string.h`. Both of these headers are described briefly in Section A.4.14.

Multi-dimensional arrays

The definition

```
int mat[2][3];
```

allocates memory for 6 `int`s, `mat[0][0]` to `mat[1][2]`, which can be thought of as a two-dimensional array, or matrix. These occupy a contiguous region of memory, with `mat[0][0]` at the lowest memory location, followed by `mat[0][1]`, `mat[0][2]`, `mat[1][0]`, `mat[1][1]`, and `mat[1][2]`. This matrix can be initialized using nested braces,

```
int mat[2][3] = {{0, 1, 2}, {3, 4, 5}};
```

Higher-dimensional arrays can be created by simply adding more indexes. In memory, a "row" of the rightmost index is completed before incrementing the next index to the left.

Static vs. dynamic memory allocation

A command of the form `float arr[5]` is called *static memory allocation*, meaning that the size of the array is known at compile time. Another option is *dynamic memory allocation*, where the size of the array can be chosen at run time.[23] With the C standard library header `stdlib.h` included using the preprocessor command `#include <stdlib.h>`, the syntax

```
float *arr; // arr is a pointer to float, but no memory has been allocated for the array
int i=5;
arr = (float *) malloc(i * sizeof(float));  // allocate the memory
```

allocates `arr[0..4]`, and

```
free(arr);
```

releases the memory when it is no longer needed. It is crucial to remember to `free` memory allocated with `malloc` when you are finished with it, so you do not run out of memory if you repeatedly allocate memory.[24] If `malloc` cannot allocate the requested memory, perhaps because the computer is out of memory, it returns a NULL pointer (i.e., `arr` will have value 0). You must **always** check that the result of `malloc` is valid before continuing with your program.

A.4.10 Relational Operators and TRUE/FALSE Expressions

`==`	equal
`!=`	not equal
`>, >=`	greater than, greater than or equal to
`<, <=`	less than, less than or equal to

Relational operators operate on two values and evaluate to 0 or 1. A 0 indicates that the expression is FALSE and a 1 indicates that the expression is TRUE. For example, the expression `(3>=2)` is TRUE, so it evaluates to 1, while `(3<2)` evaluates to 0, or FALSE.

[23] Dynamic memory is allocated from the *heap*, a portion of memory set aside for dynamic allocation (and therefore is not available for statically allocated variables and program code). You may have to adjust linker options setting the size of the heap. See Chapter 5.3.

[24] `malloc` tracks the size of the block associated with the address `arr`, so you do not need to tell `free` how much memory to release.

The most common mistake with relational operators is using = to test for equality instead of ==. For example, using the `if` conditional syntax (Section A.4.12), the test

```
int i = 2;
if (i = 3) {   // error:  this is an assignment, not a test!!
  printf("Test is true.");
}
```

always evaluates to TRUE, because the expression `(i=3)` assigns the value of 3 to `i`, and the expression evaluates to 3. Any nonzero value is treated as logical TRUE. If the condition is written `(i==3)`, it will operate as intended, evaluating to 0 (FALSE).

Be aware of potential pitfalls in checking equality of floating point numbers. Consider the following program:

```
#include <stdio.h>
#define VALUE 3.1
int main(void) {
  float  x = VALUE;
  double y = VALUE;
  if (x == VALUE) {
    printf("x is equal to %f.\n",VALUE);
  } else {
    printf("x is not equal to %f!\n",VALUE);
  }
  if (y == VALUE) {
    printf("y is equal to %f.\n",VALUE);
  } else {
    printf("y is not equal to %f!\n",VALUE);
  }
  return 0;
}
```

You might be surprised to see that your program says that `x` is not equal while `y` is! In fact, neither `x` nor `y` are exactly 3.1 due to roundoff error in the floating point representation. However, by default, the constant 3.1 is treated as a `double`, so the `double y` carries the identical (wrong) value. If you want a constant to be treated explicitly as a `float`, you can write it as 3.1F, and if you want it to be treated as a `long double`, you can write it as 3.1L.

A.4.11 Logical and Bitwise Operators

~	bitwise NOT
&	bitwise AND
|	bitwise OR
^	bitwise XOR
>>	shift bits to the right (shifting in 0's from the left)
<<	shift bits to the left (shifting in 0's from the right)

Bitwise operators act directly on the bits of the operand(s), as in the following example:

```
unsigned char a=0xC, b=0x6, c;  // in binary, a is 0b00001100 and b is 0b00000110
c = ~a;       // NOT; c is 0xF3 or 0b11110011
c = a & b;  // AND; c is 0x04 or 0b00000100
c = a | b;  // OR;  c is 0x0E or 0b00001110
c = a ^ b;  // XOR; c is 0x0A or 0b00001010
c = a >> 3; // SHIFT RT 3; c is 0x01 or 0b00000001, one 1 is shifted off the
               right end
c = a << 3; // SHIFT LT 3; c is 0x60 or 0b01100000, 1's shifted to more
               significant digits
```

Much like the math operators, we also have the assignment expressions `&=`, `|=`, `^=`, `>>=`, and `<<=`, so `a &= b` is equivalent to `a = a&b`.

A.4.12 Conditional Statements

`if-else`

The basic `if-else` construct takes this form:

```
if (<expression>) {
  // execute this code block if <expression> is TRUE, then exit
}
else {
  // execute this code block if <expression> is FALSE
}
```

If the code block is a single statement, the braces are not necessary; however, good practice dictates that you should always use braces, in case you want to add additional statements later. The `else` and the block after it can be eliminated if no action needs to be taken when `<expression>` is FALSE.

`if-else` statements can be made into arbitrarily long chains:

```
if (<expression1>) {
  // execute this code block if <expression1> is TRUE, then exit this if-else chain
}
else if (<expression2>) {
  // execute this code block if <expression2> is TRUE, then exit this if-else chain
}
else {
  // execute this code block if both expressions above are FALSE
}
```

An example `if` statement is in `getUserInput`.

switch

If you would like to check if the value of a single expression is one of several possibilities, a `switch` may be simpler, clearer, and faster than a chain of `if-else` statements. Here is an example:

```
char ch;
// ... omitting code that sets the value of ch ...
switch (ch) {
  case 'a':        // execute these statements if ch has value 'a'
    <statement>;
    <statement>;
    break;         // exit the switch statement
  case 'b':
    // ... some statements
    break;
  case 'c':
    // ... some statements
    break;
  default:         // execute this code if none of the previous cases applied
    // ... some statements
}
```

Notice the `break;` statement after each `case`. These statements are required to prevent the code from "falling through" to the next case, which is usually undesirable.

A.4.13 Loops

for loop

A `for` loop has the following syntax:

```
for (<initialization>; <test>; <update>) {
  // code block
  }
```

If the code block consists of only one statement, the surrounding braces can be eliminated, but it is a good idea to use them anyway.

The sequence is as follows: at the beginning of the loop, the `<initialization>` statement is executed. Then the `<test>` is evaluated. If it is TRUE, the code block is executed, the `<update>` is performed, and we return to the `<test>`. If it is FALSE, the `for` loop exits.

The following `for` loop is in `calculateGrowth`:

```
for (i = 1; i <= invp->years; i = i + 1) {
  invp->invarray[i] = invp->growth*invp->invarray[i-1];
}
```

The <initialization> step sets `i = 1`. The <test> is TRUE if `i` is less than or equal to the number of years we will calculate growth in the investment. If it is TRUE, the value of the investment in year `i` is calculated from the value in year `i-1` and the growth rate. The <update> adds 1 to `i`. In this example, the code block is executed for `i` values of 1 to `invp->years`.

It is possible to perform more than one statement in the <initialization> and <update> steps by separating the statements by commas. For example, we could write

```
for (i = 1, j = 10; i <= 10; i++, j--)  { /* code */ };
```

if we want `i` to count up and `j` to count down.

`while` loop

A `while` loop has the following syntax:

```
while (<test>) {
  // code block
}
```

First, the <test> is evaluated, and if it is FALSE, the `while` loop exits. If the test is TRUE, the code block is executed and we return to the <test>.

In `main` of `invest.c`, the `while` loop executes until the function `getUserInput` returns 0, i.e., FALSE. `getUserInput` collects the user's input and returns an `int` that is 0 if the user's input is invalid and 1 if it is valid.

`do-while` loop

This is similar to a `while` loop, except the <test> is executed at the end of the code block, guaranteeing that the loop is executed at least once.

```
do {
  // code block
} while (<test>);
```

`break` and `continue`

If anywhere in the loop's code block the command `break` is encountered, the program will exit the loop. If the command `continue` is encountered, the rest of the commands in the code block will be skipped, and control will return to the <update> in a `for` loop or the <test> in a `while` or `do-while` loop. Examples:

```
while (<test1>) {
  if (<test2>) {
    break;  // jump out of the while loop
  }
  // ...
}
```

```
while (<test1>) {
  if (<test2>) {
    continue;  // skip the rest of the loop and go back to <test1>
  }
  x = x+3;
}
```

Use `break` and `continue` judiciously; they can make your code difficult to read. If you find yourself relying on numerous `break` and `continue` statements in a single loop, you may want to rethink your approach.

A.4.14 The C Standard Library

C comes with a standard library, aspects of which you can use by including the appropriate `.h` header file. The header file provides data types, function prototypes and macros required for part of the library.[25] We have already seen examples of standard header files such as `stdio.h`, which contains input/output functions; `math.h` in Section A.4.7; and `stdlib.h` in Section A.4.9. Third-party libraries can also be included by including their header files and linking with them.

It is well beyond our scope to provide details on the C standard library. Here we highlight a few particularly useful functions in `stdio.h`, `string.h`, and `stdlib.h`.

***Input and output:* `stdio.h`**

```
int printf(const char *Format, ...);
```

The function `printf` is used to print to the "standard output," which, for a PC, is typically the screen. It takes a formatting string `Format` and a variable number of extra arguments, determined by the formatting string, as indicated by the ... notation. The keyword `const` means that `printf` cannot change the string `Format`.

An example comes from our program `printout.c`:

```
int i;
float f;
double d;
char c;
```

[25] Reminder: if you include `<math.h>`, you should also compile your program with the `-lm` flag, so the math library is linked during the linking stage. The math library is logically part of the C standard library, it just happens to be in a different file.

```
i = 32;
f = 4.278;
d = 4.278;
c = 'k'; // or, by ASCII table, c = 107;

printf("Formatted output:\n");
printf(" i = %4d  c = '%c'\n",i,c);
printf(" f = %19.17f\n",f);
printf(" d = %19.17f\n",d);
```

which produces the output

```
Formatted output:
i =   32  c = 'k'
f = 4.27799987792968750
d = 4.27799999999999958
```

The formatting strings consist of plain text, the special character `\n` that prints a newline, and directives of the form `%4d` and `%19.17f`. Each directive indicates that `printf` will be looking for a corresponding variable in the argument list to insert into the output. A non-exhaustive list of directives is given here:

- `%d` Print an `int`. Corresponding argument should be an `int`.
- `%u` Print an `unsigned int`. Corresponding argument should be an integer data type.
- `%ld` Print a `long int`.
- `%f` Print a `double` or a `float`. Corresponding argument should be a float or a double.
- `%c` Print a character according to the ASCII table. Argument should be `char`.
- `%s` Print a string. Argument should be a pointer to a `char` (first element of a string), terminated with a NULL character (`'\0'`).
- `%x` Print an `unsigned int` as a hexadecimal number.

The directive `%d` can be written instead as `%4d`, for example, meaning that four spaces are allocated to write the integer, which will be right-justified in that space with unused spaces blank. The directive `%f` can be written instead as `%6.3f`, indicating that six spaces are reserved to write out the variable, with one of those spaces being the decimal point and three of the spaces after the decimal point.

```
int sprintf(char *str, const char *Format, ...);
```

Instead of printing to the screen, `sprintf` prints to the string `str`. An example of this is in `sendOutput`. The string `str` must have enough memory allocated to fit the results.

```
int scanf(const char *Format, ...);
```

The function `scanf` is a formatted read from the "standard input," which is typically the keyboard. Arguments to `scanf` consist of a formatting string and pointers to variables where the input should be stored. Typically the formatting string consists of directives like `%d`, `%f`, etc., separated by whitespace. The directives are similar to those for `printf`, except they do not accept spacing modifiers (like the 5 in `%5d`).

One notable difference between formatting strings is that, unlike `printf`, `scanf` does distinguish between `float`s and `double`s. To read a `double` with `scanf` use `%lf` rather than `%f`.

For each directive, `scanf` expects a pointer to a variable of that type in the argument list. A very common mistake is the following:

```
int i;
scanf("%d",i);  // WRONG!  We need a pointer to the variable.
scanf("%d",&i); // RIGHT.
```

The pointer allows `scanf` to put the input into the right place in memory.

`getUserInput` uses the statement

```
scanf("%lf %lf %d", &(invp->inv0), &(invp->growth), &(invp->years));
```

to read in two `double`s and an `int` and place them into the appropriate spots in the investment data structure. `scanf` ignores the whitespace (tabs, newlines, spaces, etc.) between the inputs.

```
int sscanf(char *str, const char *Format, ...);
```

Instead of scanning from the keyboard, `scanf` scans the string pointed to by `str`.

```
FILE* fopen(const char *Path, const char *Mode);
int fclose(FILE *Stream);
int fscanf(FILE *Stream, const char *Format, ...);
int fprintf(FILE *Stream, const char *Format, ...);
```

These commands are for reading from and writing to files. Say you have a file named `inputfile`, sitting in the same directory as the program, with information your program needs. The following code would read from it and then write to the file `outputfile`.

```
int i;
double x;
FILE *input, *output;
input = fopen("inputfile","r");     // "r" means you will read from this file
output = fopen("outputfile","w");  // "w" means you will write to this file
fscanf(input,"%d %lf",&i,&x);
fprintf(output,"I read in an integer %d and a double %lf.\n",i,x);
fclose(input);                      // these streams should be closed ...
fclose(output);                     // ... at the end of the program
```

Normally, you would check the return value of `fopen`. If it returns NULL, than it failed to open the file.

```
int fputc(int character, FILE *stream);
int fputs(const char *str, FILE *stream);
int fgetc(FILE *stream);
char* fgets(char *str, int num, FILE *stream);
int puts(const char *str);
```

These commands write (put) a character or string to a file, get a character or string from a file, or write a string to the screen. The variable `stdin` is a `FILE *` that corresponds to keyboard input, and `stdout` is a `FILE *` that corresponds to screen output.

String manipulation: `string.h`

```
char* strcpy(char *destination, const char *source);
```

Given two strings, `char destination[100],source[100]`, we cannot simply copy one to the other using the assignment `destination = source`. Instead we use `strcpy(destination,source)`, which copies the string `source` (until reaching the string terminator character, integer value 0) to `destination`. The string `destination` must have enough memory allocated to hold the source string.

```
char* strcat(char *destination, const char *source);
```

Appends the string in `source` to the end of the string `destination`, which must be large enough to hold the concatenated string.

```
int strcmp(const char *s1, const char *s2);
```

Returns 0 if the two strings are identical, a positive integer if the first unequal character in `s1` is greater than `s2`, and a negative integer if the first unequal character in `s1` is less than `s2`.

```
size_t strlen(const char *s);
```

The type `size_t` is an unsigned integer type. `strlen` returns the length of the string `s`, where the end of the string is indicated by the NULL character ('`\0`').

```
void* memset(void *s, int c, size_t len);
```

`memset` writes `len` bytes of the value `c` (converted to an `unsigned char`) starting at the beginning of the string `s`. So

```
char s[10];
memset(s,'c',5);
```

would fill the first five characters of the string `s` with the character `'c'` (or integer value 99). This can be a convenient way to initialize a string.

General purpose functions in `stdlib.h`

```
void* malloc(size_t objectSize)
```

`malloc` is used for dynamic memory allocation. An example use is in Section A.4.9.

```
void free(void *objptr)
```

`free` is used to release memory allocated by `malloc`. An example use is in Section A.4.9.

```
int rand()
```

It is sometimes useful to generate random numbers, particularly for games. The code

```
int i;
i = rand();
```

places in `i` a pseudo-random number between 0 and `RAND_MAX`, a constant which is defined in `stdlib.h` (2,147,483,647 in our gcc installation). To convert this to an integer between 1 and 10, you could follow with

```
i = 1 + (int) ((10.0*i)/(RAND_MAX+1.0));
```

One drawback of the code above is that calling `rand` multiple times will lead to the same sequence of random numbers every time the program runs. The usual solution is to "seed" the random number algorithm with a different number each time, and this different number is often taken from a system clock. The `srand` function is used to seed `rand`, as in the example below:

```
#include <stdio.h>    // allows use of printf()
#include <stdlib.h>   // allows use of rand() and srand()
#include <time.h>     // allows use of time()

int main(void) {
  int i;
```

```
    srand(time(NULL)); // seed the random number generator with the current time
    for (i=0; i<10; i++) printf("Random number:  %d\n",rand());
    return 0;
}
```

If we take out the line with `srand`, this program produces the same ten "random" numbers every time we run it. Note that this program includes the `time.h` header to allow the use of the `time` function.

```
void exit(int status)
```

When `exit` is invoked, the program exits with the exit code `status`. `stdlib.h` defines `EXIT_SUCCESS` with value 0 and `EXIT_FAILURE` with value −1, so that a typical call to `exit` might look like

```
exit(EXIT_SUCCESS);
```

A.4.15 Multiple File Programs and Libraries

So far our programs have used the C Standard Library, which provides several functions, including `printf` and `scanf`. Accessing these functions is possible because:

1. The preprocessor command `#include <stdio.h>` inserts the header file `stdio.h` into the current compilation unit, providing function prototypes for library functions.
2. The linker links the pre-compiled library object code in the C Standard Library with your program.

Thus, using a library usually requires header files and object code. We could also loosely define a library to consist of a header file and a C file (without a `main` function) containing source code for the library functions.

The purpose of a library is to collect functions that are likely to be useful in multiple programs so you do not have to rewrite the code for each program. The same principles apply when dividing your project into multiple source files. Think about what functions may be generally useful yet related and put them into their own file. Having code in multiple files not only promotes reuse, but it also isolates your project's components making them easier to test and debug. Many libraries evolve from a collection of source files that were designed for one particular project, but prove more generally useful.

Let us look at an example.

A simple example: The rad2volume library

Pretend you want to write several programs that need to calculate the volume of a sphere given its radius. You could copy and paste the formula into all of your programs. If, however, you made some mistake and wanted to fix it, you would then need to find where you used the formula in every program and change it. By placing the formula in a function, in its own file, you only need to make one correction to fix everything.

In this example, you decide to write one helper C file, `rad2volume.c`, with a function `double radius2Volume(double r)` that can be used by other C files. For good measure, you decide to make the constant `MY_PI` available also. To test your new rad2volume library consisting of `rad2volume.c` and `rad2volume.h`, you create a `main.c` file that uses it. The three files are given below.

```
// ***** file:  rad2volume.h *****
#ifndef RAD2VOLUME_H            // "include guard"; don't include twice in one compilation
#define RAD2VOLUME_H            // second line of the "include guard"

#define MY_PI 3.1415926            // constant available to files including rad2Volume.h
double radius2Volume(double r);    // prototype available to files including rad2Volume.h

#endif                             // third line, and end, of "include guard"
```

```
// ***** file:  rad2volume.c *****
#include <math.h>                  // for the function pow
#include "rad2volume.h"            // if the header is in the same directory, use "quotes"

static double cuber(double x) {    // this function is not available externally
  return pow(x,3.0);
}

double radius2Volume(double rad) {  // function definition
  return (4.0/3.0)*MY_PI*cuber(rad);
}
```

```
// ***** file:  main.c *****
#include <stdio.h>
#include "rad2volume.h"

int main(void) {
  double radius = 3.0, volume;
  volume = radius2Volume(radius);
  printf("Pi is approximated as %25.23lf.\n",MY_PI);
  printf("The volume of the sphere is %8.4lf.\n",volume);
  return 0;
}
```

The C file `rad2volume.c` contains two functions, `cuber` and `radius2Volume`. The function `cuber` is only meant for internal, private use by `rad2volume.c`, so there is no prototype in rad2volume.h and it is also declared static so it is not visible to other source files. The function radius2Volume is meant for public use by other C files, so a prototype for radius2Volume is included in the library header file rad2volume.h. The constant MY_PI is also meant for public use, so it is defined in rad2volume.h. Now radius2Volume and MY_PI are available to any file that includes rad2volume.h. In this case, they are available to main.c and rad2volume.c. Typically, it is good practice for the implementation file to #include its own header; this prevents problems relating to a mismatch between function prototypes and function definitions.

Each of main.c and rad2volume.c is compiled independently to create the object code files main.o and rad2volume.o. The linker combines these files into the final executable. main.c compiles successfully because it has a prototype for rad2Volume from including rad2volume.h, and it expects that, during the linking stage, rad2Volume will be present. If no object code passed to the linker defines rad2Volume then a linker error occurs.

Note the three lines making up the *include guard* in rad2volume.h. During preprocessing of a C file, if rad2volume.h is included, the macro RAD2VOLUME_H is defined. If the same C file tries to include rad2volume.h again, the include guard will recognize that RAD2VOLUME_H already exists and therefore skip the prototype and constant definition, down to the #endif. Without include guards, if we wrote a .c file including both header1.h and header2.h, for example, not knowing that header2.h already includes header1.h, header1.h would be included twice, possibly causing errors.

The two C files, rad2volume.c and main.c, can be compiled into object code using the commands

```
gcc -c rad2volume.c -o rad2volume.o
gcc -c main.c -o main.o
```

where the -c flag indicates that the source code should be compiled and assembled, but not linked. The result is the object files rad2volume.o and main.o. The two object files can be linked into a final executable using

```
gcc rad2volume.o main.o -o myprog
```

Instead of typing these three lines, the single command

```
gcc rad2volume.c main.c -o myprog
```

will automatically compile and link the files.

Executing myprog, the output is

```
Pi is approximated as 3.14159260000000006840537.
The volume of the sphere is 113.0973.
```

More general multi-file projects

Generalizing from the previous example, a header file declares constants, macros, new data types, and function prototypes that are needed by the files that `#include` them. A header file can be included by C source files or other header files. Figure A.3 illustrates a project consisting of one C source file with a `main` function and two helper C source files without a `main` function. (Every C program has exactly one `.c` file with a `main` function.) Each of the helper C files has its own header file. This project also has another header file, `general.h`, without an associated C file. This header contains general constant, macro, and data type definitions that are not specific to either helper source file or the `main` C file. The arrows indicate that the pointed-to file `#include`s the pointed-from header file.

Assuming that all the files are in the same directory, the project in Figure A.3 can be built by the following four commands, which create three object files (one for each source file) and link them together into `myprog`:

```
gcc -c main.c -o main.o
gcc -c helper1.c -o helper1.o
gcc -c helper2.c -o helper2.o
gcc main.o helper1.o helper2.o -o myprog
```

The build is illustrated in Figure A.4. Each C file is compiled independently and requires the constants, macros, data types, and function prototypes needed to successfully compile and assemble into an object file. During compilation of a single C file, the compiler neither has nor needs access to the source code for functions in other C files. If `main.c` uses a function in `helper1.c`, for example, it needs only a prototype of the function, provided by `helper1.h`. The prototype tells the compiler the return type of the function and what arguments it takes, allowing the compiler to check if the code in `main.c` uses the function properly. Calls to the function from `main.o` are linked to the actual function in `helper1.o` at the linker stage.

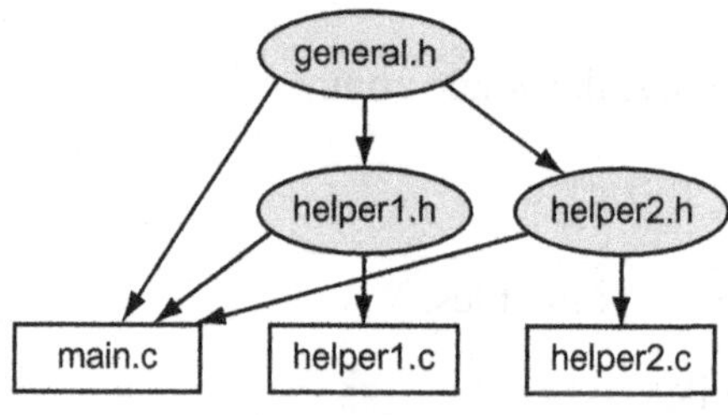

Figure A.3
An example project consisting of three C files and three header files. Arrows point from header files to files that include them.

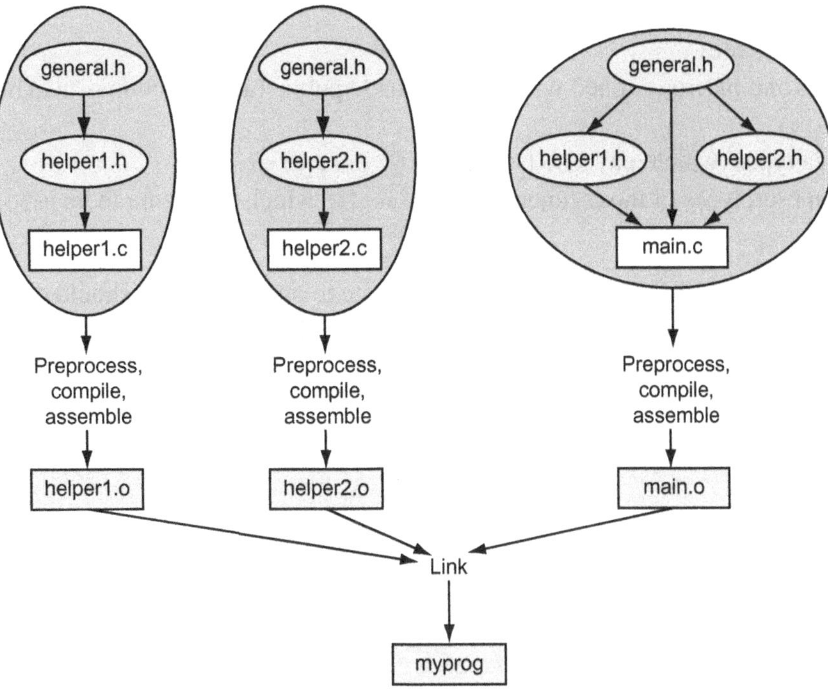

Figure A.4
The building of the project in Figure A.3.

Another benefit of splitting your project comes during compilation. If a `.c` file and the header files it includes do not change, then the `.c` file need not be compiled every time you build your project; the existing `.o` file may be used during the linking stage.

According to Figures A.3 and A.4, `main.c` has the following preprocessor commands:

```
#include "general.h"
#include "helper1.h"
#include "helper2.h"
```

The preprocessor replaces these commands with copies of the files `general.h`, `helper1.h`, and `helper2.h`. But when it includes `helper1.h`, it finds that `helper1.h` tries to include a second copy of `general.h` (see Figure A.3; `helper1.h` has a `#include "general.h"` command). Since `general.h` has already been copied in, it should not be copied again; otherwise we would have multiple copies of the same function prototypes, constant definitions, etc. The include guards prevent this duplication from happening.

In summary, the `general.h`, `helper1.h`, and `helper2.h` header files contain definitions that are made public to files including them. We might see the following items in the `helper1.h` header file, for example:

- an include guard
- other include files
- constants and macros defined with `#define` made public (and which may also be used by `helper1.c`)
- new data types (which may also be used by `helper1.c`)
- function prototypes of those functions in `helper1.c` which are meant to be used by other files

If a variable, function prototype, or constant is private to one C file, you should define it with the `static` keyword in the C file and not include it in the header file.

A header file like `helper1.h` could also have the declaration

```
extern int Helper1_Global_Var;   // no space is allocated by this declaration
```

where `helper1.c` has the global variable definition

```
int Helper1_Global_Var;          // space is allocated by this definition
```

Then any file including `helper1.h` would have access to the global variable `Helper1_Global_Var` allocated by `helper1.c`. Global variables defined in `helper1.c` with the `static` keyword are private to `helper1.c` and cannot be accessed by other files. If you have to use global variables (which should be avoided generally), declare them `static` whenever possible. Only in rare circumstances should you need to use `extern`.

Makefiles

When you are ready to build your executable, you can type the `gcc` commands at the command line, as we have seen previously. A `Makefile` simplifies the process, particularly for multi-file projects, by specifying the dependencies and commands needed to build the project. A `Makefile` for our rad2volume example is shown below, where everything after a # is a comment.

```
# ***** file:  Makefile *****
# Comment:  This is the simplest of Makefiles!

# Here is a template:
# [target]:  [dependencies]
# [tab] [command to execute]

# The thing to the left of the colon in the first line is what is created,
```

```
# and the thing(s) to the right of the colon are what it depends on.  The second
# line is the action to create the target.  If the things it depends on
# haven't changed since the target was last created, no need to do the action.
# Note:  The tab spacing in the second line is important!  You can't just use
# individual spaces.

# "make myprog" or "make" links two object codes to create the executable
myprog:  main.o rad2volume.o
	gcc main.o rad2volume.o -o myprog

# "make main.o" produces main.o object code; depends on main.c and rad2volume.h
main.o:  main.c rad2volume.h
	gcc -c main.c -o main.o

# "make rad2volume.o" produces rad2volume.o; depends on one .c and one h file
rad2volume.o:  rad2volume.c rad2volume.h
	gcc -c rad2volume -o rad2volume.o

# "make clean" throws away any object files to ensure make from scratch
clean:
	rm *.o
```

With this `Makefile` in the same directory as your other files, you should be able to type the command `make [target]`,[26] where `[target]` is `myprog`, `main.o`, `rad2volume.o`, or `clean`. If the `target` depends on other files, `make` will make sure those are up to date first, and if not, it will call the commands needed to create them. For example, `make myprog` triggers a check of `main.o`, which triggers a check of `main.c` and `rad2volume.h`. If either of those have changed since the last time `main.o` was made, then `main.c` is compiled and assembled to create a new `main.o` before the linking step.

The command `make` with no target specified will make the first target (which is `myprog` in this case).

Ensure that your `Makefile` is saved without any extensions (e.g., `.txt`) and that the commands are preceded by a tab (not spaces).

There are many more sophisticated uses of `Makefiles` which you can learn about from other sources.

A.5 Exercises

1. Install C, create the `HelloWorld.c` program, and compile and run it.
2. Explain what a pointer variable is, and how it is different from a non-pointer variable.
3. Explain the difference between interpreted and compiled code.

[26] In some C installations `make` is named differently, like `nmake` for Visual Studio or `mingw32-make`. If you can find no version of `make`, you may not have selected the `make` tools installation option when you performed the C installation.

4. Write the following hexadecimal (base-16) numbers in eight-digit binary (base-2) and three-digit decimal (base-10). Also, for each of the eight-digit binary representations, give the value of the most significant bit. (a) 0x1E. (b) 0x32. (c) 0xFE. (d) 0xC4.
5. What is 333_{10} in binary and 1011110111_2 in hexadecimal? What is the maximum value, in decimal, that a 12-bit number can hold?
6. Assume that each byte of memory can be addressed by a 16-bit address, and every 16-bit address has a corresponding byte in memory. How many total bits of memory do you have?
7. (Consult the ASCII table.) Let `ch` be of type `char`. (a) The assignment `ch = 'k'` can be written equivalently using a number on the right side. What is that number? (b) The number for `'5'`? (c) For `'='`? (d) For `'?'`?
8. What is the range of values for an `unsigned char`, `short`, and `double` data type?
9. How many bits are used to store a `char`, `short`, `int`, `float`, and `double`?
10. Explain the difference between `unsigned` and `signed` integers.
11. (a) For integer math, give the pros and cons of using `char`s vs. `int`s. (b) For floating point math, give the pros and cons of using `float`s vs. `double`s. (c) For integer math, give the pros and cons of using `char`s vs. `float`s.
12. The following `signed short int`s, written in decimal, are stored in two bytes of memory using the two's complement representation. For each, give the four-digit hexadecimal representation of those two bytes. (a) 13. (b) 27. (c) −10. (d) −17.
13. The smallest positive integer that cannot be exactly represented by a four-byte IEEE 754 `float` is $2^{24} + 1$, or 16,777,217. Explain why.
14. Give the four bytes, in hex, that represent the following decimal values: (a) 20 as an `unsigned int`. (b) −20 as a two's complement `signed int`. (c) 1.5 as an IEEE 754 `float`. (d) 0 as an IEEE 754 `float`.

    To verify your answers, use the program `typereps.c` below. This program allows the user to enter four bytes as eight hex characters, then prints the value of those four bytes when they are interpreted as an `unsigned int`, a two's complement `signed int`, an IEEE 754 `float`, or four consecutive `char`s. To do this, the program creates a new data type `four_types_t` consisting of four bytes, or 32 bits. These same 32 bits are interpreted as either an `unsigned int`, `int`, `float`, or four `char`s depending on whether we reference the bits using the fields `u`, `i`, `f`, or `char0`–`char3` in the `union`.

```
#include <stdio.h>

typedef union {    // a new data type consisting of four bytes
  unsigned int u; // the 32 bits interpreted as an unsigned int
  int i;          // the same 32 bits interpreted as a two's complement int
  float f;        // the same 32 bits interpreted as an IEEE 754 single prec float
  struct {
    char char0:8; // bits  0 -  7 interpreted as char, called char0
    char char1:8; // bits  8 - 15 interpreted as char, called char1
    char char2:8; // bits 16 - 23 interpreted as char, called char2
```

```
    char char3:8; // bits 24 - 31 interpreted as char, called char3
  };
} four_types_t;  // the new type is called four_types_t

int main(void) {
  four_types_t val;

  while (1) {      // exit the infinite loop using ctrl-c or similar
    printf("Enter four bytes as eight hex characters 0-f, e.g., abcd0123:  ");
    scanf("%x",&val.u);
    printf("\nThe 32 bits in hex:                          %x\n",val.u);
    printf("The 32 bits as an unsigned int, in decimal:  %u\n",val.u);
    printf("The 32 bits as a signed int, in decimal:     %d\n",val.i);
    printf("The 32 bits as a float:                      %.20f\n",val.f);
    printf("The 32 bits as 4 chars:                      %c %c %c %c\n\n",
           val.char3, val.char2, val.char1, val.char0);
  }
  return 0;
}
```

Below is a sample output. Note that only the ASCII values 32-126 have a visible printed representation, so the printouts as `char`s are meaningless in the first two examples.

```
Enter four bytes as eight hex characters 0-f, e.g., abcd0123: c0000000

The 32 bits in hex:                          c0000000
The 32 bits as an unsigned int, in decimal:  3221225472
The 32 bits as a signed int, in decimal:     -1073741824
The 32 bits as a float:                      -2.00000000000000000000
The 32 bits as 4 chars:                      ?

Enter four bytes as eight hex characters 0-f, e.g., abcd0123: ff800000

The 32 bits in hex:                          ff800000
The 32 bits as an unsigned int, in decimal:  4286578688
The 32 bits as a signed int, in decimal:     -8388608
The 32 bits as a float:                      -inf
The 32 bits as 4 chars:                      ? ?

Enter four bytes as eight hex characters 0-f, e.g., abcd0123: 48494a4b

The 32 bits in hex:                          48494a4b
The 32 bits as an unsigned int, in decimal:  1212762699
The 32 bits as a signed int, in decimal:     1212762699
The 32 bits as a float:                      206121.17187500000000000000
The 32 bits as 4 chars:                      H I J K
```

You can easily modify the code to allow the user to enter the four bytes as a `float` or `int` to examine their hex representations.

15. Write a program that prints out the sign, exponent, and significand bits of the IEEE 754 representation of a `float` entered by the user.
16. Technically the data type of a pointer to a `double` is "pointer to type `double`." Of the common integer and floating point data types discussed in this chapter, which is the most similar to this pointer type? Assume pointers occupy eight bytes.

17. To keep things simple, let us assume we have a microcontroller with only $2^8 = 256$ bytes of RAM, so each address is given by a single byte. Now consider the following code defining four global variables:

```
unsigned int i, j, *kp, *np;
```

Let us assume that the linker places `i` in addresses 0xB0..0xB3, `j` in 0xB4..0xB7, `kp` in 0xB8, and `np` in 0xB9. The code continues as follows:

```
                 // (a) the initial conditions, all memory contents unknown
kp = &i;         // (b)
j = *kp;         // (c)
i = 0xAE;        // (d)
np = kp;         // (e)
*np = 0x12;      // (f)
j = *kp;         // (g)
```

For each of the comments (a)-(g) above, give the contents (in hexadecimal) at the address ranges 0xB0..0xB3 (the `unsigned int i`), 0xB4..0xB7 (the `unsigned int j`), 0xB8 (the pointer `kp`), and 0xB9 (the pointer `np`), at that point in the program, after executing the line containing the comment. The contents of all memory addresses are initially unknown or random, so your answer to (a) is "unknown" for all memory locations. If it matters, assume little-endian representation.

18. Invoking the `gcc` compiler with a command like `gcc myprog.c -o myprog` actually initiates four steps. What are the four steps called, and what is the output of each step?
19. What is `main`'s return type, and what is the meaning of its return value?
20. Give the `printf` statement that will print out a `double d` with eight digits to the right of the decimal point and four spaces to the left.
21. Consider three `unsigned chars`, `i`, `j`, and `k`, with values 60, 80, and 200, respectively. Let `sum` also be an `unsigned char`. For each of the following, give the value of `sum` after performing the addition. (a) `sum = i+j;` (b) `sum = i+k;` (c) `sum = j+k;`
22. For the variables defined as

```
int a=2, b=3, c;
float d=1.0, e=3.5, f;
```

give the values of the following expressions. (a) `f = a/b;` (b) `f = ((float) a)/b;` (c) `f = (float) (a/b);` (d) `c = e/d;` (e) `c = (int) (e/d);` (f) `f = ((int) e)/d;`

23. In each snippet of code in (a)-(d), there is an arithmetic error in the final assignment of `ans`. What is the final value of `ans` in each case?

a.
```
char c = 17;
float ans = (1 / 2) * c;
```

b.
```
unsigned int ans = -4294967295;
```

c.
```
double d = pow(2, 16);
short ans = (short) d;
```

d.
```
double ans = ((double) -15 * 7) / (16 / 17) + 2.0;
```

24. Truncation is not always bad. Say you wanted to store a list of percentages rounded down to the nearest percent, but you were tight on memory and cleverly used an array of chars to store the values. For example, pretend you already had the following snippet of code:

```
char percent(int a, int b) {
  // assume a <= b
  char c;
  c = ???;
  return c;
}
```

You cannot simply write c = a / b. If $\frac{a}{b} = 0.77426$ or $\frac{a}{b} = 0.778$, then the correct return value is c = 77. Finish the function definition by writing a one-line statement to replace c = ???.

25. Explain why global variables work against modularity.
26. What are the seven sections of a typical C program?
27. You have written a large program with many functions. Your program compiles without errors, but when you run the program with input for which you know the correct output, you discover that your program returns the wrong result. What do you do next? Describe your systematic strategy for debugging.
28. Erase all the comments in invest.c, recompile, and run the program to make sure it still functions correctly. You should be able to recognize what is a comment and what is not. Turn in your modified invest.c code.
29. The following problems refer to the program invest.c. For all problems, you should modify the original code (or the code without comments from the previous problem) and run it to make sure you get the expected behavior. For each problem, turn in the modified portion of the code only.
    a. *Using* if, break *and* exit. Include the header file stdlib.h so we have access to the exit function (Section A.4.14). Change the while loop in main to be an infinite loop by inserting an expression <expr> in while(<expr>) that always evaluates to 1 (TRUE). (What is the simplest expression that evaluates to 1?) Now the first command inside the while loop gets the user's input. if the input is not valid, exit the program; otherwise continue. Next, change the exit command to a break command, and see the different behavior.
    b. *Accessing fields of a* struct. Alter main and getUserInput to set inv.invarray[0] in getUserInput, not main.
    c. *Using* printf. In main, before sendOutput, echo the user's input to the screen. For example, the program could print out You entered 100.00, 1.05, and 5.
    d. *Altering a string.* After the sprintf command of sendOutput, try setting an element of outstring to 0 before the printf command. For example, try setting the third element of outstring to 0. What happens to the output when you run the program? Now try setting it to '0' instead and see the behavior.

e. *Relational operators.* In `calculateGrowth`, eliminate the use of `<=` in favor of an equivalent expression that uses `!=`.
f. *Math.* In `calculateGrowth`, replace `i=i+1` with an equivalent statement using `+=`.
g. *Data types.* Change the fields `inv0`, `growth`, and `invarray[]` to be `float` instead of `double` in the definition of the `Investment` data type. Make sure you make the correct changes everywhere else in the program.
h. *Pointers.* Change `sendOutput` so that the second argument is of type `int *`, i.e., a pointer to an integer, instead of an integer. Make sure you make the correct changes everywhere else in the program.
i. *Conditional statements.* Use an `else` statement in `getUserInput` to print `Input is valid` if the input is valid.
j. *Loops.* Change the `for` loop in `sendOutput` to an equivalent `while` loop.
k. *Logical operators.* Change the assignment of `valid` to an equivalent statement using `||` and `!`, and no `&&`.

30. Consider this array definition and initialization:

```
int x[4] = {4, 3, 2, 1};
```

For each of the following, give the value or write "error/unknown" if the compiler will generate an error or the value is unknown. (a) `x[1]` (b) `*x` (c) `*(x+2)` (d) `(*x)+2` (e) `*x[3]` (f) `x[4]` (g) `*(&(x[1]) + 1)`

31. For the (strange) code below, what is the final value of `i`? Explain why.

```
int i,k=6;
i = 3*(5>1) + (k=2) + (k==6);
```

32. As the code below is executed, give the value of `c` in hex at the seven break points indicated, (a)-(g).

```
unsigned char a=0x0D, b=0x03, c;
c = ~a;       // (a)
c = a & b;    // (b)
c = a | b;    // (c)
c = a ^ b;    // (d)
c = a >> 3;   // (e)
c = a << 3;   // (f)
c &= b;       // (g)
```

33. In your C installation, or by searching on the web, find a listing of the header file `stdio.h`. Find the function prototype for one function provided by the library, but not mentioned in this appendix, and describe what that function does.
34. Write a program to generate the ASCII table for values 33 to 127. The output should be two columns: the left side with the number and the right side with the corresponding character. Turn in your code and the output of the program.
35. We will write a simple *bubble sort* program to sort a string of text in ascending order according to the ASCII table values of the characters. A bubble sort works as follows.

Given an array of n elements with indexes 0 to $n - 1$, we start by comparing elements 0 and 1. If element 0 is greater than element 1, we swap them. If not, leave them where they are. Then we move on to elements 1 and 2 and do the same thing, etc., until finally we compare elements $n - 2$ and $n - 1$. After this, the largest value in the array has "bubbled" to the last position. We now go back and do the whole thing again, but this time only comparing elements 0 up to $n - 2$. The next time, elements 0 to $n - 3$, etc., until the last time through we only compare elements 0 and 1.

Although this simple program `bubble.c` could be written in one function (`main`), we are going to break it into some helper functions to get used to using them. The function `getString` will get the input from the user; the function `printResult` will print the sorted result; the function `greaterThan` will check if one element is greater than another; and the function `swap` will swap two elements in the array. With these choices, we start with an outline of the program that looks like this.

```
#include <stdio.h>
#include <string.h>
#define MAXLENGTH 100        // max length of string input

void getString(char *str);  // helper prototypes
void printResult(char *str);
int greaterThan(char ch1, char ch2);
void swap(char *str, int index1, int index2);

int main(void) {
  int len;                  // length of the entered string
  char str[MAXLENGTH];      // input should be no longer than MAXLENGTH
  // here, any other variables you need

  getString(str);
  len = strlen(str);        // get length of the string, from string.h
  // put nested loops here to put the string in sorted order
  printResult(str);
  return(0);
}

// helper functions go here
```

Here's an example of the program running. Everything after the first colon is entered by the user. Blank spaces are written using an underscore character, since `scanf` assumes that the string ends at the first whitespace.

```
Enter the string you would like to sort:  This_is_a_cool_program!
Here is the sorted string:  !T____aacghiilmoooprrss
```

Complete the following steps in order. Do not move to the next step until the current step is successful.

a. Write the helper function `getString` to ask the user for a string and place it in the array passed to `getString`. You can use `scanf` to read in the string. Write a simple call in `main` to verify that `getString` works as you expect before moving on.

b. Write the helper function printResult and verify that it works correctly.
c. Write the helper function greaterThan and verify that it works correctly.
d. Write the helper function swap and verify that it works correctly.
e. Now define the other variables you need in main and write the nested loops to perform the sort. Verify that the whole program works as it should.

Turn in your final documented code and an example of the output of the program.

36. A more useful sorting program would take a series of names (e.g., Doe_John) and scores associated with them (e.g., 98) and then list the names and scores in two columns in descending order. Modify your bubble sort program to do this. The user enters a name string and a number at each prompt. The user indicates that there are no more names by entering 0 0.

    Your program should define a constant MAXRECORDS which contains the maximum number of records allowable. You should define an array, MAXRECORDS long, of struct variables, where each struct has two fields: the name string and the score. Write your program modularly so that there is at least a sort function and a readInput function of type int that returns the number of records entered.

    Turn in your code and example output.

37. Modify the previous program to read the data in from a file using fscanf and write the results out to another file using fprintf. Turn in your code and example output.

38. Consider the following lines of code:

```
int i, tmp, *ptr, arr[7] = {10, 20, 30, 40, 50, 60, 70};

ptr = &arr[6];
for(i = 0; i < 4; i++) {
  tmp = arr[i];
  arr[i] = *ptr;
  *ptr = tmp;
  ptr--;
}
```

a. How many elements does the array arr have?
b. How would you access the middle element of arr and assign its value to the variable tmp? Do this two ways, once indexing into the array using [] and the other with the dereferencing operator and some pointer arithmetic. Your answer should only be in terms of the variables arr and tmp.
c. What are the contents of the array arr before and after the loop?

39. The following questions pertain to the code below. For your responses, you only need to write down the changes you would make using valid C code. You should verify that your modifications actually compile and run correctly. Do not submit a full C program for this question. Only write the changes you would make using legitimate C syntax.

```
#include <stdio.h>
#define MAX 10

void MyFcn(int max);
```

```
int main(void) {
  MyFcn(5);
  return(0);
}

void MyFcn(int max) {
  int i;
  double arr[MAX];

  if(max > MAX) {
printf("The range requested is too large.  Max is %d.\n", MAX);
return;
  }
  for(i = 0; i < max; i++) {
    arr[i] = 0.5 * i;
    printf("The value of i is %d and %d/2 is %f.\n", i, i, arr[i]);
  }
}
```

a. `while` loops and `for` loops are essentially the same thing. How would you write an equivalent `while` loop that replicates the behavior of the `for` loop?
b. How would you modify the `main` function so that it reads in an integer value from the keyboard and then passes the result to `MyFcn`? (This replaces the statement `MyFcn(5);`.) If you need to use extra variables, make sure to define them before you use them in your snippet of code.
c. Change `main` so that if the input value from the keyboard is between −`MAX` and `MAX`, you call `MyFcn` with the absolute value of the input. If the input is outside this range, then you simply call `MyFcn` with the value `MAX`. How would you make these changes using conditional statements?
d. In C, you will often find yourself writing nested loops (a loop inside a loop) to accomplish a task. Modify the `for` loop to use nested loops to set the ith element in the array `arr` to half the sum of the first $i-1$ integers, i.e., `arr[i]` $= \frac{1}{2}\sum_{j=0}^{i-1} j$. (You can easily find a formula for this that does not require the inner loop, but you should use nested loops for this problem.) The same loops should print the value of each `arr[i]` to 2 decimal places using the `%f` formatting directive.

40. If there are n people in a room, what is the chance that two of them have the same birthday? If $n = 1$, the chance is zero, of course. If $n > 366$, the chance is 100%. Under the assumption that births are distributed uniformly over the days of the year, write a program that calculates the chances for values of $n = 2$ to 100. What is the lowest value n^* such that the chance is greater than 50%? (The surprising result is sometimes called the "birthday paradox.") If the distribution of births on days of the year is not uniform, will n^* increase or decrease? Turn in your answer to the questions as well as your C code and the output.
41. In this problem you will write a C program that solves a "puzzler" that was presented on NPR's CarTalk radio program. In a direct quote of their radio transcript, found here

`http://www.cartalk.com/content/hall-lights?question`, the problem is described as follows:

> ***RAY***: *This puzzler is from my "ceiling light" series. Imagine, if you will, that you have a long, long corridor that stretches out as far as the eye can see. In that corridor, attached to the ceiling are lights that are operated with a pull cord.*
>
> *There are gazillions of them, as far as the eye can see. Let us say there are 20,000 lights in a row.*
>
> *They're all off. Somebody comes along and pulls on each of the chains, turning on each one of the lights. Another person comes right behind, and pulls the chain on every second light.*
>
> ***TOM***: *Thereby turning off lights 2, 4, 6, 8 and so on.*
>
> ***RAY***: *Right. Now, a third person comes along and pulls the cord on every third light. That is, lights number 3, 6, 9, 12, 15, etc. Another person comes along and pulls the cord on lights number 4, 8, 12, 16 and so on. Of course, each person is turning on some lights and turning other lights off.*
>
> *If there are 20,000 lights, at some point someone is going to come skipping along and pull every 20,000th chain.*
>
> *When that happens, some lights will be on, and some will be off. Can you predict which lights will be on?*

You will write a C program that asks the user the number of lights n and then prints out which of the lights are on, and the total number of lights on, after the last (nth) person goes by. Here's an example of what the output might look like if the user enters 200:

```
How many lights are there? 200

You said 200 lights.
Here are the results:
  Light number   1 is on.
  Light number   4 is on.
  ...
  Light number 196 is on.
  There are 14 total lights on!
```

Your program `lights.c` should follow the template outlined below. Turn in your code and example output.

```
/**************************************************************************
 * lights.c
 *
 * This program solves the light puzzler.  It uses one main function
 * and two helper functions:  one that calculates which lights are on,
 * and one that prints the results.
 *
 **************************************************************************/

#include <stdio.h>
#include <stdlib.h>          // allows the use of the "exit()" function
```

```
#define MAX_LIGHTS 1000000  // maximum number of lights allowed

// here's a prototype for the light toggling function
// here's a prototype for the results printing function

int main(void) {

  // Define any variables you need, including for the lights' states

  // Get the user's input.
  // If it is not valid, say so and use "exit()" (stdlib.h, Sec 1.2.16).
  // If it is valid, echo the entry to the user.

  // Call the function that toggles the lights.
  // Call the function that prints the results.

  return(0);
}

// definition of the light toggling function
// definition of the results printing function
```

42. We have been preprocessing, compiling, assembling, and linking programs with commands like

```
gcc HelloWorld.c -o HelloWorld
```

The `gcc` command recognizes the first argument, `HelloWorld.c`, is a C file based on its `.c` extension. It knows you want to create an output file called `HelloWorld` because of the `-o` option. And since you did not specify any other options, it knows you want that output to be an executable. So it performs all four of the steps to take the C file to an executable. We could have used options to stop after each step if we wanted to see the intermediate files produced. Below is a sequence of commands you could try, starting with your `HelloWorld.c` code. Do not type the "comments" to the right of the commands!

```
> gcc HelloWorld.c -E > HW.i  // stop after preprocessing, dump into file HW.i
> gcc HW.i -S -o HW.s         // compile HW.i to assembly file HW.s and stop
> gcc HW.s -c -o HW.o         // assemble HW.s to object code HW.o and stop
> gcc HW.o -o HW              // link with stdio printf code, make executable HW
```

At the end of this process you have `HW.i`, the C code after preprocessing (`.i` is a standard extension for C code that should not be preprocessed); `HW.s`, the assembly code corresponding to `HelloWorld.c`; `HW.o`, the unreadable object code; and finally the executable code `HW`. The executable is created from linking your `HW.o` object code with object code from the `stdio` (standard input and output) library, specifically object code for `printf`.

Try this and verify that you see all the intermediate files, and that the final executable works as expected. (An easier way to generate the intermediate files is to use `gcc`

`HelloWorld.c -save-temps -o HelloWorld`, where the `-save-temps` option saves the intermediate files.)

If our program used any math functions, the final linker command would be

```
> gcc HW.o -o HW -lm          // link with stdio and math libraries, make
                                 executable HW
```

The C standard library is linked automatically, but often the math library is not, requiring the extra `-lm` option.

The `HW.i` and `HW.s` files can be inspected with a text editor, but the object code `HW.o` and executable `HW` cannot. We can try the following commands to make viewable versions:

```
> xxd HW.o v1.txt            // can't read obj code; this makes viewable v1.txt
> xxd HW v2.txt              // can't read executable; make viewable v2.txt
```

The utility `xxd` just turns the first file's string of 0's and 1's into a string of hex characters, represented as text-editor-readable ASCII characters 0..9, A..F. It also has an ASCII sidebar: when a byte (two consecutive hex characters) has a value corresponding to a printable ASCII character, that character is printed. You can even see your message "Hello world!" buried there!

Take a quick look at the `HW.i`, `HW.s`, and `v1.txt` and `v2.txt` files. No need to understand these intermediate files any further. If you do not have the `xxd` utility, you could create your own program `hexdump.c` instead:

```
#include <stdio.h>
#define BYTES_PER_LINE 16

int main(void) {
  FILE *inputp, *outputp;                        // ptrs to in and out files
  int c, count = 0;
  char asc[BYTES_PER_LINE+1], infile[100];

  printf("What binary file do you want the hex rep of? ");
  scanf("%s",infile);                            // get name of input file
  inputp = fopen(infile,"r");                    // open file as "read"
  outputp = fopen("hexdump.txt","w");            // output file is "write"

  asc[BYTES_PER_LINE] = 0;                       // last char is end-string
  while ((c=fgetc(inputp)) != EOF) {             // get byte; end of file?
    fprintf(outputp,"%x%x ",(c >> 4),(c & 0xf)); // print hex rep of byte
    if ((c>=32) && (c<=126)) asc[count] = c;     // put printable chars in asc
    else asc[count] = '.';                       // otherwise put a dot
    count++;
    if (count==BYTES_PER_LINE) {                 // if BYTES_PER_LINE reached
      fprintf(outputp,"  %s\n",asc);             // print ASCII rep, newline
      count = 0;
    }
  }
  if (count!=0) {                                // print last (short) line
    for (c=0; c<BYTES_PER_LINE-count; c++)       // print extra spaces
      fprintf(outputp,"   ");
```

```
    asc[count]=0;                                // add end-string char to asc
    fprintf(outputp,"  %s\n",asc);               // print ASCII rep, newline
  }
  fclose(inputp);                                // close files
  fclose(outputp);
  printf("Printed hexdump.txt.\n");
  return(0);
}
```

Further Reading

Bronson, G. J. (2006). *A first book of ANSI C* (4th ed.). Boston, MA: Course Technology Press.

Kernighan, B. W., & Ritchie, D. M. (1988). *The C programming language* (2nd ed.). Upper Saddle River, NJ: Prentice Hall.

APPENDIX B

Circuits Review

This appendix is meant as a brief refresher on basic analysis of circuits with resistors, capacitors, inductors, diodes, bipolar junction transistors, and operational amplifiers, at the level they are used in this book. In particular, this appendix does not cover the general frequency response of circuits with complex impedance. It also does not cover digital circuit design; in this book, most logical operations are performed by the PIC32.

B.1 Basics

Primary quantities of interest in circuit analysis and design are **voltage** and **current**.

Voltage is an effort variable, analogous to force in mechanical systems. In fact, voltage is sometimes referred to as *electromotive force*. Just as a force causes a mass to move, a voltage causes electrons (and therefore current) to flow. The unit of voltage is a Volt (V). Voltage is measured across elements (e.g., the voltage, or potential, at the positive terminal of a 9 V battery is 9 V greater than at the negative terminal). By defining the voltage at a particular point in a circuit as 0 V, or ground (GND), it is possible to refer to the voltage at a point, implicitly comparing it to ground.

Current is a flow variable, analogous to velocity in mechanical systems. The unit of current is an Ampere (A), commonly shortened to amp. Current is measured as a flow through circuit elements. Current into a circuit element must equal the current coming out of the element, and therefore current can only flow around a closed loop. It cannot, for example, flow into an element and stop there.

Just as force times velocity is power in mechanical systems, voltage times current is power in electrical systems. The unit of power is the Watt (W), and 1 W = 1 A × 1 V. For example, Figure B.1 shows a generic circuit element (perhaps a battery, resistor, capacitor, diode, etc.). The voltage V across the element is defined to be positive if the potential is higher at the end of the element labeled +; otherwise V is negative. The current I through the element is defined to be positive if it is in the direction of the arrow, from + to −; otherwise I is negative.[1] With

[1] Note: The labeling of the ends of the element as + and − does *not* necessarily indicate which end has higher potential. It just indicates the convention chosen to call the voltage positive or negative. Similarly, the arrow does not necessarily indicate which direction the current actually flows.

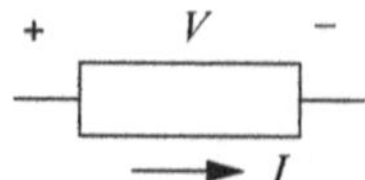

Figure B.1
Defining positive voltage V across, and current I through, a circuit element.

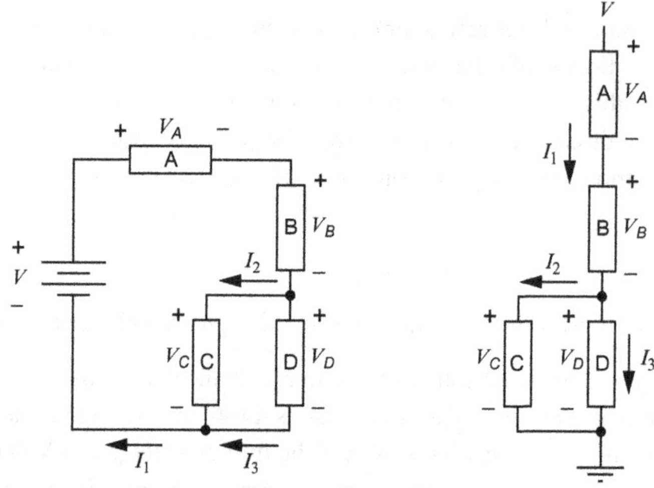

Figure B.2
The same circuit drawn two different ways, with a battery of voltage V and four generic elements, A through D.

these conventions, the power consumed by the element is $P = IV$. When $IV > 0$, the element is consuming electrical power, either dissipating it (as with a resistor) or storing it as energy (as with a capacitor or inductor). When $IV < 0$, the element is providing electrical power (e.g., a battery or a discharging capacitor). The unit of energy is the Joule (J), and 1 J = 1 W × 1 s.

Figure B.2 shows a circuit with a battery of voltage V and four generic elements, labeled A through D. The voltages across the elements are V_A through V_D. The current I_1 flows through the battery, A, and B; I_2 flows through C; and I_3 flows through D. The same circuit is drawn in two different ways. In one, a battery is drawn explicitly, allowing a closed loop for current to be clearly visualized. In the other, which is more common in circuit schematics, the closed loop through the battery is left implicit. This circuit also introduces the ground symbol the voltage level defined as zero volts (at the negative terminal of the battery in this case).

To solve for voltages and currents in this circuit, we use *Kirchhoff's current law* (KCL) and *Kirchhoff's voltage law* (KVL). KCL says that current is preserved at any node: current into the node is equal to current out of the node. In Figure B.2, there are two nodes where currents come together, indicated by dots, and each provides the same equation:

$$I_1 = I_2 + I_3.$$

KVL says that the sum of voltages around any closed loop must be zero. As you step around a loop, add voltages from elements where you proceed from the − terminal to the + terminal, and subtract voltages from elements where you proceed from the + terminal to the − terminal. For example, there are three loops in Figure B.2: through the battery, A, B, and C; through the battery, A, B, and D; and through C and D. These yield the following equations, respectively:

$$\begin{aligned} V - V_A - V_B - V_C &= 0 \\ V - V_A - V_B - V_D &= 0 \\ V_C - V_D &= 0. \end{aligned}$$

Only two of these equations are independent. For example, the third equation can be used to show that the first two are equivalent.

We now have three independent equations (one from KCL and two from KVL) to solve for seven unknowns in the circuit: the three currents I_1, I_2, and I_3, and the four voltages across the elements, V_A, V_B, V_C, and V_D. To get four more equations, we need the *constitutive laws* of the elements, relating the voltages across the elements to the currents through them. Let us begin with the constitutive laws of the common linear circuit elements: resistors, capacitors, and inductors.

B.2 Linear Elements: Resistors, Capacitors, and Inductors

Resistors, capacitors, and inductors are called *linear* circuit elements because the voltages across the elements are proportional to the current, time integral of the current, or derivative of the current, respectively. The symbols, units, constitutive laws, and information about power and energy are summarized in Table B.1. Resistors only dissipate power, as heat, while

Table B.1: The three linear circuit elements and the constitutive laws relating the current *I* through them and the voltage *V* across them

Element	Schematic	Symbol	Unit	Constitutive Law	Power (W)	Energy Stored (J)
Resistor	+ −	R	Ohm (Ω)	$V = IR$	I^2R dissipated	0
Capacitor	+ −	C	Farad (F)	$I = C\frac{dV}{dt}$	$CV\frac{dV}{dt}$	$\frac{1}{2}CV^2$
Inductor	+ −	L	Henry (H)	$V = L\frac{dI}{dt}$	$LI\frac{dI}{dt}$	$\frac{1}{2}LI^2$

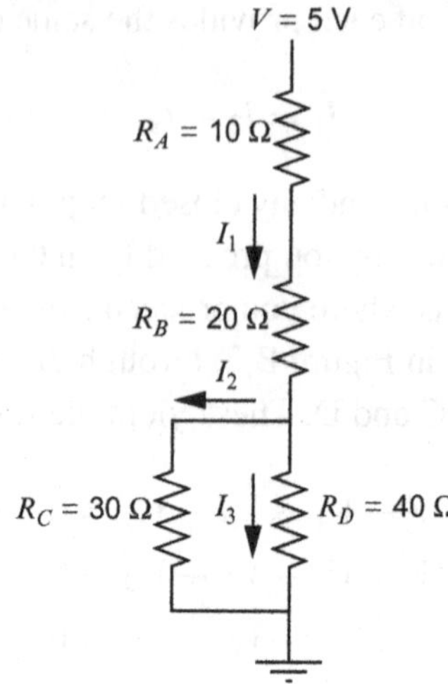

Figure B.3
A resistor network.

capacitors and inductors do not dissipate any power, but either charge (consuming electrical power and storing it as energy) or discharge (providing electrical power).

Figure B.3 shows the circuit of Figure B.2 with the generic elements replaced by resistors. We can solve for the four voltages across the resistors and the three currents by simultaneously solving the seven KCL, KVL, and constitutive law equations:

$$\begin{aligned}
\text{KCL: } I_1 &= I_2 + I_3 \\
\text{KVL: } 0 &= V - V_A - V_B - V_D \\
0 &= V_C - V_D \\
\text{Constitutive laws: } V_A &= I_1 R_A \\
V_B &= I_1 R_B \\
V_C &= I_2 R_C \\
V_D &= I_3 R_D
\end{aligned}$$

Substituting the battery voltage $V = 5$ V and the resistances $R_A = 10\ \Omega$, $R_B = 20\ \Omega$, $R_C = 30\ \Omega$, and $R_D = 40\ \Omega$, the currents and voltages can be solved as

$$\begin{aligned}
&I_1 = 0.106 \text{ A}, \quad I_2 = 0.061 \text{ A}, \quad I_3 = 0.045 \text{ A} \\
&V_A = 1.061 \text{ V}, \quad V_B = 2.121 \text{ V}, \quad V_C = 1.818 \text{ V}, \quad V_D = 1.818\ V.
\end{aligned}$$

Table B.2: Equivalent resistance, capacitance, and inductance of elements in series and parallel

Elements	In Series	In Parallel
Resistors R_1, R_2	$R_1 + R_2$	$R_1 R_2/(R_1 + R_2)$
Capacitors C_1, C_2	$C_1 C_2/(C_1 + C_2)$	$C_1 + C_2$
Inductors L_1, L_2	$L_1 + L_2$	$L_1 L_2/(L_1 + L_2)$

According to our sign convention, where I is defined as positive if it flows from the terminal labeled + to the terminal labeled −, the power consumed by the battery is $-I_1 V = -(0.106\text{ A})(5\text{ V}) = -0.53$ W. Since the power consumed is negative, the battery is providing power. The power consumed by R_A is $I_1 V_A = I_1^2 R_A = 0.112$ W. The power consumed by R_B, R_C, and R_D can be calculated as 0.225, 0.112, and 0.081 W, respectively, and the sum of the power dissipated by the resistors is 0.53 W, equal to the power provided by the battery, as we would expect.

If any of the elements were capacitors or inductors, those constitutive laws would relate the current through a capacitor to the rate of change of the voltage, or the voltage across an inductor to the rate of change of current. Instead of simply solving linear equations as above, we must now solve linear differential equations. In this book we do not delve into analysis of linear circuits with arbitrary combinations of resistors, inductors, and capacitors, but focus on circuits with resistors only, as above, as well as circuits with resistors and either a single capacitor or a single inductor (Section B.2.1). Such circuits cover many practical cases of interest in mechatronics.

In Figure B.3, the resistors R_A and R_B are said to be *in series*. The resistors R_C and R_D are said to be *in parallel*. A simple derivation shows that resistors in series act like a single resistor of greater resistance, $R_{\text{series}} = R_A + R_B$, and resistors in parallel act like a single resistor of lesser resistance (since there are now two paths for the current to follow), $R_{\text{parallel}} = R_C R_D/(R_C + R_D)$. Similar relationships can be derived for capacitors and inductors (Table B.2).

The last linear element we will use is the *potentiometer*, or *pot* for short (Figure B.4). A pot is a resistor with three connections: the terminals at either end, like a regular resistor, and a third connection called the *wiper*. The wiper is an electrical contact that can slide from one end of the resistor to the other, creating a variable resistance between the wiper and the end connections. If R_+ is the resistance between the + terminal of the resistor and the wiper, and R_- is the resistance between the − terminal and the wiper, then the sum of R_+ and R_- always equals R, where R is the total resistance of the pot between the two ends. Pots often come packaged in rotary knobs, and turning the knob moves the wiper to allow R_+ and R_- to be varied from approximately 0 to R.

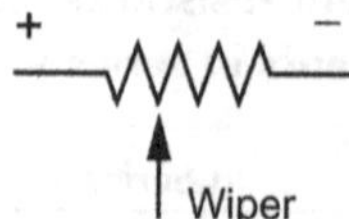

Figure B.4
Symbol for a potentiometer.

B.2.1 Time Response of RC and RL Circuits

Figure B.5(a)–(d) shows four circuits, each consisting of a resistor R and either a capacitor C or an inductor L. For mechatronics, circuits with resistors and a single inductor are important for understanding the behavior of motors, which have significant inductance. Circuits with resistors and a single capacitor are often used for signal filtering.

The circuits in Figure B.5(a)–(d) are powered by a time-varying voltage $V_{in}(t)$ that periodically switches between $V_{hi} > 0$ and 0 V relative to ground. Let us focus on the circuit in Figure B.5(a), where $V_{out} = V_C$ is the voltage across the capacitor.

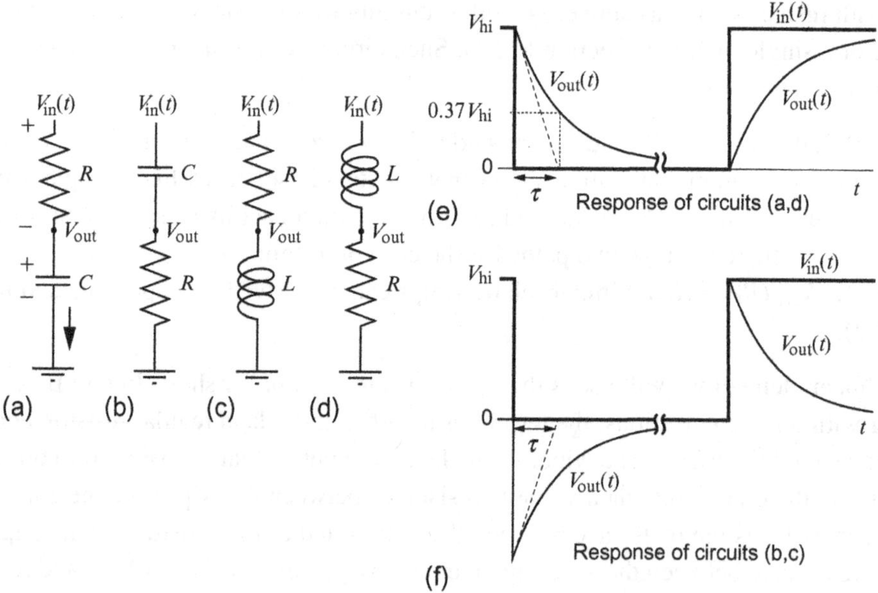

Figure B.5
(a–d) RC circuits and RL circuits. (e) Response of the circuits in (a) and (d) to a changing $V_{in}(t)$. Note the discontinuity in time in the middle of the plot, to allow the response to reach steady state. (f) Response of the circuits in (b) and (c) to a changing $V_{in}(t)$.

KVL tells us that the circuit in Figure B.5(a) satisfies

$$V_{in}(t) = V_R(t) + V_C(t) = I(t)R + \frac{1}{C}\int_0^t I(s)\,\mathrm{d}s.$$

Assume that the input voltage $V_{in}(t)$ is equal to V_{hi}, and has been for a long time so that any transients have died out. Current flowing through the capacitor has charged it up until, in steady state, the capacitor is fully charged to a voltage $V_C = V_{hi}$ and an energy $\frac{1}{2}CV_{hi}^2$. Therefore there is no voltage across the resistor, so $I = 0$. The capacitor is acting like an open circuit.

The key point is that the voltage across the capacitor $V_C(t)$ cannot be discontinuous in time if current is finite. For example, V_C cannot change from 0 to 5 V instantaneously; it takes time for the current to integrate to develop a voltage across the capacitor.

Now consider what happens when $V_{in}(t)$ instantly changes from V_{hi} to 0 V. Since the voltage across the capacitor cannot change instantaneously, just after the switch occurs, the voltage across the capacitor is still V_{hi}. This means that the voltage across the resistor R is $-V_{hi}$. Therefore current must be flowing from ground through the capacitor and resistor. By KVL, and the constitutive law of the resistor, we can calculate the current I just after the switch at time $t = 0$ (let us call this time 0^+):

$$V_{in}(0^+) = 0 = V_R(0^+) + V_C(0^+) = I(0^+)R + V_{hi} \quad \rightarrow \quad I(0^+) = -\frac{V_{hi}}{R}.$$

This negative current begins to discharge the energy stored in the capacitor, and therefore the voltage across the capacitor begins to drop. To solve for $\mathrm{d}V_C/\mathrm{d}t$, the rate of change of the voltage across the capacitor, at time 0^+, we use the constitutive law of the capacitor:

$$I(0^+) = -\frac{V_{hi}}{R} = C\frac{\mathrm{d}V_C}{\mathrm{d}t}(0^+) \quad \rightarrow \quad \frac{\mathrm{d}V_C}{\mathrm{d}t}(0^+) = -\frac{V_{hi}}{RC}.$$

If the capacitor continued to discharge at this rate, it would fully discharge in RC seconds.

Of course the capacitor does not continue to discharge at this rate; the rate slows as the voltage across the capacitor drops. To fully solve for $V_C(t)$ using KVL and the constitutive law of the capacitor, we solve the first-order linear differential equation

$$\begin{aligned} 0 &= I(t)R + V_C(t) \\ 0 &= C\frac{\mathrm{d}V_C}{\mathrm{d}t}(t)R + V_C(t) \\ \frac{\mathrm{d}V_C}{\mathrm{d}t}(t) &= -\frac{1}{RC}V_C(t) \end{aligned} \tag{B.1}$$

to get

$$V_C(t) = V_0 e^{-t/RC} \tag{B.2}$$

where the initial voltage V_0 is V_{hi}. This is an exponential decay to zero as $t \to \infty$. The *time constant* of the decay is $\tau = RC$, in seconds. One time constant after the switch, the voltage is $V_C(\tau) = V_0 e^{-1} = 0.37V_0$; the voltage has decayed by 63%. After 3τ, the voltage has decayed to 5% of its initial value.

Note that the time constant $\tau = RC$ is large if R is large, since the large resistance limits the current to charge or discharge the capacitor, or if C is large, because it takes more time for the current to charge or discharge energy in the capacitor, which is $\frac{1}{2}CV_C^2$.

If instead $V_{in}(t)$ has been at 0 V for a long time and then switches to V_{hi} at $t = 0$, a similar derivation yields

$$V_{out} = V_C(t) = V_{hi}(1 - e^{-t/RC}),$$

a rise from 0 V asymptoting at V_{hi}. The voltage across the capacitor rises to 63% (95%) of V_{hi} after time τ (3τ).

Figure B.5(e) shows a plot of the fall and rise of the voltage $V_{out}(t) = V_C(t)$ in the circuit in Figure B.5(a) in response to a $V_{in}(t)$ occasionally switching between V_{hi} and 0.

In Figure B.5(b), the positions of the capacitors and the resistors are reversed, so $V_{out} = V_R = V_{in}(t) - V_C(t)$. The response of $V_{out}(t)$ to the switching $V_{in}(t)$ is shown in Figure B.5(f). In this case, V_{out} spikes to $-V_{hi}$ on a falling edge of $V_{in}(t)$, then decays back to zero, and spikes to V_{hi} on a rising edge of $V_{in}(t)$, then decays back to zero.

In summary, the output of the circuit in Figure B.5(a) is a smoothed version of $V_{in}(t)$, where the output gets smoother as RC gets larger, while the output in circuit in Figure B.5(b) responds most strongly at the times of the switches of $V_{in}(t)$. Smoothing is characteristic of a *low-pass filter*, while strong response to signal changes is characteristic of a *high-pass filter*; see Section B.2.2.

The circuits in Figure B.5(c) and (d) can be analyzed similarly, now using the constitutive law $V_L = L\,dI/dt$ for the inductor instead of $I = C\,dV_C/dt$ for the capacitor. It is also important to realize that the inductor does not allow current to change discontinuously, as a discontinuous current implies an unbounded voltage $L\,dI/dt$ across the inductor. It takes time to charge or discharge the energy in the inductor, $\frac{1}{2}LI^2$, and therefore I cannot change instantaneously.

Based on this analysis, we see that the response of the RL circuit in Figure B.5(c) is that shown in Figure B.5(f), but now with a time constant $\tau = L/R$. Similarly, the response of the RL circuit in Figure B.5(d) is shown in Figure B.5(e), again with a time constant $\tau = L/R$.

Table B.3: Summary of capacitor and inductor behavior

Element	Rule Enforced	Discharged	Charged
Capacitor	Continuous voltage	Wire	Open circuit
Inductor	Continuous current	Open circuit	Wire

Note that the time constant is large if R is small, since the low resistance does not dissipate much power for a given current, or if L is large, since it takes longer to charge or discharge the inductor's energy $\frac{1}{2}LI^2$.

Some rules of analyzing circuits with a capacitor or inductor are summarized in Table B.3. When a capacitor is initially discharged, it lets current flow freely (like a wire), and when it is fully charged, it behaves like an open circuit (no current flows). When an inductor is initially discharged, it behaves like an open circuit (it takes time for current to build up as initially all voltage is claimed by $L\,\mathrm{d}I/\mathrm{d}t$), and when it is fully charged and $dI/dt = 0$, it lets current flow freely with no voltage across it (like a wire).

Application: Switch debouncing

Figure B.6 illustrates a closing mechanical switch, nominally generating a clean falling edge from GND to V. In practice, mechanical switches tend to *bounce*; the two metal contacts impact and bounce before coming into steady contact. The result is a $V_0(t)$ that rapidly switches between V and GND before settling at GND. Switch bounce is a common problem, and programs responding to button presses should not respond to the bounces.

To remedy the signal bounce, a debouncing circuit, as shown in Figure B.6, can be used. First, an RC filter is used to slow down the voltage variations, creating the signal $V_1(t)$. Because the

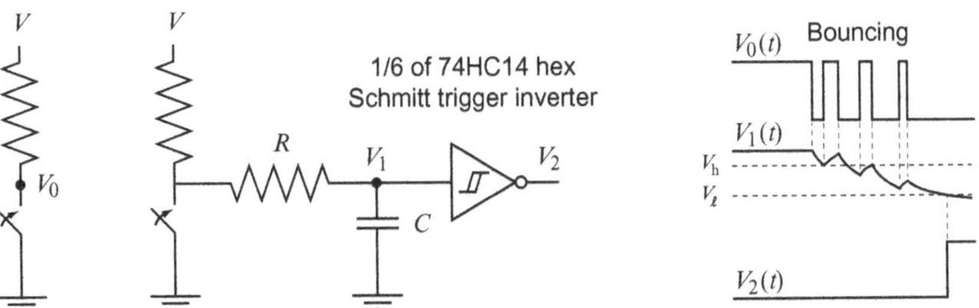

Figure B.6

(Left) Bounces on the closing of a mechanical switch generate the output signal $V_0(t)$ on the right. (Middle) A debouncing circuit. The bouncing signal is RC filtered, creating the signal $V_1(t)$. This signal then passes through a Schmitt trigger inverter, creating a single clean rising edge $V_2(t)$. (Right) The signals $V_0(t)$, $V_1(t)$, and $V_2(t)$. The Schmitt trigger hysteresis voltages V_h and V_ℓ are shown on the signal $V_1(t)$.

bouncing transitions occur quickly, the signal $V_1(t)$ changes little during the bouncing. Once the bouncing has ended, $V_1(t)$ drops steadily to zero, according to the RC time constant.

The signal V_1 is then fed to a digital output Schmitt trigger chip. The purpose of a Schmitt trigger is to implement *hysteresis*: if the input is currently low, the output does not change until the input has risen past a voltage V_h; if the input is high, the output does not change until the input has dropped below a voltage V_ℓ; and $V_h > V_\ell$. This hysteresis means that small variations of the input signal should not change the output signal. Further, a Schmitt trigger *inverter* makes the digital output opposite the input. The 74HC14 chip has six Schmitt trigger inverters on it.

Because the Schmitt trigger inverter ignores the small voltage variations at $V_1(t)$ during bouncing, it does not change its output until the switch contact is steady. The end result of the debouncing circuit, $V_2(t)$, is a single clean rising edge, after the bounces have terminated.

Since it is unlikely that you will need to press and release a button in less than 10 ms, it is not unreasonable to choose $RC \approx 10$ ms.

There are other debouncing circuits, and debouncing can instead be performed in software. See Exercise 16 of Chapter 6.

B.2.2 Frequency Response of RC and RL Circuits

In the previous section we focused on the time response of RC and RL circuits in response to step changes in voltage. The step response is helpful to understand, as microcontrollers and some sensors output digital signals. We should remember, however, that by Fourier decomposition, any periodic signal of frequency f can be represented by a sum of sinusoids of frequency f, $2f$, $3f$, etc. For example, the 50% duty cycle square wave of amplitude 1 and frequency f in Figure B.7 can be represented by an infinite sum of sinusoids at frequencies f, $3f$, $5f$, etc. Therefore it is useful to understand the behavior of circuits in response to sinusoidal inputs.

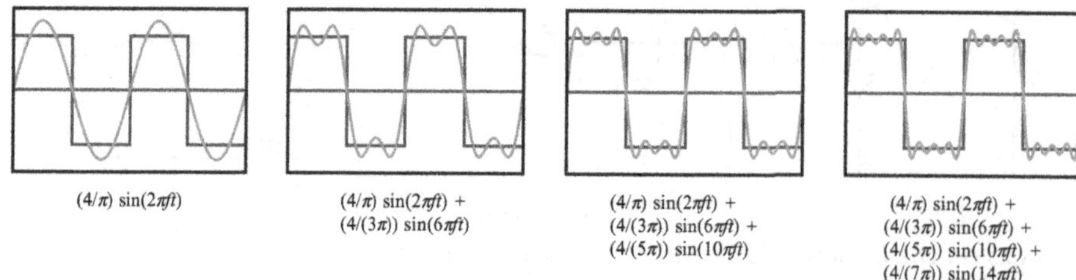

Figure B.7
The lowest four frequency components of a Fourier decomposition of a square wave of amplitude 1 and frequency f.

If the input to a linear system, like an RCL circuit, is a sinusoid of the form $V_{\text{in}}(t) = A\sin(2\pi ft)$, the output is a scaled, phase-shifted sinusoid of the same frequency, $V_{\text{out}}(t) = G(f)A\sin(2\pi ft + \phi(f))$, where the system's gain $G(f)$ and phase $\phi(f)$ are a function of the frequency f of the input (Figure B.8). Collectively the gain $G(f)$ and the phase $\phi(f)$ are called the *frequency response* of the system. For periodic non-sinusoidal input signals like the square wave in Figure B.7, the output is the sum of the individually scaled and shifted sinusoids that constitute the Fourier decomposition of the input.

Each of the RC and RL circuits in Figure B.5 is a linear system, with $V_{\text{in}}(t)$ as input and $V_{\text{out}}(t)$ as output. Without derivation (take a linear systems or circuits course!), the frequency responses of the circuits are plotted in Figure B.9, where Figure B.9(a) is the frequency response $G(f)$ and $\phi(f)$ of the circuits in Figure B.5(a) and (d), and Figure B.9(b) is the frequency response of the circuits in Figure B.5(b) and (c). Note that the frequency and gain

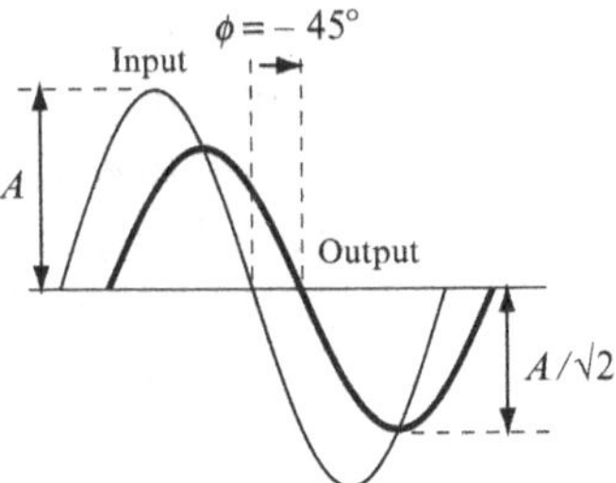

Figure B.8

An example linear system time response to a sinusoidal input of amplitude A. The output is phase shifted by $\phi = -45°$ and scaled by a gain $G = 1/\sqrt{2}$.

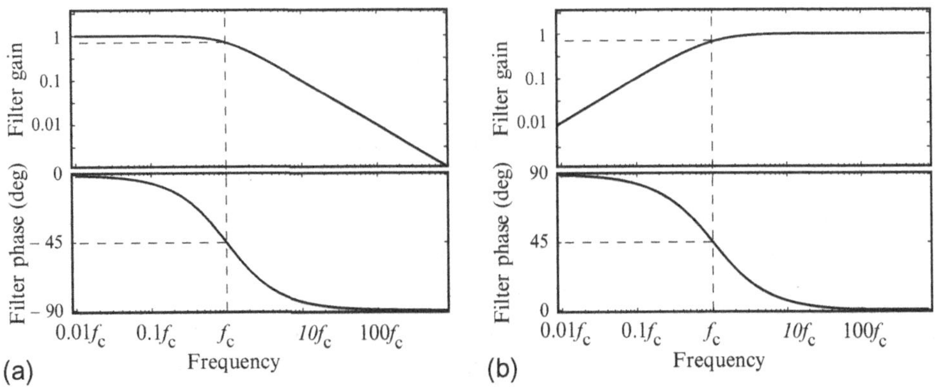

Figure B.9

(a) The frequency response of a first-order low-pass filter, e.g., the circuits in Figure B.5(a) and (d). (b) The frequency response of a first-order high-pass filter, e.g., the circuits in Figure B.5(b) and (c).

are plotted on log scales, to cover larger ranges of values and to more clearly show the essential features of the response.

The frequency response in Figure B.9(a) corresponds to a *low-pass filter* (LPF). This name comes from the fact that low-frequency sinusoids are passed from the input to the output with little change: $G \approx 1$ and $\phi \approx 0$. As the frequency increases, the gain drops and the output signal begins to lag the input signal ($\phi < 0$). At the *cutoff frequency* $f_c = 1/(2\pi\tau)$, where $\tau = RC$ for the RC circuits and $\tau = L/R$ for the RL circuits, the gain has dropped to $G(f_c) = 1/\sqrt{2}$ and the phase has dropped to $\phi(f_c) = -45°$. Beyond the cutoff frequency the gain drops by a factor of 10 for every increase in the frequency by a factor of 10, so $G(10f_c) \approx 0.1$ and $G(100f_c) \approx 0.01$. The phase ϕ continues to drop, asymptoting at $-90°$ at high frequencies.

The frequency response in Figure B.9(b) corresponds to a *high-pass filter* (HPF). This name comes from the fact that high-frequency sinusoids are passed from the input to the output with little change: $G \approx 1$ and $\phi \approx 0$. As the frequency decreases, the gain drops and the output signal begins to lead the input signal ($\phi > 0$). At the cutoff frequency $f_c = 1/(2\pi\tau)$, the gain has dropped to $G(f_c) = 1/\sqrt{2}$ and the phase has risen to $\phi(f_c) = 45°$. Beyond the cutoff frequency the gain drops by a factor of 10 for every decrease in the frequency by a factor of 10, so $G(0.1f_c) \approx 0.1$ and $G(0.01f_c) \approx 0.01$. This means that DC (constant) signals are completely suppressed by the filter. The phase ϕ continues to rise with decreasing frequency, asymptoting at 90° at low frequencies.

Low-pass and high-pass filters are useful for isolating a signal of interest from other signals summed with it. For example, LPFs can be used to smooth and suppress high-frequency noise on a sensor line. HPFs can be used to suppress DC signals and only look for sudden changes in a sensor reading, just as the output depicted in Figure B.5(f) is largest when the input suddenly changes value and drops to zero when the signal is constant.

The LPFs and HPFs illustrated in Figure B.9 are called *first order* because the circuit response is described by a first-order differential equation, e.g., (B.1). First-order filters have relatively slow *rolloff*—in the cutoff frequencies, the filter gain drops by only a factor of 10 for every factor of 10 in frequency. By using more passive elements, it is possible to design second-order filters with a gain rolloff of 100 for every factor of 10 in frequency, which are better at suppressing signals at frequencies we want to eliminate with less effect on signals at frequencies we want to preserve. Higher-order filters, with even steeper rolloff, can be constructed by putting first- and second-order filters in series. LPFs and HPFs can also be combined to create *bandpass* filters, which suppress frequency components below some f_{min} and above some f_{max}, or *bandstop* or *notch* filters, which suppress frequency components between f_{min} and f_{max}.

Filters constructed purely with resistors, capacitors, and inductors are called *passive* filters, as these circuit elements do not generate power. More sophisticated *active* filters can be created using operational amplifiers (Section B.4). These circuits have the advantage of having high input impedance (drawing very little current from V_{in}) and low output impedance (capable of supplying a lot of current at V_{out} while maintaining the desired behavior as a function of V_{in}).

The gain (or magnitude) portion of the frequency response is often written in terms of decibels as M_{dB}, where

$$M_{dB} = 20 \log_{10} G.$$

So a gain of 1 corresponds to 0 dB, a gain of 0.1 corresponds to −20 dB, and a gain of 100 corresponds to 40 dB.

B.3 Nonlinear Elements: Diodes and Transistors

Nonlinear circuit elements are critical for computing and nearly all modern circuits. Two common types of nonlinear elements are diodes and transistors. While there are many kinds of transistors, in this section we focus on bipolar junction transistors (BJTs), notably excluding field effect transistors (FETs), which are extraordinarily useful and come in many different varieties. In keeping with the spirit of this appendix, we do not get into the semiconductor physics of these nonlinear elements, but focus on simplified models that facilitate analysis.

Analyzing circuits with simplified models of nonlinear elements is quite different from circuits with only linear elements. We do not simply write a set of equations and solve them. Instead, a nonlinear element can operate in different regimes (two regimes for a diode: conducting and not conducting; and three regimes for a BJT: off, linear, and saturated), each regime with its own governing equations. In principle, we have to solve a complete set of circuit equations for each possible combination of regimes for the nonlinear elements in the circuit. All but one of these guesses at the operating regimes will be wrong, leading to equations and inequalities without valid solutions.

B.3.1 Diodes

Figure B.10 shows the circuit symbol and the simplified current-voltage behavior of a diode. When the voltage across the diode is less than the *forward bias voltage* $V_d \geq 0$, no current flows through the diode. When current flows, it is only allowed to flow in the direction indicated in Figure B.10, from anode to cathode, and the voltage across the diode is V_d. It is never possible to have a voltage greater than V_d across the diode. A typical forward bias voltage for a diode is around 0.7 V, but other values are also possible.

Figure B.11(a) shows a simple resistor-diode circuit. Assume $V_{in} = 5$ V and $V_d = 0.7$ V. To solve for V_{out}, we analyze the circuit for the two possible cases of the diode: conducting or not conducting.

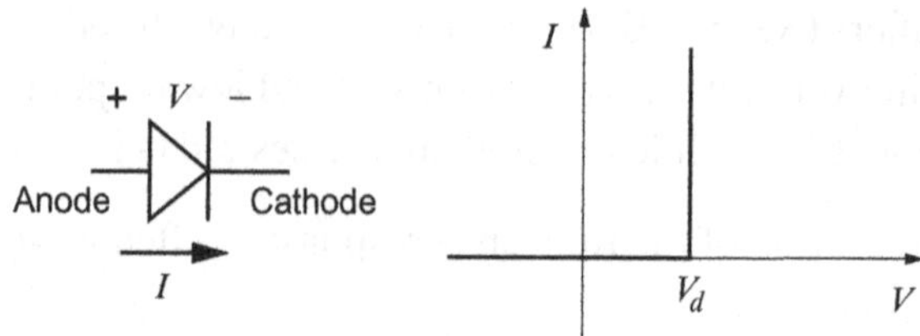

Figure B.10
(Left) The circuit symbol for a diode, indicating positive current and positive voltage. (Right) The simplified current-voltage relationship for a diode with forward bias voltage V_d.

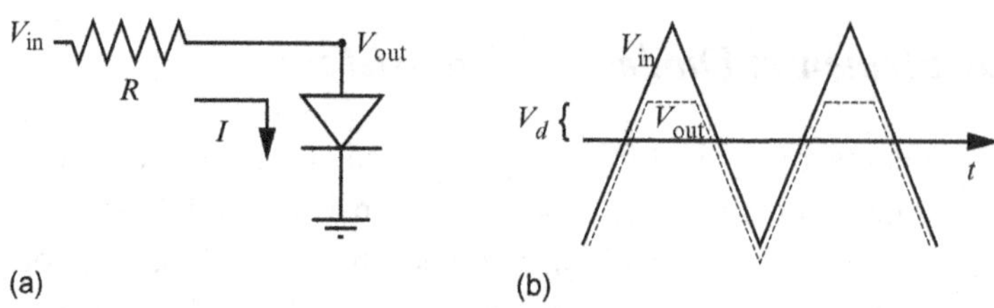

Figure B.11
(a) A resistor-diode circuit. (b) The output voltage V_{out} for a sawtooth V_{in}. The output voltage is equal to the input voltage for $V_{in} < V_d$, but the output voltage is capped at V_d by the diode. Reversing the direction of the diode would cause the output to track the input for $V_{in} > -V_d$, and the voltage would never drop below $-V_d$.

- **Case 1: diode is not conducting.** In this case, we know $I = 0$, so there is no voltage drop across the resistor, so $V_{out} = V_{in} = 5$ V. But we know that the diode can never have more than $V_d = 0.7$ V across it. Therefore this regime is not valid.
- **Case 2: diode is conducting.** Since the diode is conducting in this case, we know that the voltage across the diode is the forward bias voltage $V_d = 0.7$ V. Therefore the current I must be $(5\text{ V} - 0.7\text{ V})/R$ from the constitutive law of the resistor. This current is flowing in the right direction (positive current) and therefore does not violate the current-voltage relationship of the diode, so this is a valid solution.

Figure B.11(b) illustrates V_{out} as V_{in} follows a sawtooth profile, showing that V_{out} can never exceed V_d. The power dissipated by the diode when current I flows is IV_d.

A *light-emitting diode* (LED) is just a diode that emits visible or invisible light when current flows. A typical forward bias voltage for an LED is 1.7 V.

If a large negative voltage is placed across a diode, it may break down, allowing negative current to flow. While this is a failure mode for most diodes, for *Zener* diodes it is the intended use. Zener diodes are designed to have specific (and often relatively small) negative

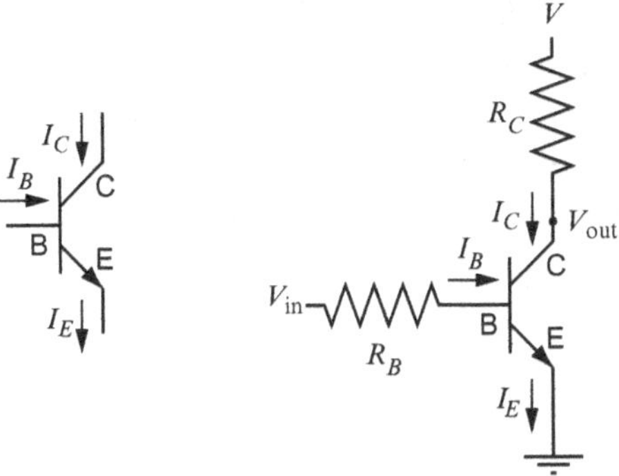

Figure B.12
(Left) The circuit symbol for an NPN transistor, showing the collector, base, and emitter. (Right) A resistor-transistor circuit.

breakdown voltages and to allow current to flow easily while at that breakdown voltage. Zener diodes can be used in voltage regulator applications.

B.3.2 Bipolar Junction Transistors

Bipolar junction transistors come in two flavors: NPN and PNP. We will focus on the NPN BJT, then return to the PNP.

Figure B.12 shows the circuit symbol for an NPN BJT. It has three connections: the collector (C), base (B), and emitter (E). Current flows into the collector and base, and the sum of those currents flows out of the emitter, $I_E = I_C + I_B$. The voltage drop from the base to the emitter is denoted V_{BE} and the voltage drop from the collector to the emitter is written V_{CE}. In normal usage, I_C, I_B, I_E, and V_{CE} are all nonnegative.

The basic function of the NPN BJT is to attempt to generate a collector current that amplifies the base current, $I_C = \beta I_B$, where β is the gain of the transistor (also commonly referred to as h_{FE}). A typical value of β is 100. Depending on the amount of base current flowing, the transistor can be in one of three modes—off, linear, or saturated—and each mode provides three equations governing the transistor voltages and currents:

- **Linear:** $I_B > 0$ and $V_{CE} > V_{CE,sat}$. In this mode, the collector current is $I_C = \beta I_B$, and the transistor is not yet saturated, so if I_B increases, I_C will also increase. Saturation occurs when V_{CE} drops to the collector-emitter saturation voltage $V_{CE,sat}$, which is commonly around 0.2 V or so. In the linear mode, V_{BE} is equal to $V_{BE,on}$, the PN junction diode voltage drop from the base to the emitter. A typical value is $V_{BE,on} = 0.7$ V. **Governing equations:** $I_C = \beta I_B$, $V_{BE} = V_{BE,on}$, and $I_E = I_C + I_B$.

- **Off:** $I_B = 0$. This means that $I_C = 0$ and therefore $I_E = 0$. V_{BE} is less than $V_{BE,on}$. **Governing equations:** $I_B = I_C = I_E = 0$.
- **Saturated:** $I_B > 0$ and $V_{CE} = V_{CE,sat}$. In this mode the collector cannot provide more current even if the base current increases; the transistor is saturated. This is because the voltage between the collector and emitter cannot drop below $V_{CE,sat}$. This means that the relationship $I_C = \beta I_B$ no longer holds. **Governing equations:** $V_{CE} = V_{CE,sat}$, $V_{BE} = V_{BE,on}$, and $I_E = I_C + I_B$.

When $I_E > 0$, the power dissipated by the transistor is $I_B V_{BE,on} + I_C V_{CE,sat}$.

Figure B.12 shows an NPN BJT in a *common emitter* circuit, so called because the emitter is attached to ground ("common"). We can determine the transistor's operating mode as a function of V_{in}:

- **Off:** $V_{\text{in}} \leq V_{BE,on}$. Input voltages in this range do not provide enough voltage to turn on the base-emitter PN junction while also providing a base current $I_B > 0$. Since $I_C = 0$, there is no voltage drop across R_C, and $V_{\text{out}} = V$.
- **Saturated:** $V_{\text{in}} \geq V_{BE,on} + R_B(V - V_{CE,sat})/(\beta R_C)$. When the transistor is saturated, the output voltage is $V_{\text{out}} = V_{CE,sat}$, and the voltage across R_C is $V - V_{CE,sat}$. This means $I_C = (V - V_{CE,sat})/R_C$. At the boundary between the linear and saturated regions, the relationship $I_C = \beta I_B$ is still satisfied, so $I_B = I_C/\beta$. So the minimum V_{in} for saturation is the sum of $V_{BE,on}$ and the voltage drop $I_B R_B$ across the base resistor.
- **Linear:** All V_{in} between the off and saturated regimes. In this regime, $I_B = (V_{\text{in}} - V_{BE,on})/R_B$ and $I_C = \beta I_B$, so

$$V_{\text{out}} = V - I_C R_C = V - \frac{\beta R_C}{R_B}(V_{\text{in}} - V_{BE,on}).$$

To increase the gain of a transistor we can use two transistors, Q1 and Q2, as a *Darlington pair* (Figure B.13(a)). The two collectors are connected and the emitter of Q1 feeds the base of Q2. The two together act like a single transistor, with the base of Q1 as the base of the pair

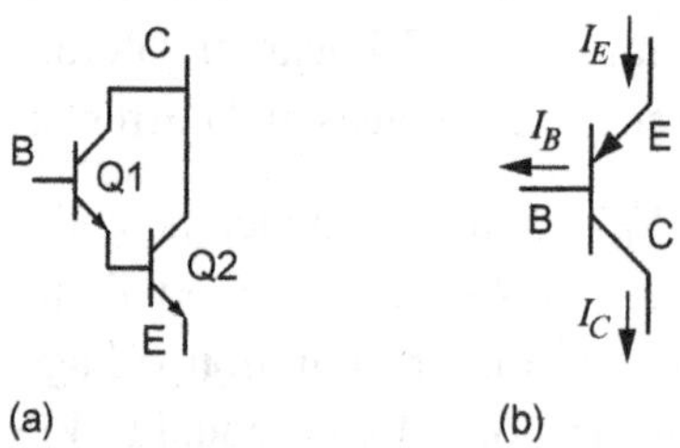

Figure B.13
(a) An NPN Darlington pair. (b) A PNP BJT.

and the emitter of Q2 as the emitter of the pair, but now the overall gain is $\beta_1\beta_2$. $V_{BE,on}$ for the Darlington pair is the sum of the individual base-emitter voltages.

Finally, Figure B.13(b) shows the circuit symbol for a PNP transistor. For a PNP BJT, $I_C = \beta I_B$ as with the NPN, but now I_B and I_C flow out of the transistor and $I_E = I_C + I_B$ flows into it. At saturation, V_E is greater than V_C (typically by about 0.2 V), and when the transistor is on, the voltage drop from V_E to V_B is approximately 0.7 V. The PNP BJT is off if $I_B = 0$; it is saturated if $I_B > 0$ and the voltage drop from V_E to V_C indicates saturation; and otherwise it is in the linear mode.

B.4 Operational Amplifiers

The circuit symbol for an operational amplifier (op amp) is shown in Figure B.14(a). Apart from the power supply inputs, the op amp has two inputs, a noninverting input labeled + and an inverting input labeled −, and one output. Figure B.14(b) shows a particular chip, the 8-pin Texas Instruments TLV272, which has two op amps.

An ideal op amp obeys the following rules:

1. Input impedance is infinite. No current flows in or out of the inputs.
2. Output impedance is zero. The op amp can produce any current necessary to satisfy the following rule.

3(a). If there is no feedback connection between the output and the inputs, then the output satisfies $V_{out} = G(V_{in+} - V_{in-})$, where the gain G is very large, effectively infinite. (The output goes to its maximum positive or negative value if V_{in+} and V_{in-} are different.)

3(b). If there is a current path from the output to the inverting input (*negative feedback*), for example through a capacitor or resistor, then the voltage at the two inputs are equal. This is because the large gain G of the op amp attempts to eliminate the voltage difference $V_{in+} - V_{in-}$.

Almost all useful op amp circuits have negative feedback, so rule 3(b) applies.

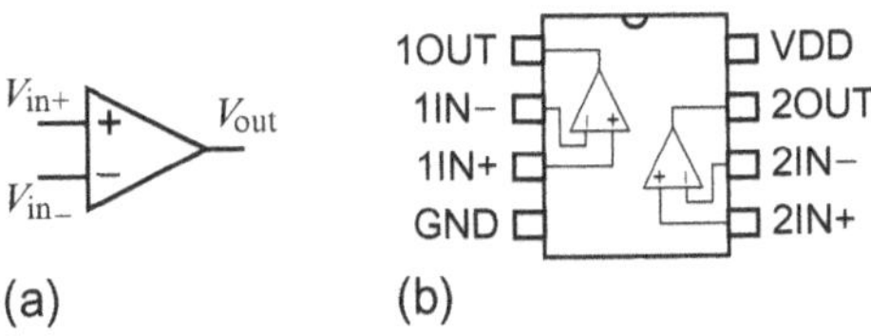

Figure B.14
(a) The op amp circuit symbol. (b) The 8-pin TLV272 integrated circuit, with two op amps.

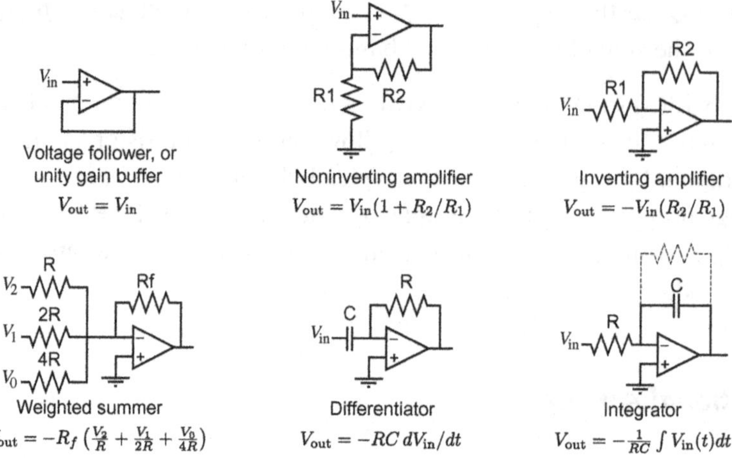

Figure B.15

Common op amp circuits. Of these circuits, note that only the unity gain buffer and the noninverting amplifier present the very high input impedance of the op amp at the V_{in} input; in the other circuits, the input impedance is dominated by external resistors or capacitors. See Section B.5 for a discussion of input impedance.

Figure B.15 shows several useful circuits built with op amps. To analyze these circuits, remember that no current flows in or out of the op amp inputs (current only flows through the external resistors and capacitors), and since there is negative feedback in each of them, the voltages at the two inputs are equal.

For example, to analyze the weighted summer circuit, we recognize that both inputs are at 0 V (ground). Therefore the currents flowing through R, $2R$, and $4R$ are simply V_2/R, $V_1/(2R)$, and $V_0/(4R)$. Since no current flows in or out of the − input, these currents sum to give the current I through the feedback resistor R_f, and the output voltage is simply $V_{out} = -IR_f$. If the input voltages V_i are binary, this circuit provides an analog voltage representation of the three-digit binary number $V_2V_1V_0$, where V_2 represents the most significant bit (since it provides the most current) and V_0 represents the least significant bit. If instead the three resistors $R, 2R, 4R$ are replaced by variable resistances set by potentiometers, the weighted summer is similar to an audio mixer.

The response of the integrator circuit is obtained by recognizing that the current flowing from V_{in} is $I = -V_{in}/R$ and that

$$V_{out} = -V_C = -\frac{1}{C}\int I(t)\,\mathrm{d}t = -\frac{1}{RC}\int V_{in}(t)\,\mathrm{d}t.$$

If V_{in} is constant at zero, ideally V_{out} should also be zero. In practice, however, the voltage across the capacitor is likely to drift due to nonidealities of the op amp (Section B.4.1), including slight input offset. For this reason, it is good practice to put another resistor in parallel with the capacitor. This resistor serves to slowly bleed off capacitor charge, counteracting voltage drift. This resistance should be much larger than R to prevent significant impact on the circuit's behavior at frequencies of interest.

The voltage follower circuit simply implements $V_{out} = V_{in}$, but it is quite useful because it draws essentially no current from the source providing V_{in} (such as a sensor) while being able to provide significant current at the output. For this reason the circuit is sometimes called a unity gain *buffer*: the op amp provides a buffer that prevents the circuit connected to the output of the op amp from affecting the behavior of the circuit connected to the input of the op amp. This allows individual circuits to be designed and tested modularly, then cascaded using buffers in between (Section B.5).

One application of the unity gain buffer is to implement an RC LPF or HPF (Section B.2.2). By cascading a unity gain buffer, then a passive RC LPF or HPF, we get the ideal frequency response of the passive filter but with high input impedance, as opposed to the relatively low input impedance of the passive filter alone. There are many more sophisticated higher-order op amp filter designs that achieve better attenuation at frequencies to be suppressed; consult any text on the design of op amp filters. In particular, for LPFs, HPFs, bandpass, and notch filters, popular filters are Butterworth, Chebyshev, and Bessel filters, each with somewhat different properties. These names refer to the form of the mathematical transfer function from input to output. To implement these transfer functions using op amps, resistors, and capacitors, there are different types of circuit designs; popular choices are the Sallen-Key circuit topology and the multiple feedback circuit topology.

B.4.1 Practical Op Amp Considerations

If you want to purchase an op amp chip, you will find that there are tens of thousands to choose from! How do you choose? Op amp data sheets can be bewildering to read, with many different characteristics, most of them depending on the particular operating condition (the power supply voltage, the load at the output, etc.). Here are a few characteristics to consider, along with the values for the flexible and inexpensive TLV272.

- **Supply voltage range.** An op amp has both a minimum and a maximum allowable voltage across the power supply lines. TLV272: 2.7-16 V.
- **Output voltage swing (rail-to-rail or not).** The maximum outputs of some op amps do not reach all the way to the power supply rails, falling short by 1 V or more. Other op amps are *rail-to-rail*, meaning that the output voltage can come close to the power supply rails. The TLV272 is rail-to-rail, with a maximum output voltage swing to within about 0.1 V of the rails.

- **Input voltage range.** Even if the output goes rail-to-rail, the differential inputs may not be allowed to approach the rails. TLV272: inputs can go rail-to-rail.
- **Output current.** The maximum amount of current that can be provided by a single output. TLV272: up to 100 mA.
- **Unity gain bandwidth.** This is a specification of how fast the op amp output can change. For an input signal at this frequency, the effective gain of the amplifier (which we assumed to be infinite) has dropped to one. TLV272: 3 MHz.
- **Slew rate.** This is another measure of how fast the output can change, in V/s. TLV272: 2.4 V/μs.
- **Input bias current.** There is actually a very small current at the inputs (we assumed it to be zero). This is the typical average of those input currents. TLV272: 1 pA.
- **Input offset current.** This is the typical difference between the two input currents. TLV272: 1 pA.
- **Input offset voltage.** Ideally zero voltage difference at the inputs would cause zero voltage at the open-loop output. In practice, the input levels may have to be slightly different to achieve a zero voltage output. TLV272: 0.5 mV.
- **Common-mode rejection ratio.** The ideal amplifier amplifies only the difference between the voltages at the + and − inputs, but there is actually a small amplification of the common voltage between them. For example, if the voltage V_+ is 5.1 V and the voltage V_- is 5.0 V, the usual amplifier gain acts on the 0.1 V difference while the common-mode gain acts on the average, 5.05 V. The CMRR specifies the ratio of the differential gain to the common-mode gain. TLV272: 80 dB (or 10,000).
- **Number of op amps.** Some chips have more than one op amp. TLV272: two op amps.
- **Packaging.** Op amps come in different types of packaging. DIP packages are easiest to work with for breadboard prototyping. TLV272: available in a variety of packages, including an 8-pin DIP.
- **Price.** Price increases for higher bandwidth and slew rates, higher output current, lower offset voltage, higher common-mode rejection ratio, rail-to-rail operation, etc. TLV272: about one dollar.

B.4.2 Single Supply Design and Virtual Ground

When using an op amp in microcontroller applications, often the only power supply available is a positive voltage rail and ground (no negative rail), and the positive voltage may be small (e.g., 3.3 V). First of all, this likely means that a rail-to-rail op amp should be used, to maximize the output voltage range, and it should be capable of being powered by the microcontroller voltage (e.g., 3.3 V). Secondly, notice that many of the standard op amp circuits (Figure B.15) provide an output voltage that has a sign opposite of the input voltage, which the op amp cannot produce if there is no negative rail.

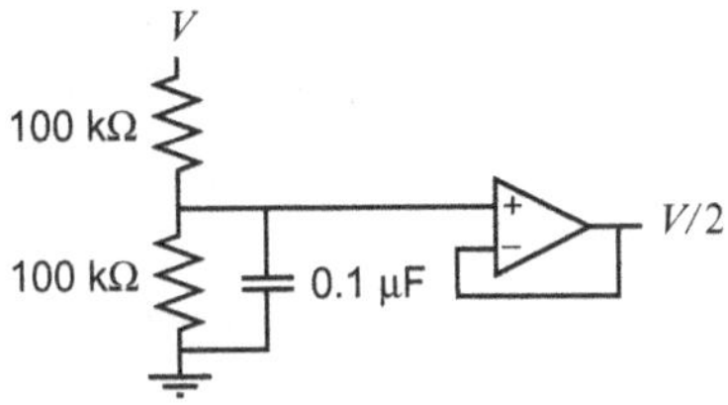

Figure B.16
Creating a virtual ground at $V/2$.

One way to handle this issue is to introduce a *virtual ground* at a voltage halfway between the positive supply rail and ground, effectively creating a bipolar supply about this virtual ground. Figure B.16 illustrates the idea. A unity gain buffer is used in combination with a voltage divider to create a virtual ground at $V/2$. The capacitor helps stabilize the reference voltage to any transients on the power supply line. This virtual ground is then used in place of ground in the inverting circuits in Figure B.15. Inverted voltages are now with respect to $V/2$ instead of 0.

Since the op amp likely sinks or sources less current than a typical power supply, care should be taken to make sure that these limits are never exceeded.

B.4.3 Instrumentation Amps

An *instrumentation amp* is a specialized amplifier designed to precisely amplify the difference in voltage between two inputs, $V_{out} = G(V_{in+} - V_{in-})$. Like an op amp, it has inputs V_{in+} and V_{in-}, but unlike an op amp, it is not used with a negative feedback path. Instead, an instrumentation amp like the Texas Instruments INA128 allows you to connect a single external resistor R_G to determine the gain G, where

$$G = 1 + \frac{50\ \text{k}\Omega}{R_G}.$$

The INA128 is typically used to implement gains G from 1 ($R_G = \infty$, i.e., no connection at the gain resistor inputs) to 10,000 ($R_G \approx 5\ \Omega$).

Other instrumentation amps allow you to choose from a fixed set of very precise gains, not dependent on an external resistor (with its associated tolerance). These are sometimes called programmable-gain instrumentation amps, and an example is the TI PGA204, which uses two digital inputs to choose gains of 1, 10, 100, or 1000. A related design is the TI INA110, offering gains of 1, 10, 100, 200, or 500.

Instrumentation amps distinguish themselves in their very high common-mode rejection ratio (120 dB for the INA128). Instrumentation amps are typically more expensive than op amps, e.g., in the range of 10 to 20 USD for quantities of one. If you are trying to save money, you can build your own difference-and-gain circuit using multiple op amps, but you will not achieve the same performance as an instrumentation amp.

B.5 Modular Circuit Design: Input and Output Impedance

One way to design a complex circuit is to design subcircuits, each with a specific function, and then cascade them so that the output of one circuit is the input of another. Modular circuit design is similar to modular code design: each subcircuit has a specific function and well-defined inputs and outputs.

Modularity requires that connecting the output of circuit A to the input of circuit B does not change the behavior of either circuit. Modularity is assured if circuit A has low output impedance (it can source or sink a lot of current with little change in the output voltage) and circuit B has high input impedance (it draws little current at the input).[2] For constant (DC) voltages and currents, if a change ΔI in the current drawn from the output of a circuit causes a change of voltage ΔV, then the DC output impedance (or simply the output resistance, since the voltage is DC) is $|\Delta V/\Delta I|$. A circuit's input resistance can be measured similarly. High input impedance means that a change in input voltage gives a very small change in input current.

Input and output impedance are generally frequency dependent. For sinusoidal signals of any frequency $\omega = 2\pi f$, the impedance of a resistor is simply its resistance R. The magnitude of the impedance of an inductor is ωL, meaning that the impedance increases linearly with ω—the impedance is zero at DC and infinite at infinite frequency. The magnitude of the impedance of a capacitor is $1/(\omega C)$, indicating that the impedance magnitude is infinite at DC and zero at infinite frequency.

As a simple DC example, consider the following design problem. We want to provide a user the ability to choose an input voltage between 0 and 3 V by turning a potentiometer knob. So we decide the circuit B will be a 10 kΩ potentiometer with one end at 3 V and the other end at 0 V, with the wiper providing the user's input signal. No 3 V supply is available, however; there is only a 6 V supply. So we decide to design a circuit A, a voltage divider consisting of two resistors of resistance R, to create the 3 V reference. The output of circuit A becomes the input for circuit B (see Figure B.17).

Let us say we choose $R = 100$ kΩ for the voltage divider. Then the currents in Figure B.17 can be calculated using

[2] It is actually the ratio of input impedance to output impedance that matters. This ratio should ideally be multiple orders of magnitude.

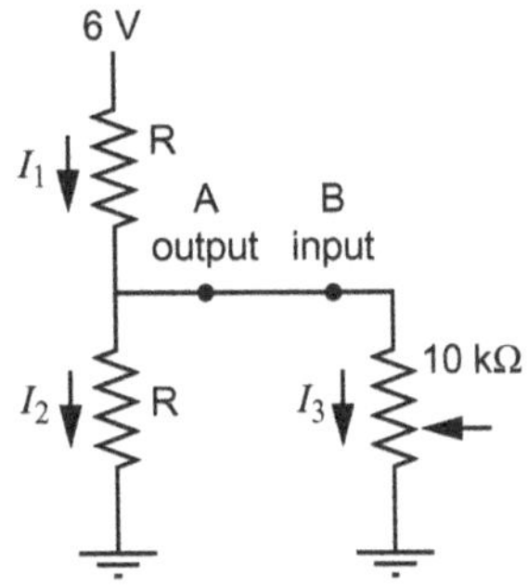

Figure B.17
A voltage divider circuit A feeding a potentiometer circuit B.

$$I_1 = I_2 + I_3$$
$$6\,\text{V} = I_1\,100\,\text{k}\Omega + I_2\,100\,\text{k}\Omega$$
$$I_2\,100\,\text{k}\Omega = I_3\,10\,\text{k}\Omega$$

to find $I_1 = 55\ \mu\text{A}$, $I_2 = 5\ \mu\text{A}$, and $I_3 = 50\ \mu\text{A}$. This means the voltage divider actually creates a voltage at the output of A of 6 V – (55 μA)(100 kΩ) = 0.5 V instead of 3 V. Circuit B "loads" or "pulls down" the output of circuit A. The output impedance R of circuit A is too high relative to the 10 kΩ input impedance of circuit B, defeating circuit modularity. Our attempt to design circuits A and B independently and put them together has failed.

On the other hand, if we choose R = 100 Ω for the voltage divider, we find that $I_1 = 30.15$ mA, so $V_A = 2.985$ V, very close to our target of 3 V. The output impedance of circuit A is much lower, so modularity is more closely achieved. This comes at the cost of greater power dissipated by the voltage divider, $V^2/R = (6\ \text{V})^2/200\ \Omega = 180$ mW vs. $(6\ \text{V})^2/200\ \text{k}\Omega = 0.18$ mW.

Op amps, with their high input impedance and low output impedance, are quite useful in achieving circuit modularity. In particular, a unity gain buffer between the output A and input B in Figure B.17 would eliminate any loading of the circuit A by circuit B, allowing us to use higher resistances for R and therefore wasting less power.

Further Reading

Hambley, A. R. (2000). *Electronics* (2nd ed.). Upper Saddle River, NJ: Prentice Hall.

Horowitz, P., & Hill, W. (2015). *The art of electronics* (3rd ed.). New York, NY: Cambridge University Press.

APPENDIX C

Other PIC32 Models

As of this writing, there are nearly 200 different PIC32 models, arranged in six major families: PIC32MX1xx/2xx, PIC32MX3xx/4xx, PIC32MX5xx/6xx/7xx (from which our PIC32MX795F512H was chosen), PIC32MX1xx/2xx/5xx 64-100 pins, PIC32MX330/350/370/430/450/470, and PIC32MZ. The five PIC32MX families all use the MIPS32 M4K processor as the CPU, at speeds of 40-120 MHz, while the MZ family has a different architecture and uses the MIPS32 microAptiv microprocessor as the CPU at up to 200 MHz. "MIPS32" refers to CPU architectures and associated assembly language instructions licensed by Microchip from Imagination Technologies.

The main differences between the families are the CPU architecture (MIPS32 M4K vs. microAptiv), CPU clock speeds, amount of RAM and flash, physical packaging, available peripherals, number of pins, and the extent to which the pins can be mapped to different functions. This appendix provides a brief introduction to the features of the different families.

C.1 The PIC32MX5xx/6xx/7xx Family

Devices in the PIC32MX5xx/6xx/7xx family have names of the form

$$\text{PIC32MX}\quad \underbrace{5, 6, \text{or } 7}_{\text{communication options}}\quad \underbrace{\text{x x}}_{\text{other model options}}\quad \text{F}\quad \underbrace{\text{yyy}}_{\text{flash in KB}}\quad \underbrace{\text{H or L}}_{\text{64 or 100 pins}}$$

The 5xx series has full-speed USB and CAN peripherals, the 6xx series has full-speed USB and Ethernet, and the 7xx series has full-speed USB, Ethernet, and CAN. The xx code can be 34, 64, 75, or 95, corresponding to other model options, but primarily indicating the amount of RAM available (16, 32, 64, and 128 KB, respectively). The yyy code indicates the amount of flash memory, in KB. (All devices in this family also have an additional 12 KB of boot flash.) Devices ending in H have 64 pins and devices ending in L have 100 pins. Therefore the PIC32MX795F512H has full-speed USB, CAN, and Ethernet; 128 KB of RAM; 512 KB of flash; and 64 pins.

The M4K CPU can operate at up to 80 MHz for all devices in the PIC32MX5xx/6xx/7xx family. All devices have full-speed USB, five 16-bit counter/timers with up to two 32-bit counter/timers, five input capture devices, five output compare modules with 16-bit resolution, 16 10-bit ADC inputs, six UARTs, two comparators, DMA, and PMP. Devices with 64 pins have three SPI and four I^{2}C peripherals; devices with 100 pins have four SPI and five I^{2}C peripherals. All devices are available only in surface mount packages.

C.2 PIC32MX3xx/4xx Family

PIC32MX3xx/4xx devices are the first to have appeared in the PIC32 line. The M4K CPU can operate at up to 80 MHz for most devices in this family, but a few devices are limited to 40 MHz. Devices in this family have up to 32 KB of RAM and 512 KB of flash and have names of the form

PIC32MX	3 or 4	xx	F	yyy	H or L
	3: no USB; 4: with USB	20, 40, or 60		flash in KB	64 or 100 pins

Thus the PIC32MX460F512H has full-speed USB, 32 KB of RAM (with the "60" option), 512 KB of flash, and 64 pins.

Devices in the PIC32MX3xx/4xx family have similar capabilities to those in the PIC32MX5xx/6xx/7xx family, except they do not offer Ethernet or CAN, have fewer UART, SPI, and I^{2}C peripherals, and have only one 32-bit counter/timer.

C.3 PIC32MX1xx/2xx Family

PIC32MX1xx/2xx devices are more recent than the 3xx/4xx and 5xx/6xx/7xx families. They are smaller devices, coming in 28-, 36-, and 44-pin devices. The 28-pin devices are available in DIP (dual inline package), convenient for breadboarding. The maximum CPU clock speed for a 1xx/2xx device is 40 or 50 MHz, depending on the model. The 1xx/2xx devices do not have a prefetch cache module; they run at full speed pulling instructions from flash.

Devices in this family have up to 64 KB of RAM and 256 KB of flash and have names of the form

PIC32MX	1 or 2	xx	F	yyy	B, C, or D
	1: no USB; 2: with USB	other model options		flash in KB	28, 36, or 44 pins

The xx code can be 10, 20, 30, 50, or 70, which correspond to 4, 8, 16, 32, or 64 KB of RAM, respectively. Thus the PIC32MX230F064B has full-speed USB, 16 KB of RAM, 64 KB of flash, and 28 pins.

Devices in this family differ from the 5xx/6xx/7xx family in that they have only 3 KB of boot flash; fewer ADC inputs; fewer UART, SPI, and I^{2}C peripherals; and no Ethernet or CAN.

The 1xx/2xx devices have three comparator modules instead of two, programmable with up to 32 reference voltages as compared to the 16 of the 5xx/6xx/7xx family. Other interesting new features, which we have not seen in the previous models, include Peripheral Pin Select (PPS), audio interface by SPI, and charge-time measurement for capacitive touch sensing, as discussed below.

PPS allows a wide range of digital-only peripherals to be assigned flexibly to different device pins, a major feature for low-pin-count devices like PIC32MX1xx/2xx devices. While a pin of our PIC32MX795F512H can support several possible peripherals, devices with PPS have much more flexibility in mapping certain peripherals to different pins. A remappable peripheral does not have default I/O pins; SFRs must be configured to assign the peripheral to specific pins before it can be used. Examples of peripherals that can be remapped by PPS include UARTs, SPI modules, counter/timer inputs, input capture, output compare, and interrupt-on-change inputs. Some peripherals cannot be remapped, such as I^2C peripherals and ADC inputs, because of special requirements on the I/O circuitry for those peripherals.

The 1xx/2xx devices' SPI modules support audio interface protocols for 16-, 24-, and 32-bit audio data. One example is the Inter-IC Sound (I^2S) protocol, which allows the transmission of two channels of digital audio data using the SPI peripheral. The I^2S capability allows a 1xx/2xx device to communicate with digital audio equipment as either the master or slave.

Finally, the 1xx/2xx's Charge-Time Measurement Unit (CTMU) provides a current source to interface with an external capacitive touch sensor, such as a capacitive on/off button or even an *x-y* touchpad. The CTMU is used with one or more ADC channels to measure the capacitance of one or more analog capacitive sensors.

C.4 PIC32MX1xx/2xx/5xx 64-100 Pin Family

The PIC32MX1xx/2xx/5xx 64-100 pin family expands on the features of the PIC32MX1xx/2xx family, which includes PPS, CTMU, and audio interface protocols. The M4K CPU operates at speeds up to 50 MHz, and the devices have 64 or 100 pins and up to 64 KB of RAM and 512 KB of flash. These devices feature more analog input channels (up to 48), more UART and SPI peripherals, and some models feature USB and CAN. Device names have the form

$$\text{PIC32MX}\quad \underbrace{\text{1, 2, or 5}}_{\text{communication options}}\quad \underbrace{\text{xx}}_{\text{other model options}}\quad \text{F}\quad \underbrace{\text{yyy}}_{\text{flash in KB}}\quad \underbrace{\text{H or L}}_{\text{64 or 100 pins}}$$

The 1xx series has neither CAN nor USB, the 2xx series features full-speed USB but no CAN, and the 5xx series has both full-speed USB and CAN. The code xx can be 20, 30, 50, or 70, corresponding to 8, 16, 32, or 64 KB of RAM, respectively. Thus the PIC32MX570F512H offers both USB and CAN and has 64 KB of RAM, 512 KB of flash, and 64 pins.

C.5 PIC32MX330/350/370/430/450/470 Family

The PIC32MX330/350/370/430/450/470 family differs from the PIC32MX1xx/2xx/5xx 64-100 pin family by offering 12 KB of boot flash, up to 128 KB of RAM, and CPU speeds up to 120 MHz, the fastest in the PIC32MX families. Devices in this family have two SPI channels (compared to the three or four of the PIC32MX1xx/2xx/5xx 64-100 pin family) and do not offer CAN. 4xx devices in this family offer full-speed USB while 3xx devices do not.

C.6 PIC32MZ Family

The most recent addition to the PIC32 line, the PIC32MZ family is the most advanced. Devices in this family have names of the form

$$\text{PIC32MZ} \quad \underbrace{\text{xxxx}}_{\text{flash in KB}} \quad \text{three-letter code} \quad \underbrace{\text{yyy}}_{\text{number of pins}}$$

The number of pins (yyy) is either 064, 100, 124, or 144. The three-letter code indicates whether the PIC32 has CAN modules; an external bus interface (EBI), described below; a floating point unit (FPU) for fast floating point operations; and a Crypto Engine, a hardware module used to accelerate applications requiring encryption, decryption, and authentication. PIC32MZs currently have up to 512 KB of RAM.

All PIC32MZs have a much larger boot flash segment (160 KB), nine 16-bit counter/timers configurable to up to four 32-bit counter/timers, nine output compares with up to 32-bit resolution, nine input captures, six UARTs, up to five I^2C modules, up to six SPI modules supporting audio interfaces, PPS for more flexible pin remapping, Ethernet, and high-speed USB, as opposed to the slower full-speed USB of the PIC32MX models. PIC32MZs do not currently have a CTMU.

Two new capabilities on the PIC32MZ are the 50 MHz External Bus Interface (EBI) and the 50 MHz Serial Quad Interface (SQI). SQI is similar to SPI, except it has four data lines and supports single lane, dual lane, and quad lane modes of operation. In single lane mode, it is identical to SPI. EBI allows a high-speed connection to external memory devices, like NOR flash devices or SRAM, allowing you to seamlessly address external memory in your C code.

The PIC32MZ family also has several different peripheral buses, each potentially clocked at different frequencies.

Certainly the biggest difference of the PIC32MZ family from PIC32MX devices is its different microprocessor, the MIPS32 microAptiv core. The microAptiv CPU can be clocked at up to 200 MHz, and it has multiple shadow register sets, a larger number of interrupt sources, and new assembly instructions and hardware to accelerate digital signal processing calculations. Other capabilities of the microAptiv core can be found in Section 50 of the Reference Manual.

C.7 Conclusion

To learn more about a specific PIC32 model, first consult the Data Sheet for the appropriate PIC32 family to learn the specific capabilities of each model. After that, you can consult the sections of the Reference Manual for more information. You will find it helpful to be armed with the knowledge of the features that your PIC32 model has or does not have, since the Reference Manual is currently written to cover all PIC32 models. After that, you can modify the sample code provided in this book for your particular PIC32, or start with sample code provided by Microchip.

Index

Note: Page numbers followed by *f* indicate figures and *t* indicate tables.

E

F

T

Printed in the United States
By Bookmasters